TEACHER'S EDITION

For use with SCIENCE 6
Third Edition

BJU PRESS
GREENVILLE, SOUTH CAROLINA

Advisory Committee
from the administration, faculty, and staff of Bob Jones University
Steve Skaggs, M.Ed., *Director of Product Development, BJU Press*
Philip D. Smith, Ed.D., *Provost*

Consultants
Verne Biddle, Ph.D., *Professor, Department of Chemistry*
Dottie Buckley, *Elementary Authors Coordinator, BJU Press*
R. Terrance Egolf, CDR, USN (retired), *Secondary Science Authors, BJU Press*
Will Gray, *Bible Integration, BJU Press*
Donald Jacobs, Ed.D, *Chairman, Department of Elementary Education*
Thomas E. Porch, D.M.D., *Secondary Science Authors, BJU Press*
Bryan Smith, Ph.D., *Bible Integration, BJU Press*

NOTE: The fact that materials produced by other publishers may be referred to in this volume does not constitute an endorsement of the content or theological position of materials produced by such publishers. Any references and ancillary materials are listed as an aid to the student or the teacher and in an attempt to maintain the accepted academic standards of the publishing industry.

SCIENCE 6 for Christian Schools™ Teacher's Edition
Third Edition

Coordinating Writers
Peggy S. Alier
Joyce Garland

Contributing Writers
Eva Autry
Adelyn Forrest
Donald Jacobs
Jocelyn Loucks
Amy Miller
Janet E. Snow
Debra White

Project Editor
Ellen M. Gildersleeve

Composition
Kelley Moore

Designer
Wendy Searles

Cover Design
John Bjerk
Elly Kalagayan

Project Manager
Vic Ludlum

Photo Acquisition
Susan Perry
Carla Thomas

Illustration Coordinator
Dave Schuppert

Illustrators
Anne Bastine
Matt Bjerk
Aaron Dickey
Cory Godbey
Preston Gravely
Jim Hargis
Dave Schuppert

Produced in cooperation with the Bob Jones University Department of Science Education of the School of Education, the College of Arts and Science, and Bob Jones Elementary School.

Photograph credits appear on pages A78–A79.

for Christian Schools is a registered trademark of BJU Press.

BJU Press grants the original purchaser a limited license to copy the reproducible pages in this book for use in his classroom. Further reproduction or distribution shall be a violation of this license. Copies may not be sold.

© 1991, 1996, 2004 BJU Press
Greenville, South Carolina 29614
First Edition © 1977 BJU Press

Printed in the United States of America
All rights reserved

ISBN 1-59166-007-6

15 14 13 12 11 10 9 8 7 6 5 4 3 2

We asked customers what they were looking for in their teaching materials.

You told us.

I am looking for Teacher's Editions that
- ☑ are easy to use
- ☑ aim at teaching comprehension
- ☑ improve test scores
- ☑ have lots of teaching ideas
- ☑ help me reach different learning styles
- ☑ have four-color throughout
- ☑ are reasonably priced

I want Student Books that have
- ☑ grade-appropriate activities
- ☑ challenging information
- ☑ engaging presentations
- ☑ photographs
- ☑ lots of color everywhere

We listened.

And we did as you asked. We think you will like what you see in this book. Enjoy your teaching. We're still listening – contact us any time.

1.800.845.5731
www.bjup.com

Table of Contents

Introduction
- Goals for SCIENCE 6 vi
- Instructional Materials vii
- Sample Text Lesson viii
- Sample Activity Lesson x
- Teaching a Lesson xii
- Bible Integration xii
- Review Lessons xii
- Science Notebook xii
- Teaching an Activity or Explorations Lesson xiii
- Science Process Skills xiv
- Assessment and Grading xv
- Scheduling xv
- Science Fairs xvi

Table of Lesson Plans xviii

Lesson Plans 1

Appendix
- Technology Lesson Plans A3
- Reproducibles A17
- Rubrics A49
- Games A56
- Materials List A58
- Bible Action Truths and Bible Promises A69
- Index A73
- Photo Credits A78

Goals for SCIENCE 6

Develop a knowledge of God
- Inculcate the understanding that God is the Creator and Sustainer of the universe (Col. 1:16–17)
- Identify the orderliness and precision of God's creation (Eccles. 3:1–7)
- Inspire curiosity, wonder, and appreciation of God's creation (Ps. 19:1)

Encourage Christian growth
- Promote disciplined and orderly approaches to problem solving (I Thess. 5:21)
- Teach cooperative skills through group activities
- Challenge students to be good stewards of God's creation (Gen. 2:15)

Promote scientific literacy
- Establish foundational science facts and skills for further science instruction
- Balance presentation of facts with active participation
- Teach processes involved in a scientific method
- Show the integration of science into the student's everyday life
- Affect a positive attitude toward science through active participation and relevant discussions

Instructional Materials

Student Text

The SCIENCE 6 *for Christian Schools* Student Text provides grade appropriate information through text, diagrams, graphs and charts, and annotated pictures and photographs. Special interest boxes and Quick Check sections are included in all text lessons. The Student Text also contains instructions for experiments, activities, and projects. Application and higher-level thinking questions are included at the end of each chapter. A glossary and index are included at the end of the book.

Teacher's Edition

The SCIENCE 6 *for Christian Schools* Teacher's Edition contains 165 lessons. These lessons are divided into six units and a Technology section that includes one technology lesson for each unit. Each unit consists of two or three interrelated chapters. Most lessons include additional background information, cross-curricular links, and scientific activities.

The Teacher's Edition also includes useful information about science fairs, science process skills, and the management of activities. Reproducibles, a materials list, and several rubrics designed for assessing Activity and Explorations lessons are located in the back of the Teacher's Edition.

Activity Manual

The SCIENCE 6 *for Christian Schools* Activity Manual provides a variety of pages to aid the student's understanding. Study Guides provide a systematic review of concepts. Pages for recording information for Activity and Explorations lessons reinforce good scientific methods. Preview, Reinforcement, Bible Integration, and Expansion pages are also included.

Answers are located in the Activity Manual Answer Key, which must be purchased separately.

Tests

The Test Packet includes one copy of each of the fifteen chapter tests. It also provides two or three short quizzes for use with each chapter. An answer key for the tests and quizzes is available separately.

Introduction vii

Sample Text Lesson

Use the **materials list** to determine the materials needed to teach the lesson.

Encourage student curiosity and science awareness by discussing the **interest boxes**.

Clarify and enhance the lesson with information from **Science Background** and **Science Misconceptions.**

Apply **integrated Bible truths** and **principles.**

Begin the lesson with an **introductory discussion** or **activity.**

Enrich the student's knowledge by incorporating **activities, cross-curricular links,** and **demonstrations.**

Any **materials** needed for enrichment activities are listed in the links.

Introduction

pH Scale

Scientists use a special scale to determine the concentration, or amount, of an acid or base in a solution. The term *pH* comes from French words meaning "the power of hydrogen." The **pH scale** is numbered from 0 to 14. Acids measure from 0 to 6.9 on the scale, with 0 being highly acidic and 6 being slightly acidic. Bases measure from 7.1 to 14 on the scale, with 14 being highly basic and 8 being slightly basic. Solutions with a pH of 7.0 are **neutral**, meaning they are neither basic nor acidic. Pure water has a pH of 7.0.

Soil pH affects how well plants absorb nutrients from the soil. Most plants grow best if the soil pH is around 6.0–7.0. People can buy pH test kits at most lawn and garden stores. Once a gardener knows the pH of his soil, he can raise or lower the pH by adding substances to the soil. To increase the pH, most gardeners add a type of a base called lime to the soil. To lower the soil pH, many gardeners add peat moss or compost. These items usually have some type of acid in them.

Discussion: pages 174–75

- What is an *acid*? {a compound that forms hydrogen ions when dissolved in water}
- What determines the strength of an acid? {the number of hydrogen ions it forms}
- How do acids taste? {sour}
- What are some common weak acids? {Possible answers: lemon juice and vinegar}
- What are examples of corrosive acids? {Possible answers: hydrochloric acid and gastric acid}
- What is a *base*? {a compound that forms hydroxide ions when dissolved in water}
- What determines the strength of a base? {the number of hydroxide ions it forms}
- What are some characteristics of bases? {They taste bitter and feel slippery.}
- What are some common weak bases? {Possible answers: baking soda and antacids}
- How do you dilute an acid or a base? {add more water to the solution}
- What is the name of the scale that is used to show the concentration of an acid or a base in a solution? {pH scale}
- How is the pH scale numbered? {0–14}
- Why is pure water considered a neutral compound? {It has a pH of 7, which is neither an acid nor a base.}
- What numbers on the scale indicate that a compound is an acid? {0–6.9}
- Which numbers indicate a strong acid? {lower numbers}
- What numbers on the scale indicate that a compound is a base? {7.1–14}
- Which numbers indicate a strong base? {higher numbers}
- How do gardeners increase the soil pH? {They add lime or another base.}
- How do gardeners decrease the soil pH? {They add compost, peat moss, or an acid.}
- Why do gardeners sometimes change the pH of their soil? {Possible answer: Not all plants grow well in the same type of soil.}
- Discuss the *pH scale* on Student Text pages 174–75. Compare and contrast acid and base characteristics that the student may know about the items pictured.

> Evaluate the understanding of science concepts through **discussion**.

> Promote **higher-level thinking skills**. Answers to these questions are not taken directly from the pages being discussed.

> Include discussion of **graphic information**.

Chapter 7

Lesson 80 185

Discussion: pages 176–77

- Explain what it means to neutralize an acid or a base. {Possible answer: When an acid and a base come in contact with each other, the reaction changes the pH closer to 7, or neutral.}
- Explain how an antacid can help your stomach feel better. {An upset stomach can be caused by too much gastric acid. The antacid contains a base. When the acid and base are combined, they neutralize each other.}
- What substances form when an acid and a base react chemically? {water and a salt}
- What is a *salt*? {a compound that contains positive ions from a base and negative ions from an acid}
- Are all salts the same? {no}
- What makes salts different from one another? {the different combinations of bases and acids which form different types of salt}
- What are some common salts? {sodium chloride—table salt; potassium chloride—salt substitute; calcium carbonate—chalk}

Answers

1. Bases form hydroxide ions in water, taste bitter, and feel slippery.
2. Acids form hydrogen ions in water and taste sour.
3. Possible answers: Acids include lemon, vinegar, tomato juice, and coffee. Bases include baking soda, toothpaste, antacid, and ammonia.
4. Salts form as a result of the neutralizing of an acid and a base by a chemical reaction.

Activity Manual

Study Guide—page 107
This page reviews Lesson 80.

Bible Integration—page 108
This page looks at the life of Robert Boyle, who was both a scientist and a Christian.

Assessment

Test Packet—Quiz 7-B
The quiz may be given any time after completion of this lesson.

> Review each text lesson with a **Quick Check**.

> Reinforce, review, and enrich student learning with pages from the **Activity Manual**.

Continued from page 186

- What are the results of the tests? {The ammonia turns the red litmus paper blue, and the blue litmus paper stays blue. The vinegar turns the blue litmus paper red, and the red litmus paper stays red.}

Fill a medicine dropper with ammonia and add five drops to the vinegar. Have another student stir the mixture and check it with blue litmus paper.

Continue adding ammonia one drop at a time and checking with litmus paper until the litmus paper no longer reacts.

- What change did you observe in the litmus paper? {At first the blue litmus paper turned red, but after adding ammonia the paper stayed blue.}
- Is this liquid still an acid? What is it? {No; it is neutral.}
- How many drops of ammonia were needed before the solution became neutral? {Answers will vary.}
- What formed when you neutralized the acid and base? {water and a salt}
- How do we know the solution is neutral? {Litmus paper does not change color when a solution is neutral.}

> Assess the student's work and knowledge by using a **quiz**, **test**, or **rubric**.

Chapter 7

Lesson 80 187

Sample Activity Lesson

Incorporate some of the **science process skills** into each activity.

Most activities state a **problem** that needs a solution.

In the Activity Manual, the student records the specific **materials and measurements** used for each activity.

Lesson 82
Student Text pages 180–81
Activity Manual pages 111–12

Objectives
- Hypothesize about the effectiveness of several antacids
- Make and use a model of "upset stomach" acid
- Infer information from the model

Materials
- See Student Text page

Introduction
Purpose for reading:
Student Text pages 180–81
Activity Manual pages 111–12

The student should read all the pages before beginning the activity.

- When you have an upset stomach, what kind of medicine do you take? *(Answers will vary.)*
- Do you prefer one type of medicine over another? Why? *(Answers will vary.)*
- Do you know wh[ich] medi[c]ine works be[st?...] do you know [...]
- Today you wi[ll...] which antacid [...] acid.

Which Antacid Is Best?

The stomach uses acid to digest food, but the stomach should not be too acidic or too basic. Many antacids can help an upset stomach caused by the presence of too much acid. Some antacids are sold as medications. Other antacids may be food or cooking ingredients found around your house. In this activity you will test the effectiveness of several different antacids on experimental "upset stomachs."

Process Skills
- Hypothesizing
- Experimenting
- Observing
- Inferring
- Recording data

Problem
Which antacid works best to neutralize an acid?

Procedure
1. Write a hypothesis in your Activity Manual explaining which antacid you think will work best to neutralize an acid.
2. Combine 80 mL of water and 40 mL of vinegar in the container to make an *upset stomach mixture*.
3. Place a pH indicator strip into the *upset stomach mixture* for 30 seconds. Immediately compare the color of the strip [to] the pH chart. Record the pH level in your [A]ctivity Manual.
4. [Pour] 20 mL of the *upset stomach mixture* [into] each plastic cup. These are your "upset [stom]achs."
5. [Place] one dose of each antacid into a [diffe]rent "upset stomach" cup and mix well. [(Tab]lets should first be crushed.)

Materials:
- metric measuring cups and spoons
- water
- vinegar
- 200 mL or larger container
- pH indicator paper (range of 1–14)
- 6 clear plastic cups
- 6 spoons or stirring sticks
- baking soda
- milk
- One dose each of four different commercial antacids (or their generic equivalents) such as:
  - Alka-Seltzer
  - Maalox
  - Pepto Bismol
  - Rolaids
  - Tums
  - Milk of magnesia

180

The **materials list** in the Student text names the basic resources needed for each activity.

Which Antacid Is Best?

Name _____

Use Student Text pages 180–81.

Problem
Which antacid works best to neutralize an acid?

Materials
- metric measuring cups and spoons
- 80 mL water
- 40 mL vinegar
- 200 mL or larger container
- pH indicator paper
- 6 clear plastic cups
- 6 spoons or stirring sticks
- 5 mL baking soda
- 80 mL milk

List the antacids that you use.
_____
_____
_____

Hypothesis (Write one sentence that explains which antacid you think works best to neutralize stomach acid.)

Procedure
1. What is the pH level of the *upset-stomach mixture*?
2. In the chart, list the four antacids that you are using.
3. Test the pH level of each "upset stomach" after adding each antacid, including the baking soda and milk.
4. Record your results.

Antacid name	pH level after adding the antacid
baking soda	
milk	

Lesson 82; pp. 180–81
Activity 111

Teacher Helps

The pH indicator paper is different from litmus paper. The activity will be most effective if you ensure that each of your antacids is made of different ingredients. Two antacids with the same ingredients will likely have the same pH reading.

Crush tablets by placing them in plastic bags and pounding them with a hammer.

Red cabbage juice can be used in place of the pH paper, but this will give a less accurate reading. The colors of the cabbage juice and their coordinating pH levels are as follows:

- pH 2–red
- pH 6–violet
- pH 10–blue-green
- pH 4–purple
- pH 8–blue
- pH 12–greenish yellow

Science Background

A normal stomach is actually slightly acidic and has a pH level of 1.6–2.4. The general principles used in this activity are valid; however, the pH numbers used do not approach the true physiologic pH values in the stomach.

An upset stomach may be caused by too much acid in the stomach rather than a lower pH level of acid in the stomach.

Not all antacids work the same way. Since this activity does not involve a real stomach, only the antacids that act as a buffer will give accurate results for this activity. Antacids that reduce the amount of acid in the stomach will appear to be less effective, although they may be more effective in a more acidic stomach.

Make teaching easier by using the management ideas and other information included in **Teacher Helps**.

The student formulates a **hypothesis** that states what he is trying to prove in the activity.

The student records observations and measurements in the **Procedure** section. Data is sometimes recorded using graphs and charts.

x   Introduction

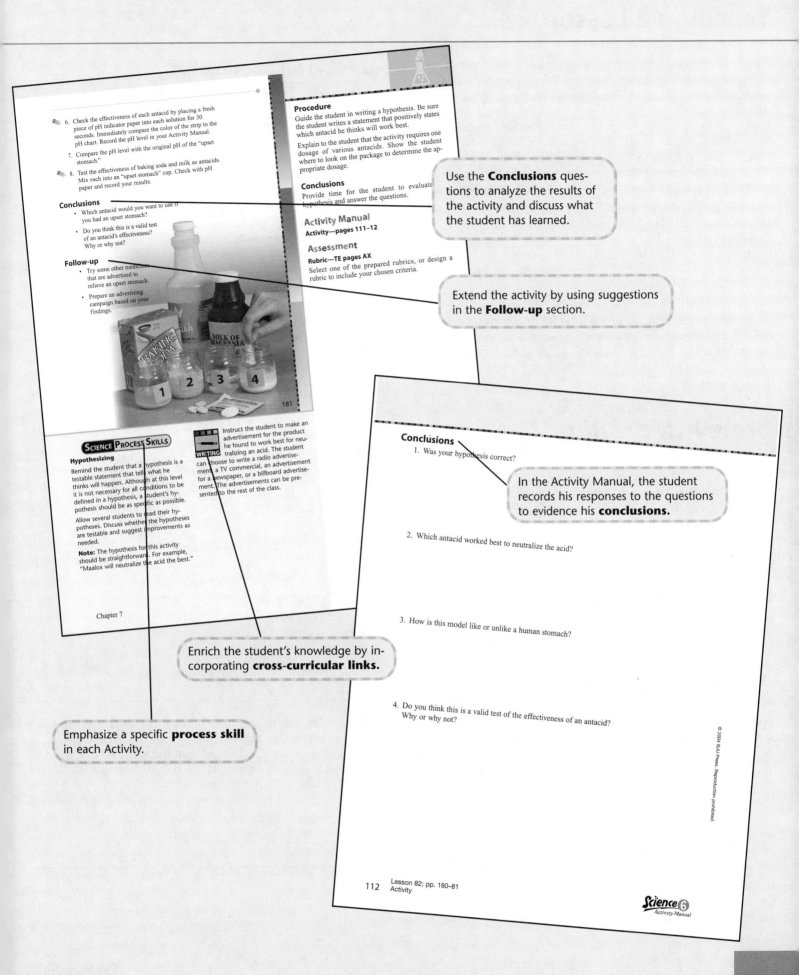

Introduction

Teaching a Lesson

Each lesson begins with relevant questions or a short activity to focus the student on the day's lesson. Most lessons cover 4–6 pages of the Student Text. The Purpose for Reading questions should be presented before the student begins to read the text. Each student should then read the material silently on his own. This may be done as a portion of the science class period; however, the teacher may choose to have the student read the text before the science lesson begins. Oral reading of the text should be reserved for short passages only as they relate to the discussion.

The lesson provides questions to help the student comprehend what he has read. This discussion provides an auditory reinforcement of key ideas. The higher-level thinking questions (marked with the light bulb icon) help the student apply the information that he has read by relating it to previous knowledge and everyday situations.

Enrichment information and activities allow the teacher to choose additional material to use with the lesson based on the class dynamics and the time available. Some enrichment material and activities reinforce a concept taught in the lesson, while other enrichment materials extend a concept. Enrichment material may also apply the science concepts to other subjects.

Reading for Information

The text lessons focus on reading for information, which is an essential skill for the student. Most of the reading required of a student on a secondary level will be informational reading.

There are several ways that the teacher can assist the student in developing and refining his informational reading skills. The PQ3R method is presented in Lesson 13. This method and other similar methods will aid the student in gathering information from the text.

The questions from the Purpose for Reading section help the student focus on the material as he reads. The Quick Check at the end of each lesson helps the student evaluate how well he has comprehended what he read.

Often graphic information, such as charts, graphs, maps, and diagrams, are part of the Student Text. Graphic information is used increasingly as a means of communicating. Learning to assimilate information from graphic sources is a reading skill that the student also needs to master.

Bible Integration

All subjects lend themselves to Bible integration. However, science by its very nature is an observation of God's revelation of Himself through His creation. SCIENCE 6 *for Christian Schools* incorporates not only Bible Action Truths and Bible Promises in the lessons, but also biblical truths that help the student develop a Christian worldview by showing God's nature and man's responsibility as revealed through creation.

Bible integration is often included as part of the text discussion. It may also be included as Bible Links in the Teacher's Edition or as Science and the Bible interest boxes in the Student Text.

Each chapter also includes a Bible Integration page in the Activity Manual. Although the page is usually placed near the lesson to which it most closely corresponds, it may be used at any time during the teaching of the chapter.

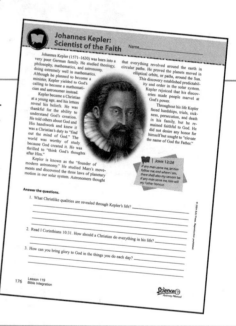

Review Lessons

A review lesson is planned for each chapter. Use of the suggested game or an alternate game from the *Game Bank* on pages A56–A57 enables the teacher to reinforce concepts in a fun way.

The review lesson also gives the teacher an opportunity to check to see that each student has accurately completed the study guides for the chapter. The material for each chapter test is taken from the Activity Manual study guides for that chapter. A student who knows the material (not just the answers) covered on the study guides will be adequately prepared to take the tests. (**Note:** The tests must be purchased separately.)

Science Notebook

It is highly recommended that students keep a looseleaf notebook for science. A notebook will allow the student to keep study guides, activity records, and other useful information organized.

Teaching an Activity or Explorations Lesson

It is important to emphasize the value of the Activity and Explorations lessons in SCIENCE 6. In order for the student to truly know science, he must be able to use it. Knowing how to apply science is the basic premise behind the term *scientific literacy*.

The Activity and Explorations lessons allow the student to demonstrate his understanding of science concepts. However, these activities also allow the student to apply knowledge from other subject areas, such as measurement skills from math and writing and communication skills from English.

Many activities in SCIENCE 6 are open-ended experiments. These require the student to incorporate various science process skills. Because the goal of these activities is to teach a mental process as opposed to specific procedures, the activities may not follow an exact format. An activity may have more than one correct result.

Explorations are project-oriented activities. They often require the student to work outside of class and usually require more creativity than the activities that involve simple experiments. While doing explorations, the student will develop and use skills such as conducting research, writing, and making oral presentations.

Most of the Activity and Explorations lessons can be completed by students working in groups. Using science groups helps students learn cooperation and management skills. Brainstorming ideas among group members is usually a good problem-solving technique.

Groups for Activities

Place students in groups of three or four. Some science groups work well with only two students, but groups of more than four tend to discourage participation by all members.

A science group should be a mixture of students from all achievement levels. You may find that students who are not good "book learners" are much better at doing hands-on activities.

At the beginning of the year, you may choose to assign tasks to each member of the groups to ensure active participation by all. Later, you should be able to allow the group to decide the tasks of each member.

Groups for Explorations

Using groups for explorations may require extra input from the teacher. Because the Explorations lessons are less structured, students may have a difficult time assigning themselves appropriate tasks. However, because these lessons often require a high level of creativity, students may be less intimidated in a group setting.

Management Tips

Being prepared is the key to a successful activity. Materials lists are given in the Student Text and in the Activity Manual. The Teacher's Edition also includes a materials chart on pages A58–A68. You will need to decide specific quantities based on the number of science groups you have in your class. You may need an additional quantity of some materials in order for each student to follow up the activity with new variables.

Having newspaper or plastic to cover work surfaces and buckets or containers to dispose of waste material will make the activities more manageable. Paper towels and damp rags are helpful also.

Introduction xiii

Science Process Skills

Science is not just a collection of facts. It is also a demonstration of processes that show understanding of how science works. Each Activity lesson in SCIENCE 6 allows the student to practice multiple science process skills. Additionally, one particular process skill is highlighted for discussion in each Activity lesson.

Although process skills are emphasized in the Activity lessons, they are also used in class discussions, demonstrations, and Explorations lessons. The following is a list of process skills that are used throughout SCIENCE 6.

Hypothesizing	formulating a statement that can be tested by experimentation
Observing	using the senses to gather information regarding objects and events
Classifying	grouping or ordering objects based on similarities or differences
Inferring	suggesting explanations based on previous knowledge or observation
Predicting	forecasting an expected result based on prior experience or knowledge
Communicating	using written, oral, or graphic means to transmit information to others
Measuring	using standard measuring devices and techniques to quantify information
Defining operationally	explaining an object or event properly but in terms of the student's observation and experience
Experimenting	setting up an investigation to test a hypothesis
Making models	creating a physical or mental representation to explain or clarify ideas, objects, or events
Recognizing and controlling variables	identifying the changing and unchanging factors in an investigation and adjusting one factor to obtain data
Collecting and recording data	gathering information about objects and events in an organized and systematic manner

Assessment and Grading

SCIENCE 6 provides a test packet that includes **chapter tests** and two or three **quizzes** per chapter. The tests are based on the study guide pages that are part of the Student Activity Manual.

To assess Activity and Explorations lessons, the Teacher's Edition includes a variety of **rubrics**. Rubrics can be used in many ways. A rubric is a valuable tool for evaluating work that is not objective. In a group activity it is often beneficial not only to give the group as a whole a grade, but also to give each student an individual grade. Sometimes allowing each member of a group to evaluate the contributions of the other members in the group will encourage all of the students to participate. Rubrics can also be given to the student for self-assessment. A rubric allows a student to see why he received a particular grade and shows him areas in which he can improve.

Each rubric in the Teacher's Edition contains a *Level of Importance* column. This allows the teacher to assign a different level of importance to each category based on the teacher's criteria.

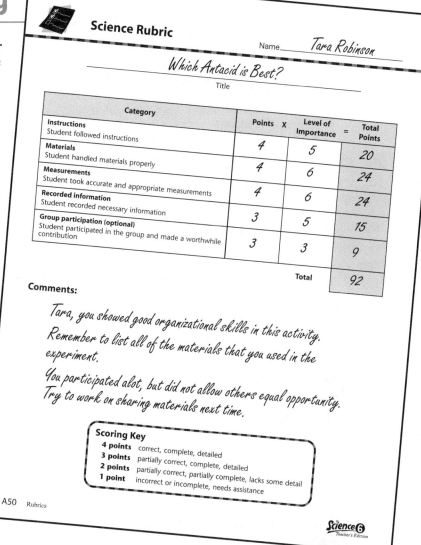

Scheduling

Full-year Plan

SCIENCE 6 provides material for a year-long program. Test and review days are included, and some activities have been allotted two lesson days for completion. The instruction of all textual information, activities, and explorations will adequately fill a full year science program.

Half-year Plan

A program that includes less than a full year of science instruction must adjust the use of the materials to best fit individual needs. The Table of Lesson Plans suggests lessons to include in a one-semester program. The last two chapters, dealing with the human body and health topics, are not included in the one-semester plan.

Tips

The following are additional ways to adjust the program to fit an alternate schedule.

- The material covered in Chapter 13, *Heredity and Genetics,* may not appear on the achievement tests. You may choose to skip this chapter in order to spend additional time on earlier chapters.

- The text lessons generally consist of four pages. You may choose to combine two text lessons or split three text lessons over a two-day period.

- Explorations are not included as part of the semester plan. However, many of the explorations can be incorporated into another subject, such as English, since they often require research and study skills.

- Many explorations and the enrichment activities included in most lessons can be adjusted for use in a learning center.

Science Fairs

The idea of coordinating a science fair for the first time may sound intimidating. The following information is meant to make the job a little easier. In addition to the suggestions here, there are numerous resources available on the Internet and in libraries. These resources go into greater detail regarding each aspect of planning, preparing, and running your fair.

Purpose of a Science Fair

Your students will benefit greatly from participating in a science fair. Completing and presenting a science fair project requires a synthesis of many skills that are important for your student to develop. The following are some of the skills sharpened through science fair participation:

- problem solving
- practice in researching
- use of the scientific method
- research writing
- public speaking

Planning Your Science Fair

If you have been placed in charge of coordinating a science fair, there are many logistical issues that you will need to address. The following checklist should help your planning:

- ✓ Choose a date for your fair well in advance and reserve a building or large room. Saturday is a good day, as it allows more parents and friends to visit the fair.
- ✓ Choose judges. One or two judges may suffice, depending on the size of your fair. The judge should be a professional in the area of science. A good choice would be a doctor, college science teacher, or college student majoring in science.
- ✓ Make a list of guidelines, hints, and judging criteria for the students and parents. Send these home three to six months in advance to allow plenty of time for preparation. Send your judges the judging criteria ahead of time as well.
- ✓ Charge a small fee per exhibit or per family to offset the cost of building rental, awards, and honorariums for the judges.
- ✓ Allow adequate space for each student's display. One half of an 8-foot table is visually sufficient to set up a project. Provide seating for the students while waiting for the judges.
- ✓ Make a map of the tables in the room ahead of time, and assign each participant to a space. Make sure there are plenty of electrical outlets in your fair area.
- ✓ Create a schedule for the day of the fair.
- ✓ Plan an awards program. Present awards for the best projects. A ribbon is a good award choice. Additionally, you may want to provide participation certificates for all the students. Provide a time for everyone to view the winning exhibits with their ribbons after the program.

Preparing for a Fair

Have a group meeting with parents, students, and teachers shortly after sending home the guidelines. The purpose of the meeting is to answer questions, explain expectations, and present project ideas to stimulate thinking. Allow time in class to pursue ideas for projects. Encourage each student to choose a topic that interests him. Brainstorm ideas with the students. If a student has an idea that interests him but would not necessarily make a good project, try suggesting branching into subtopics or related topics. The best topic asks a question that can be answered by conducting an experiment. Have available for your students some good resources on how to do a science fair project.

Presenting the Project
Display

The display should provide a clear presentation of the student's project. An upright display board should clearly show the title of the project and any graphs, charts, photos, explanations, and Bible application. The exhibit should also include visual aids, a written report, and the logbook.

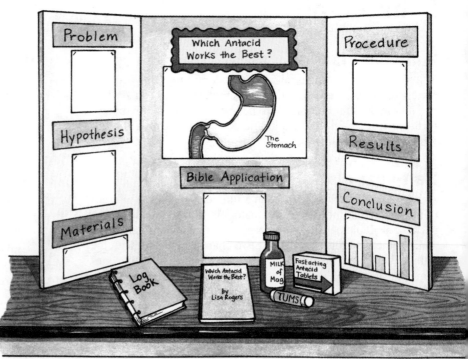

Logbook

A logbook is a journal of detailed steps and activities the student performs to accomplish the project and prepare the display. The logbook should include information regarding the observations and results of the scientific experiment.

Written report

The written report should include the following:

1. **Title Page**—gives the title of the project, the student's name, and his grade level
2. **Table of contents**—lists the sections of the paper and the corresponding page numbers
3. **Purpose**—gives the hypothesis and explains the student's idea, source of the idea, etc.
4. **Procedure**—explains the procedure step-by-step; gives instructions and materials as if someone were following the student's directions; sometimes includes pictures
5. **Results**—discusses the student's results; gives charts, graphs, pictures, etc.; identifies any errors in the experiment
6. **Conclusion**—draws conclusions from the results
7. **Bibliography**—lists references used, such as books or articles; acknowledges any help received from parents, professional individuals, businesses, etc.; acknowledges any donors of supplies

Oral report

The student should be prepared to explain his project orally to the judges. You may want to encourage the student to have a short speech prepared, or you may prefer to expect the student only to answer questions the judges ask. Each student should be prepared to explain how he came up with his idea, the procedures he followed, and the results of his experiment. Additionally, the student should be prepared to explain to the judges any problems he encountered, any help he received, what he learned, etc.

As part of his training in public speaking, the student should be encouraged to dress nicely and show proper manners while addressing the judges. The student should be encouraged to stand and use good eye contact while speaking to the judges.

Judging Criteria

- If possible, divide the grade levels so that younger students are not competing with older students.
- You may make separate judging categories for experiments, demonstrations, and models.
- The judging criteria may include the following: scientific merit (most important), originality, visual appearance, written report, oral presentation, and biblical application.
- You may choose to use the judging rubric on TE page A54 or make a rubric with your chosen criteria.

Table of Lesson Plans

Lessons marked with * are suggested for a one-semester course.

Lesson	TE Pages	ST Pages	AM Pages	Assessment	Content and Objectives
*1	1–3	1–3	1		**Unit and Chapter Opener** • Recognize the interrelationship of science concepts • Explain that ideas about science change, but that God never changes
*2	4–7	4–7	2		**Earthquakes** • Recognize that the earth's surface changes constantly • Explain the theory of plate tectonics • Use graphic information to identify the parts of the earth and the different kinds of faults • Infer that plate boundaries are unstable areas of the earth's surface
*3	8–11	8–11	2–4	Quiz 1-A	**Earthquake Waves** • Compare and contrast P, S, and two kinds of L waves • Explain the differences between the Mercalli scale and the Richter scale • Describe some ways that earthquakes cause damage
4	12–13		5–6		**Activity: A Science Experiment** • Use a scientific method
*5 6	14–15	12–13	7–8	Rubric	**Activity: Construction Site** • Make a model of a structure that can withstand the motion of an earthquake • Record and analyze information to form conclusions • Identify variables
*7	16–19	14–17	9–10		**Volcanoes** • Explain the causes of a volcanic eruption • Identify the parts of a volcano • Describe three ways volcanoes are classified
8 9	20–21	18–19	11–13	Rubric	**Activity: Create an Eruption** • Design a volcano based on one of the three kinds of volcanoes • Make a model of a volcano • Communicate the type of volcano made and the process used to make the volcano
*10	22–24	20–22	14	Quiz 1-B	**Weather Effects** • Identify possible dangers of volcanoes • List some of the meteorological effects of a volcanic eruption • List some of the products of volcanoes • Describe other kinds of thermal eruptions
11–12	25	23	15–16	Rubric	**Explorations: I.N.V.E.N.T.** • Identify the dangers and difficulties associated with exploring volcanoes • Design a piece of equipment that would help in volcano research
13	26–27 A18–A19		17–18		**PQ3R** • Use the PQ3R method to read informational text
14	28	24	19–20		**Chapter Review** • Recall concepts and terms from Chapter 1 • Apply knowledge to everyday situations

Chapter 1—Earthquakes and Volcanoes

Vocabulary	Process Skills	Enrichment	Bible Integration
			Creation under the curse of sin God's omniscience Interrelationship of the parts of creation God's use of creation for His glory Man's finite knowledge
lithosphere, plate, theory of plate tectonics, plate boundary, earthquake, fault		History Link Bible Link Demonstration	The Flood as God's judgement on sin God's omnipotence God's use of creation for His purposes
focus, seismic wave, epicenter, seismologist, seismograph, magnitude, tsunami		History Link Demonstration Bible Link Math Link Technology Link	Christ as solid foundation for life God's love shown through Christ's death Man's demonstration of God's love
scientific method, problem, material, hypothesis, procedure, conclusion	Hypothesizing Recording data		
	Predicting Experimenting Observing Making and using models Recording data Identifying and controlling variables		Christians as dependable workers Christians as faithful workers
magma, volcano, volcanologist, magma chamber, lava, vent, crater, ash, cone, Ring of Fire, hot spot, cinder, tephra, pyroclastic flow		Geography Link Math Link History Link	
	Predicting Making and using models Measuring Observing Communicating		
vog, debris flow, igneous rock, hot spring, geyser, mud pot		Language Link History Link Demonstration	God as Master of creation God's use of forces for Earth's benefit
	Communicating	Language Link	

Table of Lesson Plans

Lesson	TE Pages	ST Pages	AM Pages	Assessment	Content and Objectives
15				Chapter 1 Test	• Demonstrate knowledge of concepts taught in Chapter 1
	29	25	21		**Chapter Opener** • Recognize that scientific inferences are not always accurate
*16	30–33	26–29			**Rock Cycle and Mechanical Weathering** • Identify the three types of rocks and explain how they are formed • Differentiate between mechanical and chemical weathering • Define weathering and provide examples of mechanical weathering
*17	34–37	30–33	22	Quiz 2-A	**Chemical Weathering and Caves** • Define and give examples of chemical weathering • Describe how acid rain forms • Explain how limestone caves are formed as a result of chemical weathering
18	38–39		23–24		**Measurement** • Measure length to the nearest millimeter • Measure mass to the nearest gram • Measure volume to the nearest milliliter
*19	40–42	34–36	25–26	Quiz 2-B	**Soil** • Compare the different kinds of soil and their relative sizes • Describe the factors that determine the composition of soil • Illustrate and describe the five soil horizons
20	43 A20	37	27–28	Rubric	**Explorations: Soil Detective** • Follow the procedure of a flow chart • Analyze a soil sample
21	44–45	38–39	29–30	Rubric	**Activity: Retaining the Right Amount** • Record observations • Analyze experiment results • Predict the particles needed for a specific soil sample
22					
*23	46–49	40–43			**Erosion** • Differentiate between erosion and weathering • Distinguish among examples of mass wasting • Describe how sediments are carried and deposited by a stream
*24	50–51	44–45	31–32	Rubric	**Activity: Stream Erosion** • Record and analyze data • Measure accurately • Identify variables
*25	52–55	46–49	33–34	Quiz 2-C	**Wave, Wind, and Ice Erosion** • Recognize the real-life problems of sand erosion and deposition • Explain how wind causes erosion • Compare the effects of ice erosion with other kinds of erosion • Identify how rocks are eroded by the forces of wind and glaciers
26	56	50	35–36		**Chapter Review** • Recall concepts and terms from Chapter 2 • Apply knowledge to everyday situations

Chapter 2—Weathering and Erosion

Table of Lesson Plans

Vocabulary	Process Skills	Enrichment	Bible Integration
sedimentary, igneous, metamorphic, rock cycle, weathering, mechanical weathering, chemical weathering		Language Link Activity	
carbonic acid, speleothem, stalagmite, stalactite		Bible Link History Link Demonstration Language Link Writing Link	Christians as faithful witnesses God's perfect design God's use of forces for Earth's benefit
	Measuring	Demonstration Bible Link	Christian behavior as showing God's love to others Christians as faithful workers
soil, humus, pedologist, sand, clay, silt, texture, loam, horizon, topsoil, subsoil, bedrock		Language Link Bible Link	
	Observing		
	Hypothesizing Identifying variables Measuring Experimenting Observing Recording data		
sediment, erosion, agents of erosion, deposition, mass movement, load, delta, birdfoot delta		Language Link Technology Link Bible Link Demonstration	God as Master of creation God's use of forces for Earth's benefit
	Hypothesizing Identifying variables Measuring Experimenting Observing Recording data		
deflation, dust storm, sandstorm, glacier, plucking, moraine		Language Link Bible Link Demonstration Writing Link Technology Link	Man as steward of God's creation Man's use of God's resources

Table of Lesson Plans

Lesson	TE Pages	ST Pages	AM Pages	Assessment	Content and Objectives
27				Chapter 2 Test	• Demonstrate knowledge of concepts taught in Chapter 2
	57	51	37		**Chapter Opener** • Recognize that God plans natural events to benefit His creation
*28	58–61	52–55	38–40		**Nonrenewable Energy Resources** • Recognize the difference between renewable and nonrenewable resources • Explain how fossil fuels are formed • Identify the sources and uses of petroleum, natural gas, and coal • Describe the benefits and problems related to the use of nuclear energy
*29	62–63	56–57	41–42	Rubric	**Activity: Clean Up the Spill** • Explain the different methods of cleaning up an oil spill • Predict the best method for removing the oil • Use a model to demonstrate the different methods of cleanup • Compare the methods used in this activity with the methods used in a real oil spill
*30	64–67	58–61	43–44	Quiz 3-A	**Renewable Energy** • Describe some renewable energy resources • Compare and contrast renewable sources of energy
*31	68–71	62–65			**Minerals and Soil** • Name and identify the uses of several metals • Recognize soil as a natural resource • Identify several ways to conserve soil
32	72–73	66–67	45–46	Rubric	**Activity: Erosion Prevention** • Make models of soil with and without erosion prevention • Infer how certain materials prevent soil erosion
*33 34	74–79	68–73	47–48	Quiz 3-B	**Water Resources; Preserving Our Resources** • Identify several natural resources • Explain how the ocean is the source of most fresh water • Identify locations of fresh water • Describe the different kinds of ice
35	80–81 A21–A22		49–50		**Explorations: Water in Israel** • Compare the differences between water accessibility in Bible times and water accessibility now • Identify several ways to conserve water • Recognize the living water
36	82	74	51–52		**Chapter Review** • Recall concepts and terms from Chapter 3 • Apply knowledge to everyday situations

Chapter 3—Natural Resources

xxii Table of Lesson Plans

Vocabulary	Process Skills	Enrichment	Bible Integration
			God's use of forces for Earth's benefit
natural resource, renewable resource, nonrenewable resource, fossil fuel, petroleum, crude oil, refinery, petrochemical, natural gas, coal, uranium		Art Link Bible Link Geography Link	God's provision for man Man's uses of God's resources The Flood's effect on the earth
	Hypothesizing Predicting Making a model Observing Inferring	History Link	Man's responsibility for his actions
reservoir, hydroelectric energy, geothermal energy, wind energy, solar energy		Bible Link Demonstration History Link Geography Link Scientist Link Technology Link	
mineral, vein, ore, smelting, malleable, fallow, ground cover, contour plowing		Bible Link Technology Link	Giving God the best God's plan for worship God's refining in Christians' lives Man's use of God's resources
	Observing Making a model Recording data Inferring	Bible Link	
hydrosphere, water cycle, saline, phytoplankton, ground water, aquifer, drawdown, humidity, ice sheet, ice shelf, iceberg, sea ice, reduce, reuse, recycle		Bible Link Geography Link Technology Link Writing Link Activity	God's design for Earth's resources God's provision for man Man as a steward of God's creation
cistern	Recording data	Math Link	God's gift of eternal life Salvation as the living water

Table of Lesson Plans

Lesson	TE Pages	ST Pages	AM Pages	Assessment	Content and Objectives
37				Chapter 3 Test	• Demonstrate knowledge of concepts taught in Chapter 3
	83–85	75–77	53		**Unit and Chapter Opener** • Understand the interrelationship of science concepts • Recognize that God supplies the needs of every organism
*38	86–90	78–82			**Cells and Organisms** • Distinguish between living things and nonliving things • Identify five characteristics of living things • Identify men associated with the development of the microscope • Explain the cell theory
*39	91–92 A23		54		**Using a Microscope** • Identify the parts of a microscope • Explain how to use a microscope
*40	93–95	83–85	55–56	Quiz 4-A	**Cells** • Identify a cell as a living organism • Discuss the relationship of cells, tissues, organs, and systems • Identify cell structures • Compare and contrast plant and animal cells
*41 42	96	86		Rubric	**Activity: Cell Model** • Demonstrate knowledge of cell structure • Construct a 3-D model of a cell • Prepare a written report
43 44	97	87	57	Rubric	**Explorations: An Organized Cell** • Correlate the function of cell structure to another organization • Write and present a skit to compare a cell to an organization
*45	98–99	88–89			**Reproduction of Cells** • Describe the process of cell division—both mitosis and meiosis • Recognize when mitosis occurs and when meiosis occurs
46	100 A24	90	58	Rubric	**Activity: Classifying** • Distinguish groups according to chosen criteria • Complete a classification chart
*47	101–5	91–95	59–61	Quiz 4-B	**Living Kingdoms** • Name the six kingdoms • Identify characteristics of each kingdom • Explain how man is similar to and yet different from other living organisms
*48	106–7	96–97	62		**Naming Organisms** • Recognize that Carolus Linnaeus was responsible for the method of classification that we use • List the levels of the classification system from largest to smallest • Compare the common names and the scientific names of organisms • Write a scientific name properly
49	108	98	63–64		**Chapter Review** • Recall concepts and terms from Chapter 4 • Apply knowledge to everyday situations

Chapter 4—Cells and Classification

Table of Lesson Plans

Vocabulary	Process Skills	Enrichment	Bible Integration
organism, life span, life cycle, environment, energy, cell, cell theory, microscope		Bible Link Demonstration Scientist Link Writing Link	Creation under the curse of sin Death and decay as a result of sin God's perfect design God's perfect creation Man's finite knowledge
eyepiece, body tube, adjustment knob, nosepiece, arm, objective lens, specimen, clips, slide, stage, light source, base		Language Link Math Link	
unicellular, multicellular, tissue, organ, system, cell membrane, cytoplasm, organelle, nucleus, chromosome, mitochondria, endoplasmic reticulum, ribosome, vacuole, cell wall, chloroplast, chlorophyll		Language Link Bible Link	
cytologist	Making and using a model Communicating		
	Communicating Making and using a model		
cell division, mitosis, sexual reproduction, meiosis			God's design for man's body Purity of a Christian's heart
criteria	Observing Classifying Communicating		
classification, bacteria, colony, protozoan, algae, leaven, photosynthesis		Demonstration Language Link Activity	Effects of a little sin God's perfect design God's plan of redemption God's provision for His creation God's provision for man's sin Man as God's special creation Man's God-given dominion Man's fall Christ as sacrifice
common name, kingdom, phylum, class, order, family, genus, species, scientific name		Language Link Activity Scientist Link	God's omniscience God's orderly design God's variety in creation Man's finite knowledge

Table of Lesson Plans

Lesson	TE Pages	ST Pages	AM Pages	Assessment	Content and Objectives
				Chapter 4 Test	• Demonstrate knowledge of concepts taught in Chapter 4
50	109	99	65		**Chapter Opener** • Recognize that man is unaware of many parts of God's creation
*51	110–13	100–103	66		**Sponges, Stinging Animals, and Mollusks** • Recognize invertebrates and vertebrates as a broad way to distinguish animals • Recognize that unique animal characteristics allow classification • Describe the unique characteristics of the sponge, jellyfish, mollusk, and echinoderm groups
52	114	104	67	Rubric	**Explorations: Snail Terrarium** • Construct a terrarium • Observe land snails • Record observations
*53	115–17	105–7	68–70		**Echinoderms; Flatworms; Roundworms; Segmented Worms** • Describe three types of worms • Compare a free-living worm with a parasite • Explain why worms can be helpful to man • Explain why worms can be harmful to man
*54	118–21	108–11	71–72	Quiz 5-A	**Arthropods** • Identify crustaceans, arachnids, centipedes, millipedes, and insects as arthropods • Describe basic characteristics of each kind of arthropod
55	122–23 A25	112–13	73–74	Rubric	**Activity: Mealworm Movement** • Observe the larval stage of complete metamorphosis • Observe the pupal stage of complete metamorphosis • Collect and record observation data
*56	124–27	114–17			**Fish and Amphibians** • Identify the characteristics of fish • Identify the characteristics of amphibians • Recognize how fish and amphibians are alike • Describe the life cycle of most amphibians
*57	128–31	118–21	75	Quiz 5-B	**Reptiles and Birds** • Identify two characteristics of reptiles • Identify two characteristics of birds • Explain how birds and reptiles are similar and different
*58 *59	132–37	122–27	76	Quiz 5-C	**Mammals and Humans** • Identify four characteristics of mammals • Explain how marsupials are different from other mammals • Recognize how humans are different from mammals
60	138–39	128–29	77–78	Rubric	**Activity: Blubber Mitts** • Write a hypothesis • Record temperatures and observations • Relate the effectiveness of shortening or lard as an insulator to the effectiveness of animal blubber

Chapter 5—Animal Classification

Table of Lesson Plans

Vocabulary	Process Skills	Enrichment	Bible Integration
			God's care for His creation God's omniscience Man's finite knowledge
invertebrate, vertebrate, nematocyst, mollusk		Language Link Activity History Link	God's orderly design God's perfect design God's provision for His creation
	Observing Recording data		God's use of creation for His purposes
filter feeder, parasite, free-living, segment		Art Link Demonstration	Christians as faithful workers
arthropod, exoskeleton, molt, crustacean, arachnid, insect, metamorphosis, incomplete metamorphosis, complete metamorphosis		Language Link Bible Link	
	Experimenting Recording data Observing Identifying and controlling variables	Bible Link	
cold-blooded, cartilage, amphibian		Writing Link History Link	God's perfect design God's use of creation for His purposes
omnivore, herbivore, carnivore, warm-blooded		Language Link Bible Link History Link Demonstration Math Link	Man's imitation of creation God's perfect design God's provision for His creation God's use of creation for His purposes
insectivore, nocturnal, echolocation, pride, blubber, pod		Language Link Demonstration History Link	Man created in God's image Man as God's special creation
	Predicting Experimenting Measuring Inferring Observing Recording data		

Table of Lesson Plans

	Lesson	TE Pages	ST Pages	AM Pages	Assessment	Content and Objectives
Chapter 5	61	140–41	130–31	79	Rubric	**Explorations: Animal Robotics** • Associate animal parts with mechanical tools • Research to design a robotic animal • Prepare a drawing and description of the robotic animal
	62					
	63	142	132	80–82		**Chapter Review** • Recall concepts and terms from Chapter 5 • Apply knowledge to everyday situations
Chapter 6—Plant Classification	64				Chapter 5 Test	• Demonstrate knowledge of concepts taught in Chapter 5
		143	133	83		**Chapter Opener** • Recognize that man's knowledge must continually be re-evaluated
	*65	144–47	134–37	84		**Nonvascular Plants; Seedless Vascular Plants** • Explain the difference between vascular and nonvascular plants • Recognize that vascular plants can be classified as seed-bearing plants and seedless plants • Identify ferns, horsetails, and club mosses as seedless vascular plants • Identify the parts of a fern: rhizome, fronds, fiddleheads
	*66	148–51	138–41	85–86	Quiz 6-A	**Gymnosperms** • Recognize that seed-producing plants can be classified as gymnosperms and angiosperms • Identify four kinds of gymnosperms • Identify two kinds of conifers • Describe several ways that man uses conifers
	*67	152–55	142–45	87–88		**Angiosperms** • Recognize that angiosperms include trees, shrubs, and flowering plants • Distinguish among annuals, biennials, and perennials • Name some ways that angiosperms are used • Compare monocotyledons and dicotyledons
	*68	156	146	89–90	Rubric	**Activity: Classification Check** • Plan a visual to demonstrate how plants are classified
	*69	157	147		Rubric	**Explorations: Plant Products** • Research products made from a given plant • Prepare a display to demonstrate research results
	*70	158–61	148–51	91–92	Quiz 6-B	**Plant Parts** • Identify the two kinds of vascular tissue and their functions • List three main functions of the stem of a plant • Describe the difference between herbaceous and woody stems • List three main functions of the root system • Describe the differences between taproots and fibrous roots
	71	162–63	152–53	93–94	Rubric	**Activity: How Big is My Tree?** • Measure the circumference, height, and crown of a tree and calculate the tree's point value • Produce and read a graph to show relationships
	72	164	154	95–96		**Chapter Review** • Recall concepts and terms from Chapter 6 • Apply knowledge to everyday situations

Table of Lesson Plans

Vocabulary	Process Skills	Enrichment	Bible Integration
	Inferring	Activity	Man's imitation of creation Man's God-given curiosity
			God's orderly design Man's finite knowledge
vascular plant, nonvascular plant, rhizoid, rhizome, frond, fiddlehead		Language Link Demonstration History Link	God's love of beauty God's variety in creation
angiosperm, gymnosperm		Bible Link Demonstration Geography Link Math Link	Giving God the best
annual, biennial, perennial, cotyledon, dicotyledon, monocotyledon		Activity Language Link	Man's God-given dominion
taxonomy	Observing Classifying Communicating	Language Link	
	Communicating	History Link	
xylem, phloem, vascular bundle, cambium, tuber, herbaceous, primary root, taproot, fibrous root		Demonstration Technology Link	The Bible as final authority Christians rooted and grounded in Christ God's perfect design God's provision for His creation
	Measuring Observing Inferring Communicating Collecting, recording, and interpreting data		

Table of Lesson Plans

Lesson	TE Pages	ST Pages	AM Pages	Assessment	Content and Objectives
				Chapter 6 Test	• Demonstrate knowledge of concepts taught in Chapter 6
73	165–67	155–57	97		**Unit and Chapter Opener** • Understand the interrelationship of science concepts • Recognize that man's inferences are sometimes inaccurate
*74	168–71	158–61	98		**Atoms** • Describe and label the size, charge, and location of each part of an atom • Recognize that an element is made of only one kind of atom • Differentiate between atomic mass and atomic number
*75	172–75	162–65	99–100	Quiz 7-A	**Elements** • Identify the terms *period, group,* and *family* as they relate to a periodic chart • Understand that the periodic chart is only a classification system • Describe the process that Mendeleev used for arranging elements • Identify the types of information in a square on the periodic table
76	176	166	101–2	Rubric	**Explorations: Wanted: U or Your Element** • Research an element • Construct a visual aid
*77	177–79	167–69			**Compounds; Chemical Formulas; Chemical Reactions** • Explain that a chemical change occurs when atoms combine • Give examples of synthesis and decomposition reactions • Demonstrate how to write a chemical formula
*78	180–81	170–71	103–4		**Atomic Bonding** • Compare and contrast ionic and covalent bonding • Explain what causes an ion
79	182–83	172–73	105–6	Rubric	**Activity: Hot or Cold** • Recognize whether chemical reactions occur • Collect data to identify a reaction as endothermic or exothermic
*80	184–87	174–77	107–8	Quiz 7-B	**Acids and Bases** • Compare and contrast characteristics of acids and bases • Describe the purpose of an indicator • Identify products that are acids, bases, or salts • Explain how a salt is formed
*81	188–89	178–79	109–10	Rubric	**Activity: pH Indicator** • Identify a solution as an acid or a base by using a pH indicator solution • Estimate the strength of an acid or base solution by using a pH indicator solution • Observe the effects of acids and bases on an indicator
82	190–91	180–81	111–12	Rubric	**Activity: Which Antacid is Best?** • Hypothesize about the effectiveness of several antacids • Make and use a model of "upset stomach" acid • Infer information from the model
83	192	182	113–14		**Chapter Review** • Recall concepts and terms from Chapter 7 • Apply knowledge to everyday situations

Chapter 7—Atoms and Molecules

Table of Lesson Plans

Vocabulary	Process Skills	Enrichment	Bible Integration
			God's creation of invisible forces God's holding all creation together God's omniscience Man's finite knowledge
chemistry, atom, element, nucleus, proton, neutron, atomic mass, electron, shell, atomic number, atomic theory, particle accelerator		Language Link Scientist Link Technology Link Activity	God as Master of creation Man's finite knowledge
chemical symbol, periodic table of the elements, period, group, family		Scientist Link Art Link Activity Math Link	God's orderly design
	Communicating		
molecule, chemical reaction, compound, chemical formula		Activity Demonstration	
closed, covalent bond, ion, positive ion, negative ion, ionic bond		Bible Link Activity	Christians bonded in Christ
endothermic reaction, exothermic reaction	Predicting Observing Measuring Experimenting Recording data	Language Link	Evidences of salvation
acid, base, alkali, pH scale, neutral, indicator, neutralize, salt		Activity	God's creation for man's enjoyment
	Predicting Measuring Observing Recording data		
	Hypothesizing Experimenting Observing Inferring Recording data		

Table of Lesson Plans

	Lesson	TE Pages	ST Pages	AM Pages	Assessment	Content and Objectives
rowspan="10"	Chapter 8—Electricity and Magnetism	84				Chapter 7 Test	• Demonstrate knowledge of concepts taught in Chapter 7
		193	183	115		**Chapter Opener** • Recognize God's use of man's curiosity
	*85	194–97	184–87	116		**Static Electricity; Current Electricity** • Explain what causes static electricity • Identify the two things needed for an electric current to flow • Describe the characteristics of conductors, resistors, and insulators
	86	198–99	188–89	117–18	Rubric	**Activity: An "Unbreakable" Circuit** • Design and build an "unbreakable" circuit • Experiment to test hypotheses
	*87	200–203	190–93	119–22	Quiz 8-A	**Circuits; Measuring Electricity; Batteries** • Differentiate between parallel circuits and series circuits • Distinguish among the three basic units of electrical measurement: volt, ampere, and watt • Explain how a battery works
	*88	204–6	194–96	123–24		**Magnetism** • Describe what happens to magnets at their poles • Explain the relationship between magnetism and electricity • Identify and describe the parts of a generator • Explain how a generator works
	89	207 A26	197		Rubric	**Explorations: Famous Inventors** • Research an inventor • Present a speech honoring an inventor
	*90	208–9	198–99	125–26	Rubric	**Activity: Build an Electromagnet** • Identify ways to increase a wire's magnetism • Predict ways to strengthen an electromagnet • Experiment to test predictions
	*91	210–13	200–203	127–28	Quiz 8-B	**Electronics** • Explain the difference between electricity and electronics • Identify the benefits of an integrated circuit • Identify some of the parts of a computer
	92	214	204	129–30		**Chapter Review** • Recall concepts and terms from Chapter 8 • Apply knowledge to everyday situations
Chapter 9	93				Chapter 8 Test	• Demonstrate knowledge of concepts taught in Chapter 8
		215	205	131		**Chapter Opener** • Recognize that only God can create energy
	*94	216–19	206–9	132		**Motion** • Differentiate between speed and velocity • Explain why a reference point is needed to observe motion • Explain the relationship of mass and velocity to momentum

xxxii Table of Lesson Plans

Vocabulary	Process Skills	Enrichment	Bible Integration
			Man's finite knowledge Man's God-given curiosity
static electricity, current electricity, circuit, conductor, switch, short circuit, insulator, resistor		Language Link Demonstration Bible Link	Man as steward of God's creation Man's God-given dominion
	Hypothesizing Predicting Experimenting Inferring Identifying and controlling variables	Bible Link	
series circuit, parallel circuit, volt, watt, ampere, electric cell, electrolyte, battery		Activity Math Link Demonstration	God's perfect design God's provision for His creation
magnet, magnetism, magnetic field, electromagnet, generator		Demonstration	
	Communicating	Language Link	
	Hypothesizing Predicting Experimenting Observing Inferring Identifying and controlling variables Recording data		
electronic device, electrical signal, binary number system, integrated circuit, semi-conductor, CPU, ROM, RAM		Scientist Link Activity History Link	Discerning what is true Knowing God as greatest wisdom
			God as only Creator
energy, potential energy, kinetic energy, mechanical energy, motion, reference point, distance, speed, instantaneous speed, velocity, acceleration, force, friction, momentum		Math Link Language Link	Christ as a Christian's reference point Bible as final authority

	Lesson	TE Pages	ST Pages	AM Pages	Assessment	Content and Objectives
Chapter 9—Motion and Machines	*95	220–23	210–13	133–34	Quiz 9-A	**Laws of Motion** • Identify Newton's three laws of motion • Explain that both gravity and friction work against inertia
	96	224–25	214–15	135–36	Rubric	**Activity: Mini Cars in Motion** • Plan a demonstration to illustrate the laws of motion • Experiment to show each of the laws of motion with toy cars • Identify the laws of motion in real-life situations
	97	226	216	137–38	Rubric	**Explorations: Roller Coaster** • Design and make a model roller coaster • Discover relationships between slope, speed and momentum
	*98	227–29	217–19	139–40		**Work; Simple Machines: Levers** • Explain that *work* equals *force* times *distance* • Differentiate among the three classes of levers
	*99	230–33	220–23	141–44	Quiz 9-B	**Pulleys; Wheels and axles; Inclined planes; Wedges; Screws; Compound Machines** • Describe a pulley, wheel and axle, inclined plane, wedge, and screw • Discern between a fixed pulley, a moveable pulley, and a block and tackle • Explain what a compound machine is
	*100	234–35	224–25	145–46	Rubric	**Activity: How Much Force?** • Experiment to show that an inclined plane reduces the amount of force needed to do work • Measure metrically in newtons and centimeters • Define operationally the results of the activity
	101	236	226	147–50		**Chapter Review** • Recall concepts and terms from Chapter 9 • Apply knowledge to everyday situations
Chapter 10	102				Chapter 9 Test	• Demonstrate knowledge of concepts taught in Chapter 9
		237–39	227–29	151		**Unit and Chapter Opener** • Understand the interrelationship of science concepts • Recognize God's miraculous control over nature
	*103	240–43	230–33	152		**Our Closest Star; Characteristics of Stars** • Explain how stars produce their own light • Distinguish between apparent magnitude and absolute magnitude of stars • Recognize that stars are classified according to color • Explain ways distance is measured in space • Read a graph
	*104	244–47	234–37	153–54	Quiz 10-A	**Kinds of Stars** • Differentiate between a pulsating variable star and an eclipsing variable star • Describe the causes of a nova and of a supernova • Explain how a neutron star or black hole is formed

Table of Lesson Plans

Vocabulary	Process Skills	Enrichment	Bible Integration
first law of motion, inertia, gravity, second law of motion, third law of motion		Demonstration Writing Link Math Link Activity Bible Link	Spirit-filled Christians
	Experimenting Making and using models Observing Communicating		God's orderly design
	Making and using models Inferring		
work, newton, joule, machine, effort force, resistance force, lever, fulcrum, first-class lever, second-class lever, third-class lever		Bible Link Language Link Demonstration	God's design of man's body
pulley, fixed pulley, moveable pulley, block and tackle, mechanical advantage, wheel and axle, inclined plane, wedge, screw, threads, compound machine		Demonstration History Link Writing Link	Results of unconfessed sin
	Measuring Observing Defining operationally Recording data		
			God's use of creation for His purpose God overruling His natural laws
magnitude, apparent magnitude, absolute magnitude, dwarf, giant star, supergiant, light-year, parallax		Language Link Bible Link Demonstration Math Link	God's omniscience God's use of creation for His glory God's use of creation for His purposes
variable star, pulsating variable star, eclipsing variable star, nova, nebula, supernova, neutron star, pulsar, black hole		Language Link Writing Link Demonstration Technology Link	God's variety in creation Man's finite knowledge

Table of Lesson Plans

Lesson	TE Pages	ST Pages	AM Pages	Assessment	Content and Objectives
*105	248–51	238–41	155		**Observing the Heavens** • Identify various constellations • Explain why a Christian should not be involved in astrology • Describe the difference between a reflecting telescope and a refracting telescope • Identify instruments used to study the stars
106	252 A27–A28	242		Rubric	**Activity: Pop Can Constellations** • Make a model of a constellation. • Recognize and name some star groups and constellations
107	253 A29–A32	243		Rubric	**Explorations: A Different Look** • Make a model of a constellation • Plot points on a graph • Recognize the relative distances of stars
*108	254–59	244–49	156–58	Quiz 10-B	**Star Groups** • Identify how many stars are in a binary star group and in a multiple star group • Differentiate between an open star cluster and a globular cluster • Identify our galaxy as the Milky Way • Recognize that our galaxy is part of a cluster of galaxies called the Local Group • Describe asteroids, meteoroids, meteors, meteorites, and comets
109	260–61 A33		159–60	Rubric	**Explorations: Stargazing** • Read and use a star chart • Identify objects in the night sky • Record observations
*110	262–63	250–51	161–62	Rubric	**Activity: Crater Creations** • Measure mass and length • Use a chart to record information • Make and test predictions
111	264	252	163–64		**Chapter Review** • Review concepts and terms from chapter 10 • Apply knowledge to everyday situations
112				Chapter 10 Test	• Demonstrate knowledge of concepts taught in Chapter 10
	265	253	165		**Chapter Opener** • Recognize that God's creation is orderly
*113	266–69	254–57	166		**The Sun and the Seasons** • Identify the parts of the Sun • Describe the characteristics of a solar storm • Explain why Earth experiences seasons • Understand that the Sun's gravitational pull keeps the planets in orbit
*114	270–73	258–61	167–68	Quiz 11-A	**The Planets** • Describe similarities among the inner planets • Explain how man has gradually learned about the planets • Identify characteristics of Mercury, Venus, and Mars

Chapter 10—Stars

Chapter 11

Vocabulary	Process Skills	Enrichment	Bible Integration
constellation, circumpolar constellation, astrology, astronomy, refracting telescope, reflecting telescope, radio telescope, spectroscope, redshift		Bible Link Geography Link Activity Language Link History Link Demonstration	Faith in the Word of God for guidance God's Word as the only true source of guidance God's omnipotence God's use of creation for His glory
	Making and using models Observing		
	Measuring		
binary system, multiple star group, open star cluster, globular cluster, galaxy, Local Group, asteroid, meteoroid, meteor, meteorite, comet		Language Link Demonstration Activity History Link Writing Link	God's omnipotence God as Master of creation
			Faith in the Word of God
	Hypothesizing Measuring Observing Recording data Identifying and controlling variables Communicating		
		Bible Link	God's orderly design God's perfect design God's provision for His creation
photosphere, chromosphere, corona, sunspot, faculae, solar flare, solar prominence, solar wind, aurora, revolution, axis, rotation		Bible Link Language Link Math Link Demonstration	God's perfect design
terrestrial		Language Link Math Link Scientist Link History Link	

Table of Lesson Plans

Lesson	TE Pages	ST Pages	AM Pages	Assessment	Content and Objectives
*115	274–77	262–65			**Earth; the Moon; Project Apollo; Eclipses** • Explain some ways God made Earth unique • Describe why the same side of the Moon always faces Earth • Give details about the *Apollo 11* mission • Describe the causes of solar and lunar eclipses
*116	278–79	266–67	169–70	Rubric	**Activity: Spare Parts Solar Oven** • Construct a solar oven that will melt a marshmallow • Infer the relationship between materials used and results
*117	280–83	268–71	171–72		**The Outer Planets** • Identify characteristics of each of the outer planets • Recognize that the *Voyager* probes have explored the outer planets
118	284–85		173–74	Rubric	**Explorations: Solar Walk** • Construct a scale model of the solar system • Gain a greater understanding of the vastness of our solar system
*119	286–89	272–75	175–76	Quiz 11-B	**Space Exploration** • Explain how a rocket uses thrust to launch • Define Newton's third law of motion • Distinguish between a satellite and a probe
120	290–91		177–78	Rubric	**Explorations: Travel Brochure** • Design a travel brochure for a planet • Collect data
121	292–93	276–77	179–80	Rubric	**Activity: Rocket Race** • Hypothesize how design affects the performance of a balloon rocket • Construct a balloon rocket • Demonstrate an understanding of Newton's third law of motion
122	294	278	181–82		**Chapter Review** • Review concepts and terms from chapter 11 • Apply knowledge to everyday situations
123				Chapter 11 Test	• Demonstrate knowledge of concepts taught in Chapter 11
	295–97	279–81	183		**Unit and Chapter Opener** • Understand the interrelationships of science concepts • Recognize that man's inferences are sometimes faulty
*124	298–301	282–85	184		**Plant Reproduction** • Explain the purpose for each part of a flower • Differentiate between pollination and fertilization • Explain how scientists classify fruits • Describe the process of germination

Chapter 11—Solar System

Chapter 12

Table of Lesson Plans

Vocabulary	Process Skills	Enrichment	Bible Integration
satellite, solar eclipse, totality, lunar eclipse		Demonstration Bible Link Writing Link Technology Link Activity Language Link	Christians as a reflection of God God's provision for His creation God's loving care
	Observing Inferring Identifying variables Recording data	Bible Link Writing Link Art Link	God's orderly design
		Demonstration Language Link Math Link Writing Link	God's omnipotence
	Measuring Making and using models	Writing Link Demonstration	God's love for man God's salvation through Christ God's loving care
probe, International Space Station, dehydrated		Language Link Scientist Link Technology Link Bible Link Writing Link History Link	
	Communicating		
	Hypothesizing Measuring Making and using models Observing Inferring Recording data	Activity	
		Scientist Link	The Bible as final authority God as Master of creation Man's finite knowledge
sepal, stamen, anther, pistil, ovary, ovule, style, stigma, cross-pollination, self-pollination, fertilization, zygote, embryo, fruit, germinate, cotyledon, seed coat		Language Link Demonstration Bible Link	God's provision for His creation God's plan for salvation God's love of beauty

Table of Lesson Plans

Lesson	TE Pages	ST Pages	AM Pages	Assessment	Content and Objectives
Chapter 12—Plant and Animal Reproduction					
*125	302–3	286–87	185–86	Rubric	**Activity: Flower Dissection** • Measure the parts of a flower • Identify the parts of a flower
*126	304–7	288–91	187–88	Quiz 12-A	**Seeds in Cones; Spores** • Explain how conifers reproduce • Compare and contrast seeds and spores • Identify some organisms that reproduce by spores
*127	308–11	292–95	189		**Animal Reproduction** • Recognize that animals begin as a single cell • Compare and contrast placental and marsupial gestation • Differentiate between the different types of eggs • Understand why some animals lay many eggs
128	312–13		190		**Explorations: What Value Does God Place on Life?** • Recognize the value that God places on life • Recognize that God provides eternal life
129	314–17 A34	296–99	191–94	Quiz 12-B Rubric	**Asexual Reproduction** • Identify some methods of asexual reproduction **Activity: It's a Race** • Set up an experiment to observe and compare the rate of growth of a plant cutting and a seed
130	318	300	195–96		**Chapter Review** • Recall concepts and terms from Chapter 12 • Apply knowledge to everyday situations
Chapter 13—Heredity and Genetics					
131	319	301	197	Chapter 12 Test	• Demonstrate knowledge of concepts taught in Chapter 12 **Chapter Opener** • Recognize that a parent's acquired abilities are not part of inherited traits
*132	320–22	302–4	198–200	Rubric	**Heredity** • Explain how chromosomes, DNA, and genes are related • Identify some learned and inherited traits **Activity: It's All in the Genes** • Take a survey of a sampling group • Graph recorded survey results
*133	323–24 A35	305–6	201–4	Quiz 13-A	**DNA: the Double Helix** • Identify the structure of a DNA molecule • Recognize James Watson and Francis Crick as those who identified DNA structure • Make a model of a DNA molecule • Identify ways DNA testing is used
134	325	307	205–6	Rubric	**Explorations: DNA Extraction** • Extract DNA from organic matter
*135	326–29	308–11	207		**Father of Genetics; Dominant and Recessive Genes** • Describe some of Mendel's experimental procedures • Explain some of Mendel's conclusions • Recognize the difference between dominant genes and recessive genes

Table of Lesson Plans

Vocabulary	Process Skills	Enrichment	Bible Integration
	Measuring Observing Recording data Defining operationally		God's perfect design
spore, fruiting body		Demonstration	
gestation, placental mammal, marsupial mammal		Writing Link	God's perfect design God's variety in creation
			God's value of life God's gift of eternal life
asexual reproduction, binary fission, budding, regeneration, fragmentation, vegetative reproduction	Hypothesizing Measuring Observing Recording data Communicating		
			God's plan for heredity
trait, gene, DNA, heredity, inherited	Collecting data Interpreting data Communicating	Language Link Demonstration	God's knowledge of each individual The Bible as final authority The Holy Spirit's guidance
		History Link Scientist Link	
	Observing	Bible Link	
purebred plant, P generation, hybrid, dominant trait, recessive trait, phenotype, genotype, codominant, incomplete dominance		Scientist Link Language Link	Honesty

Table of Lesson Plans

	Lesson	TE Pages	ST Pages	AM Pages	Assessment	Content and Objectives
Chapter 13	*136	330–33	312–15	208–10		**Punnett Squares; Pedigrees** • Predict genetic probability using a Punnett square • Interpret a pedigree chart • Identify some sex-linked traits
	*137	334–35 A36–A37	316–17	211–12	Rubric	**Activity: Paper Pet Genetics** • Use Punnett squares to predict genotypes • Construct paper pets based on predicted genotypes
	*138	336–39	318–21	213	Quiz 13-B	**Genetic Disorders and Diseases; Genetic Engineering** • Identify and discuss some common genetic diseases and disorders • Explain why genetic diseases are not easy to cure • Name some examples of genetic engineering
	139	340	322	214–16		**Chapter Review** • Recall concepts and terms from chapter 13 • Apply knowledge to everyday situations
Chapter 14—Nervous System	140				Chapter 13 Test	• Demonstrate knowledge of concepts taught in Chapter 13
		341–43	323–25	217		**Unit and Chapter Opener** • Recognize the interrelationship of science concepts • Recognize that man's inferences are sometimes inaccurate
	141	344–47	326–29	218		**The Central Nervous System** • Identify the two main parts of the nervous system • Describe the parts of the central nervous system • List the four lobes of the cerebrum • Differentiate among the functions of the three parts of the brain
	142	348–51 A38–A39	330–33	219–20	Quiz 14-A	**The Peripheral Nervous System** • Identify the parts of a neuron • Explain how neurons send messages • Compare the two parts of the peripheral nervous system • Describe how a reflex occurs
	143	352–53	334–35	221–22	Rubric	**Activity: Reaction Time** • Explore variables that affect reaction time
	144	354–57	336–39	223		**The Five Senses** • Recognize how the five senses interact with the nervous system • Read diagrams for information • Identify the nerves associated with hearing, sight, and smell • Explain how the different senses communicate with the brain
	145	358–59 A40	340–41	224	Rubric	**Activity: Touch Tester** • Predict and identify areas of the body that are the most sensitive to touch
	146	360–63	342–45	225–28		**Memory and Sleep** • Differentiate between short-term memory and long-term memory • Identify two categories of long-term memory • Describe some characteristics of REM sleep

xlii Table of Lesson Plans

Vocabulary	Process Skills	Enrichment	Bible Integration
pedigree, sex-linked trait		Scientist Link Demonstration	Identified in Christ
	Making and using models Inferring Interpreting data Communicating	Bible Link	
sickle cell anemia, cystic fibrosis, Down syndrome, genetic engineering		Writing Link Technology Link	God as Master of creation God's knowledge of each individual
		Bible Link	God's perfect design God's design for man's body
central nervous system, peripheral nervous system, brain, cerebrum, lobes, cerebellum, brain stem, spinal cord		Demonstration Bible Link	God's design of man's body God's perfect design
neuron, sensory neuron, motor neuron, dendrite, impulse, axon, synapse, reflex		Language Link Activity Demonstration	God's perfect design
	Predicting Measuring Inferring Identifying & controlling variables Recording and interpreting data	Bible Link	
		Technology Link Activity Demonstration	God's perfect design Holy Spirit's guidance Faith in the Word of God Man's finite knowledge
	Predicting Measuring Inferring Recording Data		
memory, short-term memory, long-term memory			God's command to remember God's design of man's body Godly wisdom

Table of Lesson Plans

Lesson	TE Pages	ST Pages	AM Pages	Assessment	Content and Objectives
Chapter 14					
147	364–67	346–49	229–30	Quiz 14-B	**The Endocrine System; Disorders and Drugs** • Compare the nervous system and the endocrine system • Identify the function of some glands in the endocrine system • Identify some common nervous system disorders • Recognize some of the problems resulting from drug abuse
148	368–69 A41–A42				**Explorations: Effects of Drug Abuse** • Identify some common categories of drugs • Explain how some types of drugs affect the nervous system • List some biblical reasons for not taking drugs
149	370	350	231–32		**Chapter Review** • Recall concepts and terms from Chapter 14 • Apply knowledge to everyday situations
Chapter 15—Immune System					
150				Chapter 14 Test	• Demonstrate knowledge of concepts taught in Chapter 14
	371	351	233		**Chapter Opener** • Recognize that man's inferences are sometimes inaccurate
151	372–75	352–55	234		**Diseases** • Recognize that disease is a consequence of Adam's sin • Explain how diseases are classified • Identify four common pathogens • List some diseases caused by each pathogen
152	376–79	356–59	235–36	Quiz 15-A	**Pathogens and Noncommunicable Diseases** • List several ways that pathogens are spread • Differentiate between communicable diseases and noncommunicable diseases • Explain some of the jobs of an epidemiologist
153	380–81	360–61		Rubric	**Activity: Of Epidemic Proportions** • Recognize how quickly pathogens can spread • Identify the source of contamination
154	382–85	362–65			**The Immune System** • Identify several defensive barriers of the body • List two of the body's nonspecific defenses • Identify the body's specific defense against pathogens • Explain some functions of white blood cells during the immune response
155	386–89	366–69	237–38	Quiz 15-B	**Immunity; Antibodies and Antibiotics; Malfunctions of the Immune System** • Explain three ways that the body can obtain immunity • Compare and contrast antibiotics and antibodies • Identify some problems that occur when the immune system malfunctions
156	390 A43–A48	370			**Activity: Defend and Capture** • Model the interactions between the immune system and pathogens
157	391	371		Rubric	**Explorations: Extra, Extra, Read All About It** • Research and write an article about a medical discovery
158	392	372	239–40	Chapter 15 Test	**Chapter Review** • Recall concepts and terms from Chapter 15 • Apply knowledge to everyday situations

Table of Lesson Plans

Vocabulary	Process Skills	Enrichment	Bible Integration
hormone, endocrine gland, target cell, hypothalamus, pituitary gland, epilepsy, multiple sclerosis, Parkinson's disease, Alzheimer's disease		History Link Activity	God's design of man's body Consequences of sin Man's body as God's temple Man's responsibility to glorify God
			Man's body as God's temple Man's sinful nature
			God as Great Physician God's omnipotence
communicable, noncommunicable, pathogen		Scientist Link Language Link History Link Bible Link	Consequences of sin God's omnipotence God's protection of His people
vector, epidemic, airborne pathogen, contact, food-borne pathogen, water-borne pathogen, epidemiologist		History Link Demonstration Activity Geography Link Writing Link	God as Master of creation God's omniscience
	Making and using models Observing Inferring Recording data Communicating		
defensive barrier, cilia, inflammatory response, immune response		History Link	God's plan for man's body Consequences of sin Faith in the Word of God God's perfect design
antibody, memory cell, immunity, vaccine, antibiotic, allergen, autoimmune disease		Technology Link Language Link	Man's sinful nature God's power over sin God's omniscience God's omnipotence
	Observing Communicating Defining operationally		God as Master of creation
	Communicating	Geography Link	

Table of Lesson Plans

Lesson	TE Pages	ST Pages	AM Pages	Assessment	Content and Objectives
160	A4–A5		241–42		**Autonomous Underwater Vehicles** • Explain what an autonomous underwater vehicle is • Explain the advantages of an underwater observatory • Identify some ways using an AUV may benefit studying the ocean
161	A6–A7		243–44		**Fiber Optic Sponges** • Explain how the spicules of a Rossella sponge are like optic fibers • Identify ways that studying a Rossella sponge may improve fiber optic technology • Recognize man's duplication of God's creation
162	A8–A9		245–46		**Maglev Train** • Explain how electromagnets are used in maglev trains • Identify some ways a maglev train may benefit the environment and transportation
163	A10–A11		247–48		**Inflatable Spacecraft** • Describe some types of inflatable spacecraft • Understand the basics of inflatable technology • Explain the advantages of inflatable spacecraft
164	A12–A13		249–50		**Glowing Space Plants** • Identify characteristics of the thale cress • Explain how genetic engineering produces glowing plants • Recognize that scientists use the same basic methods that Mendel used
165	A14–A15		251–52		**Robotic Surgery** • Compare robotic surgery with traditional surgery • Describe some advantages and disadvantages of long-distance robotic surgery

Technology Lessons

Table of Lesson Plans

Vocabulary	Process Skills	Enrichment	Bible Integration
		History Link	Spirit-filled Christians Bible as only authority
		Demonstration	Christians as lights in the world Man's technology patterned after God's creation God's perfect design
		Demonstration	The Christian's attraction to Christ The Christian's hatred of sin
		Math Link	Man as steward of God's creation
		Activity	Attitudes as a reflection of a Christian's walk
		History Link	Technology as a demonstration of man loving his neighbor as himself

Table of Lesson Plans xlvii

Student Text pages 1–3
Activity Manual page 1

Lesson 1

A Changing Earth

Objectives
- Recognize the interrelationship of science concepts
- Explain that ideas about science change, but that God never changes

Unit Introduction

SCIENCE 6 for Christian Schools is divided into six units. Grouping chapters together in units helps students understand relationships between the science concepts.

In Unit 1 students will see how the volcanoes and earthquakes discussed in Chapter 1 move the earth's surface and expose more surface area to the processes of weathering and erosion discussed in Chapter 2.

Many minerals and ores discussed in Chapter 3 are found in volcanic areas.

Good soil management, as discussed in Chapter 3, helps prevent the wind and water erosion discussed in Chapter 2.

Use every opportunity that arises to help the student connect new information with material that he has already learned.

➤ **Look through Unit 1. What are the topics of the chapters we will study in this unit?** *{Possible answers: earthquakes and volcanoes, weathering and erosion, natural resources}*

➤ **Why do you think these chapters are organized into the same unit?** *{Answers will vary. Elicit that all of the chapters are about physical attributes of the earth.}*

➤ **One way to see the greatness of God is to look at the way all creation fits together. Although the earth is under the curse of sin, God uses all things for His glory. Whether it is a thorn bush that provides shelter for an animal or a volcano that enriches the soil, God controls all of creation.**

Teacher Helps

Science Notebook—You may choose for the student to keep a science notebook. The notebook will provide a place to keep Activity Manual pages as well as other papers used in Activities, Explorations, and other science-related projects.

You may choose to organize the pages by chapter or group them by type, such as Reinforcement pages, Study Guides, Activities, etc.

Keeping a notebook will help the student develop organizational skills. The notebook can also be used in the student's portfolio to demonstrate his progress and ability to think scientifically.

Plan to check the content, organization, and neatness of the notebook at least once per chapter.

Additional information is available on the BJU Press website.

www.bjup.com/resources

Unit 1   Lesson 1   1

Project Idea

The project idea presented at the beginning of each unit is designed to incorporate concepts of each chapter as well as information gathered from other resources. You may choose to use the project as a culminating activity at the end of the unit or as an ongoing activity while the chapters are taught.

Unit 1—A New Town

Design a map of a region that incorporates land features from the unit. Plan a new town that makes use of the land features and natural resources available. The town should include some of the following: industry, residential areas, transportation, agriculture, businesses, communication, parks, and recreation.

Technology Lesson

A correlated lesson for Unit 1 is provided on TE pages A4–A5. The lesson may be taught with this unit or with the other Technology Lessons at the end of the course.

Bulletin Board

Note: The suggested bulletin board idea may be modified for use in a learning center.

As each chapter in Unit 1 is discussed, the student makes fact cards to add to the map. Facts can be historical or may be from current events.

Ideas for fact cards include:

   Chapter 1—earthquakes and volcanic eruptions

   Chapter 2—mudslides, avalanches, or other natural disasters

   Chapter 3—locations of oil fields or other large sources of natural resources

Earthquakes and Volcanoes

"Call unto me, and I will answer thee, and shew thee great and mighty things, which thou knowest not."
Jeremiah 33:3

Great & Mighty Things

Throughout history man has tried to explain God's wonderful creation. Many of these explanations seem laughable to us today. Some people thought that earthquakes occurred because a huge turtle moved. Others believed that volcanoes erupted because a Hawaiian goddess was angry. Many beliefs in false gods came from attempts to explain natural events. Today man often uses theories or laws to explain the world around him. But only God knows all things. Man continually has to revise his ideas as God reveals more and more about His creation. There are many aspects of creation that man will never fully understand until he meets his Creator face to face.

Science Background

Turtles and earthquakes—Many stories about what causes earthquakes exist. One tale tells of the earth held on the backs of four elephants. These elephants stand on the back of a turtle that stands on a cobra. If any of the animals moves, an earthquake occurs. Another legend places the earth on the backs of seven turtles. As the turtles move away from each other, earthquakes occur.

Hawaiian goddess and volcanoes—Religions of the islands in the Pacific Ocean have many goddesses. Pele, the goddess of fire, is specific to Hawaii. People believe she lives in a crater on one of the islands. When Pele becomes angry, a volcano erupts.

Introduction

God's Word tells of many great and mighty things that man cannot understand without God's revelation. As Christians grow in Christ, God reveals more of Himself to them through His Word.

God's creation also contains great and mighty things that are beyond man's understanding. Through the ages man has tried to explain the world around him. But all knowledge and understanding come from God. As God allows man to learn more and more about His creation, man constantly must reevaluate his ideas.

At the beginning of each chapter, you will find a short selection that gives examples of man's incomplete or inaccurate conclusions. Our purpose is not to undermine the value of scientific discoveries. Instead, we desire to show that God allows man to learn more about creation in order to demonstrate man's finiteness and God's omniscience. It is important for students to understand that all truth, whether scriptural or scientific, comes from God.

Chapter Outline

I. Earthquakes
   A. Plate Tectonics
   B. Causes of Earthquakes
   C. Earthquake Waves
   D. Detecting Earthquakes
   E. Measuring Earthquakes
   F. Building for Earthquakes
   G. Related Disasters

II. Volcanoes
   A. Causes of Volcanoes
   B. Locations of Volcanoes
   C. Classifying Volcanoes

III. Effects of Volcanoes
   A. Dangers of Volcanoes
   B. Products of Volcanoes
   C. Other Thermal Eruptions

Activity Manual

Preview—page 1

The purpose of this Preview page is to generate student interest and determine prior student knowledge. The concepts are not intended to be taught in this lesson.

Lesson 2

Student Text pages 4–7
Activity Manual page 2

Objectives

- Recognize that the earth's surface changes constantly
- Explain the theory of plate tectonics
- Use graphic information to identify the parts of the earth and the different kinds of faults
- Infer that plate boundaries are unstable areas of the earth's surface

Vocabulary

lithosphere
plate
theory of plate tectonics
plate boundary
earthquake
fault

Materials

- several dry sticks, each about 1 cm in diameter
- eye protection (glasses or safety goggles)
- world map

Introduction

Give one student eye protection and a stick. Instruct the student to try to break the stick slowly. After the stick breaks, ask the following questions.

➤ **Did the stick break easily?** *{It probably did not.}*

➤ **How did the stick feel as it bent?** *{There was tension.}*

➤ **Did the stick break quickly or slowly?** *{It probably bent for a while and then broke quickly.}*

Allow several other students to try breaking sticks.

Conclude that tension built in each stick before the stick broke suddenly.

➤ **This lesson presents another situation in which tension builds and is released suddenly.**

Purpose for reading:
Student Text pages 4–7

Note: Purpose for reading questions help prepare the student for reading the lesson pages.

➤ **What beliefs do some scientists have about the surface of the earth and the formation of land masses?**

➤ **What causes earthquakes?**

In 1811–12 a series of earthquakes hit New Madrid (muh DRID), Missouri. One of these earthquakes was so strong that it rang bells in Boston's church steeples over 1,100 miles away and toppled chimneys in Virginia. It even caused the Mississippi River to run backwards for a while! An excerpt from a letter written at the time compares the movement of the earth to waves in a gentle sea. Few of us experience the ground moving like the sea. However, the ground that we think of as solid and steady is actually in continuous motion. Sometimes this movement causes large shifts in the earth's surface, resulting in earthquakes. At other times, the earth's surface splits and allows the molten rock beneath to escape as a volcano.

Earthquakes

Plate Tectonics

The earth is made up of the inner and outer core, the mantle, and the crust. The earth's crust is quite thin compared to the other parts of the earth. The portion of the crust that forms the continents is approximately 45 km (25 mi) thick. The crust under the ocean is even thinner—only about 5 to 8 km (2 to 4 mi) thick. Some scientists believe that the earth's **lithosphere** (LITH uh SFEER), the crust and upper area of the mantle, consists of large pieces called **plates**. These plates float on the partly melted rock in the earth's mantle. The idea that the earth's crust is made up of moving plates is called the **theory of plate tectonics**.

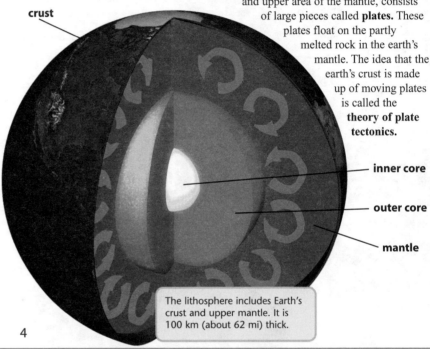

The lithosphere includes Earth's crust and upper mantle. It is 100 km (about 62 mi) thick.

Labels: crust, inner core, outer core, mantle

SCIENCE BACKGROUND

Hypothesis—A scientist starts with an educated guess at the solution to the problem, and he tests it repeatedly.

Theory—After testing the hypothesis and obtaining results from a number of observations, the scientist develops a theory.

Law—After a period of time and repeated testing, a generalization called a law is made concerning the body of theories on a specific topic.

Tectonics—Tectonics is the study of the earth's structural features.

Creationists and the Noahic Flood—Creationists have differing views about how mountains and oceans formed after the Noahic Flood.

Convection currents—Scientists think that the hot matter in the earth's mantle moves in convection currents. Hot matter rises and spreads out sidewise and then cools and falls. This circular motion is similar to the way water boils. As the hot matter spreads out, it moves the plates.

SCIENCE MISCONCEPTIONS

We are not sure how certain landforms were formed. We cannot base how the earth may have changed in the past on how it changes now.

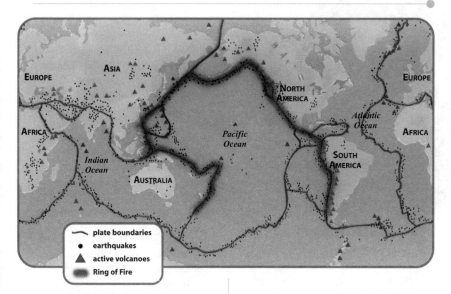

Places where the plates meet are called **plate boundaries.** Scientists think that currents in the molten rock of the earth's mantle may move the plates a few centimeters each year. This movement may cause the plates to separate, to collide, or to slide along their plate boundaries.

Some scientists believe that at one time the earth could have been a single large landmass that they call *Pangaea* (pan JEE uh). Because there was no recorded observation, we cannot be sure that such a landmass existed. We also cannot know how the landmass may have broken into pieces. However, the Bible tells us in the book of Genesis that God sent a great flood to destroy the wickedness on the earth. Genesis 7:11 states that "the fountains of the great deep [were] broken up, and the windows of heaven were opened." Many Creation scientists think that the earth's surface went through catastrophic changes during the Noahic Flood. These changes could have caused the great landmass to break and separate. The plates may have moved with such tremendous force that landforms such as mountains could have been formed as plates collided. Some scientists claim that landforms took millions of years to form, but it is likely that they formed in a much shorter period of time.

 New Madrid earthquakes— A series of large earthquakes shook the New Madrid area of Missouri from December 1811 to February 1812. The quakes during this period were estimated to have been between 7.5 and 8.3 on the Richter scale. People felt aftershocks for months afterward. The quakes caused immense damage. They pushed up some large areas of land and made others sink down, creating lakes where lakes had not been before. Movement under the Mississippi River's surface created large waves, and the river looked like it was flowing upstream. For more information on this subject, research Lorenzo Dow's Journal, which contains a firsthand account.

 Genesis 1:9–10 tells us that God gathered together the waters to let the dry land appear. The word "earth" describing the dry land is singular, but the word "seas" used for the water is plural. In 1858 the Creation scientist Antonio Snider-Pellegrini used these verses to develop the idea of a single landmass existing at one time.

Discussion:
pages 4–5

 Why is it difficult to be certain about the earth's interior and its movements? *{Answers may vary. Man has limited ability to observe the earth's interior and its movements.}*

➤ **Which is thicker, the crust under the continents or the crust under the oceans?** *{the crust under the continents}*

➤ **What is the lithosphere?** *{the crust and upper part of the earth's mantle}*

➤ **What are the large pieces of the lithosphere called?** *{plates}*

➤ **What is the theory of plate tectonics?** *{the idea that the earth's crust is made of moving plates}*

➤ **In what three ways can plates move against each other?** *{collide, separate, and slide}*

➤ **What is Pangaea?** *{a large landmass that some scientists think may have existed at one time}*

➤ **Can we know for sure that Pangaea existed? Why?** *{No; there is no recorded observation of this landmass.}*

Refer the student to the map on Student Text page 5.

 What do you notice about the locations of volcanoes and earthquakes? *{They tend to be near plate boundaries. They also tend to be near each other.}*

Locate the area of the New Madrid earthquakes. Did these earthquakes occur near a plate boundary? *{no}*

➤ **Where are the nearest plate boundaries?** *{the Pacific coast and the middle of the Atlantic Ocean}*

 From the New Madrid earthquakes, what can you conclude about the relationship of plate boundaries and earthquakes? *{Answers will vary but should include that not all earthquakes occur along plate boundaries.}*

You may wish to discuss further the catastrophic effects that the Noahic Flood had on Earth.

Discussion:
pages 6–7

➤ **What usually causes an earthquake?** *{the sudden shifting of rocks along plate boundaries}*

💡 Not all earthquakes occur along plate boundaries. Name an earthquake that was not near a plate boundary. *{the New Madrid earthquake}*

➤ **What is a fault?** *{a break in the earth's surface along which rock can move}*

➤ **What is an earthquake?** *{the released energy that causes vibrations in the earth's surface}*

A plate boundary is also a fault because it is a break in the surface of the earth.

Use the world map to locate places as they are discussed.

➤ **What are the three kinds of faults?** *{reverse, normal, strike-slip}*

➤ **What determines the kind of fault?** *{how the rocks move against each other}*

➤ **What causes a reverse fault?** *{rocks pushing together until a section of rock moves upward}*

➤ **What landforms may have been formed by reverse faults?** *{deep ocean trenches and some mountains}*

➤ **What are some examples of reverse faults?** *{the Himalaya Mountains, Marianas Trench}*

➤ **What causes a normal fault?** *{rocks moving apart}*

➤ **What sometimes fills the gap created by a normal fault at a plate boundary?** *{Possible answer: molten rock from under the crust}*

➤ **What kind of fault is the Great Rift Valley in Africa?** *{a normal fault}*

➤ **What do scientists call a normal fault that occurs under the ocean?** *{sea-floor spreading}*

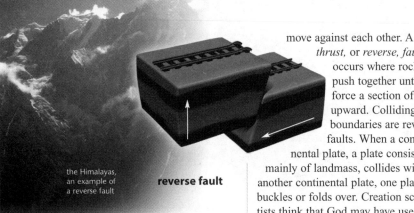

the Himalayas, an example of a reverse fault

reverse fault

Causes of Earthquakes

When two surface plates push and scrape against each other, energy builds up in the moving plates. This energy may build up for many years. But just as a stick bends only so far before it breaks, eventually the rocks along the plate boundaries shift suddenly and release their stored energy, resulting in an **earthquake**. The released energy causes vibrations that shake and rattle the earth's surface. Sometimes the release of energy is small. These little tremors are so small that no one even feels them. However, if the release of energy is great, the vibrations may cause widespread destruction.

Faults are breaks in the earth's surface along which rocks can move. There are three kinds of faults, depending on how the rocks move against each other. A *thrust*, or *reverse, fault* occurs where rocks push together until they force a section of rock upward. Colliding plate boundaries are reverse faults. When a continental plate, a plate consisting mainly of landmass, collides with another continental plate, one plate buckles or folds over. Creation scientists think that God may have used this process to form some landforms, such as the Himalaya Mountains in Asia. Where a continental plate collides with an oceanic plate, one plate often slides under the other plate, creating deep ocean trenches, such as the Marianas Trench in the western Pacific.

The second type of fault is a *normal fault*. As rocks move apart, a section of rock may fall between the separating rocks. Where normal faults occur at plate boundaries, the fault sometimes allows molten rock from under the crust to fill in the gap and form new land. The Great Rift Valley in Africa is an example of a separating plate boundary. New land is also forming along the

the Great Rift Valley, an example of a normal fault

normal fault

SCIENCE MISCONCEPTIONS

Plate boundaries are faults, but not all faults occur at plate boundaries.

 The surface of the earth is constantly changing. However, God's Word is infallible and unchangeable. His Word endures forever. [BAT: 8b Faith in the power of the Word of God]

 During the early 1900s San Francisco was the largest city on the west coast. Many people lived in closely built wooden structures. An earthquake that struck in 1906 devastated the city. Buildings and roads collapsed. Trees were uprooted. Telephone lines fell down, and water pipes broke. Fires started and burned for three days because water was not available for fighting them. It is said that more buildings were lost to the fires than to the actual earthquake.

Mid-Atlantic Ridge in the middle of the Atlantic Ocean. As molten rock pushes up between plates, it causes the plates to spread apart. The magma then cools, forming new land. This is called *sea-floor spreading*. Many scientists use sea-floor spreading to support their belief in evolution, but it is simply a fascinating characteristic of the world God created.

A *strike-slip fault* occurs as rocks move horizontally past each other. One of the most famous strike-slip faults is the San Andreas Fault in California. It is the site of many earthquakes.

There are other reasons an earthquake might happen. Adding or removing large amounts of earth may cause the earth to shift or move, resulting in an earthquake. This may occur during the construction of large buildings or dams. Sometimes as molten rock under a volcano moves, it can also cause an earthquake. All of these causes, however, usually create only small earthquakes.

Imagine what happens to pipes, gas lines, and power and telephone lines when an earthquake shakes an urban area. Even if buildings have been constructed to withstand earthquakes, many things still cannot bend and move with the shaking of the earth.

It is God who allows all things to happen in His creation, including earthquakes. Psalm 60:2 says that God has "made the earth to tremble." When earthquakes occur, it is not by chance.

the San Andreas Fault, an example of a strike-slip fault

strike-slip fault

QUICK CHECK
1. What are the main parts of the earth?
2. Give two or three reasons an earthquake may occur.
3. Describe and illustrate the three kinds of faults.

Direct a Demonstration

Demonstrate how rocks move along faults

Materials: two damp sponges, table or other smooth surface

Demonstrate a reverse fault by pushing the sponges against each other along the table. Repeat the demonstration several times.

➤ Did the sponges always move the same way when pushed together? {no} Describe how they moved. {Possible answers: Both pushed up together. One pushed under the other. One folded under the other.}

Rock in a reverse fault can respond in similar ways.

Demonstrate the two other kinds of faults.

➤ What happened to the sponges as they rubbed against each other? {They pulled out of shape.}

As rocks move along faults, they can be pulled or pushed out of shape, causing stress. The release of that stress can cause earthquakes.

Discussion:
pages 6–7

➤ What causes a strike-slip fault? {rocks moving horizontally past each other}

➤ Along which famous fault in California do many earthquakes occur? {San Andreas Fault}

➤ What human activity can cause earthquakes? {moving large quantities of earth}

➤ What size earthquake does this activity cause? {a very small earthquake}

Refer the student to the map on Activity Manual page 2.

➤ The San Andreas fault is located along the coast of California. How is the fault marked on the map? {with a purple line}

➤ What major U.S. cities are close to the San Andreas Fault? {Possible answers: Sacramento, San Francisco, San Jose, Los Angeles, and San Diego}

Imagine that an earthquake has just occurred in San Francisco. What kinds of problems would emergency personnel have to deal with? {Possible answers: Fires may occur because of broken gas lines. Water supplies may be disturbed because of pipe breaks. Communication may be difficult. Roads may be impassable. Buildings and bridges may be broken down. Electrical lines may be down.}

Some scientists use the kinds of faults to explain evolutionary processes.

➤ Could God have used earthquakes to shape the earth we see today? {yes}

➤ What are some characteristics of God demonstrated through disasters such as earthquakes? {Possible answers: providence, omnipotence, omniscience}

Answers
1. crust, mantle, inner core, and outer core
2. Possible answers: sudden shifting of the earth's crust; addition or removal of large amounts of earth; movement of molten rock under a volcano
3. reverse fault—rock is pushed up
   normal fault—rock drops down between separating rocks
   strike-slip fault—rock moves horizontally along other rock

Activity Manual
Expansion—page 2
You will use this page again in the next lesson.

Lesson 3

Student Text pages 8–11
Activity Manual pages 2–4

Objectives

- Compare and contrast P, S, and two kinds of L waves
- Explain the differences between the Mercalli scale and the Richter scale
- Describe some ways that earthquakes cause damage

Vocabulary

focus
seismic wave
epicenter
seismologist
seismograph
magnitude
tsunami

Materials

- pie plate
- medicine dropper

Introduction

➤ What things do you enjoy doing at a lake or pond?

➤ Have you ever thrown or skipped rocks on a lake or pond?

➤ What happened to the water when the rock hit it? {Ripples spread out from the place where the rock entered the water.}

Demonstrate how ripples move out from a source by filling a pie plate with water and using the medicine dropper to drop water into it one drop at a time.

📖 You, too, can have a ripple effect on others. God first loved us. He demonstrated His love to us through His Son. Christians should show godly love towards others by their words and actions. [BAT: 5 Love-Life Principle]

Purpose for reading:
Student Text pages 8–11

➤ How do earthquakes travel?

➤ What are some effects of earthquakes?

Earthquake Waves

Earthquakes take place below the surface of the earth. Waves of energy are sent out from the **focus,** or beginning point of an earthquake. These vibrations, or **seismic** (SIZE mik) **waves,** flow out from the focus in all directions, similar to the ripples caused when a pebble is tossed into a pond. The point on the surface of the earth directly above the focus is called the **epicenter.**

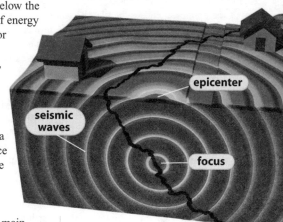

Body waves are seismic waves that occur beneath the surface of the earth. The two main kinds of body waves are P waves and S waves. The P waves, or primary waves, move quickly through both the solid and liquid in the earth's interior. These waves can actually be felt on the side of the earth opposite the focus of an earthquake.

S waves, or secondary waves, move more slowly. They cannot move through the liquid material in the earth. **Seismologists** (size MAHL uh jists), scientists who study the movement of the earth, use the difference in the speed of P and S waves to help calculate the location of the focus of an earthquake.

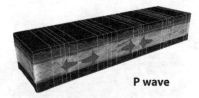

P wave

The *P wave* is the fastest-moving body wave. It travels in a straight path by a push- and pull-motion.

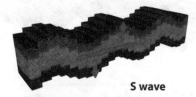

S wave

The *S wave* moves more slowly than the *P wave.* It moves in an up-and-down zigzag pattern.

8

Teacher Helps

Place a shallow clear plastic or glass plate or pan on an overhead projector to allow students to see the ripples in the water easily.

Science Background

S-waves—also called *shear waves* because of the way they move

P-waves—also called *compressional waves*

History

The earliest seismoscope, or instrument to detect an earthquake, was called an *earthquake weathercock.* Chang Heng, a Chinese geographer, invented it in AD 132. The "dragon jar" was eight feet tall and was made of bronze. When an earthquake occurred, a metal rod inside the jar fell, causing one of the dragons on the side of the jar to drop a ball from its mouth to the frog's mouth below. This indicated that an earthquake had occurred, as well as the direction the earthquake had come from. For 1,600 years this was the only method available for detecting earthquakes. The pendulum seismograph used today was invented in 1897.

8 Lesson 3

When P and S waves reach the surface of the earth, they produce land waves, or L waves. *L waves* are the slowest moving and most destructive waves. Love waves and Rayleigh (RAY lee) waves are two types of land waves.

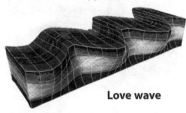

Love wave

Love waves, the fastest moving land waves, move back and forth in a zigzag pattern.

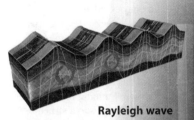

Rayleigh wave

Rayleigh waves move along the ground in a rolling motion, similar to the way ocean waves roll.

Detecting Earthquakes

Seismologists use a machine called a **seismograph** to detect, time, and measure the movements of the earth. As the earth moves, seismographs produce *seismograms,* or records of the movements. Early seismographs used rotating drums and pens that touched the paper on the drums to form seismograms. Today, most seismographs are part of an advanced computer system. These instruments measure and record up-down, east-west, and north-south movements, giving scientists an accurate record of earthquake activity.

seismograph

seismogram

Direct a DEMONSTRATION

Demonstrate the different types of waves

Materials: a plastic spring toy

Instruct two students to hold the spring toy, one at each end. Tell them to stand about five to six feet apart and lay the spring toy on the floor.

- To demonstrate the movement of the P wave, ask one student to push in his end of the spring toy. The "P wave" will travel down to the end of the toy and back.
- To demonstrate the S wave, tell the students to stand. One student should move the spring toy in an up and down motion.
- To demonstrate the Love wave, tell the students to move the spring toy in a side-to-side motion.

Ask the students if they can think of a way to demonstrate the Rayleigh wave using the spring toy.

Discussion:
pages 8–9

➤ What do we call the beginning point of an earthquake? *{focus}*

➤ What are seismic waves? *{the vibrations of an earthquake}*

💡 Energy is defined as the ability to move something through a distance. Why would seismic waves be called energy waves? *{They cause vibrations, or movements, of the earth.}*

➤ What is the point on the surface of the earth directly above the focus called? *{epicenter}*

➤ What is a seismologist? *{a scientist who studies movements of the earth}*

➤ Describe the movements of P and S waves. *{The P waves move in a straight line by a push and pull motion. The S waves are slower and travel up and down in a zigzag pattern.}*

➤ Which waves are more destructive, body waves or land waves? Why? *{land waves; They move on the surface of the earth.}*

➤ Describe the movements of Love waves and Rayleigh waves. *{Love waves move in a side-to-side zigzag pattern. Rayleigh waves move in a circular pattern, like rolling waves on the ocean.}*

➤ What machine do scientists use to measure the movement of the earth? *{a seismograph}*

➤ How have seismographs changed since their invention? *{Today most seismographs are part of a computer system.}*

💡 People have described earthquakes as sounding like trains. Why would the seismic waves from an earthquake create sound? *{Possible answer: Vibrations create sound. Seismic waves are huge vibrations in the earth's crust.}*

Discussion:
pages 10–11

- What two scales measure earthquakes? {Mercalli scale and Richter scale}
- Which scale depends on man's observations of an earthquake's destruction? {the Mercalli scale}
- Using the Mercalli scale, would it be difficult to determine the strength of an earthquake in a largely unpopulated area? Why or why not? {Yes, because the scale measures observable destruction. Without buildings to observe, determining the strength of an earthquake would be difficult.}
- Charles Richter developed the Richter scale. How is it different from the Mercalli scale? {The Richter scale measures the strength of the seismic waves rather than just the observable damage.}
- Why is the Richter scale more accurate than the Mercalli scale? {The Richter scale is not dependent on man's observations or the population of an area.}
- In what ways can architects improve a building's stability? {Possible answers: by laying the foundation in rock, using good design, and using strong building materials}
- On what foundation are Christians to build their lives? {Jesus Christ}
- Jesus Christ is a solid foundation in an unstable world.
- Which of the earthquake scales is influenced by the way buildings are constructed? {the Mercalli scale} Why? {It is based on the amount of destruction that occurs.}
- How does the way buildings are constructed influence Mercalli scale measurements? {Mercalli scale measurements for the same waves may be smaller than Richter scale measurements in areas with reinforced structures but greater in areas with weaker structures.}
- Refer the student to *Science and the Bible*.
- Where does the Bible describe the earth opening up and swallowing people? {Numbers 16}
- Why were Korah and his followers judged this way? {Guide to the conclusion that Moses asked God to do a "new thing," so that all the people would know God's specific judgment on those who opposed His chosen leaders.}

10 Lesson 3

Measuring Earthquakes

Scientists use two scales to measure and compare the strength of earthquakes. The Mercalli (MUR kah lee) scale is based on the amount of destruction that an earthquake causes to man-made structures. This scale can be used to gain a general idea of the strength of earthquakes in the past.

The Richter (RIK tuhr) scale measures the **magnitude,** or strength, of the seismic waves of an earthquake. An earthquake is assigned a decimal number based on the strength of its seismic waves. Each whole number is ten times greater than the previous one. For example, an earthquake with a magnitude of 6 has seismic waves ten times greater than an earthquake with a magnitude of 5. But an earthquake with a magnitude of 7 has seismic waves one hundred (10 × 10) times greater than an earthquake with a magnitude of 5.

Building for Earthquakes

Man cannot predict or control earthquakes, but he can try to minimize the damage they cause. Most destruction in an earthquake occurs because structures such as buildings and roads collapse. Therefore, engineers and architects work to develop structures that will sustain only minor damage in an earthquake and will remain standing.

The Bible describes a man who hears and obeys God as having his foundation on a rock—Jesus Christ (Matt. 7:24–27). Likewise, the best foundation for a building is solid rock. A building with a foundation on sand or a landfill has little stability.

In addition, building materials and design are key factors to a structure's stability. Concrete reinforced with steel rods is a common building material. A building constructed using a steel frame also tends to be stable. The frame

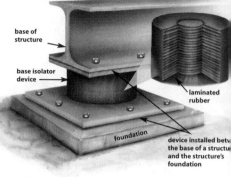

Earthquake Isolation Device

base of structure
base isolator device
laminated rubber
foundation
device installed between the base of a structure and the structure's foundation

Richter scale effects

2.0–2.9   3.0–3.9   4.0–4.9   5.0–5.9   6.0–6.9   7.0–7.9   8.0–8.9

SCIENCE BACKGROUND

Richter scale—Each number on the Richter scale represents seismic waves that are ten times greater than the previous number. However, the increase in energy released by the earthquake is actually about thirty times greater for each successive number.

Other effects of earthquakes—Some other effects earthquakes have on the earth's features include cave-ins, wells drying up as water tables shift, and seiches (the sloshing of water in lakes).

 Luke 6:46–49 compares a Christian's obedience to God's Word with the construction of a house. A house with a deep foundation built on solid rock will withstand the floods and violent shaking of storms. Likewise, a believer who has taken the time to study and obey God's Word establishes a deep foundation on which to base his life's decisions. That deep foundation prepares him to face storms of difficulty that may enter his life. [Bible Promise: D. Identified in Christ]

Science and the Bible

Some scientists who question the accuracy of the Bible doubt that the earth could actually open up and swallow something. But in Numbers 16, the Bible records just such an occurrence. When Korah rallied others against God's chosen leaders, Moses asked God to do "a new thing." That new thing was the earth opening up and swallowing Korah, his followers, and all their goods. They "went down alive into the pit, and the earth closed upon them: and they perished from among the congregation" (Num. 16:33).

connects all parts of the structure and allows it to move as one piece instead of pulling apart.

Related Disasters

Sometimes earthquakes cause or result from other catastrophic events. An earthquake, volcano, or landslide occurring under or near the ocean can cause a series of giant waves called **tsunamis** (tsoo NAH meez). These waves can move over 700 km (435 mi) an hour at sea with a height of only about 1 m. As the waves reach the shallow water near the shore, they can reach a height of over 30 m (98 ft). These waves cause much damage when they reach land.

Another event associated with an earthquake is a volcanic eruption. Seismic waves from beneath a volcano alert scientists to the immense pressure building up below the surface. For example, Mount St. Helens, in the state of Washington, began shooting steam and ash up in March 1980. In May, an earthquake caused a landslide that allowed the pressure to release. The northern side of the mountain exploded, and lava and gases poured from the mountain.

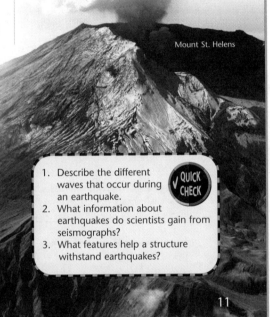
Mount St. Helens

QUICK CHECK
1. Describe the different waves that occur during an earthquake.
2. What information about earthquakes do scientists gain from seismographs?
3. What features help a structure withstand earthquakes?

MATH
The Richter scale is based on powers of ten. Each increase of one number on the scale represents an earthquake magnitude ten times greater than the magnitude of the previous number.

▶ How much greater is a magnitude 4 earthquake than a magnitude 3 earthquake? {10 times}

▶ How much greater is a magnitude 4 earthquake than a magnitude 2 earthquake? {100 times}

▶ How much greater is a magnitude 8 earthquake than a magnitude 3 earthquake. {100,000 times}

TECHNOLOGY
Triangulation—Scientists use a method called triangulation to locate the epicenter of an earthquake. Seismograph readings of the same earthquake are taken from three earthquake stations. Scientists use these measurements to determine the distance the earthquake is from each station. The point at which all three measurements meet is the epicenter of the earthquake. Methods of triangulation are also used by rescuers to find the locations of lost aircraft, boats, and even cellular phones.

Discussion:
pages 10–11

▶ What is a tsunami? {a huge wave resulting from an earthquake, landslide, or volcano that occurs under or near the ocean}

▶ Why is a tsunami so dangerous? {Possible answers: The waves are large and strong. It is a series of waves. Since a tsunami is the result of sudden movement, there are no definite warning signs that a tsunami will occur.}

Tsunamis were once called "tidal waves," but the Moon does not affect these waves as it does true tidal waves.

💡 *Tsunami* is a Japanese word meaning "harbor wave." Why would a tsunami be called a harbor wave? {Possible answer: The greatest destruction occurs when the wave is confined to an area such as a shallow beach or harbor.}

▶ What other catastrophic events are associated with earthquakes? {Possible answers: volcanic eruptions, landslides}

▶ How did a landslide caused by an earthquake affect the eruptionof Mount St. Helens? {The landslide released some pressure, and one side of the mountain exploded.}

Answers

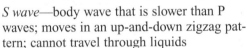

1. *P wave*—fast-moving body wave; travels in a straight line by a push and pull motion; can travel through solids and liquids

   S wave—body wave that is slower than P waves; moves in an up-and-down zigzag pattern; cannot travel through liquids

   L wave—land wave; the slowest moving and most destructive; examples are Love and Rayleigh waves

2. A seismograph detects, times, and measures movements of the earth.

3. concrete reinforced with steel rods; foundation laid in rock; steel framing

Activity Manual

Expansion—pages 2–3
Guide the students as they complete these pages. These pages may also be used as independent work for advanced students.

Study Guide—page 4
This page reviews Lessons 2 and 3.

Assessment

Test Packet—Quiz 1-A
The quiz may be given any time after completing this lesson.

Lesson 4

Activity Manual pages 5–6

Objective
- Use a scientific method

Vocabulary
scientific method   hypothesis
problem   procedure
material   conclusion

Materials
- one fettuccine noodle
- paper or foam cup
- 30 cm piece of string
- roll of pennies
- 2 stacks of books 30 cm high

Introduction

Note: The following experiment is intended as a teacher demonstration. The emphasis is to explain each step of a scientific method and practice the process skills that will be used in the other activities. Guide the student through the completion of the pages as the parts of a science experiment are discussed.

The scientific method is a systematic approach to problem solving. Used consistently, it makes science a subject of discovery through observation. Although the specific steps of the scientific method may vary, they should include using a hypothesis to identify a problem, experimenting to test the hypothesis, and drawing conclusions to test the accuracy of the hypothesis.

Discussion:
Activity Manual pages 5–6

➤ What is the problem of an experiment? {the question that the experiment will answer}

The problem is worded as a question and is specific rather than general. Write the problem for display. The problem for this experiment is "How many pennies can one strand of suspended fettuccine hold before breaking?"

➤ Write the problem on Activity Manual page 5.

Display the materials. The student will often be able to choose the materials he will use for his experiments. For this experiment, the materials are a cup, piece of string, fettuccine noodle, pennies, and stacks of books.

➤ List the materials you will use.

A Science Experiment

Name _____

Problem
Each experiment begins with a **problem**, or question that needs answering. This experiment's problem asks about the strength of uncooked fettuccine.

A. Write the problem.
How many pennies can one strand of fettuccine hold before breaking?

Materials
In the list of **materials**, include all the materials and equipment you will use to conduct the experiment. In some experiments all of the materials will be listed for you. In others you will choose some or all of the materials.

B. List the materials you will use.
1 fettuccine noodle (uncooked)
1 paper or foam cup
30 cm piece of string
1 roll of pennies
2 stacks of books 30 cm high

Hypothesis

A **hypothesis** is a scientist's idea of the answer to a problem. As you conduct an experiment, the hypothesis is a *statement* of your idea or a diagram of how the problem will be answered through that experiment. A hypothesis often includes *specific criteria* about the conditions of the experiment.

C. Four choices for a hypothesis are given. Read each and decide if it is a good hypothesis. If you answered No, explain what is wrong with that hypothesis.

1. Will ten pennies break one suspended piece of fettuccine?   ☐Yes ■No
   This is a question, not a statement. (statement)

2. Twelve pennies will break the suspended fettuccine.   ☐Yes ■No
   This hypothesis does not tell how many pieces of fettuccine are being used. (specific criteria)

3. I think six pennies will break one suspended piece of fettuccine.   ■Yes ☐No

4. One piece of fettuccine will hold no more than ten pennies without breaking.   ☐Yes ■No
   This hypothesis does not tell that the fettuccine is suspended. (the condition of the fettuccine)

D. Write a hypothesis in your own words.
Answers will vary.

 God encourages Christians to test and prove things. In Acts 17:11, the Berean believers were commended because they did not just accept what someone said, but they did their own research to find out if what they were hearing was true. [BAT: 6a Bible Study]

Procedure

The **procedure** is the steps of an experiment. The steps must be followed exactly to assure the accuracy of the results. Scientists **record** their observations and results throughout an experiment.

E. Follow the procedure.

1. Prepare the cup by making a string handle for it.
2. Hang the empty cup on the piece of fettuccine.
3. Suspend the fettuccine between the stacks of books. Place a book on each end of the fettuccine to hold it in place.
4. Support the cup with your hand and place one penny inside it. Gently lower the cup until it hangs from the fettuccine.
5. Repeat step 4 until the number of pennies reaches the number in your hypothesis or until the fettuccine breaks.
6. Record your observations and results.

Number of pennies	Did the fettuccine break?

Conclusions

The **conclusions** evaluate the accuracy of a hypothesis. They relate and apply the information to other areas.

F. Answer the questions to reveal your conclusions.

1. Was your hypothesis correct? If not, was your prediction too low or too high?   Answers will vary.
2. Would each piece of fettuccine hold the same number of pennies? Why or why not?   Answers will vary.

Follow-up

The **follow-up** usually considers variations to an experiment. Sometimes specific details or suggestions to change the variables are given.

Example: Try dropping the cup instead of lowering it gently after adding pennies.

© 2004 BJU Press. Reproduction prohibited.

A hypothesis is a scientist's testable idea of the answer to the problem. The hypothesis is written as a statement rather than as a question. The conditions for the hypothesis should be very specific. Discuss the wording of the sample hypotheses.

➤ **What are some other possible hypotheses about this problem?**

Record the student's responses. Allow the student to evaluate each idea to see if it meets the conditions for a hypothesis.

➤ **Write your chosen hypothesis.**

The procedure is the steps of an experiment that must be followed exactly.

➤ **Why is it important to follow the steps exactly?** {to ensure the accuracy of results}
➤ **Record your observations on the chart in the Activity Manual.**
➤ **What are you using as a unit of measurement in this experiment?** {pennies}

If the fettuccine does not break by the time you reach the number of pennies given in the hypothesis:

➤ **Have you answered the problem?** {No, because we still don't know if the fettuccine will break when we add another penny.}

If the fettuccine breaks before the student's prediction, help him see that the hypothesis was wrong. Allow the student to predict another number and form a new hypothesis to test.

Help the student draw conclusions based on the experiment. Generally, conclusions are inferences based on what the student has seen.

➤ **Was your hypothesis correct? If not, was your prediction too low or too high?**

The follow-up usually involves changing one variable or condition of the experiment.

➤ **What details or conditions about this experiment could we vary or change?** {Possible answers: the kind of pasta, the number of pieces of pasta, the length of the piece of pasta}
➤ **Would the experiment have produced the same results if the cup had been dropped suddenly instead of being lowered gently? Why?** {no; Possible answer: Dropping the cup suddenly may cause the fettuccine to break sooner.}

Activity Manual
Activity—pages 5–6

Lessons 5–6

Student Text pages 12–13
Activity Manual pages 7–8

Objectives

- Make a model of a structure that can withstand the motion of an earthquake
- Record and analyze information to form conclusions
- Identify variables

Materials

- See Student Text page

Introduction

Note: Two lesson days are allotted for this activity. You may choose to make the first structure on the first day and the second structure on the second day.

➤ What are some things you would associate with a construction site? *{Possible answers: workers, hard hats, huge machines and bulldozers, metal beams, noise}*

➤ In what ways is teamwork important to the construction of a building? *{Possible answers: to meet the goals and plans of the project; to make the best use of the knowledge and skills of each person; to save time; to use resources and materials wisely; for safety}*

➤ This activity will require you to work as a team as you construct a model structure.

God wants us to be dependable workers so that others can count on us to complete the job. [BATs: 2c Faithfulness; 2e Work]

Purpose for reading:
Student Text pages 12–13
Activity Manual pages 7–8

Display the materials the students will use.

➤ How do you think these materials can be used to find out what kinds of buildings will withstand an earthquake?

➤ Read the Student Text pages and the Activity Manual pages before you begin the activity. Think about what you need to do so that you do not miss any steps after you begin.

ACTIVITY: Construction Site

You are the engineer. You have been assigned the task of building in an earthquake-prone area. The task of your construction team is to design a structure that will withstand the forces of an "earthquake."

Problem
How would you design a building with features that will withstand an "earthquake"?

Process Skills
- Predicting
- Experimenting
- Making and using models
- Observing
- Identifying and controlling variables
- Recording data

Before the activity
1. Decide on the details about your "earthquake": the number of shakes it will have, its magnitude, the direction that it will shake, and the kind of seismic waves the shake will represent.
2. Discuss possible designs for your structure. The building must contain at least two stories. Think about the shape, materials, foundation, height, weight, and any other factors that may influence the stability of your structure.

Materials:
foam base, approximately 8″ × 10″ (20 cm × 25 cm)
package of large marshmallows
box of fettuccine noodles
Activity Manual

Procedure
1. Write your hypothesis about what features of your structure will help it withstand the stress of your "earthquake."
2. Draw your design in your Activity Manual.
3. Construct your structure on the foam base.
4. Following the limits set, replicate an earthquake by shaking the foam base appropriately.
5. Write the number of shakes your structure withstood and the amount of damage that occurred.
6. Think how you might improve your structure. Write your ideas for improvements.

12

Teacher Helps

Divide the class into science teams or groups of three or four. Place high, average, and low achievers in each group. Depending on the abilities and personalities of the students, you may need to assign tasks to ensure that each team member participates.

Keep the marshmallows cool to prevent them from getting sticky too quickly.

Encourage the students to use whole marshmallows rather than tear them apart.

SCIENCE BACKGROUND

Earthquake engineering—The goal of early engineers and architects of earthquake-resistant construction was to avoid or slow the collapse of buildings to provide time for occupants to escape. However, as technology improves, engineers and architects are better able to design structures that will not only remain standing, but will also sustain only minimal damage during an earthquake.

7. Rebuild your structure, adding your improvements. Predict how your improved building will withstand another "earthquake."

8. Simulate another earthquake under the original guidelines and record your observations.

Conclusions
- Did your rebuilt structure perform better than your original structure? Why or why not?

Follow-up
- Try making a structure with different materials. Test and compare its stability with the stability of those you already tested.

SCIENCE PROCESS SKILLS
Identifying and controlling variables—Variables are the changeable parts of an experiment. An effective experiment begins with the variables set to specific criteria, such as choosing the details of the earthquake. The scientist then changes only one variable at a time and compares the results.

➤ **What are some variables that could change in this activity?** {Possible answers: the details of the "earthquake"; the kinds of building materials used; the type of foundation; the design of the structure}

Discuss what information could be gained from changing one of these variables. For example, change the foundation to a tray of sand.

➤ **What information would be gained by changing the foundation variable?** {The results would show which foundation is more stable.}

➤ **In order to gain information about the foundation, what changes, if any, would need to be made to the other variables?** {Guide students to conclude that for the data to be accurate, all of the other variables would have to remain the same.}

Discuss which variable or variables were changed when the structure was rebuilt.

 You may choose to use Activity Manual page 13 with this lesson. This Bible Integration page discusses some of the occurrences of earthquakes in the Bible.

Procedure
Before beginning, decide as a class on the details of the "earthquake."
➤ How long will the shaking last?
➤ What direction will you shake the structure?
➤ How far will the structure move during each shake?

To make things more challenging for the teams, you may set a time limit.

Allow each group to plan and build its structure.

Test the stability of the structure with an "earthquake" as decided on by the class.
➤ **What are some qualities that made the structure stable?** {Possible answers: reinforcements; firm foundation in the foam}
➤ **What are some ways to improve your structure?** {Answers will vary.}

Direct each team to rebuild its structure, implementing the improvements discussed.

Guide the students as they make predictions, retest, and record their observations of the second structure.
➤ **Was your new hypothesis correct?** {Answers will vary.}
➤ **Compare the stability of your second structure to that of the first. Was the second more stable?** {Answers will vary.}
➤ **Does this activity illustrate the stability of a building in a real earthquake? Why or why not?** {Answers will vary.}

Conclusions
Provide time for the student to evaluate his hypothesis and answer the questions.

Activity Manual
Activity—pages 7–8
Guide the students as they use these pages to write a hypothesis, plan their structures, record their observations, and draw conclusions.

Assessment
Rubric—TE pages A50–A53
The rubric is a suggested tool for grading an activity. Guidelines for using a rubric are on TE page A49. Select one of the prepared rubrics, or design a rubric to include your chosen criteria.

Lesson 7

Student Text pages 14–17
Activity Manual pages 9–10

Objectives

- Explain the causes of a volcanic eruption
- Identify the parts of a volcano
- Describe three ways volcanoes are classified

Vocabulary

magma	ash
volcano	cone
volcanologist	Ring of Fire
magma chamber	hot spot
lava	cinder
vent	tephra
crater	pyroclastic flow

Materials

- world map or globe
- aerosol can with a warning label that says not to puncture or heat the can

Introduction

➤ **What are some household things that are pressurized?** {Possible answers: spray paint, can of whipped cream, shaving cream}

Show the aerosol can and read the warning on the label.

➤ **Why do you think it is dangerous to puncture a pressurized can?** {The contents would expand too quickly and erupt or explode.}

➤ **Why do you think it is dangerous to heat a pressurized can?** {Since heating causes most things to expand, heating a pressurized can causes the contents to expand, which further increases the pressure. The increased pressure may cause the can to explode suddenly.}

Purpose for reading:
Student Text pages 14–17

➤ How are volcanoes like a pressurized can?
➤ Are all volcanoes the same?

Volcanoes

According to the theory of plate tectonics, the plates of the earth's crust float on a layer of semiliquid rock. Because we do not usually see it, we rarely think about this melted rock. But sometimes it gets our attention in dramatic ways. When a volcano erupts, molten rock, or **magma,** may suddenly explode through the earth's surface, scorching and destroying everything in its path.

Causes of Volcanoes

A **volcano** forms where a crack in the earth's crust allows magma and gases to come to the surface. **Volcanologists** (VOL kuh NOL uh jists), scientists who study volcanoes, think that deep in the earth's lithosphere there are pockets of molten rock called **magma chambers.** Just as a hot air balloon rises through cooler air, the hot magma from these chambers rises and pushes its way through the denser rock of the earth's crust. When the magma breaks the surface, it is called **lava.**

Lava flows out of the earth through an opening in the surface called a **vent.** The bowl shape at the top of a main vent is called a **crater.** However, a volcano may have more than one vent. Vents sometimes develop on the sides of the volcano as well as at the top. An eruption that flows through the side vents is called a *flank eruption*.

The mineral and gas contents of the magma determine the explosiveness of a volcano. A hotter, thinner lava tends

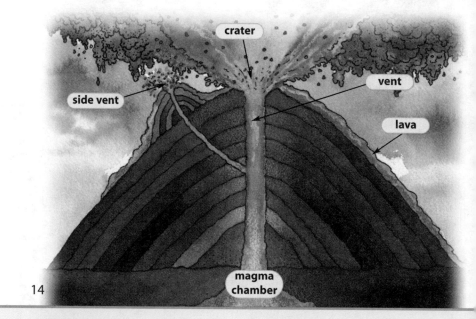

14

SCIENCE BACKGROUND

The Ring of Fire—The Ring of Fire is a zone where earthquakes and volcanoes occur often. It is located around the edges of the Pacific Ocean.

Volcano classifications—Scientists sometimes classify volcanoes into more categories than are presented in this lesson. This text presents the broadest and most common classifications.

Volcanic ash—Volcanic ash is not like the ash from burned wood or paper. It is small pieces (less than 2 mm) of crushed rock and natural glass. Exposure to a cloud of volcanic ash can cause eye, lung, and skin injuries.

Bowl-shaped craters—Scientists believe that most craters are sunken and bowl-shaped as a result of hot magma lowering as it cools.

SCIENCE MISCONCEPTIONS

Evolutionists use hot spots as proof of the great age of the earth. They claim that in order for chains of islands to have formed, the plates must have been moving for millions of years. We do not know exactly how the volcanic islands formed. Perhaps the hot spot moves instead of the plate. Whatever the case, God did not need millions of years to form any of the earth's land features. He is all-powerful and able to control the surface of our earth as well as all other aspects of our lives.

16 Lesson 7

Science and History

Mount Vesuvius (vuh SOO vee uhs) in Italy had a major eruption in A.D. 79. A heated cloud of gas, ash, and rock was forced into the air. The volcano buried the nearby city of Herculaneum. A little farther away, a poisonous gas cloud enveloped Pompeii and killed the inhabitants. Over fifteen feet of fine ash then covered the city. Archaeologists excavated this area and discovered that the ash preserved moldings of the people and places. By pouring plaster into the ash molds, archaeologists revealed details of the people who died in the eruption. These details include folds in their clothing and even the expressions on their faces.

to flow as a liquid. However, thick lava often traps gases. As the hot gases expand, they explode violently, sending globs of hot lava and rocks flying into the air. Many erupting volcanoes produce clouds filled with jagged bits of crushed rock called volcanic **ash.** A volcanic **cone,** or funnel-shaped mound, can form from layers of ash or hardened lava. The Mexican volcano Paricutin (pah REE koo TEEN) produced a cone of ash over one hundred feet high in just a few weeks.

Locations of Volcanoes

Volcanoes may occur anywhere. But two-thirds of active volcanoes are found in an area around the edges of the Pacific Ocean. These volcanoes form a ring that volcanologists call the **Ring of Fire.**

The Mediterranean Sea is another common place for volcanoes. Mount Etna is a well-known volcano of this area. It rumbles to life regularly, but it is still a popular place for sightseers who enjoy the adventure of climbing the mountain.

Volcanoes also exist under the water. Underwater eruptions called *submarine eruptions* are twenty times more frequent than eruptions on land. Many of these underwater volcanoes occur in the middle of a crustal plate rather than along a plate boundary. Scientists believe that these volcanoes are formed from **hot spots,** places where a pool of intensely hot magma rises toward the surface. Since oceanic crustal plates are usually thinner than continental crustal plates, the magma melts the rock above it and breaks through the crust. As the volcano erupts, new land forms. Some geologists believe that chains of islands, such as the Hawaiian Islands, were formed as crustal plates moved across hot spots.

15

 In *Science and History,* the pronunciation of Herculaneum is (HER kyuh LAY nee uhm) and of Pompeii is (pahm PAY).

 Use the world map to point out other volcanic islands, such as the Philippines, Iceland, the Azores, and the Galapagos Islands. Observe whether these islands occur along plate boundaries or are formed by hot spots in the earth's crust.

Discussion:
pages 14–15

- What do we call the molten rock under the earth's crust? {*magma*}
- What do we call a person who studies volcanoes? {*a volcanologist*}
- How is a volcano formed? {*A crack in the earth's crust allows magma and gases to come to the surface.*}
- What are magma chambers? {*pockets of molten rock in the earth's lithosphere*}
- What name is given to magma once it comes through the earth's surface? {*lava*}
- How do magma and gases get from a magma chamber to the earth's surface? {*through vents*}
- What shape are most craters? {*bowl-shaped*}
- What do we call an eruption that flows through a side vent rather than the main vent? {*flank eruption*}
- What determines the explosiveness of a volcano? {*the mineral and gas content of the lava*}
- How is volcanic ash different from the ash you might find in a fireplace? {*Possible answer: Volcanic ash is small bits of rock, not soft bits of burned wood.*}
- Refer back to Student Text page 5. Point out the Ring of Fire.
- What relationship can you see between plate boundaries and volcanoes? {*Volcanoes tend to occur along plate boundaries.*}
- Explain what scientists think causes a volcano to erupt. {*Rock below the earth is under pressure. It becomes hot and melts. Magma rises and pushes its way through the earth's crust, causing an eruption.*}
- Why does hot magma rise? {*because heat rises*}
- Where would you find most of the earth's volcanoes? {*under the ocean*}
- What are underwater eruptions called? {*submarine eruptions*}
- What makes hot spots different from where most volcanoes occur? {*Hot spots are not necessarily along plate boundaries.*}
- Think about the difference in thickness between the ocean crust and the continental crust. Why would hot spots be more frequent under the ocean crust? {*Answers will vary, but suggest that the ocean crust is thinner and the magma can push through more easily.*}

Chapter 1   Lesson 7   17

Discussion:
pages 16–17

➤ **What are three ways to classify volcanoes?** *{by shape, by how often they erupt, and by type of eruption}*

💡 **Which of these ways do you think is the most reliable way to classify a volcano?** *{Accept any answer, but elicit that shape is probably the most reliable classification.}*

➤ **Which kind of volcano is not cone-shaped?** *{shield volcano}*

➤ **What is an example of a shield volcano?** *{Mauna Loa}*

➤ **What are some characteristics of a cinder cone volcano?** *{Possible answers: resembles a hill; has a bowl-like crater; usually has more than one vent; made of cinders}*

➤ **Describe a composite cone volcano.** *{A composite cone volcano has steep sides and layers of lava and tephra.}*

➤ **What is tephra?** *{a mixture of cinders, ash, and rock}*

💡 **Why might a composite cone volcano be taller and steeper than a cinder cone volcano?** *{Possible answer: As the lava layers cool, they form hard rock that other layers can build upon. A cinder cone has loose rock that moves when another eruption occurs.}*

➤ **Define active, dormant, and extinct volcanoes.** *{active—erupted in recorded time and expected to erupt again; dormant—erupted in the past but currently inactive and not expected to erupt again; extinct—no recorded eruption and not expected to erupt in the future}*

💡 **Is classifying volcanoes by how often they erupt accurate? Give an example of your answer.** *{no; Possible answers: Mount Vesuvius, Mount Tambora, and Mount Katmai were all considered extinct when they erupted.}*

➤ **How does knowing the type of volcanic eruption help scientists?** *{The type of eruption gives scientists a clue as to what happens inside the volcano.}*

Classifying Volcanoes
By shape

Not all volcanoes are cone shaped. *Shield volcanoes* have gradually sloping sides and look like upside-down saucers. They are formed by a continual flow of lava. The shield volcanoes formed from hot spots are some of the largest in the world, but these volcanoes are generally not explosive.

There are two kinds of cone-shaped volcanoes. A *cinder cone volcano* is usually a volcano that resembles a hill more than a mountain. It most often has a bowl-like crater at the top and usually contains one main vent. Its eruption tends to be explosive, and it often showers bits of ash and lava, called **cinder,** into the air.

The *composite cone volcano* is the large, symmetrical, cone-shaped volcano that is commonly pictured. This volcano has steep sides that can measure several thousands of meters high. It is made of layers of hardened lava and **tephra** (TEF ruh), a mixture of cinders, ash, and rock. These volcanoes often have explosive eruptions.

Mauna Kea

cinder cone volcano

Mauna Loa
shield volcano

composite cone volcano
Mount St. Helens

SCIENCE BACKGROUND
Mauna Loa—Mauna Loa, a Hawaiian volcano, is the largest known shield volcano on Earth. Its name means "long mountain." It is 4168 m (13,677 ft) above sea level. If all of the mountain were showing, it would be the world's tallest mountain, at about 17,069 m (56,000 ft) high. This large volcano makes up over half of the land mass of the Hawaiian Islands. Thirty-three eruptions have been recorded.

MATH

Mauna Loa, a Hawaiian volcano, is 17,069 m (56,000 ft) high.

➤ If about 4168 m (13,677 ft) of Mauna Loa is above sea level, how much of this volcano is below sea level? *{12,901 m (42,323 ft)}*

➤ What fraction of Mauna Loa is above sea level? *{only about ¼}*

By how often they erupt

Volcanoes are also classified by how often they erupt. An *active* volcano is one that has erupted at some point during a recorded time period and is expected to erupt again. Some volcanoes are considered *dormant* because they have erupted in the distant past but are currently inactive and not expected to erupt again. A volcano that does not have a recorded eruption and is not expected to erupt in the future is called *extinct*. Although a volcano may be considered extinct, there is no guarantee that it will remain extinct. Mount Vesuvius, Mount Tambora (tam BOR uh), and Mount Katmai (KAT my) were all considered extinct volcanoes before they erupted suddenly, killing many people.

By the type of eruption

Not all volcanoes erupt in the same way. The type of eruption gives scientists clues to what is happening inside the volcano. A volcano with runny lava and little or no cinder, ash, and steam is called a *Hawaiian eruption*. It is a quiet eruption that may continue for long periods of time. If the volcano produces a fountain of lava that runs down the sides, volcanologists call it a *Strombolian eruption*. Neither the Hawaiian nor the Strombolian eruption is considered violent or very dangerous.

A *Vulcanian eruption*, however, is violent. It usually causes a loud explosion that sends lava, ash, cinders, and gas into the air. Similar but even more violent is a *Pelean* (puh LAY uhn) *eruption*, named after Mount Pelée on the Caribbean island of Martinique. This eruption also produces an avalanche of red-hot dust and gases called a **pyroclastic** (PIE roh KLAS tic) **flow**, which races down the side of the volcano. When Mount Pelée exploded in 1902, it destroyed a city of 30,000 people in less than two minutes.

The most powerful eruption is the *Plinian* (plin EE uhn) *eruption*. In addition to spewing out lava, this eruption blows gases, ash, and debris very high into the atmosphere. The ash can get caught in the winds of the upper atmosphere and travel for miles.

Volcanoes do not always erupt the same way every time. In fact, sometimes one eruption will change the conditions inside the volcano so that it erupts differently soon afterwards. In 1980 Mount St. Helens erupted as a Pelean eruption with a huge pyroclastic flow that leveled the side of the mountain. The eruption opened a new vent that later exploded in a Plinian eruption, showering ash for miles around the mountain.

> **QUICK CHECK**
> 1. Draw and label a diagram of a volcano.
> 2. Explain the theory of hot spots.
> 3. What are three ways to classify a volcano?

17

 Plinian eruptions are named for Pliny the Younger, a Roman statesman who observed the eruption of Mount Vesuvius in A.D. 79 from several miles away. His observations were recorded in two letters sent to a historian. The eruption of Mount Vesuvius was a Plinian eruption. Pliny the Younger's uncle, Pliny the Elder, died in the eruption.

 Find the location of each volcano on the world map as it is discussed in the Student Text.

Discussion:
pages 16–17

▶ What are characteristics of a Hawaiian eruption? *{little or no lava, runny lava, ash, steam}*

▶ Which kinds of eruptions are considered the least dangerous? *{Hawaiian, Strombolian}*

▶ What is a pyroclastic flow? *{a high-speed flow of very hot gases and dust}*

▶ Which kind of eruption produces a pyroclastic flow? *{Pelean}*

▶ How did the Pelean type of eruption get its name? *{It is named after Mt. Pelée on the island of Martinique.}*

▶ Why is a Plinian eruption considered the most powerful? *{It sends gases, ash, and debris so high that they spread over a large area.}*

▶ Why might a volcano erupt in more than one way? *{An initial eruption may create conditions that produce a different kind of eruption.}*

Answers

1.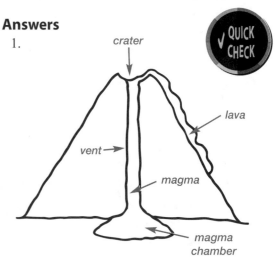

2. Scientists believe that at places on a crustal plate (other than the rim), a pool of intensely hot magma breaks through and forms new land.

3. by shape, by how often they erupt, and by the kind of eruption

Activity Manual
Expansion—page 9

This page identifies continents and the Ring of Fire on a map.

Reinforcement—page 10

A web is a graphic way to organize material. This web groups the different classifications of volcanoes by name and characteristics.

Lessons 8–9

Student Text pages 18–19
Activity Manual pages 11–13

Objectives

- Design a volcano based on one of the three kinds of volcanoes
- Make a model of a volcano
- Communicate the type of volcano made and the process used to make the volcano

Materials

- See Student Text page

Introduction

Note: Two lesson days are allotted for this activity. On the first day, introduce the activity, set guidelines and a due date for the volcanoes, and begin planning. The second lesson day may occur at a later time, when you are ready to erupt the volcanoes.

Review the three shapes of volcanoes.

▶ **What are the three shapes of volcanoes?** {shield, cinder cone, composite cone}

▶ **What are the types of eruptions?** {Hawaiian, Strombolian, Vulcanian, Pelean, and Plinian}

Discuss characteristics of each volcano shape and each type of eruption.

Purpose for reading:
Student Text pages 18–19
Activity Manual pages 11–12

▶ You will build a volcano. What do you think would be the best way to start?

▶ Read the Student Text pages and Activity Manual pages before you begin planning your volcano. What type of volcano do you want to model? What materials will you use?

The student should read all the pages before beginning the activity.

 ACTIVITY

Create an Eruption

Design a volcano model that is based on one of the three shapes of volcanoes: shield, cinder cone, and composite cone. Try to make your volcano as realistic as possible.

Process Skills
- Predicting
- Measuring
- Making and using models
- Observing
- Communicating

Procedure

1. Decide on the shape and design of your volcano. Draw your design in your Activity Manual.

2. Form your volcano according to your design. You may choose to use clay, papier-mâché, or other materials of your own. Insert a small bottle into the top of your volcano. This is where you will put your eruption materials later.

3. Record on the materials list the volcano materials you used.

4. If necessary, allow your volcano to harden. To make your model volcano look more realistic, you may choose to paint it and add other details such as dirt, rocks, and trees.

5. The eruption for your volcano will result from combining baking soda and vinegar. Experiment with different amounts of each ingredient to make the kind of eruption you want. Record the ingredients and measurements of your solution on the chart. Red food coloring added to the vinegar will produce a red "lava" flow. Be creative.

6. Record your chosen eruption materials on the materials list.

7. Use a funnel to add your ingredients to the small bottle. It is best to add the baking soda before the vinegar. Erupt your volcano using your chosen solution.

Materials:
large piece of cardboard or wood
clay or papier-mâché
small bottle or container
paint (optional)
pieces of bush, trees, or shrubs (optional)
dirt or sand (optional)
baking soda
vinegar
red food coloring (optional)
funnel
safety goggles (optional)
Activity Manual

18

 TEACHER HELPS

Constructing a volcano can be done at home before the lesson or in groups in class. If you construct the volcano in class, clay models may take less time. You will also need to allow at least one additional day for the activity.

You may want to do the eruptions outside because of the mess the volcanoes will make. Red food coloring may stain table tops and other surfaces.

Possible materials for the volcano include clay, papier-mâché, and plaster mix.

Possible design variations include making flank vents for flank eruptions, gluing real soil and plants to the slopes, and building a village nearby.

Possible eruption variations include adding liquid soap to the vinegar solution, or adding Alka-Seltzer, Jell-O, or powdered laundry detergent to the baking soda.

Conclusions
- Did your volcano erupt as you expected?
- In what ways was your eruption similar to a real volcanic eruption?
- In what ways was your eruption different from a real volcanic eruption?

Follow-up
- Try varying the quantities or using other materials to make the model volcano look or erupt more like a real volcano.

SCIENCE BACKGROUND
Chemical reaction—When baking soda and vinegar are combined, they cause a chemical reaction that results in foaming. A real volcano does not erupt as a result of this type of chemical reaction. The eruptions in this activity do not model the conditions of real volcanic eruptions. This activity focuses on modeling the shapes of volcanoes.

SCIENCE PROCESS SKILLS
Making and using models—Models help us examine and experience things that may otherwise be too small, too large, or too inconvenient to handle.

- **What parts of a model volcano are representative of a real volcano?**
  {Possible answers: the shape and design of the volcanic cone; the way the material flows from the volcano}

- **What parts of the model are not good representations of a volcano?**
  {Possible answers: the chemical makeup of the "lava"; the type of eruption; the temperature and dangers associated with the lava}

Procedure
Before erupting the volcanoes, give each student an opportunity to share the steps and materials of his project. Then have the student share the type of volcano he has modeled.

Remind the student to carefully measure and record the substances he uses to cause the eruption.

After the activity, compare eruption results.

- What materials produced a realistic eruption?
- What materials didn't work as well as expected?
- How did the amounts of eruption materials used affect the eruption results?
- How did the size of the container used in the vent affect the eruption?
- How did the size or shape of the volcano affect the eruption?

Conclusions
Provide time for the student to answer the questions.

Activity Manual
Activity—pages 11–12

Bible Integration—page 13
A Bible Integration page is included in each chapter. Although they are usually placed near lessons that contain related material, they may be used at any time.

This page discusses some of the occurrences of earthquakes in the Bible.

Assessment
Rubric—TE pages A50–A53
Select one of the prepared rubrics, or design a rubric to include your chosen criteria.

Lesson 10

Student Text pages 20–22
Activity Manual page 14

Objectives

- Identify possible dangers of volcanoes
- List some of the meteorological effects of a volcanic eruption
- List some of the products of volcanoes
- Describe other kinds of thermal eruptions

Vocabulary

vog
debris flow
igneous rock
hot spring
geyser
mud pot

Materials

- igneous rocks for display
- pictures of Yellowstone National Park

Introduction

➤ Have you ever visited Yellowstone National Park in Wyoming?

➤ What are some things you observed on your visit? *{Possible answers: Old Faithful, hot springs, animals}*

You may want to take time to share pictures from trips to Yellowstone. Research Yellowstone National Park on the Internet to view pictures of the geysers and hot springs.

Purpose for reading:
Student Text pages 20–22

➤ How can a volcano affect your everyday life even if you do not live near one?

➤ What other ways does heat escape from inside the earth?

Effects of Volcanoes

In 1815 Mount Tambora erupted in Indonesia. The Plinian eruption blasted millions of tons of ash, dust, and gases high into the atmosphere. Scientists think the gases may have reflected the Sun's rays away from the earth. This caused dramatic cooling worldwide. In fact, 1815 was known as "the year with no summer." In some places snow fell all year long. Crops were damaged and destroyed, causing a food shortage. In 1991 the eruption of Mount Pinatubo (PIN uh TOO bo) also caused a slight cooling of worldwide temperatures. This eruption helped produce brilliant sunsets and sunrises for more than a year.

The gases released from a volcano are similar to smog, the pollution caused by industry and many cars in urban areas. Scientists call volcanic gases **vog,** meaning volcanic fog. Just like city pollution, vog can aggravate respiratory problems. It can also cause acid rain as the sulfur dioxide in the gases mixes with water droplets to form an acid. Acid rain can eat away metals and stone structures and kill plant and animal life.

Dangers of Volcanoes

When you think of a life-threatening eruption, you might picture a crowd of people running from a great lava flow. Lava flows do pose a threat, but they are generally slow moving. The ash and gases released into the atmosphere are actually a greater threat to living things. The pyroclastic flow caused by a Pelean eruption probably causes more destruction than any other feature of a volcano. It was a pyroclastic flow that killed the people of Pompeii.

Another danger is a debris flow. A **debris flow** occurs when part of the mountain collapses and mud and rock fragments surge down the mountain. This debris flow can bury a city and smother the life in its path.

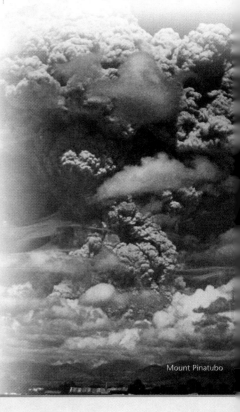
Mount Pinatubo

20

SCIENCE BACKGROUND

Volcanic rock—Igneous rock is often divided into *extrusive rock,* lava that cools quickly above ground, and *intrusive rock,* magma that cools more slowly below ground.

22 Lesson 10

pumice

Products of Volcanoes

Not all effects of volcanoes are bad. The soil around volcanoes is rich in minerals. Indonesian farmers working the land around Mount Merapi, an active volcano, can harvest three crops each year instead of one. Valuable gems are also found in and around volcanoes.

Volcanoes produce **igneous** (IG nee us) **rock** as magma and lava cool and harden. The faster the lava cools, the smoother the rock will be. Different kinds of rock can come from the same kind of lava, depending on how the lava cools. Pumice forms when the foam on the top of lava cools swiftly. Because of its many air pockets, pumice is a rock that will actually float on water! Pumice can be ground into powder and used to make abrasive soap and polish.

Obsidian (ahb SID ee uhn) is another igneous rock. It cools so quickly that it has a glassy, smooth surface. When obsidian breaks, it has sharp edges. It was once formed into weapons such as arrowheads and cutting tools. Because of its beauty, you may see obsidian in jewelry.

Some igneous rock forms when magma cools below the ground. Granite is an example of this kind of rock. Granite is composed of four minerals: quartz, feldspar, mica, and hornblende. These minerals cool slowly underground and can be seen individually in a piece of granite rock. Granite is used in buildings and monuments because it is stable and can withstand a lot of pressure.

God designed the earth. What seems like utter destruction to us is part of the Creator's mighty plan. Christians should never worry about catastrophic events, because they know that God is in control and does all things well.

Psalm 119:90 Thy faithfulness is unto all generations: thou hast established the earth, and it abideth.

obsidian

granite

Your students may enjoy reading *The Journeyman* by Elizabeth Yates, published by JourneyForth Books, a division of BJU Press. It tells the story of a young man during "the year with no summer."

Discussion:
pages 20–21

➤ **What do scientists think caused "the year with no summer?"** {the Plinian eruption of Mount Tambora}

➤ **In what ways can a volcano affect the weather?** {Possible answers: An eruption can cause changes in temperature because of the gases, ash, and dust in the atmosphere. It can cause acid rain. It can cause vog.}

➤ **What is vog?** {gases released from a volcano}

➤ **Why is vog dangerous?** {Possible answers: Vog can aggravate respiratory problems. It can cause acid rain, which eats away metal and stone and can kill plants and animals.}

➤ **Which kind of eruption can cause a cooling of the earth's temperature?** {a Plinian eruption}

➤ **Why isn't a lava flow particluarly dangerous?** {It moves slowly.}

➤ **What are two other kinds of volcanic dangers that occur quickly?** {pyroclastic flow, debris flow}

➤ **What is a debris flow?** {the surge of mud and rock fragments when part of a mountain collapses}

➤ **What kind of rock does a volcano produce?** {igneous rock}

➤ **What determines the kind of rock formed by a volcano?** {Possible answers: how quickly the lava cools; mineral content}

➤ **How does granite differ from pumice and obsidian?** {It is formed by magma cooling under the earth's surface rather than by lava cooling above the surface.}

📖 **In what way does God use volcanoes to renew the earth?** {Possible answers: Volcanoes produce new land. They bring minerals to the surface.}

➤ **Why should we not be afraid of forces such as earthquakes and volcanoes?** {We know that God is in control of all of nature's forces.} [Bible Promise: I. God as Master]

Display the igneous rocks. Allow students to infer how each rock may have cooled to form its structure.

Discussion:
page 22

- 💡 What does *thermal* mean? {Answers may vary, but elicit that *thermal* refers to heat.}
- 💡 What are some other words that use *therm* as part of the word? {Possible answers: thermometer, thermostat}
- ➤ What is a *hot spring*? {a heated pool of ground water}
- ➤ What is a *geyser*? {a hot spring that blows steam and hot water into the air periodically}
- 💡 How are a geyser and a volcano similar? {They both erupt as a result of heat and pressure.}
- ➤ What is a *mud pot*? {a hot spring that contains more mud than water}
- 💡 Why do you think these types of eruptions are called thermal eruptions? {They are heated by the earth's magma.}

Yellowstone National Park is famous for its many kinds of thermal eruptions. It appears to be located on one of the few hot spots in a continental crustal plate.

Answers

1. Its gases, ash, and dust can cause cooling. Vog pollutes the air and can cause acid rain.
2. rock that forms from cooled lava and magma
3. hot springs, geysers, mud pots

Activity Manual

Study Guide—page 14

This page reviews Lessons 7 and 10.

Assessment

Test Packet—Quiz 1-B

The quiz may be given any time after completing this lesson.

Other Thermal Eruptions

Volcanoes are not the only way that heat escapes the earth. Sometimes a body of water is located near an underground magma pool. Heat from the hot magma warms the water. Once the water is heated, it rises to the earth's surface, creating a **hot spring.** Many people enjoy these hot springs because the warm water relieves aches and pains and relaxes muscles.

A **geyser** (GY zuhr) is a hot spring that periodically blows steam and hot water into the air. The water in a geyser is under great pressure and is heated beyond the normal boiling point. As the water heats it forms steam bubbles, which become trapped in places under the earth. The pressure builds up until the water and steam explode into the air. After the eruption the geyser settles down and begins to build up pressure again. Old Faithful, located in Yellowstone National Park, is a famous geyser. Iceland and Japan also have many active geysers.

A hot spring that contains more mud than water is called a **mud pot.** As the hot water rises, it flows through the mud and warms it. Sometimes these mud pots are close to boiling, causing the mud to bubble and splatter. Mud pots usually smell bad. They give off sulfurous gases that smell like rotten eggs.

QUICK CHECK
1. Explain how a volcanic eruption can affect weather conditions.
2. What is igneous rock?
3. What are three non-volcanic types of eruptions?

 Yellowstone National Park is the oldest national park in the world. After a geological survey was done of the area in the 1870s, it was decided Yellowstone was a unique place and should not belong to any one person. On March 1, 1872, the U.S. Congress decided to make it the first national park. In addition to Old Faithful the park has approximately 10,000 other hot springs, mud pots, and geysers.

Old Faithful is not the biggest geyser at the park. It received its name by being consistent. Its eruptions are between 35 minutes and 2 hours apart, and the duration of each eruption predicts the interval to the next eruption.

Direct a DEMONSTRATION
Demonstrate how a geyser works

Materials: large metal funnel, pan, water, hot plate (or stove)

Put the wide end of the funnel in the pan. Fill the pan with water until it is level with the stem of the funnel.

Place the pan over the heat until the water boils. When the water boils it will demonstrate a geyser.

- ➤ How is the funnel like a geyser? {The hot plate heats the water like the magma heats the water pool. The hot water begins to rise. It shoots out the top of the funnel just as it shoots out the top of the geyser.}

Explorations: I.N.V.E.N.T.

Many inventions have helped scientists discover new information about volcanoes. Some information can be gathered at a distance. NASA flies radar equipment that can see through volcanic clouds over active volcanoes. This equipment tells the type and location of lava flow. Other instruments measure temperature and the gases being released. A seismograph is used to record movement in the area of the volcano. Small changes in the slope of the ground can be a warning sign. A tiltmeter registers this motion. A geodometer (JEE uh DAHM uh tuhr) is a laser beam used to assess the shape and size of a volcano.

Scientists also study volcanoes up close, but this is very dangerous. A volcanologist may wear a suit made partly of aluminum to help fireproof his clothing. Leather gloves and hiking boots are also important for protecting him from extreme heat. Sometimes he needs a gas mask. He may use a stainless steel pipe to gather lava samples. Scientists take many risks to discover information from a volcano.

What to do

1. Design an invention that will help scientists research volcanoes. Decide what information you want your invention to monitor. Where will your machine be able to gather information? Will it be mobile? If so, how does it work? Is it a piece of clothing or a machine a scientist would maneuver?
2. Follow the procedure in your Activity Manual.

Dante II

Scientists sometimes use robots to gather information from inside active volcanoes. *Dante II* made a reasonably successful trip into Mount Spurr in Alaska. However, after retrieving data, it fell to the crater floor on its way out. NASA scientists hope that developing similar robots will also aid in exploring the harsh conditions of other planets.

Student Text page 23 — Activity Manual pages 15–16 — Lessons 11–12

Objectives
- Identify the dangers and difficulties associated with exploring volcanoes
- Design a piece of equipment that would help in volcano research

Materials
- carpenter's level

Introduction

Note: Two lesson days are allotted for this exploration. On the first day, introduce the project, set a due date, and begin planning. The second lesson may occur at a later time, when the projects are presented.

Display the carpenter's level.

➤ A carpenter's level is used to indicate if an item is level horizontally or straight vertically. Volcanologists use an instrument called a *tiltmeter,* which is similar to a carpenter's level.

Purpose for reading:
Student Text page 23

The student should read all the pages before beginning the exploration.

Discussion:
page 23

➤ What dangers might scientists face when studying volcanoes? *{Possible answers: heat, uncertainty of eruption, unstable rock, poisonous gases}*

➤ What equipment might scientists use to protect themselves? *{Possible answers: protective clothing, tools to gather samples, robots, long-range sensors}*

➤ How was the *Dante II* robot used? *{It went into an active volcano.}*

💡 Why is NASA interested in the *Dante* robot design? *{Answers will vary, but elicit that a robot designed to withstand conditions in a volcano might be able to endure conditions on other planets.}*

Activity Manual
Explorations—pages 15–16

Read the scenario to the students to generate interest and excitement. Discuss some of the dangers and hazards of an active volcano.

Assessment
Rubric—TE pages A50–A53

Select one of the prepared rubrics, or design a rubric to include your chosen criteria.

SCIENCE BACKGROUND

A tiltmeter records unevenness in the earth's surface. Bulges and bubbles in the surface (crust) can indicate volcanic activity.

TEACHER HELPS — To encourage active listening from the "committee," instruct each student to record the following information about each presentation: name of inventor(s), name of invention, purpose, would they give money (yes or no and why).

Following procedures— Guide the student in learning to read procedures. Direct the student to circle all the requirements for the project in the memo on Activity Manual page 15. Encourage him to mark each step as he completes it.

Chapter 1 — Lessons 11–12 — 25

Lesson 13

Activity Manual pages 17–18

Objective
- Use the PQ3R method to read informational text

Materials
- copy of TE pages A18–A19, *Devils Postpile National Monument,* for each student

Introduction

Discuss different purposes for reading.

➤ What types of books do you read for fun?

➤ Do you expect to learn new information as you read these books? {not usually}

➤ Do you usually have to remember the information that you learn while reading for fun? {no}

➤ Sometimes you must read to gain new information. What types of books might you read for this purpose? {Possible answers: textbooks, encyclopedias, other reference books}

➤ When you read to gain information, such as from your Science textbook, do you always understand and remember what you read?

In this lesson the student will learn how to use a study skill called PQ3R to help him improve his ability to gain and remember important information as he reads.

Discussion:
Activity Manual page 17

Discuss each step of the PQ3R study skill. The discussion refers only to Activity Manual page 17 to teach each step of the PQ3R study skill. Then guide the student in using PQ3R and completing Activity Manual pages 17–18 while reading the *Devil's Postpile National Monument* article.

➤ Look at the Study Skill box on Activity Manual page 17. What does each letter of *PQ3R* stand for? {Preview, Question, Read, Recite, Review}

➤ What is a *preview*? {Possible answers: a sample look, skimming to get a general idea}

When the student reads to learn information, his first step is to *Preview* the entire selection to get a general idea of what it is about. In this step **the student *skims,* or takes a quick look at, the title, subheadings, illustrations, captions, chart titles, bold words, and italicized words.**

After the student previews the selection, he reads each section of the selection separately. The *Question, Read,* and *Recite* steps of PQ3R are repeated for each section read.

Study Skill: PQ3R   Name _____

Complete teaching instructions are located in the Teacher's Edition.

A. Practice the PQ3R study method as you read "Devils Postpile National Monument" from the page your teacher gives you. Begin by using the *Preview* step. Check each box as you complete the step.

STUDY SKILL
Preview — Skim the text.
Question — Ask questions.
Read — Look for answers.
Recite — Answer your questions.
Review — Reread and think about what you have learned.

☐ **P**review (Skim)
1. What do you look at during your preview? _title, subheadings, illustrations, captions, bold words, and italicized words_

2. What do you think this article will be about? _Devils Postpile National Monument_

B. Follow the middle three steps (*Question, Read,* and *Recite*) for the first section of the article.

☐ **Q**uestion
3. Make a question from the title of the article "Devils Postpile National Monument." _Possible answers: What is Devils Postpile National Monument? Where is Devils Postpile National Monument?_

4. Look at the first illustration. Make a question from the caption under the map. _Possible answers: Where is Mammoth Lakes? What is the monument?_

☐ **R**ead
5. Think about your questions as you read silently the first section of the article.

☐ **R**ecite
6. Tell yourself the answers to the questions you wrote in numbers 3–4.

C. Repeat the *Question, Read,* and *Recite* steps for the "Volcanic Formation" part as you read the next section.

☐ **Q**uestion
7. Make a question from the subheading "Volcanic Formation." _Possible answers: What volcanic formation is part of the monument? How was the volcanic formation formed?_

8. Make a question from the caption under the second picture. _Possible answers: What made the columns? How did the columns appear?_

9. Make a question with the word *basalt*. _Possible answer: What is basalt?_

10. Make a question with the words *columnar basalt*. _Possible answers: What is columnar basalt? How did columnar basalt form?_

Lesson 13 Expansion   17

---

26 Lesson 13

❑ **R**ead
11. Think about your questions as you read silently.

❑ **R**ecite
12. Tell yourself the answers to the questions you wrote in numbers 7–10.

D. Repeat the *Question, Read,* and *Recite* steps for the "Glacier Erosion" section.

❑ **Q**uestion
13. Make a question from the subheading "Glacier Erosion." _Possible answers: What did glaciers erode? What was formed by glacier erosion? What is glacier erosion?_

14. Make a question from the caption under the picture. _Possible answers: What is glacier polish?_

15. Make a question with the word *striations*. _Possible answer: What are striations?_

❑ **R**ead
16. Think about your questions as you read "Glacier Erosion" silently.

❑ **R**ecite
17. Tell yourself the answers to the questions you wrote in numbers 13–15.

E. Repeat the *Question, Read,* and *Recite* steps as you read the last section.

❑ **Q**uestion, **R**ead, **R**ecite Again
18. On your own paper, make questions for the next section, "Rainbow Falls." Read the section and recite your answers. _Answers will vary._

F. Complete the last step in the PQ3R study method.

❑ **R**eview
19. Look back at the title, subheadings, captions, bold words, and italicized words. Think about the information you learned.

Lesson 13
Expansion
18

➤ **What is the second step of PQ3R?** {Question}

In the *Question* step the student **forms questions from titles, headings, illustrations, captions, charts, bold words, italicized words, or other material he did not understand during the *Preview* step.** The questions may be written.

➤ **Look at the title of Activity Manual page 17. Titles usually tell you what information you will be reading. Make a question from the title on this page.** {Possible answers: What is PQ3R? What will I learn about PQ3R?}

➤ **What is the third step of PQ3R?** {Read}

After the student has formed questions about the section, he is to read the section completely. During the *Read* step he **looks for the answers to his questions.** In addition to the text, he **reads captions and examines any graphic information, such as maps and charts.** He should slow his reading or stop and reread a passage if it is especially difficult.

➤ **What is the fourth step of PQ3R?** {Recite}

The *Recite* step immediately follows each *Read* step. In this step the student **recites, thinks about, or writes the answers to his questions.** He may also find it helpful to **summarize the section in his own words or take notes.**

➤ **What is the fifth step of PQ3R?** {Review}

After the entire selection is read, the student reviews what he has learned. In the *Review* step the student **rereads the titles and headings, rereads captions and other graphic information, and thinks about the answers to his questions.**

Activity Manual
Expansion—pages 17–18
Guide the student as he applies the PQ3R study skill to read *Devils Postpile National Monument*.

PQ3R is taught in *Reading 6 for Christian Schools: As Full as the World.* If you are using this program, you may want to coordinate teaching the study skill.

Once PQ3R is learned, remind the student to use it as he reads *Science 6 for Christian Schools* or any other informational material.

SQ3R—a study method in which the first step is *Survey*. We have chosen to teach PQ3R. The **P**review step is the same as the **S**urvey step in SQ3R.

Chapter 1

Lesson 13   27

Lesson 14

Student page 24
Activity Manual pages 19–20

Objectives

- Recall concepts and terms from Chapter 1
- Apply knowledge to everyday situations

Introduction

Material for the Chapter 1 test will be taken from Activity Manual pages 4, 14, and 19–20. You may review any or all of the material during this lesson.

You may choose to review Chapter 1 by playing "Volcanic Eruptions" or a game from the *Game Bank* on TE pages A56–A57.

Diving Deep into Science

Questions similar to these may appear on the test.

Answer the Questions

Information relating to the questions can be found on the following Student Text pages:

1. page 10
2. page 15
3. pages 14 and 22

Solve the Problem

In order to solve the problem, the student must apply material he has learned. The student should attempt to answer the problem independently. The answer for this Solve the Problem can be found using information on Student Text pages 10–11. Answers will vary and may be discussed.

Activity Manual

Study Guide—pages 19–20

These pages review some of the information from Chapter 1.

Assessment

Test Packet—Chapter 1 Test

The introductory lesson for Chapter 2 has been shortened so that it may be taught following the Chapter 1 test.

Diving Deep into Science

Answer the Questions

1. Which of the earthquake scales would be more influenced by the ways buildings were constructed?
   the Mercalli scale

   Why? _This scale is based on the amount of destruction that occurs. Mercalli scale measurements for the same waves are smaller than Richter scale measurements in areas with reinforced structures, but they are greater in areas with weaker structures._

2. Think about the thickness of the ocean crust and the continental crust. Why would hot spots be more frequent under the ocean crust?
   Answers will vary, but suggest that the ocean crust is thinner, so the magma can push through the crust more easily.

3. How are a geyser and a volcano similar?
   They both erupt as a result of heat and pressure.

Solve the Problem

During the night the shaking ground awakened many residents of a small town. Big buildings that were thought sturdy began to crack and moan. When the earthquake was over, many buildings had collapsed and lay in piles of rubbish on their sandy foundations. The earthquake, however, measured only 4.8 on the Richter scale. Why do you think the earthquake caused so much damage to this city? What could people have done to prevent the massive destruction?

Answers will vary, but elicit the fact that the buildings were on sandy foundations. The sandy foundations did not provide a stable base for the buildings. The people could have put the structures' foundations deeper into a rock base.

Review Game

Volcanic Eruptions

Divide the class into teams. Draw two volcano shapes, both divided with equal horizontal lines. For each correct answer to a review question, the team colors a section of its volcano, starting at the bottom. The first team to "erupt" its volcano wins.

Weathering and Erosion

Many scientists once assumed that the Grand Canyon took millions of years to form. They were sure that the Colorado River could have eroded all the layers of rock only over a long period of time. However, in 1926 scientists realized that a canyon could form quickly. In Washington State an irrigation canal became blocked. When engineers rerouted the water into a ditch, the force of the water collapsed the underlying rock. In only six days, a small, 10-foot deep ditch became the 120-foot deep Burlingame Canyon. Creation scientists theorize that the Grand Canyon was formed in a similar way. Huge amounts of water from the Flood could have eroded layers of rock very rapidly. Through events such as the forming of Burlingame Canyon, God has given us a glimpse of His awesome power over His creation.

25

Science Background

Grand Canyon—This national park covers more than 277 miles of the Colorado River and adjacent uplands.

Burlingame Canyon—In 1904 near Walla Walla, Washington, irrigation canals were constructed to help with crop production. In 1926 one of the canals became clogged. Engineers changed the course of the water supply, sending it into a nearby irrigation ditch. This ditch was six feet wide and ten feet deep at its largest point. However, after just six days of this excessive water flow, the irrigation ditch "became a miniature Grand Canyon." Today the canyon measures 1500 feet long, 120 feet deep, and 120 feet wide!

(Student Text and Teacher's Edition information on Burlingame Canyon adapted from "How Long Does It Take for a Canyon to Form," by Dr. John D. Morris of the Institute for Creation Research.)

Student Text page 25
Activity Manual page 21

Lesson 15

Objectives
- Demonstrate knowledge of concepts taught in Chapter 1
- Recognize that scientific inferences are not always accurate

Materials
- pictures of the Grand Canyon

Introduction

Note: This introductory lesson has been shortened so that it may be taught following the Chapter 1 test.

Display the pictures of the Grand Canyon. Allow some students who have visited the Grand Canyon to share their experiences.

➤ The Grand Canyon is one of the best known examples of erosion. However, scientists disagree on how the canyon was formed. Some scientists, such as those that hold to the biblical view of a worldwide flood, believe that the canyons formed from the flow of a lot of water in a short period of time. Other scientists believe this extensive erosion was caused by a little water flowing for a very long period of time.

Chapter Outline
I. Weathering
   A. Rock Cycle
   B. Mechanical Weathering
   C. Chemical Weathering
   D. Caves
   E. Soil

II. Erosion
   A. Agents of Erosion
   B. Mass Movements
   C. Stream Erosion
   D. Wave Erosion
   E. Wind Erosion
   F. Ice Erosion
   G. Causes of Erosion

Activity Manual
Preview—page 21

The purpose of this Preview page is to generate student interest and determine prior student knowledge. The concepts are not intended to be taught in this lesson.

Lesson 16

Student Text pages 26–29

Objectives

- Identify the three types of rocks and explain how they are formed
- Differentiate between mechanical and chemical weathering
- Define weathering and provide examples of mechanical weathering

Vocabulary

sedimentary
igneous
metamorphic
rock cycle
weathering
mechanical weathering
chemical weathering

Materials

- 2 plastic film containers with lids
- water
- 2 resealable plastic bags
- a freezer

Introduction

Note: Prepare materials ahead. Water will need several hours to freeze.

Fill one film container to the top with water. Place lids on both film containers. Place each container in a plastic bag. Place in freezer.

Remove containers from the freezer and plastic bags.

➤ What affect did the freezer have on the containers? *{The empty container stayed the same. The lid came off the container filled with water.}*

➤ What can we say about the way cold affects water? *{Elicit that water expands when it freezes.}*

➤ Today we will learn about some ways water and temperature affect rocks.

Purpose for reading:
Student Text pages 26–29

➤ Are rocks still being formed today?
➤ What are some natural causes of weathering?

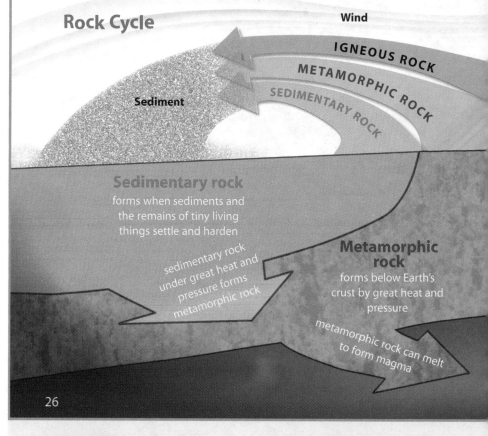

From huge granite mountains to tiny particles of sand, rocks are all around us. We use them for building, for industry, for technology, and even for pleasure. They form the foundation of the earth. But rocks are also constantly moving and changing. Forces such as earthquakes grind and shift rocks, while volcanoes spew out lava that forms new rocks. Heat and pressure deep within the earth form and transform rocks, while on the surface, wind and water break down and move rocks. All of these natural processes are continuously changing the surface of God's earth.

Weathering

Rock Cycle

Though rocks vary greatly, geologists classify them into three categories: **sedimentary, igneous,** and **metamorphic.** Scientists call the

---

SCIENCE BACKGROUND

Rock cycle—The rock cycle is an attempt to explain how rocks are formed. Some of the processes occur where they cannot be observed and are, therefore, hypothetical.

Rock formation and the Flood—Some Creation scientists believe that most sedimentary rock was formed during the Flood. The erosion caused by so much water would have moved great amounts of sediment. Pressure from the water, tectonic activity, or both at the time of the Flood may have formed the sedimentary rock. Creation scientists think it unlikely that significant amounts of sedimentary rock are still being formed today.

Uniformitarianism—People who hold to uniformitarianism believe studying the present processes opens the door to the past. It states that because things are not happening today, they could never have happened in the past. However, this belief contradicts God's Word because it leaves no room for biblical Creation or the Flood. Biblical uniformitarianism says that the processes are occurring as they have since the Flood. There is nothing in the Bible that contradicts this belief.

changing of rock the **rock cycle.** They use a diagram to show what they think happens to the rock.

Most kinds of rock are very hard. But even the hardest rock can be broken down into gravel and pebbles, sand, silt, or powdery clay. This process of breaking down rocks is called **weathering.** We usually divide weathering into two distinct processes: mechanical, or physical, weathering and chemical weathering. **Mechanical weathering** breaks rocks into smaller and smaller pieces. **Chemical weathering** transforms rocks into new substances. Both kinds of weathering take place at or near the earth's surface and are greatly affected by temperature and moisture. The process of weathering usually takes years or even centuries. But slowly, little by little, big rocks are worn away into smaller pieces.

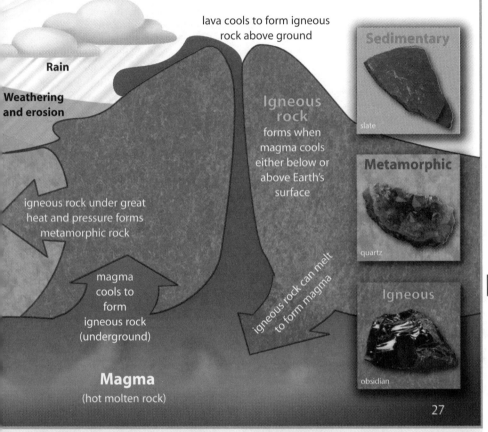

SCIENCE MISCONCEPTIONS
Weathering and erosion do not actually produce sedimentary rock. Rather, they produce the small particles that press together and harden to form sedimentary rock.

 The word *cycle* comes from the Greek word *kuklos,* which means "circle."

Discussion:
pages 26–27

- Man uses rocks in many ways. What are some examples of how rocks are used? {Possible answers: construction, landscaping, art sculptures, sandpaper}

➤ What are three classifications of rocks? {sedimentary, igneous, metamorphic}

➤ What is weathering? {processes that break rocks into smaller pieces or decompose them through chemical activity}

➤ What are two kinds of weathering? {mechanical and chemical}

➤ What affects weathering? {temperature and moisture}

➤ Does weathering usually occur quickly or slowly? {slowly}

- What is a cycle? {Elicit that a cycle is something that occurs over and over.}

- Why is the rock cycle only a theory of how rocks change? {Elicit that scientists cannot actually observe most of the processes that they think occur.}

Discuss the *Rock Cycle.*

➤ Sediment from which types of rock can form sedimentary rock? {all kinds—igneous, metamorphic, sedimentary}

➤ How does sedimentary rock form? {Sediment and remains of tiny living things settle and harden.}

- How is magma involved in the process of making rocks? {Metamorphic rock can become so hot that it turns into magma, which becomes igneous rock when it cools again.}

➤ What happens to an igneous rock to change it into a metamorphic rock? {It is heated and put under pressure.}

Discussion:
pages 28–29

➤ What is another name for mechanical weathering? {physical weathering}

➤ How do rocks change when they are weathered mechanically? {They are broken down. They become smaller.}

💡 Which rock would probably physically weather the most—a rock that is above the surface of the ground or a rock that is underground? Why? {a rock above the surface of the ground, because more of the rock's surface is exposed}

➤ What factors contribute to mechanical weathering? {Possible answers: water, rubbing, wind, plants, and animals}

➤ How does frozen water affect rocks? {When the water in the cracks of rocks freezes, it expands and cracks the rocks or pushes them farther out of the ground.}

➤ What is water freezing in a rock and cracking it called? {frost wedging}

💡 How is frost heaving different from frost wedging? {In frost heaving, the frozen water under a rock expands and lifts the rock. In frost wedging, the frozen water is in the rock, causing the rock to break.}

💡 Who do you think has more of a problem with rocks "growing" in his fields—a farmer in Alabama or a farmer in Minnesota? Why? {a farmer in Minnesota, because water in the ground freezes more often in the North, thus pushing the rocks up}

Note: If the student has difficulty estimating which state has colder weather, use a map or atlas and choose the state that is farther north as having colder weather.

➤ How does pressure release occur? {Great pressure on a rock is suddenly released, causing cracks and breaks in the rock.}

➤ What do we call the peeling away of rock in sheets? {exfoliation}

💡 What might cause a shift in the earth's surface and produce weathering by pressure release? {Possible answers: earthquake, volcano, construction}

➤ What kinds of forces weather material by abrasion? {water and wind}

💡 What types of machinery use abrasion? {Possible answers: sand blaster, pressure washer}

➤ What is characteristic of a rock weathered by water? {It has rounded, smooth edges.}

Mechanical Weathering

Mechanical, or physical, weathering is the process of breaking down rocks into smaller pieces. If more of a rock's surface is exposed, a greater amount of weathering will occur. Temperature, water, wind, and plant and animal life all contribute to mechanical weathering.

Though rocks appear solid, most actually have many small holes and cracks in them that allow water to get inside. Unlike most substances that contract as they freeze, water expands. So as the water in the rock freezes and expands during the winter, it acts like a wedge and forces the rock apart. In fact, this process is called *frost wedging* or *frost action*. Usually the process starts with small breaks, but as the cracks widen, they fill with more water, which causes larger breaks.

Another similar process, *frost heaving*, occurs when water gets underneath a rock. As the water freezes and expands, it pushes the rock farther out of the ground. At one time farmers in cold weather areas believed that rocks grew in their fields, because each spring there seemed to be new rocks that had to be cleared from their fields.

If you take a damp sponge and squeeze it in your hand, you put the sponge under pressure. When you open your hand, the sponge will expand.

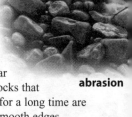
exfoliation

Some rocks are also under great pressure. A shift in the ground around them can reduce the pressure and cause the rocks to expand rapidly. This kind of mechanical weathering, called *pressure release*, creates cracks and breaks in the rocks. Often these cracks result in *exfoliation*, in which sheets of rock peel away like layers of an onion.

Abrasion is mechanical weathering that occurs when rocks rub against each other. It can be caused by water and by wind. As rivers and streams roll boulders and pebbles along their beds, the rocks gradually wear each other away. Rocks that have been abraded for a long time are rounded and have smooth edges.

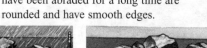
abrasion

frost heaving

28

SCIENCE BACKGROUND

Pressure release and exfoliation— Sometimes pressure release weathering and exfoliation are treated as two different weathering processes. Though exfoliation can occur because of other types of mechanical weathering, it seems to be most closely associated with rocks expanding due to pressure release.

LANGUAGE

The word *abrasion* comes from a Latin word *abrāsus*, meaning "to scrape off."

➤ What are some other ways that we use the word *abrasion*? {Possible answers: A minor cut or scrape is an abrasion. Common household cleansers are categorized as abrasive and nonabrasive.}

➤ How is an abrasion that you might get by sliding into home plate similar to an abrasion caused by wind and water? {In both cases, something is scraped off. In one case it is the rock being scraped off; in the other case, it is your skin.}

A strong wind can also abrade rocks when it picks up particles of sand and dust and blows them against the rocks. If this abrasion continues day after day, the rock will be worn away. Man has taken a lesson from nature and uses machines called sandblasters to remove grime and paint from old buildings and rust from metal by abrading the surfaces.

hoodoo

Abrasion does not affect all types of rocks equally. Soft rocks wear away faster than hard rocks. Sometimes this creates unusual rock formations called *hoodoos*. The rock on the top of a hoodoo is harder than the rock on the bottom, so the rock below abrades much faster, often leaving a vertical supporting column.

Plants and animals contribute to the weathering of some rocks. It is hard to imagine that a tiny seed could cause a rock to break apart. But just as ice expands to break rocks, the roots of a sprouting seed may grow in the cracks of a rock, causing the rock to break apart as the plant grows. Like frost heaving, huge tree roots can lift rocks out of the ground. Burrowing animals can also move rocks and expose them to additional weathering.

Fires, floods, and other catastrophic events can also cause mechanical weathering. In fact, anything that breaks rock into smaller pieces is physically weathering that rock. You can find examples of mechanical weathering all over the world. From cold northern climates to dry, sandy deserts, rock is constantly being broken down into smaller and smaller pieces.

> **QUICK CHECK**
> 1. What are the three types of rock?
> 2. What is mechanical weathering?
> 3. What are three examples of mechanical weathering?

tree splitting a rock

Discussion:
pages 28–29

💡 **Why could water be considered the single greatest factor in mechanical weathering?** *{Answers will vary but should include that water is involved in several of the weathering processes, such as frost wedging, frost heaving, and abrasion.}*

➤ **Why do rocks not weather evenly?** *{Weathering affects softer rocks more than it affects harder rocks.}*

➤ **What causes a hoodoo?** *{The rock of the lower part of a hoodoo is softer and abrades faster than the harder rock above.}*

💡 **How is weathering by plants similar to weathering by ice?** *{Roots growing in cracks act as wedges and may lift rocks out of the ground. In frost heaving, ice does the same thing.}*

➤ **How do burrowing animals contribute to mechanical weathering?** *{They help expose more of the rock.}*

Although asphalt and concrete are man-made substances, they weather much the same way as rock. Frost wedging, frost heaving, extreme temperature changes, and plants all cause these substances to crack and deteriorate.

💡 **What are some ways asphalt and concrete weather?** *{Possible answers: frost wedging in roads and sidewalks; tree roots raising}*

💡 **Why do you think upkeep on the roads in northern states is more difficult than it is in southern states?** *{Accept any answer, but elicit that extreme temperature changes and frequent freezing cause more weathering of the roads.}*

Direct an ACTIVITY
Demonstrate how sedimentary rocks are changed by heat and pressure

Materials: butter, marshmallows, crispy rice cereal, large pan, spoon, hot plate, cooking spray, plastic sandwich bags

Melt the butter and marshmallows in the large pan, stirring constantly. The cereal represents a type of rock that has been weathered into small grains. Stir the cereal into the marshmallow mixture gradually. Give each student a spoonful of the crispy rice mixture in a plastic sandwich bag.

➤ **What do you need to do to make this a solid ball?** *{squeeze it; shape it with your hands}*

The process of changing the grains of cereal into a solid ball involves applying heat and pressure, just as the application of heat and pressure changes sedimentary rocks.

Answers

1. sedimentary, igneous, and metamorphic
2. the process of breaking rocks into smaller pieces
3. Possible answers: frost heaving, frost wedging, abrasion, plant roots, burrowing animals

Lesson 17

Student Text pages 30–33
Activity Manual page 22

Objectives

- Define and give examples of chemical weathering
- Describe how acid rain forms
- Explain how limestone caves are formed as a result of chemical weathering

Vocabulary

carbonic acid   stalagmite
speleothem   stalactite

Introduction

➤ Have you ever visited an old cemetery?
➤ Did you ever notice that the statues and headstones are often worn down or crumbling?

Allow students to share experiences they have had visiting old cemeteries.

Purpose for reading:
Student Text pages 30–33

➤ How does man affect the amount of chemical weathering that occurs?
➤ What are some features of caves and caverns?

Chemical Weathering

Unlike mechanical weathering, which changes only the size and shape of a rock, chemical weathering changes the rock into a different substance. The most common types of chemical weathering are *oxidation* (OK sih DAY shun) and *reaction of acids* with minerals in the rocks. Most of us have seen oxidation. Perhaps you have noticed garden tools that have been destroyed by rust. When oxygen in the air combines with iron, iron oxide (rust) forms. Most of the metal products we use have protective coatings to slow down oxidation. But rocks have no protection, so rust can gradually eat away the metal contained in some rocks.

Another type of chemical weathering occurs when rain carries chemicals from the air onto surfaces below. Earth's atmosphere contains both water and carbon dioxide. When the carbon dioxide dissolves in the water, a weak acid called **carbonic acid** forms. Though carbonic acid is relatively weak, over long periods of time rain containing the acid can dissolve certain kinds of rock. You can see the effects of carbonic acid in old graveyards. Carbonic acid gradually wears away the limestone gravestones until the engravings are unreadable.

Carbonic acid by itself is not harmful to plant or animal life. However, man has introduced additional chemicals to the atmosphere by burning fossil fuels such as oil and coal. One of the most common chemical compounds produced by fossil fuels is sulfur dioxide. Sulfur dioxide in smoke combined with water in the atmosphere produces sulfuric acid. Rain containing sulfuric acid, carbonic acid, and other chemicals is

weathered gravestones

30

SCIENCE BACKGROUND

Oxidation—a process that occurs whenever a substance combines with oxygen to form a new substance. Iron is not the only element that can oxidize.

Carbonic acid—the weak acid that gives soft drinks their "bite"

 Chemical weathering changes rocks into different substances. Christ can change a sinner into a new creature (II Cor. 5:17).
[Bible Promise: B. Guiltless by the Blood]

trees affected by acid rain

stronger and weathers rocks much more quickly than rain containing carbonic acid alone. This stronger acid solution is called *acid rain.* Due to industrial smoke and the exhaust from cars, cities are more likely to have this kind of chemical weathering. However, upper-level winds can spread acid rain for hundreds of miles, causing damage to trees and wildlife far away from cities.

A trip to the forest reveals another kind of chemical weathering. It is not uncommon to see lichens and mosses growing on rocks. Because these organisms do not have true roots, they can survive on the little water and soil they find on some rocks. While attached to the rocks, these organisms secrete mild acids that dissolve the rocks and further break down the rocks into soil.

lichens

Whether by wind, weather, acids, or living organisms, God uses both mechanical and chemical weathering to break down rocks, forming soil and replenishing the minerals needed for plants to grow. His perfect design recycles the mineral resources that He has put on the earth.

city pollution

HISTORY

Acid rain is not a new problem. In the mid-1800s industrial smoke around London combined with the infamous London fog and caused acid fog.

Today, most developed countries have pollution controls that reduce emission of pollutants that contribute to acid rain. However, developing countries often industrialize without such pollution controls.

➤ **Why could a developing country's pollution affect other countries?** {Winds and water may carry the pollution across the countries' borders.}

Pollution is a difficult economic and environmental problem to handle. For a country to modernize, it must develop industry. Yet the negative effects of industrialization may harm the environment and citizens of the country and affect relationships with neighboring countries.

Discussion:
pages 30–31

➤ **How does chemical weathering differ from mechanical weathering?** {Chemical weathering actually changes the rock into a different substance. Mechanical weathering changes only the size and shape of the rock.}

➤ **How are chemical and mechanical weathering alike?** {Both break down rock in some way.}

➤ **What are two common types of chemical weathering?** {oxidation and reaction of acids}

➤ **What is a common form of oxidation?** {rust}

➤ **Why don't all our metal products rust away quickly?** {Most manufactured metals have protective coatings.}

💡 **Can rocks rust? Why?** {Yes; some rocks contain metal ores that can rust.}

➤ **What forms when water and carbon dioxide mix?** {carbonic acid}

➤ **How does the slight acidity of ordinary rain differ from acid rain?** {The carbonic acid in ordinary rain is not as strong as the sulfuric acid in acid rain.}

💡 **Why is acid rain more likely to occur in industrialized areas?** {The conditions, such as smoke and exhaust, which contribute to acid rain occur more frequently in industrialized areas.}

➤ **What can cause acid rain?** {burning fossil fuels such as coal and oil; volcanic gases}

➤ **Why is acid rain harmful?** {It can damage trees and wildlife.}

➤ **How do living organisms contribute to chemical weathering?** {Some organisms secrete mild acids that break down rocks.}

➤ **What are two important functions of weathering?** {forms soil; replenishes minerals for plant growth}

📖 **Weathering is an example of how God cares for His world through natural processes.**

Discussion:
pages 32–33

➤ What are some ways that mechanical weathering can form caves? {Possible answers: wind, waves, running water}

➤ What kind of weathering forms a limestone cave? {chemical weathering}

The terms *cave* and *cavern* can be used interchangeably.

➤ How does a limestone cave form? {Acidic water seeps into cracks in limestone, dissolving calcite in the limestone, which causes cavities to form.}

➤ What is the general name for the calcite formations in caves and caverns? {speleothems}

➤ How are the speleothems formed? {The calcite dissolved from the limestone is deposited out of the water.}

💡 What does "deposited out of the water" mean? {The water, in which the calcite is dissolved, evaporates, leaving the calcite to harden.}

➤ Which kind of speleothem hangs from the ceiling? {stalactite}

➤ Which kind of speleothem builds up from the ground? {stalagmite}

➤ What do we call people who explore caves and caverns? {spelunkers}

➤ Why do spelunkers have to carry mountain-climbing equipment? {Some of the features in caves include walls of rocks or chasms like you would find while climbing a mountain.}

Refer the students to *Fantastic Facts*.

➤ What huge cavern system did Jim White discover? {Carlsbad Caverns}

➤ What caused him to investigate? {seeing hundreds of thousands of bats coming out of a hole}

➤ What response did White get from others when he told of his discovery? {People laughed and did not believe that he found anything special.}

📖 Christians sometimes experience similar responses when they try to tell others about Christ or strive to live godly lives. The responses of others should not keep a Christian from saying and doing what is right. [BATs: 2c Faithfulness; 2e Work]

➤ What finally caused people to want to visit Carlsbad Caverns? {seeing black and white photos}

In 1901 a cowboy named Jim White discovered one of the greatest cave systems in the United States. As White was riding home one evening, he saw what looked like dark-gray smoke rising from the ground. He rode nearer and discovered that the "smoke" was hundreds of thousands of bats coming out of a hole in the ground. Later, White went back to explore the hole, which he named Bat Cave. What he found amazed him. He had discovered a gigantic cave. When White told his friends what he had found, they laughed. No one believed he had found anything special. White continued to explore and tell others about his discoveries.

Finally, in 1922, a photographer took some black-and-white photographs of the caverns. Suddenly, people wanted to tour the caverns and see their wonders. On October 25, 1923, New Mexico's Carlsbad Caverns was made a national monument. In 1930 it was made a national park that now covers more than seventy-three square miles.

Caves

Weathering forms many kinds of caves. Crashing waves, wind, and running water can all form caves by mechanical weathering. However, it is chemical weathering that forms limestone caves, or caverns. As water passes through the atmosphere and the ground, it combines with other substances to form acids. As this acidic water seeps into cracks in limestone, it dissolves calcite in the limestone, leaving behind large underground cavities.

Speleothems, beautiful formations in caverns, form as the dissolved calcite is deposited out of the water. Some of the most common structures in caves are stalactites (stuh LAK TYTES) and stalagmites (stuh LAG MYTES). **Stalactites** hang from the ceiling and look like stone icicles. **Stalagmites** "grow" up from the ground as a result of the dripping of dissolved calcite.

People who enjoy exploring caves are called *spelunkers* (spih LUNG kuhrz). These explorers must take their own lights with them into the caves. Since many caves have huge chasms that must be crossed or walls of rock that must be climbed, spelunkers often use mountain-climbing equipment.

32

Direct a DEMONSTRATION
Demonstrate how caverns are formed as limestone dissolves

Materials: clear plastic bottle with small opening (i.e., soda bottle), piece of aluminum foil, rubber band, jar, sand, sugar, 250 mL of warm water

Cut off the top 15 cm of the plastic bottle. Remove the bottle cap. Cover the neck opening with foil. Secure the foil with a rubber band. With the tip of a pencil, poke three or four small holes in the foil.

Place the foil end of the bottle top into the opening of the jar. Do not let the foil touch the bottom of the jar.

Gently press a 5 cm layer of sand into the prepared bottle top. Spread a 2.5 cm layer of sugar on top of the sand. Add another 5 cm layer of sand and press it down gently.

This model demonstrates layers of the earth that have limestone deposits.

Pour 125 mL of warm water into the bottle. After the water is absorbed, pour another 125 mL of warm water into the bottle.

After two or three hours, the sugar should dissolve and leave behind air pockets and gaps in the sand. This demonstrates how caverns form when limestone dissolves.

36 Lesson 17

Stalactites hang from the ceiling of a cavern.

A *column* forms when a stalactite and a stalagmite grow together.

A *drip curtain* forms when water seeps in along a crack and hardens, leaving calcite behind in a long, delicate, curtainlike sheet.

Stalagmites grow upwards from the floor of the cavern.

1. How does chemical weathering differ from mechanical weathering?
2. What causes acid rain?
3. How do speleothems form?

Luray Caverns, VA

One way to keep stalactites and stalagmites straight is to remember that stalactite has a "c" in it, as in *ceiling*, and stalagmite has a "g" in it, as in *ground*.

The word *speleothem* comes from the Greek words *speleo*, meaning "cave," and *thema*, meaning "deposit."

The word *spelunker* comes from the Latin word *spelunca*, meaning "cave" or "grotto."

The student can research other ways that caves are formed. Caves can be formed by wind and sand, by ocean waves, and by glaciers. The student may describe the process of cave formation, show pictures of different types of caves, and report any historical significance of the caves he researched (i.e., if people lived in them or hid in them).

Encourage the student to do additional research about the use of caves either historically or currently. Possible topics may include ancient wall paintings in Lascaux Cave in France; the Anasazi Indians in the southwestern United States; the caves where the Dead Sea scrolls were found; or modern uses for caves, such as for growing mushrooms.

Discussion:
pages 32–33

 What cavern is pictured on Student Text page 33? {Luray Caverns in Virginia}

➤ What is a drip curtain? {a speleothem that forms when seeping water hardens along a crack, forming a thin curtainlike sheet}

➤ What causes a column? {A stalactite and a stalagmite grow together.}

Point out the different speleothems in the photo.

Answers

1. In chemical weathering, the chemical makeup of the rock is changed, while mechanical weathering changes only the size and shape of the rock.
2. Pollutants in the air (from burning fossil fuels or from volcanoes) combine with moisture in the atmosphere to create highly acidic rain.
3. Acidic water seeps through the soil and into cracks in limestone, dissolves the limestone, and leaves behind caverns.

Activity Manual
Study Guide—page 22
This page reviews Lessons 16 and 17.

Assessment
Test Packet—Quiz 2-A
The quiz may be given any time after completion of this lesson.

Lesson 18 — Activity Manual pages 23–24

Objectives
- Measure length to the nearest millimeter
- Measure mass to the nearest gram
- Measure volume to the nearest milliliter

Materials
- assortment of 8 cm x 17 cm or smaller rocks; at least one per student
- meter stick
- centimeter ruler for each student
- mass scale
- metric beakers or measuring cups large enough to hold one rock
- water

Introduction
Note: Metric units of measure are used in this lesson and in most of the activities in SCIENCE 6 for Christian Schools.

➤ Imagine you are a geologist who has been called to an earthquake disaster site to analyze rock samples. Your job is to measure the length, mass, and volume of each rock taken from the site.

➤ The data you collect may help other scientists determine where the rock came from and may help them draw other important conclusions about the nature of the earthquake.

Accuracy is important when taking measurements, especially in science.

 Teacher Helps — Measuring correctly is an important scientific skill that the student will use in activities throughout the book. Monitor the student's measuring accuracy throughout the course. Observe the student using the various types of measurement to be sure he demonstrates the correct use of each measuring tool.

Direct a Demonstration
Demonstrate mass
Use a container, ice, and a mass scale to demonstrate the idea of the conservation of mass. Place the ice into the container and measure the mass. Record the measurement. Let the ice melt, measure the mass of the water, and compare the measurements. The mass of the ice did not change even though its shape, or state, changed. Remove some of the water and measure the mass. Since the amount of matter changed, the mass changed.

Measurement

This lesson is meant for practicing and reinforcing measurement skills and concepts learned in math.

Linear Measurement
➤ What metric tools are used to measure length? *{meter stick, centimeter ruler}*

Display the meter stick and centimeter ruler.

➤ Which tool would be the most appropriate for measuring the distance a rock tumbles down a slope? *{meter stick}*

➤ Which tool would be most appropriate for measuring your rock? *{centimeter ruler}*

Review the markings on the centimeter ruler.

➤ What do the numbered marks on the ruler indicate? *{centimeters}*
➤ What do the smaller markings indicate? *{millimeters}*
➤ How many millimeters are in one centimeter? *{10}*

Demonstrate marking the width and length of a rock. Place the rock on a sheet of paper or on an overhead projector. With a pencil (or overhead marker) held perpendicular to the paper, make two dots showing the width and two dots showing the length of the rock. Remove the rock from the paper.

Demonstrate measuring the width and length between the dots.

➤ Place your rock on your Activity Manual page. Make dots marking the width and length of your rock. Remove your rock from the page.

➤ Measure the distance between the two marks showing the width of your rock in centimeters. Record and label your answer.

➤ Measure the distance again in millimeters. Record and label your answer.

➤ Which measurement is more accurate? Why? *{millimeters; Since millimeters are smaller than centimeters, the measurement will be more accurate.}*

➤ Measure and record the length of your rock in millimeters.

Mass Measurement
➤ What is meant when we talk about the *mass* of an object? *{Mass is the quantity of matter in an object.}*

Although the size, shape, or state of an object may change, its mass does not change unless some of the matter is removed or more is added.

➤ Mass is measured using a balance scale to compare the mass of the object to be measured with the known mass of the weights.

➤ What unit is used to measure mass? *{gram}*

Demonstrate the use of the scale by measuring a rock.

➤ Measure the mass of your rock and record the measurement in your Activity Manual.

38 Lesson 18

Volume Measurement

▶ **What is the volume of an object?** *{Guide students to conclude that volume is the amount of space that an object takes up.}*

▶ **What are some measuring tools used to measure volume?** *{Possible answers: measuring cups, measuring spoons, beakers, graduated cylinders}*

The unit of measure for volume is the liter. A graduated cylinder or metric beaker can be used. Always read the markings at eye level.

▶ **Since your rock has an irregular shape, how can you use water and a beaker (measuring cup) to measure the volume of your rock?** *{Guide the student to conclude that he can drop the rock into a measured amount of water and calculate the difference between the amounts of water to determine the number of milliliters the rock takes up.}*

Demonstrate measuring the volume of a rock with water in a beaker.

▶ Measure the volume of your rock and record the measurement in your Activity Manual.

▶ Volume can also be measured in cubic units. You can find the volume of a box by multiplying the measurements of the length, width, and height of the box.

▶ **Why did we measure the volume of the rock in milliliters (mL) rather than in centimeters (cm)?** *{Draw the conclusion that because the rock has an irregular shape, using a linear measurement (centimeter) would be difficult.}*

Discuss that one milliliter is equal to one cubic centimeter. Because the measurements have this relationship, they can often be used interchangeably. You may want to calculate the volume of your rock in cubic centimeters.

As time permits, have the student measure rocks that have already been measured by others.

▶ **Were your measurements the same as the other measurements for the same rock? Why?** *{Answers will vary.}*

Discuss the importance of accurate measuring. Scientists must repeat experiments many times and compare their measurements and findings as a check to make sure their results are accurate.

Guide the student in completing the bar graph with the mass measurements for several rocks. You may choose to number all the rocks used to make identification easier.

SCIENCE BACKGROUND

Using an equal arm balance—Make sure the balance indicator points to zero. Place the object to be measured in one pan. Add weights to the other pan until the indicator points to zero, showing that the two sides are balanced. Add the weights to determine the mass of the object.

Using a triple beam balance—Make sure all the weights are pushed to the left and point to zero. (If the balance indicator does not point to zero, adjust the knob next to the pan until it does.) Place the object to be measured on the pan. Move the weights until the indicator points to zero again. As you move the weights, make sure each rests in the notch next to the number you intend for it to mark.

1. Start by moving the largest weight to the largest increment of 100 without letting the indicator go below zero.
2. Move the second largest weight to the largest increment of 10 without letting the indicator go below zero.
3. Move the smallest weight until the indicator rests at zero. Total the numbers marked by the weights to determine the mass.

Meniscus—The curve of the surface of a liquid in a container is called a *meniscus*. Look at the bottom of the meniscus to read the volume of a liquid.

SCIENCE MISCONCEPTIONS

Mass and weight—The words *mass* and *weight* are often confused. *Mass* is the amount of matter in an object. It is measured by comparing the amount of matter in an object to the amount of known matter in an object, such as a weight. Balance scales are used to measure mass in grams or pounds. *Weight* is not the measure of matter. It is the measure of the force of gravity on an object. If the gravity changes, the weight changes. Spring scales are used to measure weight in newtons (1 newton = 0.225 lb).

 Christians need to show God's love to others. How does the godly love that you demonstrate measure up?

▶ Are your words and actions toward others loving and kind?
▶ Do you give to others unselfishly?
▶ Are you friendly towards others?
▶ When was the last time that you told someone else about God's love and gift of salvation? [BAT: 5 Love-Life Principle]

Activity Manual
Activity—pages 23–24

Lesson 19

Student Text pages 34–36
Activity Manual pages 25–26

Objectives

- Compare the different kinds of soil and their relative sizes
- Describe the factors that determine the composition of soil
- Illustrate and describe the five soil horizons

Vocabulary

soil	texture
humus	loam
pedologist	horizon
sand	topsoil
clay	subsoil
silt	bedrock

Materials

- potting soil
- centimeter ruler

Introduction

Display a handful of potting soil.

➤ What do we use potting soil for? *{planting}*

➤ Why do we use potting soil instead of clay?
 {Possible answers: It is looser. It has more nutrients.}

Allow several students to share experiences they have had planting and gardening.

Purpose for reading:
Student Text pages 34–36

➤ How do different soil particles affect the texture of the soil?

➤ Which soil horizons contain nutrients from decayed materials?

Discussion:
pages 34–35

➤ What is soil? *{the loose material at the surface of the earth}*

➤ What materials are found in soil? *{weathered particles, decayed organic material, humus, air, and water}*

Organic means material from living things, such as plants and animals. Decayed organic material is composed of living things that have died.

➤ What is a pedologist? *{a scientist who studies soil}*

➤ What are the three kinds of weathered particles that make up soil? *{sand, clay, and silt}*

➤ Which kind of soil particle is the largest? *{sand}*

40   Lesson 19

Soil

Soil is the loose material at the surface of the earth. Weathering produces small particles of rocks and minerals. These small particles, along with water, air, and decayed organic material called **humus** (HYOOM uhs), make up the soil.

Soil particles

Though weathered particles greatly range in size, **pedologists** (puh DOLL uh jists), scientists who study soil, generally separate soil particles into three basic sizes.

Sand is the largest kind of particle. Some sand particles are big enough to see without a magnifying glass. Even so, they are very tiny, ranging in size from 0.06 mm to 2 mm. Sand particles are rough, causing sandy soil to have a grainy feel. Sand particles do not fit together tightly. Water can easily get between the particles, allowing the soil to drain easily and quickly. However, sometimes soil drains so quickly that it does not retain the water necessary for certain kinds of plants to grow in it.

The smallest kind of particle is **clay.** It would take approximately 100,000 particles of clay to make one particle of sand. When dry, clay has a smooth texture, but it is sticky when wet. Clay holds nutrients and water well, but its particles are so close that little air can get between them. Without air, plants may rot in the moist soil.

Silt is the third kind of particle. Its particles are very tiny, yet they are actually larger than clay. Dry silt feels powdery like flour. Silt allows water and air to mix in the soil.

Soil texture and formation

Most soil is a mixture of the three different kinds of soil particles. When scientists discuss the **texture** of a soil, they are referring to how much of each kind of particle is in the soil sample. For example, a sandy soil texture would contain a large amount of sand. When all three kinds of particles are equally evident in the soil, the soil is called **loam** (LOME). In loam, the properties of all three kinds of particles combine to form an especially fertile soil.

Relationship of sand, silt, and clay

sand   silt   clay

34

SCIENCE BACKGROUND

Potting soil—Most commercial potting soils are a mixture of soil, humus, sand, clay, and other materials.

LANGUAGE

The word *organic* is derived from the Latin word *organicus*, meaning "instrument," and has the idea of being from organized living beings.

The word *humus* is derived from the Latin word meaning "soil."

Texture Triangle

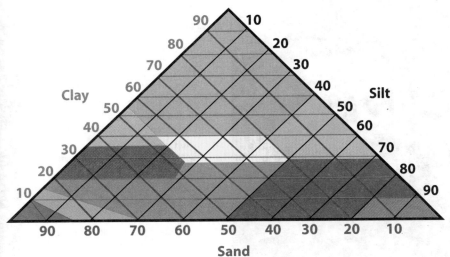

Soil texture is very important to farming. Most plants grow best in a specific kind of soil suited to the plants' design. For example, potatoes do not grow well in hard, sticky, clay soil.

The composition and fertility of soil depend on the climate, kinds of weathered rock, and types of vegetation in the area. Climate affects the rate and kind of weathering that occurs. For example, moist climates usually experience more chemical weathering than dry climates. Different rocks weather to produce different kinds of soil particles. When plants and other organic materials decay, they form humus and add nutrients to the soil.

35

 Soil—God has made the earth to be an inhabitable place for man to live (Isaiah 45:18). Part of the success of that habitation depends on the plants that are growing around us. They provide food, oxygen, energy, and many other necessary functions. God has promised to care for us just as He has cared for the plants by giving them what they need to survive (Luke 12:28). This includes giving them the correct type of soil for their survival. [Bible Promise: H. God as Father; I. God as Master]

 You may choose to review the mathematical basics of percentages. Relating percentages to cent values may help some students understand.

Discussion:
pages 34–35

▸ Look at the marking for 2 mm on the ruler. Would you consider two millimeters to be very large? {no}

▸ Look at the picture of the soil particles on Student Text page 34. A particle of sand can be two millimeters or smaller. Imagine how small the other particles are if the sand particle is two millimeters.

▸ Why does sand not make very good soil? {Water drains through too quickly, not allowing plants to get the water and nutrients that they need.}

▸ How many clay particles would it take to make one particle of sand? {about 100,000}

▸ Why does clay not make very good soil? {There is not enough space for air between the particles. Plant roots will rot without enough air.}

💡 Why might silt be a better soil than sand for growing plants? {Silt allows both water and air to mix in the soil.}

▸ What do we mean by the *texture* of a soil? {the amount of each kind of particle found in the soil}

▸ What kind of soil has all three kinds of particles? {loam}

"Equally evident" means that all three particles are present. It does not mean that there are equal amounts of each particle present.

▸ Why is the texture of soil important to farming? {Certain crops grow best in certain soil textures.}

▸ What determines the composition and fertility of soil? {kinds of weathered rock, climate, vegetation}

▸ What kind of climate would probably cause the greatest chemical weathering? Why? {Possible answer: rainy or moist climate, because moisture is needed to produce acid rain}

 Look at the *Texture Triangle* on Student Text page 35. Point out the heading and percentages along each side.

▸ What kind of particle appears to have the most influence on the texture of soil? {clay}

💡 Why are there so many kinds of loam? {Most soils probably contain a variety of weathered materials. Loam is the combination of the three particles. It can occur in many combinations.}

Discussion:
page 36

➤ What do we call the layers of soil? {horizons}

➤ What is the O horizon made of? {leaf litter and humus}

💡 What is humus? {decayed organic material}

➤ In which horizon do plants grow? {topsoil, or A horizon}

➤ What makes up topsoil? {minerals from weathered rock and humus}

➤ What is another name for the B horizon? {subsoil}

➤ How does the B horizon benefit from the layers of soil both above it and below it? {From above, it gets nutrients from the humus. From below, it gets minerals.}

💡 Why do you think the weathered fragments in the C horizon are larger than the weathered materials in the layers above? {Possible answer: The C horizon has less exposure to weathering.}

➤ What is bedrock? {unweathered parent material}

💡 Why do you think bedrock is called "parent material"? {Possible answer: Minerals and rock in other layers come from the bedrock.}

➤ What is another name for bedrock? {R horizon}

➤ What does the bedrock affect in the layers of soil above it? {the texture of the soil}

💡 Why is it difficult for most vegetation to grow where there is little topsoil and humus? {Possible answers: Plants get nutrients from decayed material. Most plants germinate and grow in topsoil.}

Answers
1. weathered particles, water, air, and humus
2. sand, silt, clay
3. the different layers of soil

Activity Manual
Expansion—page 25
You may have the students use different colored pencils to trace along the lines for each soil sample.

Study Guide—page 26
This page reviews Lesson 19.

Assessment
Test Packet—Quiz 2-B
The quiz may be given any time after completion of this lesson.

Soil horizons
Most of us think of soil as the "dirt" where we plant gardens. But soil actually consists of multiple layers called **horizons**.

Leaf litter and humus compose the top layer of soil. This layer is called the O horizon.

The next layer is the **topsoil,** or A horizon. Most plants germinate and grow roots in this layer. In addition to minerals from weathered rocks, topsoil has a high proportion of humus.

The B horizon, or **subsoil,** contains a few nutrients from the humus that have washed down through the upper layers of soil. Subsoil consists mostly of weathered minerals from the bedrock.

The C horizon consists mainly of larger weathered fragments of the bedrock. It contains clay and sand particles but very little organic material. The C horizon, however, is rich in minerals.

Underneath all the horizons is the **bedrock,** or unweathered parent material. Bedrock is the rock that greatly influences the texture of the soil above it. Sometimes it is called the R horizon, or *regolith*.

Some areas may lack layers of humus and topsoil. Very little vegetation can grow in these conditions.

1. What makes up soil?
2. List the particles of soil from the largest to the smallest.
3. What are soil horizons?

SCIENCE BACKGROUND

Horizons—Horizons may be described in different ways. However, the basic structure of the layers is usually the same regardless of the ways the layers are named.

Compost piles—Some people collect dead leaves, grass clippings, and other organic materials to form compost piles. The decaying material forms humus. Composting is simply an effort to speed up the process of producing humus to enrich the soil.

Explorations: Soil Detective

Pedologists, or soil scientists, use many scientific instruments and methods to determine the exact texture of soil. However, when they work outside of the laboratory, they often use the "feel method" to determine the texture of soil. By following a step-by-step process, pedologists can closely approximate the texture of soil simply by feeling it.

One way of showing a step-by-step process is by using a flow chart. A flow chart is a graphic representation of a procedure. It uses symbols and arrows to show what steps to take to complete a process from start to finish.

Using a flow chart you can learn to determine the texture of soil and become a "soil detective."

What to do

1. Collect a sample of soil from the ground around your school or home.
2. Use the flow chart in your Activity Manual to analyze the soil sample.
3. Record your information.
4. Repeat with soil from another location.
5. Compare your findings.

Teacher Helps

A flow chart is useful for teaching step-by-step directions. It can help a student analyze and evaluate each step taken as he progresses through an activity.

The appearance of the flow charts in the Activity Manual may confuse the student. To help him focus, cover most of the chart and expose only one or two steps at a time.

Student Text page 37
Activity Manual pages 27–28

Lesson 20

Objectives
- Follow the procedure of a flow chart
- Analyze a soil sample

Materials
- transparency of TE page A20, *How to Read a Flow Chart*

Introduction

Display the transparency *How to Read a Flow Chart*.

➤ **A flow chart is one way to show a process that requires multiple steps.**

Follow the chart step by step. Be especially alert to places where decisions are made.

➤ **Pedologists use the feel of a soil sample to place it on a texture triangle, such as the one on Student Text page 35.**

Discuss that pedologists follow a systematic method to analyze soil samples.

Purpose for reading:
Student Text page 37
Activity Manual pages 27–28

➤ How can you determine the texture of a sample of soil?

Procedure

Guide the student through the procedure.

Instruct the student to use one flowchart for each sample. Tell him to circle the decision that he makes each time.

Activity Manual
Explorations—pages 27–28

Assessment
Rubric—TE pages A50–A53

Select one of the prepared rubrics, or design a rubric to include your chosen criteria.

Lessons 21–22

Student Text pages 38–39
Activity Manual pages 29–30

Objectives

- Record observations
- Analyze experiment results
- Predict the particles needed for a specific soil sample

Materials

- several seed packets or seed catalogs that tell in what kind of soil (well-drained, dry, moist, etc.) the plant grows best
- See Student Text page
- buckets or large containers to collect used soils and water

Introduction

Note: Two lesson days are allotted for this activity. You may choose to do Part 1 on the first day and Part 2 on the second day.

Distribute the seed packets. Direct the student to read the packets to find the kind of soil recommended. Allow several students to choose plants from the seed catalog.

➤ Suppose you want to grow a plant that needs well-drained soil, but your soil is clay and does not drain well. What could you do to change the texture of your soil so that the soil would drain better? {Answers will vary, but elicit that the soil needs some sand, silt, or both to loosen it up and allow better drainage.}

➤ Suppose your soil is sandy. What could you do to the soil to help it retain water better? {Answers will vary, but elicit that adding some clay, silt, or both would help the soil retain water better.}

Refer back to the *Texture Triangle* on Student Text page 35. Notice that clay has more effect on soil than the other kinds of particles have.

Purpose for reading:
Student Text pages 38–39
Activity Manual pages 29–30

➤ What are you trying to accomplish in this experiment?

The student should read all the pages before beginning the activity.

ACTIVITY: Retaining the Right Amount

The amount of water that different types of soil can hold is one of the factors that determines how well crops can grow in that soil.

Process Skills
- Hypothesizing
- Measuring
- Experimenting
- Observing
- Identifying variables
- Recording data

Problem
How can I mix clay, sand, and potting soil to obtain a soil sample that will retain 50 percent of the water it receives after 2 minutes?

Procedure—Part 1

1. With the nail, punch 10 tiny holes in the bottom of each foam cup. (Set one foam cup aside for Part 2.)
2. Measure 250 mL of clay. Put the clay in one of the foam cups.
3. Follow the same procedure with the sand and the potting soil.
4. Place each foam cup in a smaller plastic cup. The water that seeps through the holes will collect in the plastic cup.
5. Pour 160 mL of water over the clay in the foam cup.
6. Time the experiment for 2 minutes. (You may need to periodically pull the foam cup out of the plastic cup to ensure that a vacuum has not occurred. A vacuum will prevent the water from draining into the plastic cup.)
7. Remove the foam cup from the plastic cup. Measure the water that drained into the plastic cup.
8. Record your observations and measurements in the chart in your Activity Manual. (For the soil to have a retention of 50 percent, the water seepage will have to measure 80 mL.)
9. Repeat steps 5–8 with the sand and the potting soil.

Conclusions—Part 1
- Which kind of soil allowed the most water to run through?
- Which soil allowed the least water to run through?

Materials:
sand
dry clay or clay cat litter (crushed)
potting soil
metric measuring cups
four 12- or 16-oz foam cups
small nail
four 9-oz clear plastic cups
stopwatch
water
Activity Manual

Teacher Helps: If a set of metric measuring cups for each group is unavailable, prepare clear plastic cups to use instead. Pour measured water into a cup and use a permanent marker to mark the water level. Repeat until all needed measurements are marked on the cup.

Make sure the holes in the foam cups are open. You may need to unclog the holes in the cup of sand while it is draining. If the holes are clogged, the results will be inaccurate.

Preparing the work area
- Provide large containers of water for students to measure from.
- Cover the work surface with newspaper or plastic.
- Provide buckets or containers to collect used water and soil.
- Provide a place to set the wet cups.
- Have paper or cloth towels on hand for spills.

Procedure—Part 2

1. Based on your previous observations, formulate your hypothesis. Write on the chart how much of each of the three types of soil you will mix to make a sample that will retain 50 percent of the water. The amounts need to total 250 mL. Record your hypothesis in your Activity Manual.

2. Measure your own combination of the three soils as listed on the chart. Mix the soils and put the mixture in the fourth foam cup. Place the foam cup in a plastic cup.

3. Pour 160 mL of water over your soil mixture and time the experiment for 2 minutes.

4. Measure the amount of water in the plastic cup.

5. Record your information in your Activity Manual.

6. Repeat steps 2–5 as needed.

Conclusions—Part 2

- Was your hypothesis correct?
- What mixture of soil had a water retention closest to 50 percent?
- If you had a plant that needed only 20 percent water retention, which type of soil would you use the most of?

Follow-up

- How would you set up the experiment to see whether the temperature of the water affects the soil's retention level?

SCIENCE PROCESS SKILLS

Predicting—Predictions are made based on experience, study, or knowledge. Careful observations are a key to good predictions. In the first part of this activity, the student observes the relationships between the kinds of soil and the amounts of water they can hold. The student then uses this information to predict the results of his soil mixture.

➤ How did your observations and results of the first part of the activity influence your predictions for the second part?

Chapter 2

Procedure—Part 1

➤ Which soil do you expect to drain the fastest? {sand}

➤ Which soil will likely drain the slowest? {clay}

Remind the student about the importance of accuracy in measuring.

Guide the student as he tests each container. Remind him to mark the water level in each cup after the two minutes and label what kind of soil the water seeped through.

Conclusions—Part 1

Guide the student as he uses his measurements to determine the results.

Procedure—Part 2

➤ Using your observations and results from Part 1 to form your hypothesis is called predicting. Scientists make predictions about further study based on previously gained knowledge and experiences.

The student's hypothesis should include the amount of each kind of soil that he intends to use.

Repeat trials as time permits until 50 percent of the added water is retained in the mixture.

Conclusions—Part 2

Provide time for the student to evaluate his hypothesis and answer the questions.

Activity Manual
Activity—pages 29–30

Assessment
Rubric—TE pages A50–A53

Select one of the prepared rubrics, or design a rubric to include your chosen criteria.

Lessons 21–22

Lesson 23

Student Text pages 40–43

Objectives

- Differentiate between erosion and weathering
- Distinguish among examples of mass wasting
- Describe how sediments are carried and deposited by a stream

Vocabulary

sediment
erosion
agents of erosion
deposition
mass movement
load
delta
birdfoot delta

Materials

- world map

Introduction

➤ Have you ever seen fences, netting, or other barriers between the side of a road and the base of a rocky hillside?

➤ Why do you think highway engineers use these barriers? *{Possible answer: to prevent any loose or falling rock from going on the road}*

➤ Do you think these barriers work? *{Possible answer: The barriers probably work for individual rocks, but are probably not effective to catch debris from larger mass movements.}*

Purpose for reading:
Student Text pages 40–43

➤ What do we call erosion that gravity causes?

➤ How can flooded streams help farmers and harm people living in towns?

Erosion

Agents of Erosion

Weathering produces small particles called **sediment**. Some of this sediment combines with decayed organic materials (humus), air, and water to form soil. Other sediment lies loosely on the weathered rock. But sediment seldom stays in the same location for long. When weathered material moves from one location to another, **erosion** takes place. Weathering and erosion often occur together, but they are not exactly the same. Weathering breaks down rocks, but erosion moves the broken-down material from one place to another.

The primary force behind erosion is gravity. Sometimes gravity pulls weathered material from a higher location to a lower location without the aid of other factors. However, other factors, called **agents of erosion,** often are involved in the transportation of weathered material. These agents include water, wind, and ice. Sometimes these agents work together to erode the earth's surface. Though agents like water, wind, and ice may seem to work on their own, God is their Master. Psalm 147:17–18 says, "He casteth forth his ice like morsels: who can stand before his cold? He sendeth out his word, and melteth them: he causeth his wind to blow, and the waters flow."

Weathered material that moves from one location to another must eventually stop. **Deposition** occurs when wind, water, or ice drops sediments and rocks in a new location. Usually sediment drops according to its weight. The heaviest sediment drops first, and the lightest drops last. As a result, depositions often have a layered look.

wind erosion

gravity erosion

ice erosion

water erosion

 LANGUAGE The word *deposition* can be related to the more common form, *deposit*. Examples: One may deposit money in a bank (it is carried and left there). One may deposit his books on the floor. The idea is the same, in that something is carried and then left behind.

SCIENCE BACKGROUND

mudflow—also called *mudslide*
earth flow—also called *landslide*

Mass Movements

When gravity is the primary force that moves rocks and sediments, **mass movement,** or mass wasting, occurs. Some forms of mass movement occur slowly. *Soil creep* happens as gravity slowly pulls soil down the slope of a hill. Sometimes this causes trees and fences on the hillside to lean at awkward angles.

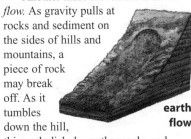
soil creep

Another mass movement is an *earth flow.* As gravity pulls at rocks and sediment on the sides of hills and mountains, a piece of rock may break off. As it tumbles down the hill, this rock dislodges other rocks and sediment, which in turn dislodge other material. The result is a pile of soil and rocks at the bottom of an incline. Sometimes an earth flow occurs slowly, but it can also happen very quickly.

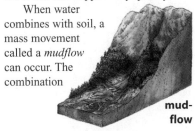
earth flow

When water combines with soil, a mass movement called a *mudflow* can occur. The combination of soil and water produces an extremely heavy and unstable area. The greater the mass of material at the top of the hill the greater the chance that gravity will start it moving downhill. A mudflow is one of the fastest and most devastating mass movements.

Soil creep and land flows are usually not very deep. The main movement is only at the surface. In *rockslides,* however, huge slabs of rock break off along cracks and faults. Sometimes an entire side of a mountain is part of a rockslide. You have probably seen examples of smaller rockslides where roads have been built through mountains. Often large rocks lie along the sides of roads where they have broken off from the slope.

rockslide

Another kind of mass movement is an *avalanche.* An avalanche occurs when a mass of snow along the side of a mountain becomes unstable. The pull of gravity starts the snow moving. Suddenly, huge amounts of snow rumble down the mountain, burying everything in their paths.

avalanche

mudflow

Although man cannot stop avalanches, he has developed technology to help detect and channel the path of an avalanche. Scientists use temperature, weather conditions, land terrain, and snowfall to help predict areas likely to have avalanches. Based on their predictions steps can be taken to prevent serious damage or danger.

Ski rangers and snow patrols constantly test and monitor snow over acres of land looking for potential problems. Sometimes they use huge guns to cause small controlled avalanches in order to break up larger, more dangerous masses of snow.

Discussion:
pages 40–41

- What is sediment? {small weathered particles}
- What is the difference between *weathering* and *erosion?* {Weathering breaks down rock, and erosion moves the weathered rock from one place to another.}
- What is the primary force in erosion? {gravity}
- What are some agents of erosion? {water, wind, and ice}
- Who controls agents of erosion, such as water, wind, and ice? {God} [Bible Promise: I. God as Master]
- What is the dropping of sediment called? {deposition}
- Why do depositions often have a layered look? {Sediments drop according to weight, so some drop farther than others.}
- What is mass movement? {erosion that is primarily caused by gravity}
- What are five kinds of mass movements? {soil creep, earth flow, mudflow, rockslides, avalanche}
- What kind of mass movement might cause the trees or poles on a hillside to lean at an angle? {soil creep}
- What happens during an earth flow? {Tumbling rock loosens other rocks, resulting in a pile of soil and rocks at the bottom of an incline.}
- Why do you think a mudflow is such a devastating mass movement? {Possible answer: Since the mud flows like water, it can fill spaces that dry rocks and soil would not fill.}
- How is a rockslide different from an earth flow? {A rockslide is the movement of many large slabs of rock rather than of a few rocks and soil, as in an earth flow.}
- What causes an avalanche? {Gravity pulls an unstable mass of snow down a mountain.}
- What tectonic forces can start a sudden mass movement? {earthquakes, volcanoes}
- How does erosion show God's preservation of the earth? {Possible answer: Erosion moves minerals and nutrients to less fertile land.}

Discussion
pages 42–43

➤ What name is given to the sediments carried in moving water? *{load}*

➤ Can you see the dissolved load in a stream? Why? *{No; it is dissolved.}*

➤ How is a suspended load different from a dissolved load? *{The sediment of a suspended load is not dissolved.}*

➤ How can the suspended load cause more mechanical weathering than water alone causes? *{The sediment abrades rocks along the sides and bottom of the stream channel.}*

💡 What determines how large a load a stream can carry? *{Possible answer: the speed and volume of the water flow}*

➤ Why is it not safe to cross through a flooded stream? *{The increased volume and speed of the water can move large objects, including cars and people.}*

➤ Why do fast streams cause deeper and wider channels than slow streams? *{The load the stream carries acts as an abrasive force, eroding more and more of the channel as the water flows faster.}*

➤ What happens when a stream slows down? *{It starts to deposit its sediments.}*

➤ Which sediments drop first? *{the heaviest}*

➤ What is it called when a stream drops its sediment? *{deposition}*

➤ How does the deposition affect the stream? *{It causes the stream to become shallower.}*

➤ What are some places where a stream might deposit its sediments? *{in the streambed; in the flooded countryside; in flooded buildings}*

➤ When is sediment welcome? *{when it fertilizes farmland}* When is it unwelcome? *{when it is deposited in homes and other buildings}*

 Point out on the map the Dead Sea in Israel and the Great Salt Lake in Utah.

➤ How do these lakes get so salty? *{Dissolved minerals and salts flow in, but not out of, the lakes.}*

➤ Where do most dissolved minerals and salts eventually end up? *{in oceans and seas}*

Science and History

In the 1960s the Aswan High Dam was built on the Nile River. The dam controls floods, produces hydroelectricity, and stores water. Though the Egyptian people enjoy the benefits of the dam, it has also caused problems. The Nile used to flood annually. Without the sediment deposited by the Nile, farmers have more difficulty maintaining the fertility of their soil.

As the dam prevents new sediment from reaching the Mediterranean Sea, the delta, Egypt's most fertile farmland, is being eaten away by the sea waves. In an effort to keep the force of the waves from carrying away this precious land, the Egyptian government has built several protective barriers offshore.

Where is all the sediment that once formed the delta? It collects behind the dam. The dam must constantly have the sediment removed so that it will not get into the generating equipment.

Aswan High Dam

Stream Erosion

Moving water transports huge amounts of sediment from one location to another. Even the tiniest raindrop moves loose material a little. As the rain washes across the surface of the earth, it carries sediment into streams. The sediment that a stream carries is called its **load**.

Some sediment, such as minerals, dissolves in the stream and is transported to larger bodies of water. This is called the stream's *dissolved load*. Most streams eventually flow to the ocean, but a few flow into large inland lakes or seas. The dissolved minerals make these bodies of water so salty that few things can live in them. The Dead Sea in Israel and the Great Salt Lake in the United States are examples of such bodies of water.

Other sediment particles cannot dissolve in water. Any sediment that is carried by a stream but is not dissolved is the stream's *suspended load*. The faster a stream flows, the more sediment it can pick up and carry along. The sediment acts as an abrasive substance, weathering and eroding additional rock along the sides and bottom of the stream channel. A rapidly flowing stream rolls larger particles

42

SCIENCE BACKGROUND

In the context of this lesson, the term *stream* means any form of water moving as a result of gravity. It might be a small creek or an enormous river.

 BIBLE Sediment deposited on farmlands fertilizes the fields, thus enabling the fields to produce more crops.

➤ What factors can help a Christian bear more fruit? *{Possible answers: Bible study, prayer, giving, telling others about Christ, friendliness, communicating with kindness}* [BATs: 5 Love-Life Principle; 6a Bible Study; 6b Prayer; 6c Spirit-filled]

along its streambed, further eroding and deepening its channel. During a flash flood, when the volume and speed of a stream are unusually great, rushing water can move large boulders. This is why it is never safe to cross a flooded stream. Just as the water can move boulders, it can also pick up cars and people and sweep them away.

When a stream slows down, it begins to drop its sediments. The heaviest sediments are dropped first. This deposition may be in the streambed. Instead of deepening a channel as before, the sediment fills it, causing the stream to be shallower. If the stream floods outside its normal channel, the sediment may be deposited in homes and buildings in the surrounding countryside.

Not all deposited sediment is destructive, however. Farmers in many places depend on the yearly flooding of their land to improve its productivity. An area that commonly floods is called a *floodplain*. Some sediment is deposited in the floodplain, and some is carried to the mouth of the river. So much sediment can deposit at a river's mouth that a new area of land forms. The new land is rich in nutrients and is often used for farming. An area of sediment at the mouth of a river is called a **delta**. Deltas were named after the triangular shape that is formed where the Nile River flows into the Mediterranean Sea. The area looks like the Greek letter *delta* (Δ).

The Mississippi River in the United States has a delta that extends into the Gulf of Mexico. Every year, the delta gets a little bigger as more sediment is added to it. Much of this sediment is topsoil washed off farmland. Unlike the neat outline of the Nile delta, the Mississippi delta sprawls in all directions. The currents and waves of the Gulf of Mexico shape the delta differently. The shape of the Mississippi Delta is so unusual that it has been named a **birdfoot delta.**

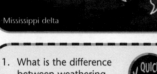
Mississippi delta

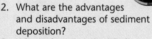
Nile delta

1. What is the difference between weathering and erosion?
2. What are the advantages and disadvantages of sediment deposition?
3. Why is the Mississippi Delta different from the Nile Delta?

Discussion
pages 42–43

➤ **What is a floodplain?** *{an area where a river or stream commonly floods}*

➤ **What happens when the same amount of water flows through a channel that has become shallower?** *{Answers will vary, but elicit that the water must go somewhere, so the river or stream will probably widen as it gets shallower.}*

➤ **What place other than a floodplain can sediment be deposited?** *{at the mouth of a river}*

➤ **What can result from sediment collecting at the mouth of a river?** *{New land can form.}*

➤ **What do we call new land formed at the mouth of a river?** *{a delta}*

➤ **Why is it called a delta?** *{The land formed at the mouth of the Nile River has a triangular shape like the Greek letter* delta.*}*

➤ **Why is the shape of the Mississippi River delta different from that of the Nile River delta?** *{The currents of the Gulf of Mexico shape the delta differently.}*

Refer the student to *Science and History.*

➤ **Why did the Egyptians build a dam across the Nile River?** *{to control floods, produce electricity, and store water}*

➤ **How is the sediment carried by the Nile River beneficial?** *{It makes the land fertile.}*

➤ **What problem does the sediment cause for the Aswan High Dam?** *{The sediment collects at the dam and can damage the generating equipment.}*

On a world map, point out where the Nile River flows into the Mediterranean Sea. Point out where the Mississippi River flows into the Gulf of Mexico.

➤ **Why is the Mississippi Delta called a birdfoot delta?** *{The erosion has not formed a uniform triangular shape. The many channels running in different directions resemble a bird's foot.}*

Answers

1. Weathering is breaking down rock. Erosion is moving weathered rock.
2. When sediments are deposited, they fertilize the land. When sediments are deposited in towns and in buildings, the effects can be devastating.
3. The shape is different because the currents and waves in the Gulf of Mexico are different from those in the Mediterranean Sea.

Direct a DEMONSTRATION
Demonstrate stream erosion

Materials: modeling clay, water, small aluminum pan, sand, potting soil

Direct a student to cover the bottom of the aluminum pan with about 5 cm of clay. Tell him to make a channel about 2.5 cm deep and 2.5 cm across and then fill the channel with potting soil.

Pour just enough water to fill the channel to the top.

➤ **What do you think will happen when we put more water into the channel?** *{The water will overflow.}*

➤ **What will happen to the sediment (soil) that is in the channel?** *{Answers will vary.}*

Pour more water into the channel until the soil floats out of the channel.

Explain that this is what happens when additional water fills a stream or river. Empty and clean out the channel. Again fill it with water.

➤ **What is another way we could get the water to overflow its banks?** *{Answers will vary. Try some of the ideas.}*

If students do not come up with the idea that increased deposition might also cause a river to overflow its banks, start filling the channel with sand until the water overflows.

➤ **How does this model show what happens to a river?** *{Elicit that as rivers drop their loads, they widen to accommodate the same amount of water.}*

Lesson 24

Student Text pages 44–45
Activity Manual pages 31–32

Objectives
- Record and analyze data
- Measure accurately
- Identify variables

Materials
- See Student Text page

Introduction
➤ Have you ever looked at a calm stream and tried to figure out which way it was flowing?

➤ Because of gravity, water flows from a higher elevation to a lower elevation. The steepness of the slope can affect the speed at which the water flows.

➤ In this activity you will investigate the effects of slope on the amount of sediment that erodes.

Purpose for reading:
Student Text pages 44–45
Activity Manual pages 31–32

➤ What are you trying to find out by doing this experiment?

The student should read all the pages before beginning the activity.

ACTIVITY

Stream Erosion

Pedro recently bought a new house. In one corner of the yard is a steep slope. After the first hard rain, Pedro was surprised at the erosion that occurred on that slope. Using the materials below, investigate the effect of a slope on erosion.

Process Skills
- Hypothesizing
- Measuring
- Experimenting
- Observing
- Identifying variables
- Recording data

Problem
How does the steepness of a slope affect the amount of erosion that is caused by a stream?

Procedure
1. Formulate and record your hypothesis in your Activity Manual.
2. Prepare a foil loaf pan by cutting it at the corners of one of the smaller ends. Fold the piece down so it is even with the bottom of the pan. Repeat with the other loaf pan. (Safety tip: The cut edges of the pan may be sharp.)
3. Measure 250 mL (about 1 cup) of dirt. Place the dirt at the end opposite the open end of one of the foil loaf pans. Follow the same procedure with the other foil loaf pan.
4. Place the open end of each loaf pan into a larger pan. This second pan will catch the runoff water. Raise the opposite end of each loaf pan and set it on a stack of books or another raised surface.
5. Using a protractor, adjust the angle of one of the loaf pans to 45 degrees. Add or take away books to keep the pan elevated at this angle. Follow the same procedure with the other pan of dirt, elevating it to a 20-degree angle.
6. Measure 120 mL of water into the spray bottle.

Materials:
2 foil loaf pans
2 small rectangular baking pans
dirt
protractor
cheesecloth
scale
water
two 12-oz clear plastic cups
spray bottle
metric measuring cups
scissors
Activity Manual

44

If students are unfamiliar with using a protractor, you may need to draw a 45° angle and a 20° angle on an index card for each group.

See Teacher Helps on page 44 for a suggested way to prepare the work area.

7. Spray the dirt in one of the pans until all the water is gone.

8. Repeat the procedure with the other pan of dirt using the same amount of water.

9. Place a double layer of cheesecloth over a plastic cup to act as a filter, and pour the contents of the container through the filter to remove the water. Put the eroded material from the 45° slope in a plastic cup and weigh it on the scale. Record your measurements.

10. Repeat step 7 with the eroded material from the 20° sloped pan.

11. Compare the results from the two containers.

Conclusions
- Why would changing the angle of the slope affect the amount of erosion?
- How is this experiment like a real stream? How is it different?

Follow-up
- What variables could you change to find out more about stream erosion?

SCIENCE PROCESS SKILLS

Measuring—Discuss some of the kinds of measurements used in the experiment, such as measuring the angle, measuring milliliters, and measuring milligrams.

➤ **Why is it important to measure the amount of water that we put in the spray bottle?** {Answers will vary, but elicit that it keeps the variable constant.}

➤ **Why is it important to measure the amount of soil that erodes?** {Answers will vary, but elicit that without measuring we are simply guessing how much soil actually erodes}.

Emphasize the importance of accurate measurement.

➤ If we fail to measure accurately, what happens to the information that we gain from the experiment? {The information becomes unreliable because it has no specific data that can be quantified.}

Procedure
Discuss with the student that we are making a simulation, or model, of what may happen in a stream. The pan is the streambed. The soil is the sediment that has settled.

Help the student prepare the aluminum pans for the experiment.

Discuss measuring using *Science Process Skills*.

➤ **Why would a steeper incline cause more sediment to erode?** {Answers will vary, but elicit that a steeper incline causes the water to run faster because of gravity. The faster the water flows, the more it erodes.}

Conclusions
Provide time for the student to evaluate his hypothesis and answer the questions.

Activity Manual
Activity—pages 31–32

Assessment
Rubric—TE pages A50–A53
Select one of the prepared rubrics, or design a rubric to include your chosen criteria.

Lesson 25

Student Text pages 46–49
Activity Manual pages 33–34

Objectives

- Recognize the real-life problems of sand erosion and deposition
- Explain how wind causes erosion
- Compare the effects of ice erosion with other kinds of erosion
- Identify how rocks are eroded by the forces of wind and glaciers

Vocabulary

deflation	glacier
dust storm	plucking
sandstorm	moraine

Materials

- world or United States map

Introduction

➤ Have you ever built a sandcastle at a beach?

➤ What happened to your sandcastle when the tide came in? *{Possible answers: The waves eroded the sandcastle. The waves left deposits of sand in and around the sandcastle.}*

Purpose for reading:
Student Text pages 46–49

➤ What are three causes of erosion other than stream erosion?

➤ Which kind of erosion can move the largest rocks?

Wave Erosion

Water is also a force of erosion along shorelines. The pounding of ocean waves exerts tremendous pressure on the rocks along the coast. As the rocks weather and erode, the ocean waves move the sediment to new locations. The shoreline constantly changes as this erosion and deposition takes place.

Sometimes waves carve out caves and sea arches from rocky cliffs. Places where land was once connected may be eroded to form islands. On the other hand, sand deposits may fill in channels and bays, forming new areas of land.

Sand deposits called sandbars may create shoals, shallow places along the coast. Shallow water can be dangerous to ships. A ship may not notice a sandbar and run aground. Sandbars constantly shift positions, which adds to the danger.

Commerce can be greatly affected when channels that are used for shipping become shallower because of sand deposits. Often large sums of money are spent to dredge (remove sand) in order to maintain shipping lanes.

Science and History

On Cape Hatteras, in North Carolina, a lighthouse stood for many years to warn ships of the shallow sandbars off the coast. But over time the shore at the lighthouse eroded, becoming narrower and narrower. It appeared that the ocean would eventually destroy this famous landmark. Because of its historical significance, a decision was made to move the lighthouse. So in 1999, the lighthouse was slowly moved farther away from the shore and repositioned to continue shining for many more years.

Cape Hatteras

Storms such as hurricanes increase the erosion and deposition caused by waves. High winds create bigger and more powerful waves. Those waves may erode a beach so much that buildings along the shore are destroyed.

sea arch

 LANGUAGE The word *deflation* comes from the Latin word that means "to blow away."

 BIBLE The Bible instructs Christians to guard their hearts (Prov. 4:23). The heart is the seat of the thoughts, will, and emotions. Christians are to put a guard around their hearts to prevent the erosion of their moral character. If Christians are not consistently guarding their hearts, Satan seeks to ensnare them and keep them from being the productive Christians that they should be (1 Pet. 5:8). [BATs 3b Mind; 3c Emotional control]

desert sand dunes

Wind Erosion

Wind is a powerful agent of erosion in dry areas such as deserts. It can also erode areas where land has become very dry due to a drought. When wind blows, picks up loose sediment, and carries it away, the process of **deflation** takes place. The wind cannot move large particles like water can. Nevertheless, a strong wind can carry tons of sediment at a time. **Dust storms** occur when the wind blows small, loose particles such as clay and silt. These dust storms can reach hundreds of meters into the air. However, since sand particles are heavier, **sandstorms** tend to be closer to the ground. Some of the sand moves as the wind bounces it along the ground.

Along the shore where sand dries out, wind often blows sand into piles called sand dunes. Deserts also have sand dunes caused by wind. Some of these are small. However, others are hundreds of meters tall. In a desert with few obstacles to stop the erosion, sand dunes can move as much as twenty-five meters per year. The prevailing, or most constant, wind determines the size and shape of the sand dunes.

beach sand dunes

47

Direct a DEMONSTRATION
Demonstrate moving wet and dry particles

Materials: a pile of paper hole punches, a saltshaker filled with water, a small aluminum pan

Place the paper punches in the middle of the pan. Blow gently. Observe what happens to the paper punches. Return the paper punches to the middle of the pan. Sprinkle water on the paper punches. Blow gently. Observe what happens this time.

➤ **How does sprinkling water on the paper punches change the effects of the wind?** {Possible answer: The water adds weight to the punches and causes them to stick together.}

WRITING

Moving the Cape Hatteras Lighthouse was an enormous and expensive undertaking.

➤ **Do you think moving the lighthouse was worth the time and money spent?**

The student should research to find details about moving the lighthouse. Then the student should write a one-page essay for or against spending the money to move the lighthouse. The argument can include information concerning the historical value of the lighthouse, the current need for a functioning lighthouse, technology that could possibly replace the lighthouse, and the impact on nature of moving the lighthouse.

Discussion:
pages 46–47

➤ **How does wave erosion affect shorelines?** {It causes the shores to change constantly.}

➤ **Why does a changing shoreline have a tremendous effect on commerce?** {Shipping lanes sometimes clog with sediment and have to be dredged in order to stay open. Sandbars create hazards for ships along the coast.}

➤ **What can cause wave erosion to be greater?** {storms, such as hurricanes}

➤ **In what kind of climate is wind the greatest erosion agent?** {dry areas}

💡 **Why would farmland normally not be eroded much by wind?** {Answers will vary, but elicit that in farming areas the soil and air is usually moist.}

➤ **What might happen to cause farmland to be more easily eroded?** {a drought}

💡 **What is a drought?** {when there is not enough rain}

➤ **What is deflation?** {wind picking up loose sediment and carrying it away}

➤ **Name two kinds of wind erosion.** {dust storms and sandstorms}

➤ **Which kind of storm is closest to the ground?** {a sandstorm} **Why?** {The particles are heavier.}

➤ **What determines the size and shape of sand dunes?** {the prevailing wind}

➤ **How might waves and wind combine to change the shoreline?** {The waves bring sand to the beach. As the sand dries, the wind picks it up and moves it into dunes.}

Refer the student to *Science and History*.

➤ **What was the purpose of the Cape Hatteras lighthouse?** {to warn ships of the shallow sandbars off the coast}

➤ **Why was the lighthouse in danger?** {The shoreline had eroded almost to the lighthouse.}

➤ **What happened to the lighthouse?** {It was moved farther inland.}

Discussion:
pages 48–49

➤ **How is a glacier formed?** {A glacier forms when layers of unmelted snow compact and turn to ice.}

➤ **What happens when a glacier starts moving?** {It slides downhill, gouging the ground underneath it.}

➤ **What happens when *plucking* occurs?** {Large pieces of bedrock are pulled out of the ground and carried along the mountainside.}

💡 **How is the erosion caused by a glacier similar to erosion caused by a stream?** {Answers will vary, but elicit that for both the sediment that is carried (suspended load) abrades the surface under it.}

💡 **How is the erosion caused by a glacier different from the erosion caused by a stream?** {Possible answer: Glaciers can carry much larger sediments and can cause much more erosion than streams can.}

💡 **Usually lower soil horizons are not exposed to extensive weathering or erosion. How is this different in areas where glaciers move? Explain.** {Possible answer: A glacier probably erodes all layers of soil since it can move bedrock.}

➤ **What happens to the soil and rock when a glacier begins to melt at the bottom of a mountain?** {The soil and rock are deposited, forming piles called moraines.}

➤ **What do we call rock that has been ground into fine powder?** {rock flour}

➤ **Why do you think it is called rock flour?** {Answers will vary, but elicit that it looks similar to flour used for baking.}

➤ **What are receding glaciers?** {glaciers that melt faster than snow falls}

➤ **What kind of valley does a receding glacier create?** {a U-shaped valley}

➤ **What lakes do scientists think were caused by a receding glacier?** {the Great Lakes in North America}

 Point out on the map the Great Lakes and states as they are discussed.

You can show how a glacier erodes. Spread a layer of clay in the bottom of an aluminum pan. Press an ice cube on the clay and move it back and forth. Remove the ice cube and observe the clay.

Place a small pile of sand on the pile of clay. Place the ice cube on the pile of sand for one minute. Pick up the ice cube and observe the bottom.

Place the ice cube on the clay again and move it back and forth. Wipe the sand off the clay and observe the clay.

Ice Erosion

Ice formations called **glaciers** erode huge amounts of all kinds of sediments and rocks. Glaciers form where snow fallen in the winter does not completely melt in the summer. When more snow falls the next winter, that snow presses the first layer down and compacts it into ice. Eventually the mass of ice becomes so heavy that gravity causes it to begin sliding slowly downhill.

As glaciers slide down a mountain, they do not slide smoothly like an ice cube across a table. Instead, they take pieces of mountain with them. Rocks become caught in the bottom of the glacier as it slides. The rocks and rough ice gouge out the ground underneath them. Where there are weaknesses in the bedrock, a glacier can pull a huge

Sioux quartzite rock left by a glacier

piece of bedrock loose and carry it along, a process called **plucking**. As the glacier continues to slide down the mountainside, the rock already picked up by the glacier scrapes soil and bedrock from the ground. After a while the glacier may erode the rock underneath it and form its own valley.

moraine

As long as more snow is accumulating than is melting, a glacier will continue to move downhill. Toward the bottom of the mountain, the temperature rises and the glacier begins to melt. As the glacier melts, it deposits the soil and rock it picked up on the way down the mountain into piles called **moraines** (muh RAINS). These piles of debris can be hundreds of meters deep and hundreds of kilometers long. Moraines often consist of *rock flour*—rocks that the glacier has ground into fine powder—and huge rocks that were never broken up. They are left behind as the glacier recedes.

A glacier that melts faster than new snow falls is called a *receding glacier*. A glacier that has completely melted often leaves behind a beautiful U-shaped valley. It may also leave lakes behind.

Many large moraines exist along the southern edges of the Great Lakes. The Valparaiso Moraine wraps around the southern tip of Lake Michigan through the states of Michigan, Indiana, Illinois, and Wisconsin. The continental glacier that helped shape Lake Michigan may have left this moraine behind.

Causes of Erosion

Most erosion is part of the normal processes that God has planned for the renewing of the earth. However, people can change the surface of the earth and sometimes cause greater-than-normal erosion. As people develop land and industry, careful planning is important. By thinking ahead about erosion and other possible results of our activities, we can be wise stewards of the resources God has entrusted to our care.

1. How can waves change a shoreline?
2. Where does most wind erosion occur?
3. What is a moraine? What causes it?

TECHNOLOGY
In many places builders who pave large areas are required to provide a retaining pond, a place where water from the paved area can run into and slow down before it flows into creeks and ditches.

💡 How would water running off a large paved area create problems for ditches and creeks? *{Accept any answer, but elicit that there would be a lot of water running at a fast rate. This water would likely cause a lot of erosion.}*

💡 Do you think builders should be required to have places for water to run off? *{Answers will vary.}*

Discussion:
pages 48–49

💡 Water, wind, and ice are called agents of erosion. Do the agents of erosion actually cause the material to weather? *{Answers will vary, but elicit that water may cause some materials to weather, but most weathering is done by the sediment and loads carried by water, wind, and ice. The agents of erosion are mainly the means to transport weathered material.}*

➤ Does all erosion occur because of natural processes? *{no}*

💡 How does man sometimes cause greater erosion than normal? *{Possible answers: cutting down all vegetation in an area; cutting through rocks to make roads; paving over land so that water runs off too quickly}*

📖 What title does God give to a man who cares for the earth? *{steward}*

➤ A Christian's goals should include plans to be a wise steward of the resources that God has entrusted to him. [BATs: 2c Faithfulness; 2d Goal setting]

Answers
1. They erode away the beach and deposit the sand in new places.
2. in dry areas
3. As a glacier recedes, it may leave behind piles of the rocks and soil that were picked up as the glacier formed.

Activity Manual
Study Guide—page 33
This page reviews Lessons 23 and 25.

Bible Integration—page 34
This page discusses the use of salt in the Bible and its application to the life of a Christian.

Assessment
Test Packet—Quiz 2-C
The quiz may be given any time after completion of this lesson.

Lesson 26

Student Text page 50
Activity Manual pages 35–36

Objectives

- Recall concepts and terms from Chapter 2
- Apply knowledge to everyday situations

Introduction

Material for the Chapter 2 test will be taken from Activity Manual pages 22, 26, 33, and 35–36. You may review any or all of the material during this lesson.

You may choose to review Chapter 2 by playing "To the Ocean" or a game from the *Game Bank* on TE pages A56–A57.

Diving Deep into Science

Questions similar to these may appear on the test.

Answer the Questions

Information relating to the questions can be found on the following Student Text pages:

1. pages 42–43
2. pages 46–47
3. pages 28–29

Solve the Problem

In order to solve the problem, the student must apply material he has learned. The answer for this Solve the Problem is based on the material on Student Text pages 34–35. The student should attempt the problem independently. Answers will vary and may be discussed.

Activity Manual

Study Guide—pages 35–36

These pages review some of the information from Chapter 2.

Assessment

Test Packet—Chapter 2 Test

The introductory lesson for Chapter 3 has been shortened so that it may be taught following the Chapter 2 test.

Diving Deep into Science

Answer the Questions

1. How would the erosion of a stream change if the water was channeled into a narrower streambed?
   Accept any reasonable answer, but elicit that the narrower streambed would cause the water to flow faster. Erosion would be greater as the water and load moved faster. Because there would be less sediment deposition, the streambed would probably deepen.

2. How might waves and wind combine to change the shoreline?
   Accept any reasonable answer, but elicit that waves move sand on and off the beach. Wind takes sand that the water has deposited and moves it farther from the beach. The wind might also move sand close to the water, where the waves can pick the sand up and move it.

3. How is the abrasion that you might get by sliding into home plate similar to the abrasion caused by wind and water?
   Possible answer: The abrasion you get may remove a layer of fabric or skin. Wind and water remove tiny layers of rock.

Solve the Problem

When you start to dig up a flower bed in your new yard, you find thick, red clay and lots of rocks. You notice that when rain falls, a small stream runs right across the area where you want to plant your flowers. Would your flowers likely grow well in this area? What could you do to the soil in the area to increase the likelihood of having a blooming garden?
Accept any reasonable answer, but elicit that flowers would probably not grow well. You could improve the soil by adding some sand and silt to help open spaces in the clay soil.

Review Game
To the Ocean

Divide the class into two teams. Each student represents a piece of sediment. The goal is to get all of the sediment to the ocean. Each time a student correctly answers a question, he moves to the designated "ocean" spot. The first team to have all its members enter the ocean wins the game.

Variation: Set a certain number of correctly answered questions as a completed trip to the ocean. See how many trips to the ocean can be completed in a certain amount of time.

Natural Resources

GREAT & MIGHTY Things

For many years man tried to prevent all forest fires. It seemed such a waste to allow one of our natural resources, forests, to burn. But researchers are now finding that fire can be beneficial to a forest. Fire returns nutrients to the soil. It clears out underbrush and opens areas of sunlight on the forest floor. This allows smaller plants to get the sunlight they need to grow. Some trees even need the heat of a fire to open their seeds. Sometimes when man tries to control natural events, he actually damages what he is trying to protect. The benefits of natural fires show the infinite wisdom of God, our great Creator.

Student Text page 51
Activity Manual page 37

Lesson 27

Objectives
- Demonstrate knowledge of concepts taught in Chapter 2
- Recognize that God plans natural events to benefit His creation

Introduction

Note: This introductory lesson has been shortened so that it may be taught following the Chapter 2 test.

In the days before modern firefighting equipment, a forest fire generally had to burn itself out. Man did not have the means to control or even contain a forest fire. However, as man's technology improves and his efforts to manage fires become more effective, man sometimes must choose whether to fight a fire or let it burn out on its own.

Chapter Outline

I. Energy Resources
   A. Fossil Fuels
   B. Nuclear Energy
   C. Renewable Energy

II. Other Resources
   A. Minerals
   B. Soil
   C. Water

III. Preserving Our Resources

Activity Manual

Preview—page 37

The purpose of this Preview page is to generate student interest and determine prior student knowledge. The concepts are not intended to be taught in this lesson.

SCIENCE BACKGROUND

A 1988 fire consumed 38 percent of Yellowstone National Park. Naturalists thought the devastation was so severe that the park would take decades to recover. However, they now believe that the fire may have been beneficial. The research that resulted from the fire caused park and forestry personnel to reevaluate their policies on natural fires.

Lesson 28

Student Text pages 52–55
Activity Manual pages 38–40

Objectives

- Recognize the difference between renewable and nonrenewable resources
- Explain how fossil fuels formed
- Identify the sources and uses of petroleum, natural gas, and coal
- Describe the benefits and problems related to the use of nuclear energy

Vocabulary

natural resource
renewable resource
nonrenewable resource
fossil fuel
petroleum
crude oil
refinery
petrochemical
natural gas
coal
uranium

Materials

- 25 checkers
- clock with a second hand

Introduction

Hide the checkers around the room. Allow one person 20 seconds to find as many as he can. Record how many he found. Allow a second and then a third person to each search for 20 seconds, and record the number of checkers each person finds.

➤ Who found the most checkers? Why was it easier for the first person to find checkers?

➤ Why is it getting more and more difficult to find new supplies of energy resources? {because there are fewer places to look}

Purpose for reading:
Student Text pages 52–55

➤ What is a nonrenewable energy resource?
➤ What are some advantages of using fossil fuels?

From the fresh water that plants need to grow to hidden veins of gold and silver, God has provided everything we need to live on Earth. We call the materials on Earth that are available for our use **natural resources.** Some of these resources we use for producing energy. Other resources we use to provide food and other products.

Energy Resources

Earth contains both nonrenewable and renewable resources. **Renewable resources** can be replaced by natural means in a relatively short amount of time. **Nonrenewable resources** cannot be replaced easily. Some of our energy resources are considered nonrenewable. We are gradually using up the earth's supply. However, no one knows how many of these energy resources lie beneath the earth's surface.

Fossil Fuels

Petroleum, natural gas, and coal are all **fossil fuels.** Fossil fuels are formed when the remains of plants and animals are buried quickly. These fossil fuels provide most of the energy we use.

Petroleum

Petroleum (puh TRO lee uhm) means "liquid rock." Petroleum is the liquid form of fossil fuel. We use petroleum products to heat our homes and produce the electricity that we use every day. Because petroleum is a nonrenewable resource, geologists

oil drilling rig

constantly search for new oil fields. One of the biggest oil finds in the United States was Alaska's Prudhoe Bay Field. It is the largest known oil field in the United States.

Wells are built to retrieve the petroleum, or **crude oil.** Different types of drilling machines are used, depending on the location of the oil site. During the oil boom in Texas in the early 1900s, drilling units dotted the landscape. But many other oil fields are located under oceans and seas. Special drilling units, or drill ships, are used to retrieve the oil from these fields.

Oil drills pump up crude oil, but the crude oil has to be refined before it can be used. Pipelines and oceangoing tankers transport the crude oil to refineries. A **refinery** is a factory that separates crude oil into different products. The Alaskan Pipeline, completed in 1977, is 1290 km (800 mi) long. It transports Alaskan oil from Prudhoe Bay in northern Alaska to the port city of Valdez. Ships at Valdez transport the oil to refineries in the lower forty-eight states.

52

SCIENCE BACKGROUND

Noahic Flood—Some scientists believe that fossil fuels were formed as a result of the Noahic Flood.

Exhaustible and non-exhaustible—Some people use the terms *exhaustible* and *non-exhaustible resources* instead of or in addition to *renewable* and *nonrenewable resources*. An exhaustible resource is something that is renewable if used properly, such as soil or forest land. A non-exhaustible resource is something such as the wind or solar energy, which cannot be exhausted no matter how much of it is used.

Refinery temperatures—The temperatures used in refining fuels may vary depending on the grade of fuel being processed. These values are not exact and may vary from one source to another.

SCIENCE MISCONCEPTIONS

Crude or refined—Crude oil is not the same oil that is put into automobiles or on door hinges. Crude oil is refined and eventually made into oil, gasoline, and kerosene.

Often people associate high fuel prices with a shortage of crude oil. But high prices may be a result of a shortage of processed fuel.

In its original form, crude oil is not a very useful product. However, by refining it we can get many different fuels and products. Heating the crude oil causes it to separate and vaporize. The vapors are collected, and when they condense they form many of the products we associate with petroleum—gasoline, kerosene, diesel, and others.

Most of us recognize that gasoline and heating oil are crude oil products. But we may not realize that **petrochemicals,** chemicals produced from oil, are used in many other ways. People use petrochemicals for making plastics, paint, fabrics, make-up, and cologne. Some petrochemicals are even added to certain foods to help keep them from spoiling.

The use of crude oil products can cause pollution. Spills sometimes occur at oil wells and during transportation. Ships that run aground, hit other ships, or catch fire may cause spills as well. Air pollution from using petroleum products is also a problem. Some cities have poor air quality because of the waste materials thousands of cars and dozens of factories produce as they burn petroleum products for energy.

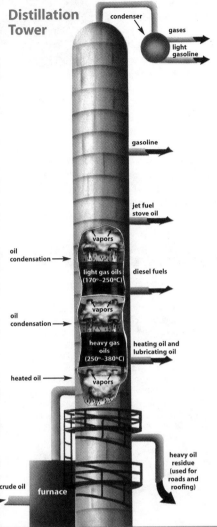

Distillation Tower

oil spill

Discussion:
pages 52–53

➤ **What are natural resources?** {materials on Earth that God has made available for our use}

➤ **What is the difference between a renewable resource and a nonrenewable resource?** {Nature can replace renewable resources easily. Nonrenewable resources cannot be replaced easily by nature.}

➤ **What are fossil fuels formed from?** {remains of plants and animals}

💡 **Why are these resources called fossil fuels?** {They are formed in much the same way as fossils are formed—from the remains of plants and animals that were buried quickly.}

➤ **Why do geologists always search for more sources of petroleum?** {Petroleum is a nonrenewable source of energy.}

➤ **Where is the largest known oil field in the United States?** {Prudhoe Bay Field in Alaska}

➤ **What happens to crude oil after it is pumped up from the earth?** {It is sent to refineries.}

➤ **What does a refinery do to the crude oil?** {separates it into different products}

➤ **What are petrochemicals?** {chemicals produced by oil}

➤ **What are some problems that can occur by using oil?** {oil spills, air pollution}

Refer the student to the diagram called *Distillation Tower*.

➤ **Does diesel fuel or heating oil vaporize at a higher temperature?** {heating oil}

➤ **What is the residue from the refinery used for?** {roads and roofing}

💡 **Do you think gasoline vaporizes at a higher or lower temperature than jet fuel? Why?** {lower; By looking at the placement of temperatures on the tower, we see that the higher the fuel is on the tower, the lower is the temperature needed for the fuel to vaporize.}

 Direct the student to make a collage showing the uses of petroleum. Provide materials for the student to make the collage, and make resource books available to him. If printed pictures are not available, the student may draw illustrations. He should label each petroleum product pictured.

Discussion:
pages 54–55

- Where is natural gas often found? {near deposits of oil}
- How do we get most of the natural gas we use? {offshore drilling}
- What is an advantage of natural gas over oil? {Natural gas burns cleaner because it does not contain sulfur.}
- What is coal? {plant material that was buried quickly}
- What biblical event can you think of that would have buried plants quickly? {the Noahic Flood}
- How much of our energy needs are supplied by coal? {about 25 percent}
- What are the three types of coal? {anthracite, bituminous, lignite}
- If anthracite is the cleanest coal to burn, why is it not used more often? {It is not common, and it is more expensive.}
- Which grade of coal is used most often? Why? {bituminous coal, because it is plentiful and fairly inexpensive}
- What has to happen to lignite before it can be used? {It has to be dried.}
- What problems result from coal burning? {Burning coal produces soot and sulfur gases that pollute the air.}
- What condition do many scientists think is caused by an excess of sulfur gases in the atmosphere? {acid rain (see Chapter 2)}
- Are pollution problems solvable? {Accept any answer, but point out that some progress has already been made; however, most solutions are very expensive.}
- Energy resources are a gift from God. What are some ways that we can use energy resources to glorify God? {Answers will vary.}
- Why is nuclear energy considered a nonrenewable resource? {It uses uranium, which is nonrenewable.}
- What are some uses of nuclear reactors? {Possible answers: to produce electricity; to power some ships or submarines}
- What are some benefits of nuclear energy? {Possible answer: It is efficient and clean.}
- Why is uranium not a fossil fuel? {Uranium is a mineral. It was never plant or animal material.}

Natural gas

Natural gas, a fossil fuel found in a gaseous state, is often found close to deposits of oil. About one-fifth of the world's natural gas comes from offshore drilling. When natural gas was first discovered, no one had a use for it. It was burned off as a waste product. Now natural gas is used to produce heat and light. It is much cleaner to burn than oil because it does not contain sulfur.

Coal

Coal was formed from plant material that was quickly buried in sediment. Coal provides just under 25 percent of our energy needs. Most of the coal mined today is used to generate electricity.

Coal comes in several grades. The best and cleanest grade is *anthracite* (AN thruh SITE) coal. Anthracite burns without smoke and produces almost no pollutants. However, it is not common and is quite expensive.

The next grade is *bituminous* (bih TOO muh nuhs) coal. It is the most common type of coal. Bituminous coal is soft and contains sulfur. Large amounts of ash are produced when bituminous coal is burned. Although bituminous coal causes

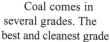
anthracite

bituminous

pollution, it is used often because it is plentiful and fairly inexpensive.

Another grade of coal is made of partially decayed plant material. *Lignite* (LIG NITE) is brown and often has pieces of wood in it. It is formed in bogs and must be dried out before it can be burned.

lignite

Burning coal for fuel produces soot and sulfur gases. Most of the soot can be cleaned from the air. Some sulfur can be removed after coal is crushed, but other forms of sulfur are much harder and more expensive to remove. However, since petroleum is more difficult to find, coal is an important energy source.

Science and History

Coal mining has always been a dangerous job. Even with today's technology, it continues to carry heavy risks. On July 24, 2002, in Somerset, Pennsylvania, nine miners were trapped after drilling too close to an abandoned mine shaft. Water from the old shaft poured in, trapping the men in a four-foot-high passageway 240 feet underground. Rescuing the miners seemed impossible. Rescuers drilled shafts from the surface in an attempt to get the miners out. Work slowed when one of the 1,500-pound drill bits broke in a shaft. To keep the miners from freezing, rescuers piped hot air down to them. It took 150 rescue workers seventy-seven hours to successfully rescue the miners.

SCIENCE BACKGROUND

Coal—Burning coal provides over 50 percent of the electricity generated in America. Coal is also useful in steel production.

Sulfur dioxide—Burning coal, especially bituminous coal, releases sulfur dioxide into the atmosphere. Sulfur gases are pollutants that can cause respiratory and other health problems. Sulfur dioxide is also a main component of acid rain.

Natural gas—Natural gas is the cleanest fossil fuel. Because it does not have an odor, color, or taste, natural gas is a popular source of fuel for homes. Many homes use gas-powered stoves, furnaces, and water heaters.

But natural gas can be deadly. It is poisonous if inhaled by living creatures. Gas companies add a rotten egg fragrance to the gas. If the gas leaks, the rotten egg smell should alert people to the danger. A natural gas detector is also a good safeguard against poisoning. Natural gas is highly flammable also. In the case of a natural gas leak, even a small spark or flame can cause a violent explosion.

Nuclear Energy

Nuclear energy does not use a fossil fuel. However, it does depend on a nonrenewable resource. Most nuclear energy depends on the mineral **uranium** (yoo RAY nee uhm). Uranium is used in nuclear reactors to produce electricity and to power some ships and submarines. In a nuclear reactor the uranium atoms are split, producing energy that heats water and turns it into steam. The steam is then used to produce electricity.

About 8 percent of the energy produced in the United States is nuclear. A nuclear energy plant is both efficient and clean. It does not pollute the air with gases or soot. However, uncontrolled nuclear energy is very dangerous. The use of nuclear power in the United States has declined because some people fear what might happen if an accident were to occur at one of the nuclear plants. They point to a reactor accident in 1979 at Three Mile Island in Pennsylvania. Because the reactor had enough backup protection, very little radiation escaped the plant. But the accident did serve as a warning that nuclear plants must be well maintained by alert and highly trained personnel.

People living in Ukraine were not so fortunate. In 1986, in the small town of Chernobyl (chur NO buhl), one of three nuclear reactors underwent a core meltdown. The nuclear reaction became so hot that it caused an explosion that blew the top off the structure intended to contain it. Radioactive material reached the atmosphere, and winds carried radioactive pollution for thousands of miles. Some of the soil around Chernobyl is still considered unsafe for farming because of the presence of radioactive particles.

The storage of used radioactive nuclear fuel is another problem. The used fuel needs special containment and burial to keep it from harming the environment.

> **QUICK CHECK**
> 1. What are natural resources?
> 2. Why are petroleum, gas, and coal called fossil fuels?
> 3. Why is nuclear energy a nonrenewable source of energy?

nuclear plant

55

 Coal deposits demonstrate evidence of a worldwide flood as described in the Bible. Coal often contains plant fossils such as leaves and stems. Sometimes whole tree trunks are found in deposits of coal. Often the tree trunks are standing, going through several strata layers. These trees, as well as other fossils, indicate that they were buried quickly and that the strata formed rapidly.

 Use maps and other resources to look at the areas affected by the Chernobyl disaster. Encourage the student to research the range of effects the disaster had on the region.

Discussion:
pages 54–55

➤ **Why has the use of nuclear power declined?** *{because of the fear of accidents and the difficulty in disposing of the waste material}*

➤ **Why was the accident at Chernobyl so much more harmful than the accident at Three Mile Island?** *{At Three Mile Island backup protection contained the radiation, but the explosion at Chernobyl destroyed the building meant to contain it.}*

💡 **Do you think it is wise to give up on nuclear power plants because of these accidents? Why?** *{Answers will vary.}*

💡 **What are some precautions that could help prevent future accidents?** *{Possible answers: well-built plants, proper waste disposal, well-trained workers}*

Answers

1. the materials on Earth that are available for our use
2. Petroleum, natural gas, and coal are called fossil fuels because they are formed when the remains of plants and animals are buried quickly.
3. Nuclear energy is a nonrenewable source of energy because it depends on a nonrenewable mineral.

Activity Manual

Bible Integration—page 38
This page discusses the uses of gold in the Bible and applies the refining process to the lives of Christians. You may choose to use this page with Lesson 31.

Study Guide—page 39
This page reviews Lesson 28.

Expansion—page 40
The student will practice reading bar graphs to learn about energy sources used around the world.

Lesson 29

Student Text pages 56–57
Activity Manual pages 41–42

Objectives

- Explain the different methods of cleaning up an oil spill
- Predict the best method for removing the oil
- Use a model to demonstrate the different methods of cleanup
- Compare the methods used in this activity with the methods used in a real oil spill

Materials

- See Student Text page

Introduction

💡 Name some characteristics of oil. *{Possible answers: does not mix with water; burns; floats; slimy or slippery; unusual smell; is used to get heat or energy}*

➤ Some properties of oil may be useful when trying to clean up a spill, while other properties make the job more difficult. Today we are going to try to clean up an oil spill.

Purpose for reading:
Student Text pages 56–57
Activity Manual pages 41–42

➤ What is an oil slick?
➤ What is the best solution for oil spills?

The student should read all the pages before beginning the activity.

ACTIVITY: Clean Up the Spill

Oil is a very important resource. Sometimes while it is being transported by ship, oil spills occur. Oil spills can cause great harm to wildlife, so people try to clean them up quickly.

Because oil floats on salt water and usually floats on fresh water, an oil spill can be easy to locate. The spilled oil spreads out rapidly, forming a thick layer called an *oil slick* on the surface of the water.

There are several ways to handle oil spills. Sometimes it is better to leave the oil alone and let it break down naturally. This is true when there is little risk to human or animal life.

Another way to handle an oil spill is to contain and remove the oil. A *boom*, or flotation foam, is used to gather the oil into a contained area. Then a *skimmer* or *sorbant* can remove the spill. A skimmer either sucks the oil off the water like a huge vacuum cleaner or adheres to the oil and lifts it off the water. A sorbant is a material that absorbs the oil like a sponge. This method is used mainly for small spills or for the later stages of cleanup of a larger spill.

Chemical agents are sometimes used to disperse and break down the oil, but they can be harmful to underwater animals. Other methods of removing the oil include burning it, washing it off beaches with hoses, vacuuming the oil, and removing sand and gravel that have been contaminated with oil.

Process Skills
- Hypothesizing
- Predicting
- Making a model
- Observing
- Inferring

Materials:
baking pan
cooking oil
water
waste pan for oil
paper towels
spoon
10–15 cotton balls
liquid dish soap
medicine dropper
Activity Manual

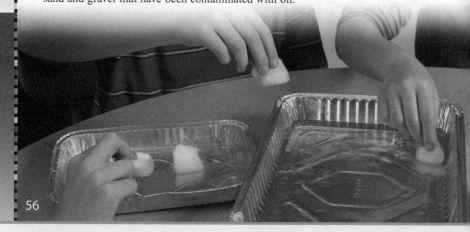

56

SCIENCE BACKGROUND

Cleaning up oil spills—The methods used to clean up oil depend on the type of oil, how much is spilled, the location of the spill, what human and animal life is affected, and the weather conditions. While containing the oil with booms and then removing it to reuse it is the best way to clean up a spill, it is also the most expensive. People who cause oil spills are fined severely. People who clean up the spills practice by performing drills. Those who transport the oil must also have a plan in case a spill should occur. If a response is quick, there will be fewer complications and often an easier cleanup process.

 TEACHER HELPS

The materials used in this activity model methods used to clean up real oil spills.

boom: rolled paper towel
skimmer: medicine dropper or spoon
sorbant: cotton balls or bits of paper towels
chemical agent: dish soap

62   Lesson 29

Problem
What is the most effective method of removing an oil spill?

Procedure
1. In your Activity Manual, list the materials you will use. Rank the five materials that will be used to clean up the spill in order from 1 to 5, 1 being the material that you think will work best for removing the oil from the water.
2. Write a hypothesis explaining which material and method you think would be best for cleaning up your oil spill. The materials listed above represent materials actually used for cleaning an oil spill.
3. Add water to the pan until it is half full.
4. Add 100 mL of cooking oil to simulate an oil spill.
5. Use the method you ranked as number one to try to clean up the oil from the water. Record the success of your method in your Activity Manual.
6. Try using the materials and methods you ranked from 2 to 5 and compare those with the method you had predicted would work best. Record your observations.

Conclusions
- Which method and materials were most effective?
- Which method and materials were least effective?
- Which method took the most time?

Follow-up
- Try other methods to clean up the oil.
- Research to find out if different types of oil require different methods of cleanup.

57

Exxon Valdez oil spill—In 1989 the *Exxon Valdez*, an American oil tanker, ran into the Bligh Reef while transporting oil from Alaska. The punctured ship spilled 11,000,000 gallons of crude oil into the water. The oil spread quickly and soon affected approximately 1,100 miles of Alaskan shoreline. People tried several traditional methods of cleanup, which proved to be ineffective.

A new, experimental method was tried. In areas where the oil was not very thick, a method called *bioremediation* was used. Microscopic bacteria that eat the hydrocarbons found in oil were put into the area. The bacteria also produced materials helpful to the environment. It took four summers to clean up the oil spill.

SCIENCE PROCESS SKILLS
Making a model

➤ What are the advantages of making and using models? *{Possible answers: A model can help us see and handle things that are otherwise too large or too small to see and handle easily. A model allows us to try new procedures on a different scale.}*

➤ Do you think scientists used models to try methods of cleaning up oil? *{probably}*

Computer simulations are a type of model.

Discussion:
Student Text pages 56–57

➤ Why should oil spills be cleaned up quickly? *{They can cause great harm to wildlife.}*

➤ What is an oil slick? *{a thick layer of oil on the surface of the water}*

➤ What are some ways of cleaning up after an oil spill? *{Possible answers: burning, containing and removing, chemicals that break up the oil, power hoses}*

Accidents do happen. God wants us to be honest and faithful workers who take responsibility for our actions. [BATs: 2c Faithfulness; 2e Work; 4c Honesty]

➤ God commands us to love and care for others. How can cleaning up an oil spill demonstrate that love and care? *{Possible answers: Removing the oil creates a healthier environment and can reduce property damage.}*

Procedure
➤ How is this model similar to a real oil spill? *{Possible answers: The oil floats on water. The oil responds to chemical agents.}*

➤ How is it not like a real oil spill? *{Possible answers: The type and thickness of the oil is different. Real oil spills are much larger. No weather conditions are involved.}*

Conclusions
Provide time for the student to evaluate his hypothesis and answer the questions.

Activity Manual
Activity—pages 41–42

Assessment
Rubric—TE pages A50–A53
Select one of the prepared rubrics, or design a rubric to include your chosen criteria.

Chapter 3

Lesson 29 63

Lesson 30

Student Text pages 58–61
Activity Manual pages 43–44

Objectives

- Describe some renewable energy resources
- Compare and contrast renewable sources of energy

Vocabulary

reservoir
hydroelectric energy
geothermal energy
wind energy
solar energy

Introduction

➤ **What are some types of nonrenewable energy?** *{Possible answers: petroleum, natural gas, coal, nuclear energy}*

There are also other sources of energy.

➤ Have you seen solar panels on a house or other building?

➤ Do you have a calculator or toy that is powered by a solar cell?

➤ Have you visited a mill that uses a water wheel or seen a field of windmills?

All of these energy resources share a common characteristic—they are renewable.

Purpose for reading:
Student Text pages 58–61

➤ What are some renewable energy resources?
➤ Why are these resources not used by everyone?

Renewable Energy

Oil, coal, natural gas, and uranium are all examples of nonrenewable sources of energy. There are also renewable sources of energy that can be replaced by natural means. Most are clean, and some are fairly efficient. Unfortunately, many of them are still too expensive to be used on a large scale.

Hydroelectric energy

Water power is the most common form of renewable energy. Even before it was used to produce electricity, rushing water powered wheels for mills that ground grain. Today about 4 percent of the energy needs of the United States is met by using hydroelectric power. By building dams across rivers to form **reservoirs,** or holding areas, behind the dams, engineers can control the flow of water. **Hydroelectric energy** is produced as water flows from the reservoirs and turns turbines. The turning of the turbines generates electricity.

Hydroelectric energy is a great renewable energy resource. In addition to energy production, reservoirs are often designed as recreational lakes. The reservoirs may also provide water for irrigation. However, hydroelectric power has potential difficulties. In order to create a reservoir, large areas of land must be flooded. Also, producing enough water to keep a reservoir filled requires a relatively large river. Additionally, engineers must consider the downstream impact of building a dam across a river.

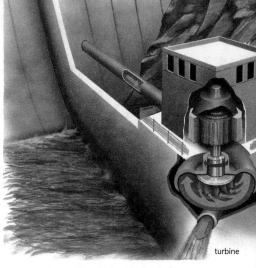
turbine

Hoover Dam

58

SCIENCE BACKGROUND

Hoover Dam—The 726.4 foot tall Hoover Dam is a concrete arch dam located on the Colorado River. The Hoover power plant located at the base of the dam houses seventeen turbines, or water wheels. This power plant produces an average of four billion kilowatts of electricity per year. A kilowatt is 1,000 watts of electricity.

America's first hydropower plant—The first American hydropower plant was at Niagara Falls. It is still a source of power today. In 1881, the first station generated electricity for nearby mills and villages. By 1896 the electricity was being transmitted 26 miles to Buffalo, New York.

BIBLE — Some resources are renewable. Christians need to be renewed daily through Bible study, prayer, and dependence on the Holy Spirit. [BATs: 6a Bible study; 6b Prayer; 6c Spirit-filled]

Geothermal energy

To produce energy from fossil fuels, the fuel must be used up, or consumed. The consumption of this fuel heats water and produces steam that turns a turbine. In contrast, **geothermal energy** uses heat from the earth to produce steam. There is no consumption of the heat source. This allows geothermal energy to be renewable.

One method of using geothermal energy is to use the steam or hot water that already comes out of the ground. Another method is to drill into a place in the earth where very hot water or steam is under pressure. Releasing the pressure causes the super-hot water to change into steam. The steam is then channeled to turn turbines and to generate electricity.

Water heated by geothermal energy can also be channeled through pipes. These hot-water pipes heat many things, such as schools, swimming pools, and greenhouses.

Geothermal plants are fairly inexpensive to build and operate. This type of heat produces no air pollution or radioactive hazards. But this resource is not perfect. Water deep in the earth carries many pollutants. When the water becomes steam, those pollutants remain behind. They must be properly disposed of to avoid contaminating streams and rivers. Also, only certain areas where there is magma close to the earth's surface are suitable for using geothermal energy.

geothermal plant

Discussion:
pages 58–59

- What are some advantages of using renewable energy resources? *{They are clean and fairly efficient.}*
- What is the biggest disadvantage of renewable energy resources? *{Many are expensive.}*
- What is the most common form of renewable energy? *{water power}*
- What are reservoirs? *{holding areas for water}*
- How does a hydroelectric dam make electricity? *{Water flows through the dam and turns a turbine that generates electricity.}*
- What are some difficulties of hydroelectric power? *{A large area of land must be flooded to form the reservoir. It requires a large river. It may create problems downstream from the dam.}*
- 💡 What are some possible problems that may occur downstream from a dam? *{Possible answers: less water in the river than before the dam; risk of flooding if the dam breaks; affects the ecosystems along the river and reservoir}*
- What does geothermal energy use to produce steam? *{heat from the earth}*
- How does geothermal energy produce electricity? *{Hot water or steam from underground sources is used to turn turbines and produce electricity.}*
- What is another way geothermal energy is used? *{It can be piped to heat structures.}*
- What are some advantages of geothermal energy? *{Possible answers: It is renewable. It is relatively inexpensive. It does not produce air pollution or radioactive waste.}*
- What problems does geothermal energy have? *{Possible answers: pollutants from the earth; available only in certain areas}*
- 💡 Is geothermal energy available in all parts of the world? Why? *{No; geothermal energy is available only where heat sources, such as magma, are close to the earth's surface.}*
- 💡 What earth feature is associated with magma? *{volcano}*

Direct a DEMONSTRATION
Demonstrate different energy sources

Materials: pinwheel, electric fan or hairdryer, container of water, and a source of steam, such as a teakettle on a hot plate

Blow on the pinwheel, or hold it in front of the fan or hair dryer to make it turn.
- What is the source of energy? *{wind}*

Pour water down the edge of the pinwheel to make it turn.
- What is the source of energy? *{moving water}*

Place the pinwheel in the path of the steam to make it turn.
- What is the source of energy? *{steam}*

Water wheels—Mabry Mill, located in Virginia, and Newlin Grist Mills, located in Pennsylvania, have two of the oldest water wheels in America. They have been preserved for future generations. At designated times, tourists and visitors can view demonstrations of the grist mills and other Early American crafts.

Point out Iceland on a world map. Iceland has over 250 hot springs, which is more than any country in the world. This form of geothermal energy is used to heat homes and greenhouses.

Discussion:
pages 60–61

➤ **What is a wind farm?** *{an area containing many windmills that turn air movements into electrical energy}*

➤ **What is wind energy?** *{energy produced by air movements}*

➤ **What are some difficulties with wind energy?** *{Possible answers: It requires a steady wind. The windmills take up a lot of space. There must be a way to store the energy.}*

➤ **What are some advantages of wind energy?** *{It is nonpolluting and quiet.}*

➤ **How does the Sun indirectly provide energy for us?** *{We eventually eat plants that have used the Sun for photosynthesis.}*

➤ **How do solar panels provide energy to heat a house?** *{They heat water that is stored to be used in the house.}*

Wind energy

Since the 1960s, a new type of farm, called a *wind farm,* has appeared. A wind farm is not covered with waving wheat but rather with giant windmills. Like water wheels, windmills have long been used to turn wheels to grind grain. But these new windmills use **wind energy,** or air movements, to generate electrical energy. The windmills often have blades that are over 30 m (100 ft) long. They take up much space, but they do not cause air or water pollution. Windmills operate with very little maintenance.

However, these windmills require steady wind. There are not many places on Earth where wind blows constantly. Also, scientists have yet to solve the problem of storing electrical energy produced by the windmills.

Solar energy

Every day the Sun provides energy for us indirectly, as plants use it to produce food through photosynthesis. Through the food chain we use that energy to power and heat our bodies. However, we can also use the Sun's energy, or **solar energy,** to power and heat other things.

One way of using the Sun's energy is to directly raise the temperature of water and air. In a house with solar collectors, sunlight raises the temperature of water. Solar panels contain tubes of water that are heated by the Sun and carried to a storage tank to be used in the house. Water heated in this way can reach temperatures of around 74°C (165°F).

With special reflectors, water temperatures may be raised to 288°C (550°F). This is another way of using

wind farm

 Windmills—Windmills seem to have originated in Persia. The oldest ones are in Afghanistan. Historians think either conquerors or crusaders brought information about windmills to Europe. From Europe, this invention spread to the rest of the known world.

 Due to Ukraine's windy terrain, wind is one of this nation's most promising natural resources. In the United States, the Midwest and Great Plains are ideal locations for wind farms. Wind farms are not limited to land. Some countries have offshore wind farms. Point out these areas on a world map.

house with solar panels

Fantastic FACTS

Solar cells are an essential part of the space program. They were used to power the moon buggy that transported astronauts on the Moon during the 1970s. But what about using solar cells on the Moon to produce electricity on Earth? Some physicists think that we should consider putting a huge solar cell grid on the Moon. As the grid collects solar energy, the energy would be converted to microwave energy and beamed to receivers on Earth that would convert it to electrical energy. Moon rocks collected during lunar missions show that the Moon has the basic materials for producing solar cells. Perhaps one day the Moon may be the means of supplying our energy needs.

solar energy. These high temperatures produce the steam needed to turn a turbine. Some scientists believe that all of the electricity needs of the United States can be met with one solar collection area, called a *solar field*. The solar field would use reflective mirrors to heat water in a tower. That hot water could produce steam that would turn a turbine to produce electricity. Such a solar field would have to be large, about 280 sq km (174 sq mi). The solar field would produce no pollutants and would use almost no water.

The biggest problems with solar energy are the expense of the solar collection cells and the large amount of land area required. Also, these systems have to store electricity since they cannot work after sunset.

 QUICK CHECK
1. What is a renewable resource?
2. Compare the advantages and disadvantages of the different types of renewable energy.

 Charles Fritz—Charles Fritz invented the first solar cell in the 1880s. Other inventors improved this invention by making it more efficient and practical. The first solar panel was built in 1954 as a joint effort by Gerald Pearson, Calvin Fuller, and Darrel Chapin. A solar panel is a battery powered by the Sun.

 Use of solar cells—In addition to producing electricity at power plants, solar cells have powered vehicles, calculators, toys, emergency phones, roadside message boards, traffic lights, and satellites. Solar cells have also been used to cook food.

Discussion:
pages 60–61

➤ How do some scientists think a solar field could provide energy? *{Reflective mirrors would heat water to produce steam that turns a turbine.}*

➤ What are some advantages of a solar field? *{Possible answers: There are no pollutants. It uses very little water.}*

➤ What are some disadvantages of solar energy? *{Possible answers: expense; amount of land required; having a way to store electricity}*

Refer the student to *Fantastic Facts*.

➤ At what unusual place do some scientists think they could put a solar grid? *{the Moon}*

➤ How might the Moon be used to produce electricity on Earth? *{A solar grid on the Moon would collect solar energy and send it to Earth to be converted to electricity.}*

Answers

1. a resource that can be replaced by natural means
2. **Hydroelectricity**—Advantage: provides reservoirs and irrigation water
   Disadvantage: most of the suitable rivers have already been dammed up

   Wind—Advantages: quiet, no pollutants
   Disadvantages: need to store energy; takes up a lot of space

   Geothermal—Advantages: can never be used up; no air pollutants or radioactive hazards
   Disadvantages: waste water can have pollutants; requires magma close to the surface

   Solar—Advantages: provides heat and electricity; no pollutants
   Disadvantages: expensive; problem of storing energy; uses a lot of land

Activity Manual
Study Guide—pages 43–44
These pages review Lessons 28 and 30.

Assessment
Test Packet—Quiz 3-A
The quiz may be given any time after completion of this lesson.

Lesson 31

Student Text pages 62–65

Objectives
- Name and identify the uses of several metals
- Recognize soil as a natural resource
- Identify several ways to conserve soil

Vocabulary
mineral   malleable
vein   fallow
ore   ground cover
smelting   contour plowing

Materials
- Bible
- samples of metals, such as aluminum (cans or foil), lead (curtain weights or fishing sinkers), iron, and copper (pieces of wire or pipe)
- hammer
- magnet

Introduction
Lead a discussion concerning the properties of metals. Suggested metals include aluminum, lead, iron, gold, silver, copper, nickel, bronze, brass, a hammer, and a magnet.

Identify some physical characteristics of each metal and test each sample to see if it can be easily dented, scratched, bent, or attracted to the magnet.

➤ Is it heavy or lightweight?
➤ Does the metal look tarnished or rusted?
➤ Does the metal dent or bend easily?
➤ Is it magnetic?

Each metal has properties that make it useful.

Purpose for reading:
Student Text pages 62–65

➤ What are some ways to conserve the soil?
➤ What is smelting?

Discussion:
pages 62–63

➤ What are two characteristics of minerals? *{Possible answers: They occur naturally in the earth. They are solid substances. They can never have been organic substances.}*

➤ How are veins of minerals formed? *{Material containing minerals melts. The minerals have different properties, so they separate into layers. As they cool, they form concentrated areas of the minerals.}*

Other Resources

Minerals

Though some minerals are very abundant, most scientists consider minerals nonrenewable natural resources. A **mineral** is a solid substance found naturally in the earth's surface. It never has been a living organism. Minerals include such common substances as salt, sand, and iron, and such rare substances as gold and diamonds.

Concentrated areas of specific minerals are called **veins.** Veins of minerals are often found near volcanic areas. The intense heat from the volcano causes the materials containing minerals to melt. The different properties and densities of these melted minerals cause many of them to settle into layers as they cool.

Veins often contain **ores,** materials with usable amounts of metal in them. Sometimes a process called **smelting** is used to separate metal in the ore from the other materials. The ore is crushed and heated until it is a liquid. The *dross,* or nonmetal part, floats to the top, where it can be removed easily. The remaining material has a much higher concentration of the metal. Often this process is repeated several times. Each time, the resulting product is purer than the time before. God uses the process of refining, or purifying, metal to illustrate the process of testing Christians to make them more like Christ (Mal. 3:3).

Precious metals

Some metals are called precious metals because of their rarity. When the wise men went to worship Jesus, they took Him gifts that were of great value. Only the best that they had would be right for the King of kings. One of the gifts that they offered was gold (Matt. 2:11). Throughout the ages, a man's wealth has often been judged by the amount of gold he possesses.

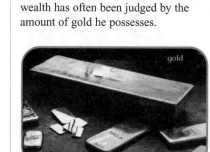

Gold is a soft and shapeable, or **malleable,** metal. People used to test gold coins by biting them. If a person's teeth dented the coins, the coins were real gold. Though gold is often used for jewelry, it must be mixed with other metals, such as copper or nickel, to strengthen it. Gold is also used in dentistry and glassmaking, and for treating arthritis and cancer.

 Heat is a purifier. In Psalm 19:6–14, the writer reminds us of the purity of God and His law. We can trust God's Word to be true. The truth of the Bible shows the sin in our lives. The Bible also provides God's plan for how we might be acceptable in His sight. [BAT: 4b Purity]

tarnished silverware

Silver is another metal used mainly for money and jewelry. Also a malleable substance, silver is often mixed with other metals. Silver is easy to scratch, and it tarnishes, or turns black. Silver may be found around your house in forms such as electrical parts, ink, glass, mirrors, or medicines. Silver is even necessary for making and processing photographic film.

copper

Other metals

The red-orange metal called *copper* is also used for money. If you have an old penny, then you have copper. One of copper's greatest values, however, is that it is a good conductor of electricity. Copper can be shaped or formed easily, yet it is sturdy and does not rust as iron does. Because of its properties, copper is used frequently around the house in wiring and plumbing. Bronze and brass are made from mixing copper with other materials.

Iron is a very plentiful and useful metal. Because iron is strong and durable, it has been used to make tools and weapons throughout history. Iron is used for many household items, such as pans, paint, and appliances. It is also used in the form of steel for constructing buildings. You can find things that contain iron by using a magnet. If the magnet is attracted to an item, the item probably contains iron.

hot roll of steel

You have probably seen aluminum foil used to wrap up leftovers before they are stored in the refrigerator. *Aluminum* is very practical because it is strong and lightweight. The transportation industry uses it to make cars and airplanes. Aluminum is the most abundant metal in the earth's crust. Even though this resource is plentiful, we can help preserve it. Recycling things like soda cans is a way to use our resources wisely.

aluminum cans

aluminum foil

63

SCIENCE MISCONCEPTIONS

The term *organic* can mean "containing carbon," since all living things contain carbon. This does not mean that all carbon was once part of living organisms. In the Student Text, *organic* is used only to refer to living or dead organisms.

Discussion:
pages 62–63

💡 **Why are minerals usually found near volcanic areas?** {Answers will vary but might include that intense heat helps to separate the minerals from each other.}

▶ **What are ores?** {materials with usable amounts of metal in them}

▶ **Describe the process of smelting.** {Ore is crushed and heated until it is a liquid. The nonmetal part, or dross, floats to the top so it can be removed. The remaining metal is purer.}

📖 Read Malachi 3:3.

▶ **How is the process of smelting similar to how God works in the life of a Christian to cause him to be more Christlike?** {Just as metal is put to the test of fire so it will be purer, a Christian may be taken through hard times of testing to help purify his life.}

▶ **Why do we call something a precious metal?** {because it is rare}

▶ **Why did biting a gold coin help test whether the coin was real or not?** {Because gold is soft, the coin would dent or move. If it had been made of something else, it would have probably been too hard to dent.}

▶ **Why do gold and silver have to be mixed with other metals for most purposes?** {They are too soft by themselves.}

▶ **What are some uses for silver?** {Possible answers: money, electrical parts, ink, glass, mirrors, medicines, and developing photographs}

▶ **What is the red-orange metal?** {copper}

▶ **Why is copper such an important metal?** {It is a good conductor of electricity.}

▶ **Why has iron been used to make tools and weapons?** {It is strong and durable.}

▶ **What is a way of finding out if something has iron in it?** {See if a magnet is attracted to it.}

▶ **What is the most abundant metal on Earth?** {aluminum}

▶ **Why is aluminum such a practical metal?** {It is strong and lightweight.}

💡 **Why would these properties make aluminum a good choice for cars and airplanes?** {Because aluminum is lightweight, less fuel is required to move things made of aluminum.}

Chapter 3   Lesson 31   69

Discussion:
pages 64–65

▶ **Why is soil considered a natural resource?** *{God gave it for our use.}*

▶ **Why were early settlers in America not careful with the soil?** *{They thought there would always be more land if they wore out the land they were using.}*

▶ **What are some ways that farmers can help their land be more fertile?** *{Possible answers: let fields lie fallow, rotate crops, plant ground cover crops, add fertilizer, use contour plowing}*

▶ **What did farmers in the Middle Ages learn about planting crops?** *{If they planted the same crop in the same field year after year, the soil wore out as a result of the plants' using all the nutrients.}*

▶ **What does it mean to leave a field fallow?** *{let the field rest and not plant in it}*

▶ **How does a field benefit from remaining fallow?** *{The rest allows natural processes to replace nutrients.}*

▶ **What are some substances that farmers can use to replenish nutrients in the soil?** *{Possible answers: compost, manure, mineral and chemical fertilizers}*

▶ **Why might a farmer plant soybeans in a field?** *{Soybeans are a crop that helps return nutrients to the soil.}*

▶ **What events in the 1930s caused scientists and farmers to work together to stop erosion?** *{dust storms}*

📖 Refer the student to *Science and the Bible*.

▶ **Why do you think God has given man the example to rest one day each week?** *{Possible answers: to worship; because our bodies need rest to be healthy}*

▶ **According to the law given Moses, what was supposed to happen during the seventh year?** *{Nothing was to be planted or harvested, in order to give the fields a year of rest.}*

▶ **What land-use principle was a result of the Sabbath year?** *{allowing the land to lie fallow}*

▶ **Was land conservation the main reason God established the Sabbath year?** *{no}* **What was the reason?** *{God intended it as a time to remind the Israelites of Him.}*

Soil

Another important natural resource is soil. America is a land blessed with much fertile soil. Many early settlers thought that no matter how they treated the soil, there would always be enough land to provide their needed timber, food, and fiber products. When fields no longer produced good crops, new fields were cleared and the old fields were abandoned. Today, Americans are much wiser in their treatment of this important natural resource.

Conservation

Fortunately, soil is a renewable resource. Even old fields can be brought back to fertility if they are properly cared for. There are several methods farmers can use to keep soil fertile and productive.

As early as the Middle Ages, farmers learned to rotate crops in their fields to help maintain soil fertility. The farmers might plant corn one year and beans the next year. They discovered that if they planted the same crop year after year, the soil wore out and became unproductive. They also let fields lie **fallow,** or resting, for a season or year. This period of rest let natural processes replace the nutrients that had been used up. Many farmers today do the same. They rotate crops and, when possible, leave fields unplanted so the soil can be replenished.

One way of replenishing the lost nutrients of the soil is by adding fertilizer. The use of compost or manure adds organic material that breaks down and releases nutrients into the soil. The addition of minerals and chemical fertilizers also produces more fertile ground. Many farmers plant crops that help enrich the soil. Soybeans are good for this, since they return needed nutrients to the soil as they grow.

During the dust storms of the 1930s, scientists and farmers worked together to stop the wearing away of the soil. They discovered that fields with vegetation had less erosion from the

When God finished His creation, He rested on the seventh day. He established the seventh day as a day of rest and worship. When God gave the Law to Moses, He established a seventh year of rest (Lev. 25:4). In that year, nothing was to be planted or harvested. Though the purpose of the Sabbath year was to remind the Israelites of God, it also established a sound principle for using the land efficiently.

64

SCIENCE BACKGROUND

Soybeans—Soybeans have a practical purpose other than just enriching the soil. Soybeans provide oil and protein and are an important food source for man and animals. Soybean sprouts are rich in Vitamin C. Soybean oil is in margarine, shortening, mayonnaise, salad oil, plastics, paint, and glue. Some of the many uses for soybeans include soy flour, soy grits, soymilk, imitation ham, hot dogs, hamburgers, soy sauce, tofu, and animal food.

devastating winds than barren fields. Unused fields were planted with a **ground cover,** a low-growing crop, such as clover. Instead of plowing immediately after a harvest, farmers began leaving the old stalks in the fields until the next spring planting. The stalks helped protect the soil and slow the evaporation of water from the ground. Today farmers use many of these same methods to protect their fields when they are not in use.

contour plowing

Contour plowing is a method used by farmers to help keep water from washing soil away. Farmers plow furrows horizontally around a hill instead of up and down it. The furrows slow the flow of rainwater downhill so that it doesn't carry away as much topsoil. The furrows also help keep the water on the fields longer, providing more moisture for the crops. By using these methods, farmers can make the land more profitable and useable for future planting.

Modern farms

Farmers closely monitor the conditions of their fields in order to ensure the best use of their land. In many parts of the world, small family farms are still the most common kind of agriculture. Because the fields are small, a single field can be irrigated and fertilized the same way. But on large industrial farms, like many in the United States, the needs of the soil in different areas of one field may vary. To help meet these needs, many large farming operations turn to technology to improve the use of their soil. A farmer may use satellites to provide information about the conditions of the soil. That information and a Global Positioning System (GPS) are then used to provide the right amounts of fertilizers and irrigation to the areas in need.

1. What are the characteristics of a mineral?
2. How is dross removed from metal ores?
3. List several uses for copper and iron.
4. What are some ways to conserve soil?

irrigation system

 TECHNOLOGY The Global Positioning System is really a network of satellites that orbit the earth and relay information to receivers on Earth. It can tell you the exact latitude, longitude, and altitude of the location of a receiver.

A farmer's use of the GPS and satellites is called precision farming. Information from satellites helps a farmer pinpoint the location of soil samples and crop yield. With this data, a farmer can make intricate grids and determine which parts of his field need more fertilizer and which parts of his field are producing as desired. By mounting a GPS receiver on a tractor, a farmer can plow or harvest a field more efficiently. The GPS helps the farmer navigate through the field without missing spots or overlapping rows.

Discussion:
pages 64–65

▶ **What did farmers and scientists discover about barren fields?** *{Barren fields have greater erosion.}*

💡 **Why would keeping topsoil from eroding be important to a farmer?** *{Answers will vary but should include that most growing occurs in the topsoil.}*

▶ **How did farmers adjust their farming methods to prevent barren fields?** *{Possible answers: They planted ground covers in unused fields. They left old stalks in the fields until the next planting.}*

▶ **What is a ground cover?** *{a low-growing crop}*

▶ **In addition to decreasing erosion, how did leaving stalks in a field help the land?** *{The stalks slowed evaporation of water from the ground.}*

▶ **What is contour plowing?** *{plowing horizontal furrows around a hill}*

▶ **What is a benefit of contour plowing?** *{The horizontal furrows slow the downhill flow of water, reducing erosion and keeping water on the fields longer.}*

▶ **What modern technology is sometimes used on large farms?** *{satellites and the Global Positioning System (GPS)}*

▶ **How does the Global Positioning System (GPS) help farmers?** *{It allows farmers to irrigate and fertilize only the areas of the field that need it rather than the whole field.}*

Answers

1. A mineral is a solid substance found naturally in the earth's surface. It never has been a living organism.
2. The ore is crushed and heated until it is a liquid. The dross floats to the top of the liquid and is removed.
3. Possible answers: copper—money, electrical parts, plumbing
   iron—tools, weapons, pans, paint, appliances, steel
4. let fields lie fallow, rotate crops, add fertilizer, plant ground cover crops, use contour plowing

Lesson 32

Student Text pages 66–67
Activity Manual pages 45–46

Objectives

- Make models of soil without erosion prevention and soil with erosion prevention
- Infer how certain materials prevent soil erosion

Materials

- See Student Text page
- bucket or large containers to dispose of used soil and water

Introduction

💡 What are some methods farmers use to conserve soil? *{Possible answers: let fields lie fallow, rotate crops, fertilize, plant ground cover crops, use contour plowing}*

➤ Which methods may help prevent soil erosion? *{plant ground cover, contour plowing}*

In this activity you may try some of these methods to help prevent soil erosion.

Purpose for reading:
Student Text pages 66–67
Activity Manual pages 45–46

➤ What problem is Farmer Brown having?
➤ How can you help Farmer Brown?

The student should read all the pages before beginning the activity.

ACTIVITY: Erosion Prevention

Process Skills
- Observing
- Making a model
- Recording data
- Inferring

You have been given the challenge of helping Farmer Brown figure out a way to prevent erosion on his farm. His land is situated on a hill with a 20-degree angle, and every year he loses valuable soil from rain runoff. He has hired you to be his land engineer. Using the materials below, find a solution to Farmer Brown's problem.

Problem
How can you reduce the amount of erosion on a 20-degree slope?

Procedure

1. In your Activity Manual, write or draw your plan for helping Farmer Brown reduce erosion.

2. Prepare the aluminum pans by cutting them at the corners of one of the smaller ends. Fold the piece down so it will not interfere with the erosion. (Safety tip: The cut edges of the pan may be sharp.)

3. Place 500 mL of potting soil at the end opposite the open end of each baking pan. One pan will have erosion protection that you construct, and the other pan will not have erosion protection.

4. Put each baking pan at a 20-degree incline with the soil at the top. Books can be used to make the incline.

5. In the pan that will have erosion protection, use the grass tufts, sticks, leaves, pebbles, and any other materials you'd like to use to slow down the erosion of the soil.

6. Slowly sprinkle 250 mL of water onto the soil with the erosion protection. Be sure to sprinkle the water over all areas of the soil. Allow the eroded material to run into the runoff container.

Materials:
- two aluminum baking pans (9"×13")
- potting soil
- container to sprinkle water (large salt shaker or a watering can)
- water
- container for runoff
- scale
- grass tufts
- sticks
- leaves
- pebbles
- cheesecloth (to act as a filter)
- plastic cup
- protractor
- books
- scissors
- Activity Manual

66

TEACHER HELPS

Filter ideas—Nylon hose could be substituted for the cheesecloth filter. Coffee filters and paper towels do not work well as filters for this activity.

Measuring the slope—Rest one end of the pan on several books. Place the flat side of the protractor on the table. Align the center point of the protractor with the point with which the pan touches the table. Measure the angle formed between the pan and the table. Adjust the books as needed to get the correct angle.

7. Place a double layer of cheesecloth over a plastic cup to act as a filter, and pour the contents of the container through the filter to remove the water. Put the eroded material in a plastic cup and weigh it on the scale. Record your measurements.

8. Repeat steps 6 and 7 with the pan without erosion protection.

9. Compare the results from the two containers.

Conclusions
- What prevented the soil in one pan from eroding as much as the soil in the other pan?
- How might your observations help Farmer Brown?

Follow-up
- Try using a different method to prevent erosion.
- Plant some grass seed in the soil. After it has sprouted, try the experiment again.
- Try other methods of pouring the water.

Farmers often try to protect their fields from erosion. With the help of the Holy Spirit, Christians can protect themselves against the temptations of Satan. [BATs: 4b Purity; 4d Victory]

SCIENCE PROCESS SKILLS
Inferring—Inferring involves interpreting observations and drawing conclusions based on previous knowledge and experience.

➤ What could you infer about soil conservation from this activity? *{Possible answer: Erosion protection slows erosion.}*

➤ How might the results affect how you set up other related experiments? *{Answers will vary, but elicit that knowledge gained from one experiment can influence later experiments.}*

Procedure
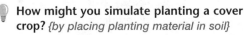 How might you simulate planting a cover crop? *{by placing planting material in soil}*

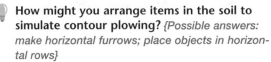 How might you arrange items in the soil to simulate contour plowing? *{Possible answers: make horizontal furrows; place objects in horizontal rows}*

Conclusions
💡 How did the items you placed in the soil slow the amount of erosion? *{Answers will vary.}*

💡 How do you think your results will help a farmer with fields on hillsides? *{Answers will vary.}*

Provide time for the student to evaluate his hypothesis and answer the questions.

Activity Manual
Activity—pages 45–46

Assessment
Rubric—TE pages A50–A53
Select one of the prepared rubrics, or design a rubric to include your chosen criteria.

Lessons 33–34

Student Text pages 68–73
Activity Manual pages 47–48

Objectives

- Identify several natural resources
- Explain how the ocean is the source of most fresh water
- Identify locations of fresh water
- Describe the different kinds of ice

Vocabulary

hydrosphere	ice sheet
water cycle	ice shelf
saline	iceberg
phytoplankton	sea ice
ground water	reduce
aquifer	reuse
drawdown	recycle
humidity	

Materials

- world map

Introduction

Note: Lessons 33–34 cover six Student Text pages. Divide the material into two lessons as best fits your schedule.

▶ **What are some things we use to produce energy?** {Possible answers: water, coal, oil, natural gas, wind, nuclear power, the Sun}

▶ **What are some things we produce from these resources?** {Possible answers: electricity, gas, heat, plastics}

Purpose for reading:
Student Text pages 68–73

▶ What is the hydrosphere?
▶ Why are oceans salty?

Water

Water is one of Earth's most valuable renewable resources. Three-fourths of the earth's surface is covered by water. Earth is the only planet known to have water in its liquid form. Water is one way God provides for the needs of His creation. All of Earth's water found in lakes, oceans, streams, rivers, soil, underground, and in the air is referred to as the **hydrosphere.** This chart shows where most of our water resources are located. Without the hydrosphere, living things could not survive.

We can see how water is replaced and reused by looking at the path it takes as it travels from land to sky and back to land. This path is called a **water cycle.** God created this process to continually replenish the earth and its living things with fresh water (Job 36:27–28).

EARTH'S WATER

Water source	% of total water
Oceans	97.2%
All icecaps and glaciers	2.0%
Ground water	0.62%
Freshwater lakes	0.009%
Inland seas, salt lakes	0.008%
Atmosphere	0.001%
All rivers	0.0001%

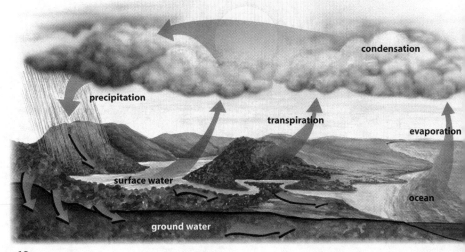

Water Cycle

68

Teacher Helps

The water cycle has been covered in previous grades, so little emphasis is put on it in this chapter. If your students have not had sufficient background on the water cycle in previous grades, you may want to spend some additional time going over the diagram and explaining the water cycle to your students.

Science Background

Transpiration—This is the part of the water cycle in which water is released by plants as they carry on photosynthesis.

Phytoplankton—These organisms live near the surface of the ocean because they need sunlight. The chlorophyll in phytoplankton gives the organisms and the ocean a greenish tint.

Photosynthesis—Plants use a process called photosynthesis to convert the energy in sunlight into a useable source of energy, sugar.

Ocean currents and weather—Water warms and cools more slowly than land. An ocean current can carry warm or cool water from one area of the ocean to another.

The Gulf Stream carries warm water from the Caribbean to the coasts of Europe. The warm air over the ocean then flows toward the cool air over the land, causing breezes and winds. These warm winds affect the weather and climate of Europe. Other ocean currents have similar effects on landmasses around the world.

74 Lessons 33–34

Oceans

Oceans contain most of Earth's water. Though we cannot drink their salty water, the oceans are a key factor in providing fresh water for us. Through the water cycle, ocean water evaporates, condenses, and returns to Earth as fresh water. The dissolved minerals and salts that are left behind cause the oceans to be **saline,** or salty.

Many parts of the world do not have enough fresh water. In the past, people in these places depended on deep wells. Now desalination (dee SAL uh NA shun) facilities are able to remove the salts and other minerals from ocean water in order to help produce fresh water. Some countries and areas depend heavily on this process for water.

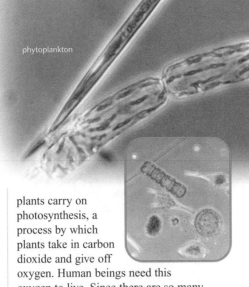
phytoplankton

TRY IT Yourself

You can demonstrate one way that desalination works. Stir a teaspoon of salt into a half cup of water. Place the cup in a dark plastic bag and set it in the sun for 1 to 2 days. The water in the cup will evaporate, leaving the salt behind. The water that evaporates and condenses on the plastic bag will be fresh.

Oceans also play a key role in the carbon dioxide-oxygen cycle. **Phytoplankton** (FYE toh PLANK ton), small plants that make up the first link in the ocean's food chain, are the chief contributors to this cycle. Millions of these tiny ocean plants can fit in a single teaspoon of water. These tiny plants carry on photosynthesis, a process by which plants take in carbon dioxide and give off oxygen. Human beings need this oxygen to live. Since there are so many of these plants, scientists estimate that phytoplankton carry out over 50 percent of the exchange of oxygen and carbon dioxide on Earth.

Oceans also influence the climates around the world. The air above ocean currents produces winds that warm or cool the land nearby. Instead of being completely frigid, northern Europe has a moderate climate because of the Atlantic Ocean's warm Gulf Stream. Even normal winds from the ocean can greatly affect the climate of a particular place.

Though man is just starting to appreciate the value of the oceans, God planned them for our use from the very beginning. In Genesis 1:9 the Bible says, "And God said, Let the waters under the heavens be gathered together unto one place, and let dry land appear: and it was so."

69

 BIBLE God the Father has provided for man's physical well-being by supplying shelter, food, and water. He has provided for man's spiritual well-being through Christ's death on the cross. [Bible Promises: E. Christ as Sacrifice; I. God as Father]

 GEOGRAPHY It is estimated that one-fifth of the world's population does not have a steady source of fresh water. These countries are located mainly in the Middle East and Africa. Saudi Arabia uses desalination to provide about 70 percent of its drinking water. Israel has approximately 30 desalination facilities.

Use a world map to show the location of Saudi Arabia.

➤ Why might Saudi Arabia be a country that could need and use desalination facilities? *{Accept any answer, but elicit that Saudi Arabia is mostly desert, but it has much coastland.}*

Discussion:
pages 68–69

➤ What is the hydrosphere? *{all of the water on, above, and under the earth}*

 Refer the student to the chart entitled *Earth's Water* and to the diagram entitled *Water Cycle*.

➤ What percentage of water is in the oceans? *{97.2%}*

➤ Which has a greater percentage of Earth's water—freshwater lakes or inland seas and salt lakes? *{freshwater lakes}*

➤ Describe the path of a drop of water through the water cycle. *{Answers will vary but should follow the diagram on page 68.}*

➤ What makes oceans saline? *{As water evaporates, dissolved minerals and salts remain behind.}*

💡 What would happen if only fresh water sources were part of the water cycle? *{Accept any answer, but elicit that there would not be enough water for our needs.}*

➤ On what other natural cycle does the ocean have a major impact? *{the exchange of carbon dioxide and oxygen, or the carbon dioxide-oxygen cycle}*

➤ What effect do phytoplankton have on Earth's carbon dioxide-oxygen cycle? *{They produce more than half of the oxygen on Earth.}*

➤ How does the ocean affect the weather? *{The air above ocean currents produces wind that warms or cools the land.}*

➤ What ocean current affects the climate of the eastern United States and northern Europe? *{the Gulf Stream}*

Discussion:
pages 70–71

➤ **Where does most of the fresh water we use come from?** {surface waters, such as rivers, streams, and lakes}

➤ **Where is the majority of liquid fresh water?** {underground}

➤ **What do we call this water?** {ground water}

➤ **What are aquifers?** {layers of sand, gravel, or bedrock that hold or move ground water}

➤ **How is most ground water retrieved?** {through a deep well in an aquifer}

➤ **What is drawdown?** {dropping of the underground water level due to water use}

💡 **If drawdown has occurred during the day, why would the water level rise during the night?** {Possible answer: Water use is less at night than during the day.}

➤ **What might cause water use to be restricted?** {drought or poor water management}

➤ **What helps purify ground water?** {soil and organisms in the soil}

➤ **What can happen if chemical contaminants get into the soil?** {Ground water can become contaminated.}

 Refer the student to the illustration on Student Text page 70.

➤ **This illustration shows that ground water can flow in several different layers, or aquifers.**

Fresh water

Only a small part of Earth's water is fresh water, and most of that water is frozen. Most of the fresh water that we use comes from rivers, streams, and lakes. But these surface waters only contain a small part of Earth's fresh water. The majority of liquid fresh water is underground. When rain falls, some of the rainwater flows into the soil and is stored beneath the surface of the earth. This water is called **ground water**. Layers of sand, gravel, or bedrock that hold and move ground water are called **aquifers** (AHK wuh furz). Aquifers have enough air space to absorb and hold water.

The most common way to retrieve ground water is through a deep well in the aquifer. During the day, when a lot of water is drawn from the well, the water level may fall. This is called **drawdown**. At night or at times when little water is taken from the well, the slowly flowing ground water refills the well. Sometimes, due to drought or poor water management, the water remains low. When this occurs, water use may be restricted to help ensure adequate water in the aquifer.

Keeping our water resources pure is very important. You might think that water becomes dirty as it flows slowly through the soil and rock of the earth. Actually, soil and the organisms living in it purify ground water by filtering out many organic contaminants. But if chemical pollution gets into the soil, it

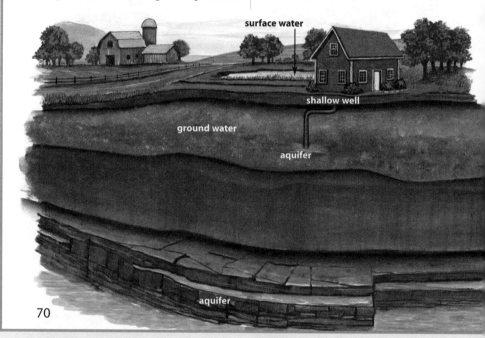

SCIENCE BACKGROUND

Drilling wells—In Bible times, men dug wells by hand. Today, drilling is a common way to find water. A giant drill is able to dig quickly and efficiently deep into the earth. Some drill rigs can drill a hole about 305 meters (1000 feet) deep.

 BIBLE Genesis 16 contains the first Bible reference to a well. The Lord provided this well or fountain of water to sustain Hagar. Hagar named the well Beeriahairoi, a name that reminded her that God lives and sees.

Fantastic Facts

The Florida Aquifer System

Perhaps you imagine an aquifer as a small area of contained underground water. While some aquifers do supply water for only a small area, other aquifers and aquifer systems cover thousands of kilometers. A huge aquifer system lies under about 260,000 sq km of the southeastern United States. It is not actually one aquifer but many aquifers that sometimes lie in layers under the surface. The deepest aquifers may reach to a depth of almost a kilometer. Florida's aquifers are among the most productive in the world. They provide more than 30 billion liters (8 billion gallons) of water every day.

can contaminate the water. Though it takes a lot of money and time to do so, chemically polluted water must be pumped out of the ground and clean water pumped in. In some places, such as near landfills, people monitor ground water for contamination.

Atmosphere

A small amount of Earth's water is held as water vapor in the atmosphere. Water becomes part of the atmosphere through evaporation. On some days we can feel the moisture in the air, and we say the day is humid. **Humidity** is the term we use to refer to water vapor in the air. When water vapor condenses, it falls to the earth and provides the water we need.

Cleaning water involves removing sediment, parasites, viruses, bacteria, and certain chemicals, such as lead and poisons, from the water. One of the oldest ways to purify water is to boil it. Today, water treatment plants use distillation, filters, chlorine, and ultraviolet light to purify water.

Distillation involves boiling water and capturing the steam. The steam is free of unwanted minerals, sediment, and chemicals.

Some water filters remove dirt particles, and other filters absorb pollutants. Carbon filters are very popular because they can absorb some pollutants and remove the bad taste from water. Adding some chlorine to water can kill many but not all pathogens.

Exposing water to ultraviolet light is a way to clean water without using chemicals or heat. The light prevents most pathogens, such as parasites, viruses, and bacteria, from causing harm by destroying the pathogens' ability to reproduce.

People must make an effort to maintain pure water resources. Christians need to make an effort to maintain lives that are pure and free from sin. A conscious effort should be made to develop godly living habits. [BAT: 4b Purity]

Discussion:
pages 70–71

➤ Can chemically polluted water be purified in the ground? {no} What must be done to clean the water? {The water must be pumped out of the ground, and then clean water must be pumped back into the ground.}

💡 What locations would require close monitoring for possible ground water contamination? {Possible answers: the ground near landfills, factories, or power plants}

💡 What are some ways that man can protect fresh water? {Possible answers: keep it free from contaminants; not use up water faster than it can be replaced}

➤ In what state is water in the atmosphere? {vapor}

➤ How does water become part of the atmosphere? {through evaporation}

➤ What name is given to the moisture in the air? {humidity}

💡 What are some examples of contamination in the atmosphere? {Possible answers: acid rain, smog, vog}

Refer the student to *Fantastic Facts*.

➤ How large is the aquifer in the southeastern United States? {about 260,000 sq km}

➤ What states are part of this aquifer? {Alabama, Florida, Georgia, Mississippi, and South Carolina}

➤ How deep is the deepest aquifer in this region? {almost one kilometer}

➤ How much water does Florida's aquifer provide? {30 billion liters per day}

💡 What impact might Florida's growing population and the state's closeness to the ocean have on the fresh water supply? {Possible answers: A greater population means possible overtaxing of aquifers as fresh water is depleted and increased chances of aquifers becoming polluted.}

Discussion:
page 72

- Where is seventy percent of Earth's fresh water located? {in Antarctica's ice sheet}
- How are glaciers and ice sheets similar? {Layers of unmelted snow form them, and they move slowly.}
- What distinguishes a glacier from an ice sheet? {Glaciers tend to be in mountains, whereas ice sheets are in relatively level land.}
- How thick is the ice sheet covering Antartica? {almost 5 km}
- What happens when an ice sheet flows into the ocean? {It forms an ice shelf.}
- How do you think the thickness of an ice shelf compares to that of an ice sheet? {Elicit that an ice shelf would probably be thinner than an ice sheet but thicker than sea ice.}
- What is the most famous ice shelf in Antarctica? {Ross Ice Shelf}

Use a world map to point out the Ross Ice Shelf. You may want to also point out a familiar state or country the size of France to give the student a comparison of the size of this very large ice shelf.

- What do we call pieces of ice that break off and float in the ocean? {icebergs}
- What important geographical point is not actually on land? {the North Pole}
- On what is the North Pole located? {on packed sea ice in the Arctic Ocean}
- What is the difference between sea ice and an ice sheet? {Possible answers: Sea ice is relatively thin, forms from salt water, and has no land under it. An ice sheet is very thick, forms from fresh water, and covers land.}
- Which do you think an ice breaker ship can move through more easily—an ice shelf or sea ice? Why? {sea ice; Sea ice is usually thinner than an ice shelf.}

Arctic Ocean and North Pole

Antarctica and South Pole

Frozen water

Seventy percent of the world's fresh water is in Antarctica, Earth's southernmost continent. This huge expanse of ice is called an **ice sheet,** and it is essentially a glacier on relatively level land. Like a glacier, an ice sheet forms when layers of snow build up. Since the weather is too cold for the snow to melt, the snow becomes deeper and deeper. The ice sheet covering Antarctica is 4776 m (15,670 ft) deep—almost 5 km (3 mi)!

Like a glacier, an ice sheet moves slowly. When an ice sheet reaches the ocean, it continues to float out over the water. This floating ice is called an **ice shelf.** The most famous of Antarctica's ice shelves is the Ross Ice Shelf. When pieces of glaciers, ice sheets, or ice shelves break off into the ocean and float independently, they are called **icebergs.**

Another kind of ice does not come from fresh water. **Sea ice** forms when ocean water freezes. At the poles, sea ice completely covers the ocean in the coldest winter months. The geographic North Pole is located in the middle of the Arctic Ocean, where there is no land, only packed sea ice. Sea ice is seldom more than 5 m (about 15 ft) thick, and it is easily broken up by winds and ocean currents.

Ross Ice Shelf

SCIENCE BACKGROUND

Ross Ice Shelf—This ice shelf, found in Antarctica, is famous because it is one of the largest ice shelves. It is 180–910 meters (600–3,000 ft) thick. The tallest points are 60 meters (200 ft) above the water. This ice shelf stretches about 965 km (600 mi) long and 804 km (500 mi) wide.

WRITING Both Frederick Cook and Commander Robert Peary claim to have been the first to reach the North Pole. Direct the student to research and write a paragraph about the North Pole expedition of one of the men. Tell him to take a position presenting evidence either for or against his person reaching the North Pole first.

Preserving Our Resources

There are things that you can do today to help conserve the resources that God has given us to use. Some of the solutions are simple. First, **reduce** the amount of resources you use. This can be as easy as turning off the lights when you leave a room or turning off the water when you're not using it.

Another thing you could do to help is to **reuse** materials that would sometimes get thrown away. Find new ways to use containers. Use newspaper to wrap things for moving or to put under a messy project. There are many things that you can reuse if you put some thought behind it.

A very important way to conserve these resources is through **recycling**. Take plastics, paper, aluminum and glass to recycling centers. They will take these resources and remake them into other products. Our landfills will last longer, and we will get the benefit of recycled products.

The most important reason of all to take the best care we can of our world is that the earth was created by God as man's home. We are to be good stewards of all that God has provided for us. Adam had the duty of taking care of Eden (Gen. 2:15). It is right and proper that we work to keep our earth beautiful and not squander the resources given to us.

> **✓ QUICK CHECK**
> 1. Diagram the path water takes through the water cycle.
> 2. Name at least three ways that the oceans provide resources for us.
> 3. What is the difference between sea ice and ice sheets?
> 4. How can we preserve our resources?

Direct an Activity

Direct the student to look around the classroom for items that are being reused instead of being discarded.

Things he may find are:

Cylindrical containers from oatmeal, powdered drink mixes, or potato chips used to store pencils, markers, rulers, etc.

Cardboard tubes from gift-wrap used to store rolled-up posters, maps, and artwork.

Interesting or unusual pictures or art from package labels, advertisements, and magazines used as writing prompts.

 Teacher Helps

An easy way to remember ways to conserve our resources is to remember the 3 Rs: reduce, reuse, and recycle.

Review recycling policies for your school and community. Identify the locations of recycling centers and the types of items collected for recycling.

Discussion:
page 73

➤ **What three things can you do to help preserve Earth's resources?** {reduce what you use, reuse what you can, and recycle}

💡 **Reduce means "to make smaller." What are some specific things you do or can do to reduce your use of resources?** {Answers will vary.}

💡 **Look around the room. What do you see being reused differently than what was intended originally?** {Answers will vary.}

📖 **What is the most important reason for taking care of the earth?** {God gave us the earth as our home. It is our privilege to care for the earth.}

➤ **What are the advantages and disadvantages of using renewable resources and nonrenewable resources?** {Answers will vary but should include ideas such as the availability of resources, the pollution that is a result of the use of certain resources, the expense of the resource, the amount of land required, etc.}

Answers

1. See illustration on Student page 68.
2. Oceans influence climates. They are part of the water cycle and the carbon dioxide-oxygen cycle.
3. Possible answer: Sea ice is relatively thin, forms from salt water, and has no land under it. An ice sheet is very thick, forms from fresh water, and covers land.
4. by reducing what we use, reusing the existing resources when possible, and recycling

Activity Manual

Reinforcement—page 47

Study Guide—page 48

This page reviews Lessons 31, 33, and 34.

Assessment

Test Packet—Quiz 3-B

The quiz may be given any time after completion of this lesson.

Lesson 35

TE pages A21–A22
Activity Manual pages 49–50

Objectives

- Compare the differences between water accessibility in Bible times and water accessibility now
- Identify several ways to conserve water
- Recognize the living water

Vocabulary

cistern

Materials

- copy of TE pages A21–A22, *Water in Israel*, for each student

Introduction

Note: The student should complete Activity Manual page 49 before the lesson. Discuss the answers from the chart.

➤ How many gallons of water does your family use?
➤ What total did you get for washing clothes?
➤ What total did you get for taking a shower?
➤ What would cause the water usage of one family to be different from that of another family? *{Possible answers: the number of people in the family; priorities of water usage}*
➤ What are some other things that you use water for? *{Possible answers: washing the car, watering the lawn, cooking}*

Purpose for reading:
TE pages A21–A22

➤ How has the use of water changed since Bible times?

Water in Israel

Name _____

Because some of the country of Israel is hot and dry, water is an especially important resource. The country's usable water, from both rivers and aquifers, is already being used to its limit. As the country's population grows, the stress on its water system grows as well. Consequently, Israel has a number of desalination plants to help supplement its fresh water supply.

Conflicts Over Water

Israel shares many water sources with its neighboring countries—Syria, Lebanon, and Jordan—as well as with its own Palestinian occupied areas. Conflicts sometimes develop over how much water each area is allotted. But disputes over water in the Middle East are not new. As far back as Abraham's time, wells were often a source of conflict. Digging wells in Bible times was an important but difficult task. Wells were considered the property of the person who dug them, and usually they were passed on as an inheritance just as land was. To take over a well that someone else had dug was considered stealing. In Genesis 21 Abraham reproves the leader of the Philistines, Abimelech, because Abimelech's servants had violently taken over a well that Abraham's servants had dug. Abraham and Abimelech made a covenant to establish Abraham's ownership of the well. Abraham called the place of the well Beersheba.

After Abraham died, the Philistines filled with dirt the wells that he had dug. When Abraham's son Isaac opened each of the wells and found water, the herdsmen of the area fought to gain control of the wells. Isaac moved on to another well that was fought over also. Finally he came to Beersheba. At Beersheba, Isaac's servants again dug a well and found water. Abimelech recognized that Isaac was blessed by God and reestablished a covenant with Isaac as he had with Abraham.

Modern Middle East

Teacher Helps: Direct the student to complete Activity Manual page 49 the night before the lesson. Parents may help the student with the chart. There are no right or wrong answers.

Sources of Water

Wells were the cleanest sources of water in ancient Israel. A short wall of limestone or stone often surrounded wells. The wall helped protect people and animals from falling into the well. Often the wells would have a stone cover that would have to be removed to draw water from the well. Because the wells were often used for watering flocks, they usually had a trough of wood or stone into which water could be poured to allow the flocks to drink.

Usually the women in a household were responsible for drawing water for the family. Typically, women would draw water early in the day and toward evening. Water was drawn by dropping a vase or waterskin attached to a rope into the well. Some wells were dug into the limestone and had steps descending into them.

Another source of water in ancient Israel was cisterns. A cistern is a large basin made of stone. The cistern collected rainwater that ran off rooftops during the rainy months. Because of the dirt that rainwater collects, water from cisterns was not suitable for drinking or cooking. However, in ancient Israel, cisterns provided water for washing and bathing.

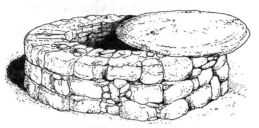

A large stone protected the well from dirt and smaller stones.

Living Water

Probably one of the most well-known Bible stories concerning a well is found in John 4. Jesus met a Samaritan woman at the well. He asked her to draw water for him to drink. Surprised that a Jew would ask a Samaritan for water, she questioned Jesus. He described for her living water that she could obtain. She would never thirst again. What was this water? The gift of salvation that God offers to all men. No conservation is needed for the living water. It is freely available in abundance for all those who seek for Christ.

© 2004 BJU Press. Limited license to copy granted on copyright page.

A22 For use with Lesson 35

To determine the weight of a gallon of water, weigh a student on a bathroom scale. Record his weight. Weigh the student again while he is holding a gallon of water. Record the weight. Allow a student to calculate the difference to find the weight of the water.

How much does the gallon of water weigh? *{A gallon of water weighs approximately 8.33 lb.}*

If you carried only two gallons of water at a time, how many trips would you need to make to carry the water your household uses in a day?

How much would all the water that your household uses weigh?

Discussion:
TE pages A21–A22

➤ **Which person in the family went to the well during Bible times?** *{the mother, or woman of the family}*

➤ **How many times did she go to the well each day?** *{two}*

➤ **What method did she use to draw water?** *{She tied a rope around a pitcher, drew the water out, and carried the water on her shoulders.}*

💡 **Look at the number of gallons of water your family used. Do you think you could carry that much water to your home each day?** *{no}*

💡 **What things would you not use as much water for if you had to carry water from a well each day?** *{Answers will vary.}*

➤ **How did the people in Bible times protect their wells?** *{They used large stones to keep out dirt and debris.}*

➤ **What often happened to water wells during times of war?** *{Enemies filled the wells with dirt and stones.}*

➤ **What is a cistern?** *{a large hole lined with stones}*

➤ **How was a cistern important to the people?** *{It provided extra water for washing and bathing.}*

➤ **How can we help protect our water supply today?** *{We can help keep trash and other items out of the water systems.}*

💡 **What are some ways you can help conserve water?** *{Possible answers: turn off water while brushing teeth; wait for a full load of laundry}*

📖 Read John 4:4–26.

➤ **What was the Samaritan woman doing at the well?** *{drawing water for her family}*

➤ **What did Jesus ask her to do?** *{give Him a drink of water}*

💡 **What is the difference between the water in the well and the water that Jesus was offering the woman?** *{Lead the students to understand that the water from the well was physical water that would quench her thirst temporarily. The water that Jesus was offering would quench her spiritual thirst for truth and give her eternal life.}*

Activity Manual
Explorations—pages 49–50

Lesson 36

Student Text page 74
Activity Manual pages 51–52

Objectives

- Recall concepts and terms from Chapter 3
- Apply knowledge to everyday situations

Introduction

Material for the Chapter 3 test will be taken from Activity Manual pages 39, 43–44, 48, and 51–52. You may review any or all of the material during this lesson.

You may choose to review Chapter 3 by playing "Mining for Gold" or a game from the *Game Bank* on TE pages A56–A57.

Diving Deep into Science

Questions similar to these may appear on the test.

Answer the Questions

Information relating to the questions can be found on the following Student Text pages:

1. page 58
2. page 68
3. page 59

Solve the Problem

In order to solve the problem, the student must apply material he has learned. The answer for this Solve the Problem is based on the material on Student Text page 70. The student should attempt the problem independently. Answers will vary and may be discussed.

Activity Manual

Study Guide—pages 51–52

These pages review some of the information from Chapter 3.

Assessment

Test Packet—Chapter 3 Test

The introductory lesson for Chapter 4 has been shortened so that it may be taught following the Chapter 3 test.

Diving Deep into Science

Answer the Questions

1. Hydroelectric power is a clean, renewable source of energy. Why is hydroelectric power not used for all of our energy needs?
   Accept any reasonable answers, but elicit that hydroelectric energy requires a large river, which is not always available. It also requires flooding an area. Flooding can be done only in a non-urban area.

2. Why are salty oceans so important to the fresh water on Earth?
   Possible answer: The salty oceans are part of the water cycle. The majority of evaporation occurs from the huge expanse of the ocean's surface area. As the water evaporates, the minerals remain in the ocean, and the evaporated water is fresh.

3. What is the relationship between geothermal energy and volcanic activity?
   Possible answer: Volcanic activity occurs where magma rises to the surface. Geothermal energy is most often available where magma is closer to the surface. Consequently, geothermal energy tends to occur in areas where there is volcanic activity.

Solve the Problem

Ian's family owns a large farm that depends on an aquifer under their land for their water supply. All of their water needs, including the well water for their household, comes from the aquifer. Last year Ian's father planted a new field. He irrigates and fertilizes it on a regular basis. Ian's mother, however, has started to notice that the water pressure at the house is low. What might be happening to the water supply, and what could be done to solve the problem?

The aquifer may not be adequate to supply both the well and the irrigation needs. Precision farming might help. Also, adjusting the times water is used to minimize drawdown may help.

Review Game

Mining for Gold

Divide the class into two teams. Have available a set number of tokens or gold candy pieces. The goal is to "find" the most gold. Each time a team answers a question correctly, they have "found" some gold and receive a token. The game continues until the vein has yielded all of its resources (all the tokens are given out). The team with the most tokens wins.

Variation: The class could see how much of a vein could be mined before it comes up empty (how many tokens in a row they can collect before someone answers a question incorrectly).

Student Text pages 75–77
Activity Manual page 53

Lesson 37

Objectives
- Demonstrate knowledge of concepts taught in Chapter 3
- Understand the interrelationship of science concepts
- Recognize that God supplies the needs of every organism

Unit Introduction

Note: This introductory lesson has been shortened so that it may be taught following the Chapter 3 test.

The purpose of this unit is to show the patterns of classification. The unit is not meant to be an exhaustive exploration of each category. It is an introductory survey of cells and classification, which will be developed further in secondary science classes.

Chapter 4 discusses the basic structures and functions of living things. The chapter also explains how living things are classified scientifically.

Chapter 5 explores how organisms in the kingdom Animalia are grouped further according to similar characteristics.

Chapter 6 explains how plants are divided further according to similar characteristics. This chapter also discusses some of the basic structures of plants.

➤ Look through Unit 2. What kinds of topics do you think you will be studying in this unit? *{Possible answers: cells, classification, animals, plants}*

➤ Why do you think these chapters are organized into the same unit? *{Answers will vary. Elicit that God's living creation has patterns that allow the organisms to be classified.}*

2
God's Living Creation

Additional information is available on the BJU Press website.
www.bjup.com/resources

Unit 2   Lesson 37   83

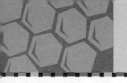

Project Idea

The project idea presented at the beginning of each unit is designed to incorporate elements of each chapter as well as information gathered from other resources. You may choose to use the project as a culminating activity at the end of the unit or as an ongoing activity while the chapters are taught.

Unit 2—A Zoo Habitat

The students design a new zoo exhibit. They choose a habitat and include the plants and animals that will be part of that habitat. They also should include the scientific names for all the plants and animals included in the exhibit.

Variation: As a class, design an entire zoo, with each group of students being responsible for a specific habitat.

Technology Lesson

A correlated lesson for Unit 2 is provided on TE pages A6–A7. This lesson may be taught with this unit or with the other Technology Lessons at the end of the course.

Bulletin Board

Note: This suggested bulletin board idea may be modified for use in a learning center.

Throughout Chapter 5, the student should gather pictures of vertebrates. As pictures are collected, the student should place each picture under the correct heading on the bulletin board. Use file folder labels to attach the common name for each animal with its corresponding picture. Use resources to find the scientific names.

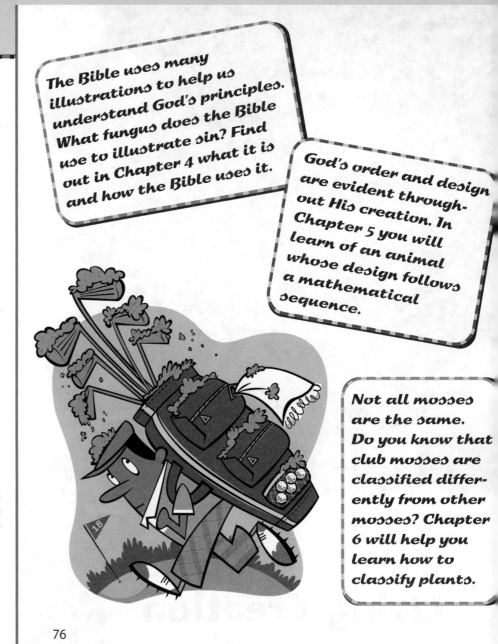

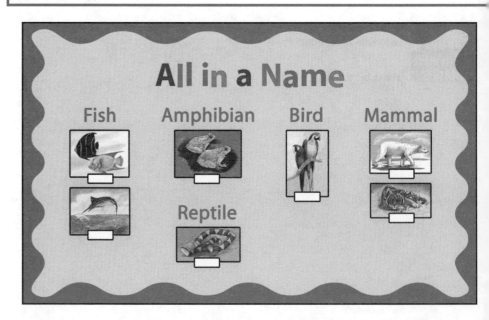

84   Lesson 37

4 Cells and Classification

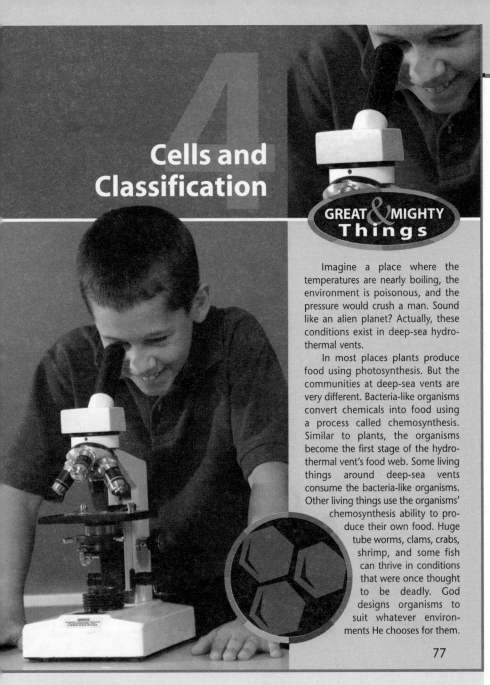

GREAT & MIGHTY Things

Imagine a place where the temperatures are nearly boiling, the environment is poisonous, and the pressure would crush a man. Sound like an alien planet? Actually, these conditions exist in deep-sea hydrothermal vents.

In most places plants produce food using photosynthesis. But the communities at deep-sea vents are very different. Bacteria-like organisms convert chemicals into food using a process called chemosynthesis. Similar to plants, the organisms become the first stage of the hydrothermal vent's food web. Some living things around deep-sea vents consume the bacteria-like organisms. Other living things use the organisms' chemosynthesis ability to produce their own food. Huge tube worms, clams, crabs, shrimp, and some fish can thrive in conditions that were once thought to be deadly. God designs organisms to suit whatever environments He chooses for them.

77

Introduction

Some environments seem so harsh, so extreme, or so poisonous that nothing should be able to live there. But God tells us that all of His creation is good. Places that are inhospitable to most living organisms provide a perfect environment for the organisms designed by God to live there.

Chapter Outline

I. Cells and Organisms
   A. Living Things
   B. Microscopes
   C. Cells
   D. Reproduction of Cells
   E. Reproduction of Multicellular Organisms

II. Classification
   A. The Living Kingdoms
   B. Naming Organisms

Activity Manual

Preview—page 53

The purpose of this Preview page is to generate student interest and determine prior student knowledge. The concepts are not intended to be taught in this lesson.

SCIENCE BACKGROUND

Hydrothermal vents—These vents are much like geysers on the ocean floor. Ocean water that has seeped beneath Earth's crust is heated by magma and hot rocks. This hot water rises through vents, or cracks in the crust. Often chimneys of dissolved metals form around the vents. Some of the dissolved metals form compounds that look like clouds of black or white smoke spewing from the vents. Because of the tremendous pressures under the ocean, the boiling water (steam) coming from a vent may reach 350°C (660°F).

Chemosynthesis—In chemosynthesis, organisms produce carbohydrates from water, carbon dioxide, and the oxidation of inorganic chemicals, such as those found near hydrothermal vents.

Deep-sea exploration—Much of the research of hydrothermal vents is conducted with the use of specially equipped deep submergence vehicles, such as *Alvin*. Deployed from the 60-man research vessel *Atlantis*, *Alvin* is a little over 23 feet long and almost 12 feet high. Its three-person crew has access to many tools aboard the submersible, including computers, video recording equipment, sonar, and manipulators, long arms that can reach up to 75 in. and lift up to 250 lb.

Chapter 4

Lesson 37   85

Lesson 38

Student Text pages 78–82

Objectives

- Distinguish between living things and nonliving things
- Identify five characteristics of living things
- Identify men associated with the development of the microscope
- Explain the cell theory

Vocabulary

organism
life span
life cycle
environment
energy
cell
cell theory
microscope

Materials

- toy animal or doll

Introduction

Show a toy animal or doll.

▶ Is this object alive?

▶ How do you know this object is not alive?

Record the reasons that the students give.

▶ In this lesson we are going to learn some characteristics that distinguish living things from nonliving things.

Purpose for reading:
Student Text pages 78–82

▶ What is a *life cycle*?

▶ What are some examples of a living organism's response to its environment?

▶ What is one characteristic of living things that nonliving things never have?

▶ Which instrument allows men to see cells?

Perhaps you have seen a little girl playing with a doll. She cleans and dresses it, talks to it, and tells it to take a nap. She may even try to feed it. But we all know that the doll is not alive.

Cells and Organisms

An **organism** is a complete living thing. You are an organism. Animals and plants are organisms. Even the mold that grows on old bread is an organism! All organisms have some things in common. These characteristics distinguish living things from nonliving things.

Living Things

Living things grow and develop

Living things are able to grow, maintain, and often repair themselves. Plants grow by using sunlight, water, and nutrients in the soil to make new leaves and stems. People and animals grow by eating food, which is then transformed into energy for growth.

living / nonliving

The Bible tells us in Genesis 1 that God created all things perfectly, both living and nonliving. As a result of sin, decay and death entered into the world (I Cor. 15:21–22). For living things, this process of birth, growth, reproduction, and death is called the **life span**. Some organisms, such as bristlecone pine trees, can live for thousands of years. On the other hand, the mayfly may live for only a few hours as an adult insect. Sometimes we refer to the life span as the **life cycle**. After an organism is born, it grows into an adult and is able to produce offspring. Finally, the organism ages and eventually dies. When an organism dies, it no longer shows the characteristics of life.

life span

SCIENCE BACKGROUND

Cell theory—The idea that living organisms are composed of cells is called a theory because even though it is based on many observations, it cannot be proven as absolutely true.

SCIENCE MISCONCEPTIONS

For something to be classified as a living organism, it must demonstrate all five characteristics of living things. Nonliving things may demonstrate some of the characteristics, but they are not alive because they do not demonstrate all five characteristics.

Living things reproduce

One of the most important scientific principles is that life comes only from life. Living things do not simply "appear" out of thin air. They must come from other living things. Since the time of Creation, all living things have come from other living things.

Birth is the first stage of life. As an organism eats, it grows and matures until it reaches a stage where it is able to reproduce itself and form new organisms. We often call organisms that reproduce themselves the *parents*. Reproduction allows the life cycle to begin again as new organisms replace the organisms that age and die.

Living things respond to their environments

When an animal senses that it is in danger, it can readily respond. Sometimes the animal will flee. At other times it will assume a defensive posture and maybe even attack. Living things respond to their **environments** (surroundings) to protect themselves or to gain an advantage, such as obtaining food. An important part of this response is movement. All living things, including plants, can move. Most plant movement is slow and is not as dramatic as animal movement. However, some plants actually move quite rapidly. A Venus flytrap can close quickly to trap an insect for lunch. Plants also respond by growing toward sources of light and water. Nonliving things do not respond to their surroundings. A piece of paper or a dead animal cannot run away or protect itself from danger.

Venus flytrap

 God commanded Adam and Eve not to eat of the tree of the knowledge of good and evil. If they ate of that tree, they would "surely die" (Gen. 2:16–17).

When Adam, the father of all people, sinned, he brought sin into the world. The punishment for sin is death. Because Adam sinned, every descendant of Adam is born a sinner and will receive the punishment for sin (Rom. 5:12).

Yet Jesus Christ, the Son of God, obeyed God by dying on the cross. He made it possible for all men to be saved from sin and the punishment for sin (Rom. 5:19; I Cor. 15:12). [BAT: 1a Understanding Jesus Christ]

Discussion:
pages 78–79

➤ **What is an *organism*?** {a complete living thing}

➤ **How do plants grow?** {by using sunlight, water, and nutrients in the soil}

➤ **How do people and animals grow?** {by eating food, which is then transformed into energy for growth}

📖 **Why did decay and death enter the world?** {because of sin}

📖 **How do we know that God created the world perfectly?** {Genesis 1 states that everything was good.}

➤ **What do we call the process of birth, growth, reproduction, and death?** {life span}

➤ **What is another term that means "life span"?** {life cycle}

➤ **Where do all living things come from?** {other living things}

💡 **Why is it necessary for organisms to reproduce?** {Eventually all organisms die. Reproduction allows new organisms to replace those that die.}

➤ **What do we call the organisms that reproduce?** {parents}

➤ **How might an animal respond to danger?** {flee, assume a defensive posture, or attack}

➤ **What is an *environment*?** {surroundings}

➤ **Do plants respond to their environments? Give an example to support your answer.** {yes; Possible answer: They grow toward sources of light and water.}

➤ **Most plants respond to their environments slowly. Give an example of a plant that responds quickly.** {Venus flytrap}

Discussion:
pages 80–81

➤ What is *energy*? {the ability to do work}

➤ How does a living thing get its energy? {from food}

➤ How might an organism use energy? {Many answers are possible, but students should recognize that everything an organism does requires energy.}

➤ When does an organism stop using energy? {only when it dies}

➤ What is produced when living organisms convert energy from one form to another? {waste}

➤ Our cells use oxygen to produce energy. What is produced as a waste product? {carbon dioxide}

➤ What are five characteristics of living organisms? {Living things grow and develop, reproduce, respond to their environments, use energy, and are made of cells.}

➤ Which characteristic is true only for living things? {made of cells}

Living things use energy

Have you ever been in a car that has run out of gas? One minute you are traveling down the freeway and the next minute you are sitting on the side of the road. In order for a car to move, it needs energy. **Energy** is the ability to do work. The energy a car needs comes in the form of gasoline, but the energy a living thing needs comes from food.

At no point during an organism's life does it stop using energy—even when it is resting. You may not be moving your arms or legs, but your heart is still pumping and your brain is still functioning. Only when an organism dies does it no longer need or use energy.

One of the laws of nature is that changing one form of energy into another is never completely efficient. As living things convert energy from one form to another, they produce waste. A living organism must be able to get rid of the waste it produces. One way you perform this task is by breathing out. As your cells use oxygen to produce energy, they also produce carbon dioxide as a waste product. Your blood carries the carbon dioxide to your lungs, where you get rid of it by breathing out.

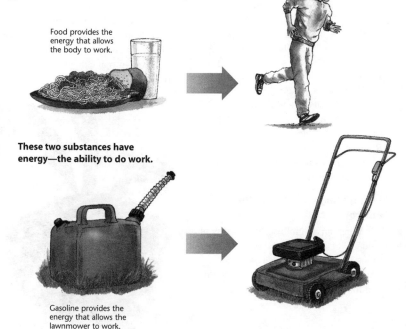

Food provides the energy that allows the body to work.

These two substances have energy—the ability to do work.

Gasoline provides the energy that allows the lawnmower to work.

80

Some characteristics of living things are also characteristics of people that are spiritually alive in Christ. To be spiritually alive, a person must accept Jesus Christ as his Savior from sin and be born into God's family.

➤ Living things grow and develop.

Spiritually alive people grow and develop in Christ (Gal. 5:22–23; I Pet. 2:2).

➤ Living things reproduce.

Spiritually alive people spread the gospel to others (Prov. 11:30; Matt. 28:19–20). [BAT: 5a Evangelism and missions]

➤ Living things respond to their environments.

Spiritually alive people respond to their environments by fleeing temptation and doing what is right (I Cor. 10:13; James 4:7).

➤ Living things use energy.

Spiritually alive people need strength (energy) from the Lord (Phil. 4:13; Ps. 27:1).

Living things are made of cells

We can find nonliving things that exhibit some of the characteristics of living organisms. For example, a mineral crystal may grow in size. Rocks move during an earthquake. And most machines use energy to do work. But there is one characteristic that applies only to living things—living things are made of cells.

In 1665 Robert Hooke, an English scientist, examined a small piece of dried cork with a microscope that he had invented. He observed small, empty chambers that he called cells. Scientists soon discovered that what Hooke had actually seen was the cell walls of the dead cork. Living cells are not empty; instead, they are filled with a watery substance. Nevertheless, the name **cell** became the term for describing the smallest unit of a living organism.

However, it was over 250 years later that scientists proposed a theory about the relationship between cells and living organisms. In 1938 two German scientists, Theodor Schwann (SHVAHN) and Matthias Schleiden (SHLY duhn), identified certain observations they had made while studying plant and animal cells. They found that all the living things they observed were made of cells, but the nonliving things were not made of cells. They also theorized that cells can function as individual living organisms or as the smallest units in a larger organism. Over the years, other scientists have observed these same things and have drawn similar conclusions. With only a few small changes and additions, Schwann's and Schleiden's conclusions form the basis for what we call the **cell theory**.

Meet the SCIENTIST — ROBERT HOOKE

Robert Hooke (1635–1703) was a man of considerable talent. His ideas and inventions covered a wide range of science topics. Hooke was a firm believer in experimental investigation. As he experimented, he carefully recorded his observations. One of his lasting effects on science was his science picture book, *Micrographia*. The finely detailed drawings in the book demonstrate Hooke's artistic ability and scientific accuracy. Chemist, physicist, naturalist, inventor, architect—Robert Hooke was all of these and more.

cork from *Micrographia*

Discussion:
pages 80–81

➤ Who discovered small, empty chambers in a dried piece of cork? {Robert Hooke}

➤ What did Hooke name these empty chambers? {cells}

➤ What was Hooke actually looking at? {cell walls of dead cork}

➤ How are living cells different from dead cells? {Living cells are not empty. They are filled with a watery substance.}

➤ What is a *cell?* {the smallest unit of a living organism}

➤ What two important observations did Theodor Schwann and Matthias Schleiden make? {Living things are made of cells, but nonliving things are not made of cells. A cell can function as an individual living organism or as the smallest unit in a living organism.}

💡 Why are these observations important? {because they form the basis for the cell theory}

💡 Why could Schwann and Schleiden not say that their ideas were absolutely true? {Accept reasonable answers, but elicit that they could not test all possible substances, so their ideas might have been disproven by something they had not tested.}

➤ Why did their conclusions eventually become accepted as the cell theory? {Over the years, other scientists tested substances and gained similar results.}

Refer the students to *Meet the Scientist*.

➤ When did Robert Hooke live? {1635–1703}

➤ Did Hooke use good scientific practices? {Yes; he experimented to prove things and kept careful records.}

➤ What is *Micrographia?* {a science picture book}

Direct a DEMONSTRATION
Demonstrate the characteristics of living things

Materials: clear container of water, food coloring, balloon, a dead plant or flower, an electric or battery-operated pencil sharpener, a student assistant or pet

Direct the student to determine whether each of the above items is living or nonliving. Set the container of water on an overhead projector or in another easily visible place in the classroom. Drop in a few drops of food coloring. Point out how the colored drops move and spread throughout the water.

➤ Is the food coloring living or nonliving? {nonliving} Why? {The coloring moves, but it does not demonstrate other characteristics of living things.}

➤ Blow up the balloon. Is it living or nonliving? {nonliving} Why? {It grew larger, but it does not demonstrate other characteristics of living things.}

➤ Is this plant a living thing? {no} Has it ever been a living thing? {yes} How do you know? {When alive, it had all the characteristics of living things.}

➤ The pencil sharpener uses energy. Is it living or nonliving? {nonliving} What is its source of energy? {a battery or electricity} How do living things get their energy? {from food}

➤ Is this student (or pet) a living thing or a nonliving thing? {living} How do we know? {He or she has all the characteristics of living things.}

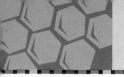

Discussion:
page 82

➤ How do our eyes help us learn about science? *{They allow us to observe the world around us.}*

📖 Can we know everything about God's creation, even with instruments such as the microscope? *{No; mankind cannot see or understand all of God's creation.}*

➤ What is a *microscope*? *{an instrument that uses lenses to magnify objects hundreds or thousands of times}*

➤ Who is credited with inventing the first microscope? *{Zacharias Jansen}*

➤ Why is Anton van Leeuwenhoek also famous for his work with microscopes? *{He made microscopes that were more powerful than the one that Jansen made. Using the microscope, he noticed tiny creatures swimming in the water and studied them.}*

➤ How powerful was van Leeuwenhoek's microscope? *{It could magnify an object to about 300 times its original size.}*

➤ Before the 1930s, how powerful were most microscopes? *{They could magnify an object about 2,000 times.}*

➤ How strong are modern electron microscopes? *{They can magnify objects 500,000 times or more.}*

Answers

1. An organism is a complete living thing.
2. Living things grow and develop, reproduce, respond to their environments, use energy, and are made of cells.
3. It is a theory that says living things are made of cells, but nonliving things are not made of cells. Cells can function as individual living organisms or as the smallest units in a living organism.
4. Zacharias Jansen

Microscopes

Without the microscope, Robert Hooke would not have been able to see the tiny chambers in the cork. Although God made our eyes to be very important tools for observing and learning about creation, our eyes cannot see everything. Some things are too small or too far away for us to see. Since science is based on observation, scientists constantly search for tools that will improve human senses. One of these tools is the microscope. The **microscope** is an instrument that uses lenses to magnify objects hundreds or thousands of times.

Many people credit the Dutch inventor Zacharias Jansen with inventing the first microscope. (He made one in the late 1500s.) However, Anton van Leeuwenhoek (LAY vun HOOK), another Dutch scientist, is also famous for his work with microscopes. He spent many hours grinding and polishing lenses to make more and more powerful microscopes. In the 1660s he used his microscope to study tiny creatures swimming in water. He called the creatures "animalcules."

Leeuwenhoek's most powerful microscope magnified an object to about 300 times its original size. Until the invention of the electron microscope in the 1930s, the most powerful microscope could magnify about 2,000 times. However, modern electron microscopes can magnify 500,000 times or more.

Hooke's microscope

QUICK CHECK
1. What is an organism?
2. What are five characteristics of living things?
3. What is the cell theory?
4. Who is credited with inventing the microscope?

82

 Zacharias Jansen (Janssen)— His father Hans, a lens maker, probably began making the first compound microscope. This microscope used more than one lens. Zacharias, when he was older, completed his father's work. There is still confusion over whether the father or the son was the inventor. Therefore, many prefer to say that both Hans and Zacharias invented the microscope.

 Students can research and prepare a short report about Robert Hooke, Zacharias Jansen (also spelled Janssen), or Anton van Leeuwenhoek.

Using a Microscope   Name

Activity Manual page 54

Lesson 39

A. Study the parts of the microscope. Put a check mark in each box as you study that part. *Complete teacher instructions are located in the Teacher's Edition.*

B. Match the descriptions with the microscope parts.
Note: This page is not intended to be graded. Students should correct with teacher's guidance.

A. adjustment knobs
B. base
C. clips
D. eyepiece
E. mirror
F. nosepiece
G. objectives
H. slide

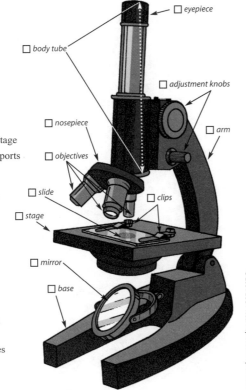

____ 1. hold the slide in place on the stage
__B__ 2. bottom of the microscope; supports the microscope
__A__ 3. turn these to move the body tube up and down to focus the image of the object
____ 4. reflects light up to the stage
__F__ 5. a rotating disc that holds objective lenses of different magnification
____ 6. a thin, small piece of glass where a sample is placed
____ 7. located at the top of the body tube; holds magnification lens
____ 8. extensions under a revolving disc; contain microscope lenses

Objectives
- Identify the parts of a microscope
- Explain how to use a microscope

Vocabulary
eyepiece   specimen
body tube   clips
adjustment knob   slide
nosepiece   stage
arm   light source
objective lens   base

Materials
- magnifying glass
- microscope
- transparency of TE page A23, *The Microscope*
- prepared slides

Introduction

Many people wear eyeglasses or contact lenses.

➤ **What do we call the part that you look through on a pair of glasses?** {lenses}

Some lenses help a person's vision by magnifying the objects being looked at.

Display the magnifying glass.

➤ **Have you ever used a magnifying glass?**

A magnifying glass also uses a lens to magnify objects. A microscope is similar to a magnifying glass. A microscope uses several lenses to magnify an object many times.

Discussion:
Activity Manual page 54

Display the transparency, *The Microsope*. Identify and discuss the parts of a microscope and how they are used.

Eyepiece—The part of the microscope that you look through. It holds the ocular lens system. Most eyepieces are 10X. This means they magnify the images 10 times.

Body tube—The adjustable cylinder that connects the eyepiece and objectives.

Adjustment knobs—Most microscopes have two adjustment knobs located on the arm. These knobs move the eyepiece and the objective lens up and down to bring the image into focus. The larger knob is used to begin bringing the image into focus. The smaller knob is for finely adjusting the focus of the image.

Nosepiece—The rotating disc that holds the objective lenses. It is turned to align an objective lens with the eyepiece.

Arm—The curved part of the microscope that connects the lenses to the base.

Objectives—These lenses are used to magnify the image of the specimen. Many microscopes have more than one objective lens. The objective lenses on many microscopes are 10X, 40X, and 100X powers.

Clips—These hold the slide in place on the stage.

Stage—The place where the specimen (slide) is placed for viewing. The stage has a hole for light to pass through to the specimen.

Light source—The light source focuses light to shine through the hole in the stage. Many microscopes use a small mirror to focus reflected light. Some microscopes use a small electric or battery-operated light bulb to produce the needed light.

Base—The bottom, or foot, of the microscope.

If desired, set up a place for the student to view specimens with a microscope.

Chapter 4   Lesson 39   91

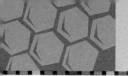

Discuss how a microscope works.

The eyepiece and each objective lens may contain two or more lenses each. These lenses work together to focus light and magnify the image of the specimen.

The symbol for magnification power is X. This means the image seen in a 10X lens appears 10 times larger than the original specimen.

The light and the magnified image first pass through the objective lens. The magnification power of an objective lens varies for each microscope. Many microscopes include three objective lenses. Often these lenses are 10X, 40X, and 100X.

The light and magnified image then pass through the tube to the eyepiece lenses. Many eyepieces are 10X. This means that the magnified image from the objective lens is magnified again as it passes through the eyepiece.

The total magnification of the image is the power of the objective lens multiplied by the power of the eyepiece. For example, an image viewed with a 40X objective lens and a 10X eyepiece will be magnified 400X (40X $\times$ 10X = 400X).

Activity Manual

Expansion—page 54

Use this page to reinforce the identification of microscope parts. You may choose to have the student complete the page as the parts of the microscope are discussed.

Using a microscope

It is important to have clean hands and materials when viewing specimens. The microscope will magnify any lint, fingerprints, or smudges that appear on the slide.

- Turn the nosepiece to the lowest power objective lens.
- Use the adjustment knob to lower the tube so the objective lens is close to the stage.
- Look through the eyepiece and adjust the light source (mirror) so the light is seen.
- Place a prepared slide under the clips on the stage. Place the slide so the specimen is centered over the hole in the stage.
- Move the adjustment knobs until the image is focused. The slide may also be moved to center the specimen under the objective lens.
- To view the specimen at a stronger power of magnification, turn the nosepiece to one of the other objective lenses and refocus the image using the adjustment knobs.

Making a slide

Materials: slide, tweezers, specimen, eyedropper, water, cover slip

- Place the slide on a clean surface. Use the tweezers to place the specimen on the center of the slide.
- Place a drop of water on the specimen.
- Hold the cover slip by the edges and gently place one edge next to the specimen. Carefully close the cover slip over the specimen

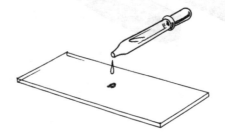 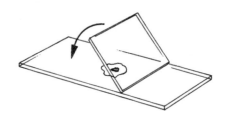

 The word *specimen* comes from the Latin word *specere*, which means "to look at." *Spectacles, spy,* and *species* also come from this same Latin word.

 The student may practice calculating the powers of magnification available with the microscope. Direct the student to multiply the power of an objective lens (usually printed on it) and the power of the eyepiece lens (usually 10).

 Carrying a microscope— The safest way to carry a microscope is to grasp the arm of the microscope with one hand and support the base with the other hand.

Slide alternative—Materials: thin cardboard, hole punch, plastic wrap, tape

With the hole punch, make a hole in the center of a 3 cm x 8 cm piece of cardboard. Tape plastic wrap across the hole on the underside of the cardboard. Pull the plastic wrap tight so there are no wrinkles. Place the specimen and drop of water in the center of the hole in the cardboard. Tap out any air bubbles that might be caught using the tweezers or the tip of the eyedropper. Tape another piece of plastic wrap over the specimen and hole.

Cells

Cells working together

A **cell** is a tiny unit of living material surrounded by a thin membrane. It does all of the things that living organisms do. It grows, reproduces, responds to its environment, uses energy, and produces and gets rid of waste. Eventually it dies.

Many organisms consist of only one cell. They are **unicellular.** The majority of unicellular organisms can be seen only through a microscope. However, most of the living things that we see without a microscope are made of millions and millions of cells. These living things are called **multicellular.** In multicellular organisms, cells become specialized. We could compare them to a baseball team. Every member of the team must individually be able to catch and throw and run. But some members specialize in pitching or hitting or playing a specific position. The goal is team performance, not individual accomplishment.

You are a multicellular organism. Your body is made of trillions of cells. Each cell in your body is designed to perform a specific function. However, some cells work together to perform a task. A **tissue** is a group of cells working together. If someone speaks of heart tissue, he is talking about the cells that function together as part of a heart. Other organisms also have tissues. For example, bark is a tissue that performs a particular function in trees. All of the tissues of an organism must successfully if the organism is to survive.

Like cells that work together to form tissues, some different kinds of tissues work together to form **organs.** Your heart is an organ. Muscles and nerves are some of the tissues that work together to form the organ that pumps blood throughout your body.

Organs also work together to form **systems.** Without our blood vessels and blood to complete the circulatory system, our heart organs would do us little good. But no matter how different the tasks that systems, organs, and tissues have, they all start with individual cells.

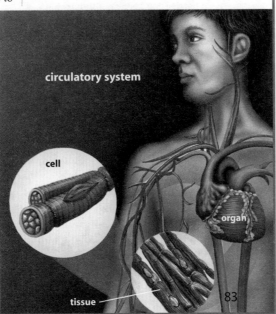

Objectives
- Identify a cell as a living organism
- Discuss the relationship of cells, tissues, organs, and systems
- Identify cell structures
- Compare and contrast plant and animal cells

Vocabulary
unicellular
multicellular
tissue
organ
system
cell membrane
cytoplasm
organelle
nucleus
chromosome
mitochondria
endoplasmic reticulum
ribosome
vacuole
cell wall
chloroplast
chlorophyll

Introduction
- Have you ever looked at the engine of a car?
- Is the engine only one piece?
- Do all parts of the engine look the same?
- Does each part have the same job?
- What happens when the parts work together to do their jobs?
- Today's lesson discusses cell structure and how cells work together in some organisms.

Purpose for reading:
Student Text pages 83–85

- What are specialized cells working together called?
- What do plant cells have that animal cells do not have?

Discussion:
page 83

- What is a *cell*? {a tiny unit of living material}
- How are unicellular and multicellular organisms different? {Unicellular organisms have only one cell. Multicellular organisms are made of many cells that work together.}
- What is meant when we say that cells in multicellular organisms are specialized? {Each has a specific purpose and task.}
- Are you a unicellular organism or a multicellular organism? {multicellular}
- What do we call a group of cells working together to perform a specific task? {tissue}
- What are some tissues in your body? {Possible answers: heart tissue, nerve tissue, muscle tissue}
- What is formed when a group of tissues work together on a specific task? {an organ}
- Which organ pumps blood throughout your body? {heart}
- What are some other organs found in your body? {Possible answers: brain, liver, kidneys}
- What is a *system*? {a group of organs that work together to complete bodily functions}
- List these terms—*cells, organs, systems, tissues*—in order from smallest to largest. {cells, tissues, organs, systems}

Discussion:
pages 84–85

Refer to the diagrams on Student Text pages 84 and 85 as the parts of cells are discussed.

➤ How is each cell shaped? *{according to its function}*

➤ What does the cell membrane provide for the cell? *{It gives the external boundary for each cell and provides a barrier to keep harmful things out while allowing necessary things into the cell.}*

➤ What is the jellylike substance inside the cell membrane? *{cytoplasm}*

➤ Cytoplasm is mostly which substance? *{water}*

➤ What are *organelles,* and where are they found? *{Organelles are the structures inside the cytoplasm of the cell that help carry out cell functions.}*

➤ What is the most recognizable organelle? *{the nucleus}*

➤ What important information is found inside the nucleus? *{the DNA}*

💡 Why is DNA important to the cell? *{DNA is the coded instructions that the cell follows for growth, reproduction, and the building of substances for the organism.}*

➤ DNA is found in little bundles throughout the nucleus of a cell. What are these bundles called? *{chromosomes}*

➤ Which organelles help to break down food and release energy? *{mitochondria}*

➤ What is the name of the organelle that serves as the cell's transportation system? *{the endoplasmic reticulum}*

➤ Which organelles are responsible for making proteins for the cell? *{ribosomes}*

Cell structures

Cells come in an amazing assortment of sizes and shapes. Some, such as red blood cells, are round and disclike. Muscle and nerve cells, on the other hand, can be long and thin. Cells that provide the outside covering of plants are often flat. Each cell is shaped according to its function.

Although cells are very different, certain structures are common to all of them. A cell membrane surrounds every cell. The **cell membrane** provides the external boundary for the material inside the cell. It also serves as a barrier that keeps out things that do not belong in the cell while allowing necessary things, such as food and oxygen, to enter the cell.

Inside the cell membrane is a jellylike substance called the **cytoplasm** (SITE uh PLAZ uhm). Cytoplasm is mostly water, but it also contains many substances, such as proteins and fats, that are essential to the cell.

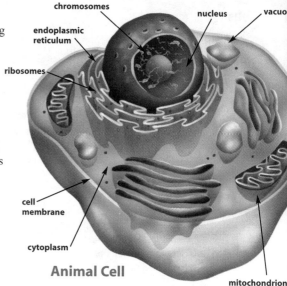

Animal Cell

Inside the cytoplasm of most cells are many tiny structures called **organelles** (OR guh NELZ), which help carry out the functions of the cell. The most obvious organelle, the **nucleus** (NOO klee uhs), is a large, circular structure separated from the rest of the cytoplasm by its own membrane. The nucleus contains DNA, the coded instructions the cell follows. The DNA is packed into tight little bundles called **chromosomes.** The cell follows the DNA code as it grows, reproduces, and builds substances for the organism. DNA eventually determines what the organism looks like.

Other organelles also help provide needs of the cell. The **mitochondria** (MY tuh KAHN dree uh) are the cell's

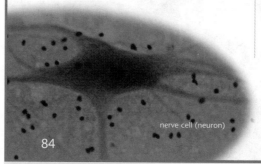

nerve cell (neuron)

SCIENCE BACKGROUND

Cells, chromosomes, and DNA—Cells have more organelles than just those mentioned in this chapter. Chromosomes and DNA will be discussed again in Chapter 13, *Heredity and Genetics.*

LANGUAGE — *Organ* and *organelle* come from the same Latin word, *organum,* meaning "tool or implement." Organelles function within cells in a similar way as organs function within systems of the body.

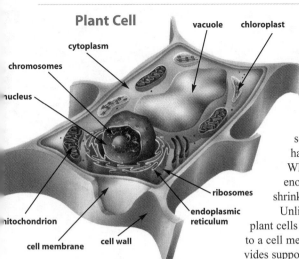

Plant Cell
- vacuole
- chloroplast
- cytoplasm
- chromosomes
- nucleus
- ribosomes
- endoplasmic reticulum
- mitochondrion
- cell membrane
- cell wall

engines. They are responsible for breaking down the cell's food and releasing energy. The **endoplasmic reticulum** (EN-duh-PLAZ-mik rih-TIK-yuh-lum), or ER, is the cell's transportation system. It is a system of passageways that allows material to move from one part of the cell to another. Along some of the ER are small organelles called ribosomes. **Ribosomes** (RY buh SOHMZ) are responsible for making the proteins that the cell needs. They also carry out the instructions given to the cell by the DNA.

Bubble-like organelles in the cytoplasm of plant, animal, and human cells are called **vacuoles** (VAK yoo OHLZ). In animal and human cells vacuoles are usually small and often temporary. They store material until it can be released outside the membrane or used by the cell. Plant cells, however, usually have one central vacuole in addition to other vacuoles. Vacuoles in plant cells sometimes hold more than half of the cell's volume. When a plant cell fails to get enough water, the vacuole shrinks and the plant droops.

Unlike animal and human cells, plant cells have a cell wall in addition to a cell membrane. The **cell wall** provides support for the plant cell. Along with the vacuoles, the cell wall helps the plant cell stay rigid and firm.

Plant cells also have large structures called **chloroplasts** (KLOR uh PLASTS), which have an abundant amount of a green pigment called **chlorophyll** (KLOR uh FILL). The chlorophyll absorbs energy from sunlight to use in photosynthesis and produces food and energy for the plant. Animal cells do not need chloroplasts because they do not use sunlight for energy.

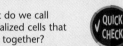

1. What do we call specialized cells that work together?
2. What are the structures that carry out the cell's functions called?
3. Why would a plant cell but not an animal or human cell need chlorophyll?

85

Just as cells, tissues, organs, and systems work together in the body to perform specific functions, Christians should work together in Christ's church to glorify His name and do His will. [BAT: 3e Unity of Christ and the church]

Discussion:
pages 84–85

➤ What are the bubble-like organelles called? {vacuoles}

➤ What is the purpose of vacuoles within a cell? {They are storage areas.}

➤ How are animal vacuoles different from plant vacuoles? {Animal vacuoles are usually small and temporary. Plant cells usually have one central vacuole in addition to other vacuoles.}

➤ Which part of the plant cell provides support for the plant? {the cell wall}

➤ What two things work together to help a plant cell stay rigid and firm? {vacuoles and cell walls}

➤ What do chloroplasts contain? {chlorophyll}

➤ Why do plant cells have chloroplasts while animal cells do not? {Plants use the chlorophyll in the chloroplasts to make food from sunlight. Animals use other methods of getting food.}

Compare the diagrams of the animal cell and the plant cell. Which parts do both types of cells have in common? {cell membrane, chromosomes, endoplasmic reticulum, mitochondrion, nucleus, ribosomes, vacuole}

Answers
1. tissues
2. organelles
3. Plants manufacture their own food, so they need the chlorophyll to carry out photosynthesis.

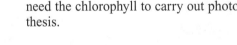

Activity Manual
Reinforcement—page 55

Study Guide—page 56
This page reviews Lessons 38 and 40.

Assessment

Test Packet—Quiz 4-A
The quiz may be given any time after completion of this lesson.

Chapter 4 — Lesson 40 — 95

Lessons 41–42

Student Text page 86

Objectives
- Demonstrate knowledge of cell structure
- Construct a 3-D model of a cell
- Prepare a written report

Vocabulary
cytologist

Materials
- picture of an object such as a car and a corresponding 3-D model
- See Student Text page

Introduction

Note: Two lesson days are allotted for this activity. On the first day, introduce the activity, set guidelines and a due date for the cell models to be completed, and begin planning. The second lesson day may occur after the cell models are finished.

Display the picture and the 3-D model.

➤ Which of these best represents what the object really looks like? {the model}

➤ Why is the model better? {Possible answers: You can see all sides. You can see inside for additional details. You can see size relationships better.}

Usually a model provides a better representation of a structure.

Purpose for reading:
Student Text page 86

➤ Which materials would you use to show the parts of a cell?

Assessment

Rubric—TE pages A50–A53

Select one of the prepared rubrics, or design a rubric to include your chosen criteria.

Cell Model

ACTIVITY

A **cytologist** (sye TOL uh JIST) is a scientist who studies cells. A cytologist does most of his work with a microscope. However, seeing only flat pictures or photographs from a microscope makes it difficult to imagine what a cell really looks like. Three-dimensional (3-D) models help people see the parts of cells better.

Your assignment is to make a 3-D model of a plant or animal cell according to the requirements given.

Process Skills
- Making a model
- Using a model
- Communicating

Procedure
1. Choose to make either a plant cell or an animal cell.
2. Decide on the outward structure of your cell. Remember that animal cells are usually round shaped, while plant cells are usually more square shaped.
3. Design your 3-D model to include all the parts of the cell shown on page 84 or 85 of your student text. (Use the diagram for the type of cell you have chosen.) You may use any materials that you choose to represent the structures in your cell.
4. Prepare a key to show what you used to represent each part of the cell.
5. Prepare a report describing the function of each organelle you placed in your cell, what you used to represent each organelle, and why you used it.
6. Be prepared to explain your model and answer questions.

Materials:
You will choose your own materials.

Follow-up
- Prepare an edible model of a cell.

Teacher Helps

You may want to prepare and display an example.

This activity may be done at home by each student or by science groups in the classroom. If you use groups, you may need to allow a few additional days to complete the activity.

Possible choices for pictures and models include cars, planes, trains, and 3-D puzzles.

SCIENCE PROCESS SKILLS

Communicating

➤ What would happen if you constructed your model and displayed it with no key? {Answers will vary, but elicit that the model would be of limited use because others would not know what the various parts represented.}

➤ How does preparing a report make your model more useful to others? {Answers will vary, but elicit that others can learn from the report.}

➤ How does preparing a report help the preparer? {Possible answers: It forces the preparer to have a logical reason for choices. It reinforces the concepts being modeled.}

➤ Most scientists' work is largely ignored until the scientists prepare extensive reports on their procedures and findings. This allows other scientists to read and evaluate their work.

Explorations: An Organized Cell

Student Text page 87
Activity Manual page 57
Lessons 43–44

Cells are very organized, and each structure in a cell has a specific purpose. You can think of a cell as an organization such as a city. Cities use multiple departments to accomplish tasks. In the same way a mayor's office manages and regulates the city, the chromosomes and DNA molecules of the nucleus regulate the work of the cell. A city also needs departments for transportation, waste management, communication, and various other tasks.

Your task is to prepare a skit comparing a cell and its internal structures to a city, business, country, factory, school, or any other organization that uses multiple departments to function.

What to do

1. Use the planning page in the Activity Manual to decide what organization your cell will be compared to.
2. Decide what part of the organization each part of the cell will represent. Research the parts of the cells as needed to gather information to aid you in your decisions.
3. Prepare a short (five-minute) skit about a day in the organization. Each department (cell structure) should be included in the skit. The skit should demonstrate the jobs of each cell structure.
4. Prepare large name cards listing each cell structure and the department it represents for the organization.
5. Present your skit to an audience.

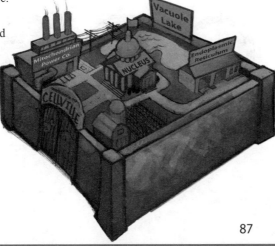

Objectives
- Correlate the function of cell structure to another organization
- Write and present a skit to compare a cell to an organization

Materials
- paper for name cards

Introduction

Note: Two lesson days are allotted for this activity. On the first day, introduce the activity, set guidelines, and allow students to begin planning. Use the second lesson day for the presentation of the skits.

➤ **The nucleus of a cell contains the DNA that determines the functions of the other parts of the cell. If we compare a cell to a school, what person would be the nucleus?** *{Possible answers: principal or administrator}*

➤ **What is the function of ribosomes in a cell?** *{They carry out instructions from the DNA and make the proteins that the cell needs.}*

➤ **Which people would be the ribosomes?** *{teachers}* **Why?** *{They carry out the instructions of the principal or administrator.}*

Purpose for Reading:
Student Text page 87
Activity Manual page 57

➤ Which group or organization does the text use to illustrate the parts of a cell and their functions?
➤ How could you use a skit to show the comparison of a cell to an organization?

Activity Manual
Explorations—page 57

Assessment
Rubric—TE pages A50–A53
Select one of the prepared rubrics, or design a rubric to include your chosen criteria.

Teacher Helps: There are no right or wrong answers to this activity. The purpose is to help students associate the cell's functions with everyday life. Even at a cellular level, God has a very purposeful design.

This activity can be done by science groups, or the entire class could work together.

Lesson 45

Student Text pages 88–89

Objectives
- Describe the process of cell division—both mitosis and meiosis
- Recognize when mitosis occurs and when meiosis occurs

Vocabulary
cell division
mitosis
sexual reproduction
meiosis

Introduction

➤ What are the five characteristics of living things? {*Living things grow and develop, reproduce, respond to their environments, use energy, and are made of cells.*}

➤ We have learned about the structure and function of cells. Now we will see some ways that cells and organisms reproduce.

Purpose for reading:
Student Text pages 88–89

➤ How do all cells reproduce?
➤ How many chromosomes are in reproductive cells?

Discussion:
pages 88–89

➤ How does the body replace dead skin cells? {*Living cells reproduce through cell division.*}
➤ What happens first when a cell gets ready to reproduce itself? {*It duplicates the chromosomes in the nucleus.*}
➤ What are two types of cell division? {*mitosis and meiosis*}
➤ If one cell undergoes mitosis, how many new cells will form? {*two*}
➤ Will these new cells be identical to the original cell? {*yes*}
➤ What type of cell division forms the reproductive cells needed to make a new life? {*meiosis*}

Reproduction of Cells

Single-celled organisms and most cells of multicellular organisms must be able to reproduce in order for the organism to survive. Imagine what would happen if the skin cells in your body were not able to reproduce themselves. As your skin cells died and were washed or rubbed away, there would be no new skin cells to replace them. Before long all of your skin would be gone! God designed cells and organisms to replace themselves through reproduction.

An individual cell reproduces itself by dividing into two cells through a process called **cell division**. The idea that one cell divides into two seems surprisingly simple. But the process is actually very complicated. First each chromosome is copied. Then the process of **mitosis** (my TOH sis) begins. Mitosis is a step-by-step process that ensures that the two new cells will be the same as the original, or parent, cell. As the cell divides, the chromosome pairs separate and move to opposite ends of the cell. Later new nuclei form and the cell splits into two new, identical cells.

Did you know that your body replaces the entire outer layer of your skin about once a month? Stomach lining cells last only a few days. Some cells, however, need to last a lifetime. Your heart muscle cells and most of your nerve cells never reproduce. You were born with all the heart and nerve cells you will ever have.

Mitosis

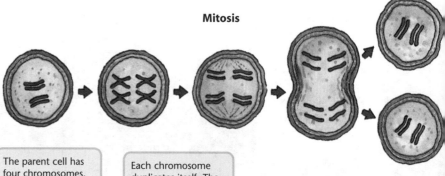

The parent cell has four chromosomes.

Each chromosome duplicates itself. The chromosomes move to opposite ends of the cell.

Two new daughter cells with exactly the same chromosomes as the parent cell are produced.

88

SCIENCE BACKGROUND

Mitosis—The new cells that form during mitosis are identical to the parent cell unless something goes wrong during cell division.

Meiosis—During meiosis, the chromosomes combine differently when the cell divides for the first time. Those chromosomes are not duplicated when the new cells divide for the second time. This leaves each of the four new cells with half of the number of chromosomes that were present in the original cell. Because reproductive cells form in this manner, each new organism is a unique combination of its father and mother.

Other forms of reproduction—Some cells reproduce in other ways. For example, some one-celled organisms divide into two organisms through binary fission. Reproduction is discussed further in Chapter 12, *Plant and Animal Reproduction*.

98 Lesson 45

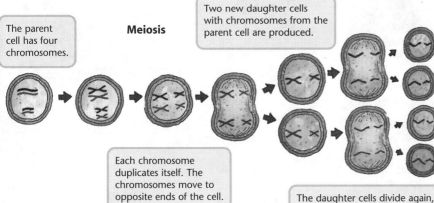

- The parent cell has four chromosomes.
- Meiosis
- Two new daughter cells with chromosomes from the parent cell are produced.
- Each chromosome duplicates itself. The chromosomes move to opposite ends of the cell.
- The daughter cells divide again, producing four cells with only two chromosomes each.

Reproduction of Multicellular Organisms

Think about the growth of a young puppy. Every day many of the puppy's cells multiply through the cell division process of mitosis. These additional cells cause the puppy's body to grow. Soon she becomes an "adult" dog. Cell division causes one puppy to grow and mature. But how do we get more puppies? More than only new cells is needed to get new puppies.

The process of creating new life using cells from male and female organisms is called **sexual reproduction**. The cells needed for this kind of reproduction are very different from other cells. Each cell contains only half the number of chromosomes as the parent cells. These special reproductive cells form through a process of cell division called **meiosis** (my OH sis).

Instead of simply dividing once as in mitosis, cells that undergo meiosis divide a second time. When they divide the second time, the chromosomes do not duplicate. Each reproductive cell ends up with only half as many chromosomes as the parent cell. Some of these reproductive cells are male and some are female. When a male reproductive cell and a female reproductive cell join, the new cell will again have the proper number of chromosomes. Once a new cell is formed, it continues to grow and multiply through the process of mitosis.

1. What process ensures that two new cells will be the same as the parent cell?
2. Name the process by which special reproductive cells form.

89

Discussion:
pages 88–89

- If one cell undergoes meiosis, how many new cells will be produced? {four} Why? {Cells that go through meiosis divide twice.}
- Are these four cells identical to the parent cell? {no} Why? {They have only half the number of chromosomes that the parent cell had.}
- What is the main difference between mitosis and meiosis? {In mitosis, chromosomes duplicate once and the cell divides once. In meiosis, chromosomes duplicate once but the cell divides twice.}
- What forms when a male reproductive cell and a female reproductive cell combine? {a new cell}
- How many chromosomes will this new cell have? {the same number as the original parent cells}
- Will the new cell be exactly the same as either parent cell? {no} Why? {The new cell is a combination of chromosomes from two different parent cells.}
- After a new cell is formed, how does that cell grow and multiply? {through mitosis}

Refer the student to *Fantastic Facts*.

- What are some parts of your body that replace cells frequently? {Possible answers: outer layer of skin, stomach lining}
- Why do you think keeping your heart healthy is important? {Possible answer: Special care must be taken to not damage or destroy heart cells, since they are not reproduced.}

As special care is given to the physical heart, care should also be given to the spiritual heart. Christians should desire to have a clean heart and a right attitude (Ps. 51:10). [BAT: 4b Purity]

Answers
1. mitosis
2. meiosis

The explanation of mitosis and meiosis is simplified purposely. This is an introduction to the processes that will be more thoroughly explained in later grades.

mitosis—Sixth-grade students should understand that in mitosis, the chromosomes duplicate once and the cell divides once. This process produces two cells identical to the parent cell. Mitosis helps multicellular organisms grow, as old cells are replaced with new cells.

meiosis—In meiosis, the chromosomes duplicate once and the original cell divides twice. The process of meiosis produces four reproductive cells.

Remembering the difference—To help the student remember the difference, tell him that the **t** in mitosis can remind him that the mitosis cell division ends with only **t**wo new cells.

Lesson 46

Student Text page 90
Activity Manual page 58

Objectives
- Distinguish groups according to chosen criteria
- Complete a classification chart

Vocabulary
criteria

Materials
- transparency of TE page A24, *Flow Chart*
- See Student Text page

Introduction

▶ Systems of classification are all around us. In a library, the books are classified as fiction or nonfiction. Most vehicles can be classified as cars or trucks.

Complete the transparency entitled *Flow Chart* with answers from the following discussion.

▶ Think about how you classify the clothing in your bedroom.

Into which two large groups can you divide all your clothing? {clean and dirty}

Into which two groups can you divide your clean clothing? {folded and hanging}

How do you sort your dirty laundry to be washed? {light clothes or dark clothes}

Add more boxes to the chart as needed. The sections of the chart may not be equal. Additional answers: Folded clothing is kept *in drawers* or *on shelves*. Hanging clothing *needs ironing* or *does not need ironing*.

Point out the organization of the chart and the way information is added with short answers.

Purpose for reading:
Student Text page 90
Activity Manual page 58

▶ What is another name for a standard used to decide how items are classified?

Procedure

Guide the student as he chooses criteria for classifying his pasta. Challenge the student to divide his pasta into many levels.

Activity Manual
Activity—page 58

Assessment

Rubric—TE pages A50–A53
Select one of the prepared rubrics, or design a rubric to include your chosen criteria.

Classifying

Every day we use systems to classify things. Often charts, graphs, or tables are used to illustrate or show systems of classification. In this activity you will decide the *criteria*, or standards, for classifying types of pasta.

Procedure

1. List your materials in your Activity Manual.
2. Write *Pasta* in the blue box at the top of the web.
3. Examine and sort your pasta into two groups. You must decide the criteria to use to make your groups. Remember that members of one group should not have the main feature of the other group.
4. Record the two groups in the yellow boxes of the web.
5. Continue dividing each group of pasta into groups according to your chosen criteria. Continue the web using your decisions.

Materials:
8 or more varieties of uncooked pasta
Activity Manual

Conclusions
- What other ways might you classify your pasta?
- Compare your chart to the chart of another science group. Were the same criteria used?

Follow-up
- Take a poll to find your classmates' favorite types of pasta. Display the results using a graph.

SCIENCE PROCESS SKILLS

Classifying

▶ What criteria did you choose to begin your classification system?

▶ Does your chart look the same as the chart from another group? Why?

▶ How would your chart be different if your criteria were based on the uses of the pasta in recipes?

▶ What do you think a scientist does if he discovers an organism that fits into more than one category? {Elicit that he may consult other scientists, and together they will make the best choice they can.}

▶ Sometimes names or the organization of whole categories change because of new organisms.

 Try to use a wide variety of pasta. Including one or more colored pastas will help the student see that some things do not always fit the specified categories.

Classification

Putting organisms with similar characteristics into groups is called **classification.** Classifying is not an exact science, but grouping and ordering organisms help scientists study the common traits of a group. As God has allowed man to learn more about various organisms, man has changed his ideas about many classifications. When the Swedish scientist Carolus Linnaeus (lih NEE uhs) originally proposed his method of classification, there were only two broad categories. He called these categories kingdoms—Animalae and Plantae. Currently, most scientists identify six kingdoms.

1. Eubacteria (YOO bak TEER ee uh)
2. Archaebacteria (AR kee bak TEER ee uh)
3. Protista (pruh TIST uh)
4. Fungi (FUN jye)
5. Plantae (PLAN tee)
6. Animalia (an uh MAY lee uh)

The Living Kingdoms
Kingdom Eubacteria

The organisms in this kingdom are called **bacteria.** These microscopic organisms are almost everywhere. On your desk right now there are probably several thousand bacterial cells. In fact, bacteria are the smallest living things known to man. Though they are unicellular, bacteria tend to live in groups called **colonies.** Unlike most other organisms, bacteria do not have well-defined nuclei. The DNA usually floats in the cytoplasm. Bacteria also do not have some of the organelles that are part of most cells.

Though you cannot see bacteria, you probably have felt the results of them. Sometimes a colony of bacteria takes up residence in a person's throat or ear. The bacteria irritate the sensitive tissue in the throat and ear and can cause a great deal of pain.

Not all bacteria species are bad. Most are harmless, and some can actually be helpful. Your intestines contain bacteria that help digest your food. Some food items, such as yogurt, contain bacteria that give them a unique taste. Bacteria can grow rapidly as long as they have warmth, water, a food source, and room to grow.

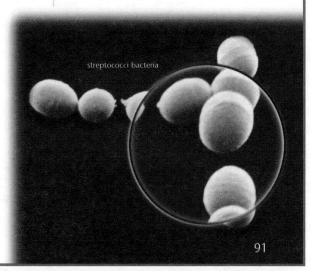
streptococci bacteria

Student Text pages 91–95
Activity Manual pages 59–61

Lesson 47

Objectives
- Name the six kingdoms
- Identify characteristics of each kingdom
- Explain how man is similar to and yet different from other living organisms

Vocabulary
classification   algae
bacteria   leaven
colony   photosynthesis
protozoan

Introduction

Note: The *Purpose for Reading* prepares and instructs the student to read all the pages in the lesson. Because the concepts and terms in this lesson may be unfamiliar to most students, the material is divided into one- or two-page sections. Each section begins with *Purpose for Reading.*

➤ **If you were trying to classify mammals into sea mammals and land mammals, where would you put a sea lion? It lives on land, but it also spends a great deal of time in the sea.**

Although classifying is not always easy, scientists have developed a reliable system.

Purpose for reading:
Student Text page 91

➤ How many kingdoms are there?
➤ What size are bacteria?

Discussion:
page 91

➤ **What is classification?** *{putting things with similar characteristics into groups}*
➤ **What name is given to the broadest category of classification?** *{kingdom}*
➤ **What are the names of the six kingdoms?** *{Eubacteria, Archaebacteria, Protista, Fungi, Plantae, Animalia}*
➤ **Where would you find bacteria?** *{almost everywhere}*
➤ **Bacteria are unicellular. What does this mean?** *{They are made of only one cell.}*
➤ **What is unusual about a bacteria cell?** *{Possible answers: A bacteria cell does not have a well-defined nucleus. The DNA floats in the cytoplasm. It does not have some of the organelles that are part of most cells.}*
➤ **How small are bacteria?** *{microscopic}*
➤ **What is a group of bacteria called?** *{a colony}*
💡 **Give some examples of when bacteria may cause you pain?** *{Possible answers: sore throat, earache}*
➤ **What are some helpful bacteria?** *{Possible answers: bacteria that help digest food, bacteria in yogurt}*
➤ **What is required for bacteria to grow?** *{a food source, water, warmth, and room to grow}*

Purpose for reading:
Student Text pages 92–93

➤ What is the most remarkable feature of Archaebacteria?

➤ What are two major kinds of Protista?

➤ What kind of fungus does a baker use?

Discussion:
pages 92–93

➤ Why did scientists divide the bacteria-like organisms into two kingdoms? *{The organisms were so different that each kind needed its own classification.}*

➤ What are some characteristics of Archaebacteria that are different from those of Eubacteria? *{Possible answers: Archaebacteria have a unique chromosome structure. They have cell walls that can withstand the extreme environments where they live. Some can survive without oxygen.}*

💡 What is another term for a sulphur spring? *{Possible answer: hot spring}*

💡 Do you think you would find Archaebacteria in mud pots? Why? *{Yes; mud pots are hot springs with mud in them.}*

➤ How are Archaebacteria similar to Eubacteria? *{Possible answers: They both have no true nuclei and do not have all the organelles that are in most cells.}*

📖 How do the Archaebacteria show God's design? *{Although their environment is harsh, they have the perfect design for the environment in which they live.}*

➤ What are some environments where you might find Archaebacteria? *{Possible answers: sulfur springs, salt lakes}*

💡 What is meant by "concentrated saline level"? *{very salty}*

💡 What other body of water might also have Archaebacteria growing in it? *{the Great Salt Lake}*

➤ What kinds of organisms are included in the kingdom Protista? *{all unicellular organisms that do not fit into any other category}*

➤ What are two groups in the kingdom Protista? *{protozoans and algae}*

➤ What does a paramecium use to move around? *{hairlike structures called cilia}*

➤ How does an amoeba eat its food? *{It extends its cell membrane to surround the food and engulfs it.}*

Kingdom Archaebacteria

Archaebacteria are unicellular organisms similar to "normal" bacteria. In fact, Eubacteria and Archaebacteria used to be classified together. But as scientists have discovered more and more about bacteria, they have concluded that some of these organisms are so distinctive that they need a classification kingdom of their own. Archaebacteria have a unique chromosome structure. They also have cell walls that are specially designed for the extreme environments where they live. Some archaebacteria are able to survive without oxygen.

Archaebacteria live in conditions that are poisonous to other living things. To most organisms the hot and acidic sulfur springs near volcanoes are deadly, but scientists have found archaebacteria growing in these springs. Archaebacteria also survive in places such as the Dead Sea, where the concentrated saline levels cause most organisms to die.

Like Eubacteria, Archaebacteria do not have true nuclei. They also may not have some of the organelles that typically make up a cell. However, God has given them a design perfectly suited to the harsh environments in which they live.

Kingdom Protista

Kingdom Protista includes all of the unicellular organisms that do not fit into any other category. These organisms have cell membranes and true nuclei. The organisms in this kingdom usually reproduce by cell division.

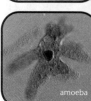
paramecium

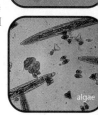
amoeba

Two kinds of organisms in this kingdom are protozoans and algae. The **protozoans** (PRO tuh ZOH uhnz) can move around and often live in water—especially pond water. One common protozoan, a *paramecium* (PEAR uh MEE see uhm), uses tiny hairlike structures called *cilia* (SILL ee uh) to propel it through the water. *Amoebas* (uh MEE buhz) are very unusual protozoans because they move around by constantly changing their cell shapes. Amoebas eat by extending their cell membranes and surrounding and engulfing food.

archaebacteria in hot springs

92

SCIENCE BACKGROUND

Hydrothermal vents—The bacteria-like organisms mentioned in the opening of this chapter are a type of Archaebacteria. Many scientists believe that the Archaebacteria are the main organisms responsible for performing chemosynthesis.

Oxygen—No known organism other than Archaebacteria has the ability to live without oxygen.

Mushrooms—Frequently, mushrooms grow in a circle called a fairy ring. Once people believed that fairies would hold meetings at these rings. Today scientists know that an underground root system from one mushroom causes the mushroom to create a ring. As this mushroom grows, it sends out roots. These roots expand to cover the area of a circle. The boundary of the roots forms the perimeter of a circle. The original mushroom resides in the center of this circle. From the root system, other mushrooms grow. Eventually, the mushrooms in the center of the circle die out, but the mushrooms on the perimeter continue to live. As they grow, the perimeter expands. Fairy rings usually reach a diameter of about six feet. A few have diameters of several hundred feet, although these make distorted circles.

Algae (AL jee) are not as mobile as the protozoans. Algae often grow in colonies that can be seen easily without the aid of a microscope. You have probably seen algae growing in ponds or fish tanks. Like plants, algae have chlorophyll in their cells, so they are able to use sunlight for energy.

Kingdom Fungi

Much variety exists among members of this kingdom. Some species of fungi are unicellular, and some exist in colonies. The cells of most organisms in the kingdom Fungi do not have cell walls. Unlike algae, a fungus cannot make its own food.

The most common example of a fungus is a mushroom. A mushroom is a complex fungus that lives off of

mushroom colony

mold on cheese

decaying plant and animal matter. The mold that grows on stale bread is another well-known example of a fungus. Fungus cells are everywhere and can reproduce rapidly in moist, warm conditions.

You may have seen your mother put fungus in homemade bread. Yeast is a special fungus that grows rapidly in bread dough. Living yeast cells produce a gas called carbon dioxide that becomes trapped in the dough. Bread rises when these small pockets of gas expand.

Most fungus species are harmless, but some, such as poisonous mushroom species, are very dangerous. Others, such as the type that causes athlete's foot, can affect humans and cause great discomfort.

Science and the BIBLE

The first mention of yeast, or **leaven**, in the Bible is in Exodus 12:15. During the first Passover feast, God commanded the children of Israel to sacrifice a lamb and eat unleavened bread, or bread without yeast. This bread was somewhat similar to the crackers we eat today. The Bible sometimes uses leaven as a picture of sin. In the New Testament, Paul warned the church in Corinth by saying, "Know ye not that a little leaven leaveneth the whole lump?" (I Cor. 5:6). The believers in that church were tolerating sin among their members. Like yeast that grows and makes bread rise, so a little sin will grow and affect other believers.

93

Direct a Demonstration
Observe yeast growth

Materials: yeast, sugar, warm water

Stir together 15 mL of yeast and 15 mL of sugar. Add 30 mL of warm water. Stir. Set aside and observe.

➤ **What gas causes bubbles to form?** {carbon dioxide}

Discussion:
pages 92–93

➤ **Why can we sometimes see algae even though they are unicellular?** {They live in colonies.}

➤ **How are algae similar to plants?** {They contain chlorophyll and are able to use sunlight for energy.}

💡 **Do you think that the term *colony* means the same when describing fungi as it does when describing bacteria?** {Answers will vary, but elicit that in either usage colony means a group of organisms living together.}

➤ **What common fungus might you find in your yard?** {a mushroom}

💡 **Why do people who are allergic to mold often have such a difficult time managing their allergy?** {Answers may vary but should include that fungus cells are everywhere.}

➤ **What conditions are best for a fungus to reproduce?** {warm, moist conditions}

💡 **Why does putting food in a refrigerator slow down the growth of a fungus?** {A refrigerator is not a warm environment.}

💡 **Do you think fungus would grow well in a desert?** {Answers may vary but should include that the dry conditions would discourage growth.}

➤ **What fungus is used to make bread?** {yeast}

➤ **How does yeast cause bread dough to rise?** {As yeast grows it gives off carbon dioxide, which forms into bubbles when trapped in the dough. The bread rises as the gas bubbles expand.}

➤ **Give an example of a fungus that causes discomfort to humans.** {Possible answers: athlete's foot, ringworm}

➤ **Fungi require certain conditions to grow. What do you think you can do to avoid getting athlete's foot?** {Possible answer: Keep your feet dry.}

📖 Refer the student to *Science and the Bible*.

➤ **What name does the Bible use for yeast?** {leaven}

➤ **What does leaven often represent in the Bible?** {sin}

💡 **Why do you think God commanded the children of Israel to eat unleavened bread at Passover?** {It showed God's desire for them to live clean, pure lives without sin.} [BAT: 4b Purity]

➤ **What did Paul mean by "a little leaven leaveneth the whole lump"?** {Just as a little yeast can cause a whole loaf of bread to rise, a little sin is able to grow and affect others.}

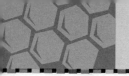

Purpose for reading:
Student Text pages 94–95

➤ Why are cell walls important to plants?
➤ How is man different from animals?

Discussion:
pages 94–95

 What important role has God designed for plants? *{to provide food}*

➤ What do plants have that helps them make their own food? *{chloroplasts that contain chlorophyll}*

➤ What process do plants use to produce food? *{photosynthesis}*

➤ What usable source of energy do plants produce through photosynthesis? *{sugar}*

➤ How do humans and animals benefit from photosynthesis? *{When they eat plants, their bodies use the sugar produced by those plants.}*

➤ In addition to food, what benefit does photosynthesis provide? *{oxygen}*

➤ Why is a cell wall more important to plants than it would be to animals? *{Animals often have skeletons to support their structure. Plants do not, so they need firmer cell boundaries for support.}*

➤ Which kinds of plants have very thick cell walls? *{shrubs and trees}*

💡 Why do these plants need very thick cell walls? *{Their sizes put a lot of pressure on the plants.}*

➤ What are some types of tissues in plants? *{Possible answers: bark, leaves, roots}*

➤ Are animals unicellular or multicellular? *{multicellular}*

➤ Although animals do not have cell walls like plants, many of them have another means of support. What is it? *{skeletal system}*

➤ What is another way that animals are different from plants? *{Animals cannot make their own food.}*

Kingdom Plantae

God has designed organisms in the plant kingdom to provide a crucial need for all living things—food. This group of multicellular organisms has chloroplasts as part of the cell structure. The chlorophyll inside the chloroplasts helps a plant make its own food. You may have noticed that most plants require sunlight to stay alive. Plants use a process called **photosynthesis** (FOH toh SIN thuh sis) to convert the energy in sunlight into a usable source of energy, sugar. Animals and humans that eat the plant also use this sugar.

Through photosynthesis plants also release oxygen into the atmosphere. Almost all living things require oxygen to stay alive. Without plants, there would not be enough oxygen for living things to survive on the earth.

In addition to cell membranes, plants have cell walls. This firmer cell boundary is very important to the plant. Some small organisms can survive without much support. Many larger organisms, such as animals, have internal or external skeletons that provide support. However, plants do not have skeletons, and not all plants are small. In fact, some plants are quite large. Therefore, plants need a way to support their structures. Cell walls give the support that plants need. Some plants, especially large plants such as shrubs and trees, have very thick cell walls.

Plants also have various tissues, such as bark, leaves, and roots, that help keep them alive. These tissues come in a wide range of shapes and sizes. God has provided a variety of plants for our use and enjoyment.

plant cells

94

Direct a DEMONSTRATION
Demonstrate cell walls

Materials: large marshmallows, empty toilet tissue tube, ruler, weight (such as a book)

In this demonstration each marshmallow is a cell in the stem of a plant. The weight is the weight of the branches, leaves, and blossoms the stem must hold.

Stack six or seven marshmallows. Measure and record their height. Place the weight on the marshmallows. Measure and record the height of the stack with the weight in place.

Place the same number of different marshmallows in the tissue tube. Measure and record the height of the tube.

Place the weight on the tube. Measure and record the height of the tube with the weight in place.

➤ What effect did the weight have on each stack of marshmallows? *{The weight squished the first stack but had little or no effect on the stack in the tube.}*

➤ What does the tissue tube represent? *{the cell wall}*

➤ How does the cell wall help the cell? *{It gives strength and support.}*

 The word *photosynthesis* comes from the Greek words *photo*, meaning "light" or "radiant energy" and *synthesis*, meaning "to put together."

Kingdom Animalia

You are probably most familiar with the animal kingdom. All animals are multicellular. Animal cells do not have cell walls, but many animals have skeletal systems to support their tissues and organs.

Unlike plants, animals cannot manufacture their own food. They are dependent on the ability to move to find and gather food for their needs. Their nervous systems help them detect food as well as respond to other conditions in their environment.

The structure, size, and overall characteristics of organisms in the kingdom Animalia vary greatly. Some animals, such as jellyfish and flatworms, use only a few organs and tissues to do small tasks such as eating and removing waste. Other animals, such as mammals, require complex body systems to live.

Some people ask the question "Are humans animals?" Physically, human bodies have many traits common to mammals. But man is in a class by himself. God created man in His own image (Gen. 1:27) and gave him specific instructions about how to live and rule over the earth. As a result of Adam's disobedience, man inherited a sinful nature. Only by faith in Jesus Christ can man gain a new nature and have eternal life. Some scientists believe that man came from animals through the process of evolution. But the Bible tells us that God created animals and humans separately by direct acts. While man may be physically similar to animals, the Bible teaches us that man is God's special creation.

Quick Check
1. Name the six kingdoms of organisms.
2. Give an example of a fungus that humans use.
3. What process do plants use to make sugar?
4. How is man different from animals?

Discussion:
pages 94–95

➤ What can animals do to obtain food? *{They move to find and gather food.}*

➤ What helps animals detect food? *{their nervous systems}*

➤ What else do their nervous systems help them do? *{respond to other conditions in their environments}*

➤ What are some animals that require only a few organs to live? *{Possible answers: jellyfish, flatworms}*

➤ How are humans like animals? *{Many physical traits of humans are common to all mammals.}*

Even with these common traits, scientists are not certain that man is physically similar to mammals.

➤ How are humans different from animals? *{Man is created in the image of God.}* [BAT: 3a Self-concept]

➤ How did we inherit a sinful nature? *{by Adam's disobedience}*

➤ What has God provided for us? *{eternal life through faith in Jesus Christ}* [Bible Promise: E. Christ as Sacrifice]

Answers
1. Eubacteria, Archaebacteria, Protista, Fungi, Plantae, Animalia
2. Possible answers: mushrooms, yeast, mold
3. photosynthesis
4. Man is different from animals because he is created in the image of God. God gave man specific directions about how to live and rule over the earth. Man inherited a sinful nature and can receive a new nature only through Jesus Christ.

Activity Manual

Bible Integration—page 59

This page illustrates that joy comes by applying God's criteria and promises.

Reinforcement—page 60

Study Guide—page 61

This page reviews Lessons 45 and 47.

Assessment

Test Packet—Quiz 4-B

The quiz may be given any time after completion of this lesson.

Direct an Activity

Make and display a 3 X 6 grid as shown.

Plantae					
Animalia					

Randomly choose and write five letters at the top of each of the five empty columns.

Give the student one minute to complete the chart with names of plants and animals that start with the corresponding letters. At the end of the time, discuss the answers.

Lesson 48

Student Text pages 96–97
Activity Manual page 62

Objectives

- Recognize that Carolus Linnaeus was responsible for the method of classification that we use
- List the levels of the classification system from largest to smallest
- Compare the common names and scientific names of organisms
- Write a scientific name properly

Vocabulary

common name
kingdom
phylum
class
order
family
genus
species
scientific name

Materials

- picture of a mountain lion and/or bison

Introduction

Show a picture of a mountain lion or bison.

➤ **What do we call this animal?** *{Possible answers: mountain lion, cougar, panther; bison, buffalo}*

➤ **Why is this animal called by different names?** *{Accept reasonable answers.}*

➤ **Do you think this might cause confusion?** *{yes}*

Scientists also had this problem. Because of the confusion, scientists developed a system to standardize the names of living organisms.

Purpose for reading:
Student Text pages 96–97

➤ Who helped classify the different organisms into groups?

➤ What two levels of classification are part of the scientific name?

Naming Organisms

Your name is what makes you unique—or so you think. Many people have very common first and last names. If you look in the telephone directory under "Davis," you may find a hundred people with that last name. Sometimes people go by their middle names or make up new first names for themselves. Naming and identifying people can be a very confusing task.

Identifying and naming organisms can be confusing too. Consider the name *spider*. Most of us know what a spider is, but the name is too general. The word *spider* could refer to a tiny, harmless spider or a large, poisonous tarantula. Over the years, people have tried to be more specific as they name organisms, so names such as black widow, roadrunner, and bald eagle have become **common names** that are widely recognized. But common names still have their problems. Sometimes an organism may have more than one common name, or a common name may apply to more than one organism. Common names are also different in different languages.

Carolus Linnaeus (1707–1778) decided to do something about this confusion. He proposed an ordering system to help classify plants and animals according to common characteristics.

96

Though some of his original system has been changed, his ideas and much of his research are still used today to classify organisms. Currently scientists use seven levels of classification. Every living thing falls into one of the *kingdoms* discussed earlier. Each kingdom is divided into phyla (FYE luh—singular, *phylum*). Members of each phylum have similar characteristics that set them apart from members of the other phyla. The

Classification of the African Lion

Kingdom: Animalia
Phylum: Chordata
Class: Mammalia
Order: Carnivora
Family: Felidae
Genus: Panthera
Species: Leo

SCIENCE BACKGROUND

Linnaen classification—Because Carolus Linnaeus developed the basis of the modern system of classification, the system is called Linnaen classification.

LANGUAGE

Allow groups to try to make a sentence using the first letter of each of the seven levels of classification.

Example: **K**ate **p**oured **c**offee **o**n **F**ather's **g**ood **s**uit.

Allow groups to share their ideas. As a class, choose the best sentence and use it repeatedly to reinforce the classification system.

phyla are split into *classes,* and every class is separated into *orders.* Orders are broken down into *families,* families into *genera* (JUHN air uh—singular, *genus*), and genera into *species.*

Linnaeus used **scientific names** for each specific type of organism. These names are unique and are not attached to any other organism. We still use many of his ideas for scientific names today.

1. A scientific name is made up of *two names*. The first name is the *genus* name and the second name is the *species* name. For example, the scientific name for dogs is *Canis familiaris. Canis* is the genus name and *familiaris* is the species name.

2. The scientific name is in Latin. Linnaeus chose this language because it was widely known. Many common words come from ancient Latin terms. For example, *canis* (as in "canine") is the Latin term for "dog," and *famil* (as in "familiar") is Latin for "friendly."

Over the years scientists have established certain rules for writing scientific names. When the scientific name is written, it must be underlined and the genus name always capitalized. When typed, it is acceptable to italicize the scientific name. Here are some examples of correct and incorrect ways of doing this:

Correct	Incorrect
<u>Canis familiaris</u>	canis familiaris
Canis familiaris	Canis Familiaris

The classification system that Linnaeus invented solved many problems for scientists. Although it is not perfect, this system makes learning about living things much easier.

Looking at the way living things are classified should help us appreciate the orderliness of God's creation. Not only are there many types of living things, but also there are many characteristics that group and set apart those living things. God's design may not always be understandable to man, but we can be assured that every creature is exactly how God planned it to be.

1. List the classification system from the largest group to the smallest.
2. What parts of the classification system make up the scientific name of an organism?

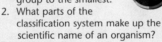

Canis familiaris

Carolus Linnaeus believed that through nature man could learn more about God. His system of classification was a way to demonstrate the orderliness found in God's creation.

In addition to being a botanist, Linnaeus was a physician and a teacher. He arranged for many of his students to travel on expeditions to faraway lands.

Linnaeus constantly reworked and expanded the first edition of his classification of living things, *Systema Natura*, as additional plant and animal specimens were sent to him.

Direct an Activity

Materials: Prepared cards with the scientific names of common animals written or typed either correctly or incorrectly according to the guidelines on Student Text page 97.

Distribute one card to each student. Each student decides if his card is correct or incorrect. If the card is incorrect, the student must tell how to correct the way the name is written.

Discussion:
pages 96–97

- What difficulties might scientists have if they use common names when identifying organisms? *{Different scientists may call the same organism by different names.}*

- Name some living things that you know more than one name for? *{Possible answers: Dianthus plants are sometimes called pinks. Chipmunks may be called ground squirrels.}*

➤ What scientist proposed a system of classification for plants and animals? *{Carolus Linnaeus}*

➤ What are the seven levels of classification of living organisms? *{kingdom, phylum, class, order, family, genus, species}*

Refer the student to *Classification of the African Lion* on Student Text page 96.

➤ When are animals most similar—in the same genus or in the same kingdom? *{in the same genus}*

➤ What two parts of the classification system make up the scientific name? *{genus and species}*

➤ What language is used to write scientific names? *{Latin}*

➤ Why was Latin chosen? *{Possible answers include that many words have their roots (or beginnings) in Latin, and that Latin was a language that all scientists could use no matter what their native tongues were.}*

➤ How should a scientific name be written? *{The genus is first and then the species. All of it is either underlined or italicized. The genus is capitalized.}*

- What does classification show us about God's creation? *{Possible answer: God's plan for orderliness, variety, and design in His creation}*

Answers
1. kingdom, phylum, class, order, family, genus, species
2. genus and species

Activity Manual
Reinforcement—page 62

Canis unfamiliaris

Lesson 49

Student Text page 98
Activity Manual pages 63–64

Objectives
- Recall concepts and terms from Chapter 4
- Apply knowledge to everyday situations

Introduction

Material for the Chapter 4 test will be taken from Activity Manual pages 56, 61, and 63–64. You may review any or all of the material during the lesson.

You may choose to review Chapter 4 by playing "Kingdom to Species" or a game from the *Game Bank* on TE pages A56–A57.

Diving Deep into Science
Questions similar to these may appear on the test.

Answer the Questions
Information relating to the questions can be found on the following Student Text pages:
1. page 93
2. page 81
3. pages 96–97

Solve the Problem
In order to solve the problem, the student must apply material he has learned. The answer for this Solve the Problem is based on the material on Student Text pages 96–97. The student should attempt the problem independently. Answers will vary and may be discussed.

Activity Manual
Study Guide—pages 63–64
These pages review some of the information from Chapter 4.

Assessment
Test Packet—Chapter 4 Test
The introductory lesson for Chapter 5 has been shortened so that it may be taught following the Chapter 4 test.

Diving Deep into Science

Answer the Questions

1. Why does putting food in the refrigerator help slow down the growth of a fungus?
   Fungus grows best in warm, moist areas. The refrigerator slows down fungus growth by keeping the food cool.

2. Why are the conclusions made by Schwann and Schleiden called a theory?
   Their conclusions are called a theory because although the results of experimentation have been consistent over the years, there is no way to know for certain that no exceptions or additions to the conclusions can be made.

3. Which animals are more alike—those only in the same class or those in the same family? Why?
   Animals in families are more alike because they have more characteristics that are similar than animals in classes. For example, all dogs (but only dogs) are in one family, but all dogs, whales, and a large variety of other animals are in the same class.

Solve the Problem

Your cousin in Wyoming keeps writing you about the problems his family is having with ground squirrels in their yard. You write back with suggestions about how you controlled the problem in your yard in North Carolina. Your cousin is not impressed. He says that you do not have "ground squirrels" in North Carolina. What animal do you think each of you is talking about? How could you make sure that you are talking about the same animal?
Answers will vary. Your cousin is probably talking about prairie dogs. You are probably talking about chipmunks. If you each would give the scientific name for the animal, you would be able to determine if the "ground squirrels" are the same animal.

98

Review Game
Kingdom to Species

Prepare the classification of several animals, similar to the one on Student Text page 96. List each level of classification on separate sentence strips. Place one animal's classification strips face down. Each time a student answers a review question correctly, he chooses one of the classification strips. Once all of the classification strips for that animal have been chosen, the students who have the strips should arrange their strips in the correct order from kingdom to species.

Read the scientific name (genus and species) aloud. Allow students to guess the common name of the animal. Repeat with other animals as time allows. Students could be divided into teams with points given for correctly organizing the classification strips and identifying the scientific name.

108   Lesson 49

Student Text page 99
Activity Manual page 65

Lesson 50

5
Animal Classification

GREAT & MIGHTY Things

Fossil after fossil of a large, bony fish with unusual fins had been found. Because no one had ever seen this fish alive, many scientists thought that it was extinct. But in 1938, much to everyone's surprise, a living coelacanth (SEE luh kanth) was caught near the Comoros Islands. Since then, many of these dark blue fish have been caught and studied. Coelacanths are not the only animals that have surprised scientists. For more than 100 years, people had also thought that another fish, the robust redhorse, was extinct. However, in the 1990s scientists noticed that many of these fish live in the Oconee River in Georgia. Even though man mistakenly thought these animals were extinct, God knew about them. He cares for all animals in His creation whether man has discovered them or not.

SCIENCE BACKGROUND

Coelacanth—These brown or bluish-colored fish can grow to 1.5 meters in length. They are found in the western Indian Ocean.

Comoros Islands—This group of islands lies in the Indian Ocean between the coast of Mozambique, Africa, and the large island of Madagascar.

Robust redhorse fish—Urbanization and the introduction of a catfish from Mexico had an effect on the decline of this fish. It had not been sighted for about 100 years but was rediscovered in 1991. The robust redhorse can live up to 25 years and can grow to be 20 pounds.

Objectives

- Demonstrate knowledge of concepts taught in Chapter 4
- Recognize that man is unaware of many parts of God's creation

Introduction

Note: This introductory lesson has been shortened so that it may be taught following the Chapter 4 test.

➤ If you have ever played hide and seek, you know that one of the best ways to keep from getting caught is to continually move to places that have already been searched.

➤ Man sometimes assumes that because he cannot find something, it does not exist. But since man cannot be everywhere at all times, his knowledge is limited and sometimes flawed. Just because an organism is not in a particular location at the moment does not mean that it never was or will not be there again.

➤ God, however, is both all-knowing and everywhere at once. He knows about all of His creatures and cares for each of them. [Bible Promise: I. God as Master]

Chapter Outline

I. Invertebrates
   A. Sponges and Stinging Animals
   B. Mollusks
   C. Echinoderms
   D. Flatworms
   E. Roundworms
   F. Segmented worms
   G. Arthropods

II. Vertebrates
   A. Fish
   B. Amphibians
   C. Reptiles
   D. Birds
   E. Mammals
   F. Humans

Activity Manual

Preview—page 65

The purpose of this Preview page is to generate student interest and determine prior student knowledge. The concepts are not intended to be taught in this lesson.

Chapter 5 — Lesson 50 — 109

Lesson 51

Student Text pages 100–103
Activity Manual page 66

Objectives

- Recognize invertebrates and vertebrates as a broad way to distinguish animals
- Recognize that unique animal characteristics allow classification
- Describe the unique characteristics of the sponge, jellyfish, mollusk, and echinoderm groups

Vocabulary

invertebrate   nematocyst
vertebrate   mollusk

Materials

- pictures of a jellyfish, snail, octopus, and sea star

Introduction

Display pictures of the animals.

➤ What similarities do you see or know about these animals? *{Answers will vary.}*

➤ Do you think scientists classify these animals together? *{Answers will vary.}*

➤ What are some ways you would classify these animals? *{Answers will vary.}*

Purpose for reading:
Student Text pages 100–103

➤ Why do scientists classify animals?

➤ What is a distinctive feature of a mollusk?

Discussion:
pages 100–101

➤ About how many species of organisms live on Earth? *{10–15 million}*

➤ What are the seven levels of scientific classification used to group living organisms? *{kingdom, phylum, class, order, family, genus, species}*

➤ In the kingdom Animalia, what obvious characteristic do scientists use to classify animals? *{whether or not each animal has a backbone}*

➤ What are animals without backbones called? *{invertebrates}*

➤ What are animals with backbones called? *{vertebrates}*

There are perhaps as many as 10–15 million living organisms on Earth. Though animals make up only a small percentage of these organisms, the kingdom Animalia still consists of millions of species. Because there are so many animals, scientists further group them by their distinctive characteristics. One of the most obvious distinguishing characteristics of animals is whether or not they have backbones. With only a few exceptions, scientists can divide animals into **invertebrates,** animals without backbones, and **vertebrates,** animals with backbones.

Invertebrates

Even though we are more familiar with vertebrates, there are actually many more invertebrates than vertebrates. In fact, 95 percent of animals are invertebrates. Since the invertebrate group is quite large, scientists split invertebrates into smaller groups based on their unique characteristics.

Sponges and Stinging Animals

Sponges belong to the phylum of animals called *Porifera* (puh RIF uh ruh). Animals in this group catch their food in an unusual way. They sit on the ocean floor and pump water through their bodies. The water goes through tiny pores, or holes, in the outside of the sponge. When the water flows through the sponge, the sponge extracts nutrients and small organisms that it needs. Then the water is pushed out through the top of the sponge.

Like sponges, jellyfish are also classified by how they get their food. A jellyfish is neither jelly nor a fish. It is an aquatic animal that has a top that looks like a blob of petroleum jelly. But underneath that top are tentacles lined with tiny stinging organelles called **nematocysts** (nih MAT uh SISTZ). Jellyfish and other animals in the phylum *Cnidaria* (nye DAIR ee uh) use nematocysts to capture their food. These stinging organelles can paralyze any small, unsuspecting animal that brushes against them. Then the jellyfish can digest the paralyzed animal.

SCIENCE BACKGROUND

Coral reefs—Scientists study coral to help the reefs continue to exist. One way scientists study coral is by slicing open dead coral to study the rings inside. The rings inside a coral are much like the rings inside a tree. Not only can the age of the coral be determined by counting the rings, but the size of each ring shows the general environmental conditions of each year.

Invertebrate and vertebrate—The term *invertebrate* is not part of the Linnaen classification system. It is a characteristic of all the phyla except Chordata. The term *vertebrate* is part of the Linnaen classification system and is the name of a subphylum of Chordata.

SCIENCE MISCONCEPTIONS

Be sure that students understand that when the Student Text refers to 10–15 million living organisms, it is referring to 10–15 million species, not individuals.

The organisms pictured on Student Text pages 100–101 are from left to right: clown fish in a sea anemone, rainbow runners, sea nettles, jellyfish, crescent tail bigeyes, French angelfish, batfish, barrel sponge, bannerfish, and southern stingray.

110 Lesson 51

Though most jellyfish pose little serious danger to people, the jellyfish's nematocysts can leave painful welts on a swimmer's body. Only a few jellyfish are venomous enough to seriously harm humans.

Though they seem very different from jellyfish, sea anemones (uh NEM uh neez) are part of the same group. Instead of floating like jellyfish, sea anemones move slowly along the ocean floor. They were once mistaken for plants because of their "petals." These "petals," or tentacles, are equipped with nematocysts that poison any small prey that might come near. The anemone then draws the prey into its mouth.

Corals are also in the same group as jellyfish. Most corals anchor themselves to the ocean floor and wait for food to come within range of their tentacles. Usually when people think of coral, they think of coral reefs. Stony corals make limestone skeletons to protect their soft bodies. The coral reefs in warm ocean waters are a buildup of the dead skeletons of these animals.

The nematocysts of these Cnidarians work extremely quickly. In some cases they work in less than five seconds. Even the fastest of creatures that wanders into the waiting tentacles of these animals may become the next meal. Although these animals cannot pursue their prey, God has provided a creative mechanism for them to get food.

Science and History

The Great Barrier Reef is the largest coral reef system in the world. Made up of around 2,900 smaller reefs, this reef wraps about 2500 km (about 1,550 mi) around the northeastern Australian coast. This reef holds some of the most deadly wildlife in the world. The box jellyfish, the most venomous jellyfish in the world, lives there. The cone snail, which has a poisonous bite, lives there also. The Portuguese man-of-war, similar to a jellyfish, also lives within this massive reef. But deadliness does not mean ugliness. Many of the venomous creatures in the reef are brightly colored, as if to say, "Danger! I am poisonous!" God's design for the animals of the Great Barrier Reef benefits both the predator and the prey.

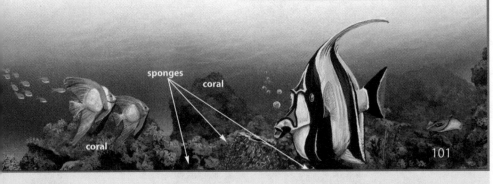

 LANGUAGE

Poriferous is an adjective meaning "to have tiny holes or openings." Animals with holes (e.g., sponges) belong to the phylum Porifera.

Nemato and nemat—Both prefixes mean "thread" and have both Latin and Greek roots. For example, *nematodes* are types of worms that have bodies shaped in long, threadlike cylinders.

➤ Which word on page 100 begins with the prefix *nemato*? {nematocysts}

Direct an Activity

Materials: natural sponge, man-made sponge, and magnifying glass

Allow the student to touch and examine the different sponges.

➤ How does the natural sponge feel? How does the man-made sponge feel?

➤ Use a magnifying glass to find the similarities and differences between the natural and man-made sponges.

The skeletons of natural sponges are made from silicon or soft and pliable spongin. Divers gather sponges from the bottom of such waters as the Gulf of Mexico, parts of the Mediterranean Sea, and the Pacific Ocean. People may use natural sponges for decorative painting or to wash vehicles.

Discussion:
pages 100–101

➤ **Is a sponge a vertebrate or an invertebrate?** {an invertebrate}

➤ **To what phylum do sponges belong?** {Porifera}

💡 **What is a *phylum*?** {the second highest level of classification (See TE pages 106–7.)}

➤ **How does a sponge get food?** {Water flows through tiny pores in the sponge. The sponge extracts the nutrients and small organisms that it needs and pushes the water out of its top.}

➤ **Which term describes the tiny stinging organelles used by some animals to capture food?** {nematocysts}

➤ **Animals in the phylum Cnidaria use nematocysts. Name some of these animals.** {Possible answers: jellyfish, sea anemones, coral}

➤ **Why were sea anemones often mistaken for plants?** {Their tentacles look like petals.}

➤ **What are coral reefs made of?** {the skeletons of dead stony coral}

💡 **Why is it important for animals in the phylum Cnidaria to have fast-acting nematocysts?** {Possible answer: Because these animals cannot pursue prey, it is important that prey not escape when it comes within their reach.}

📖 God has provided for His creatures. He tells us that just as He has provided for the animals, He will provide for us (Matt. 6:26) [Bible Promise: H. God as Father]

Refer the student to *Science and History*.

➤ **How does God's design benefit both the predator and the prey of the Great Barrier Reef?** {Accept any reasonable answer, but elicit that while bright colors attract prey, they can also warn predators that eating one of these organisms would be harmful.}

➤ **What is the name and location of the world's largest barrier reef?** {Great Barrier Reef; off the northeast coast of Australia}

➤ **What are some animals that live in the Great Barrier Reef?** {colorful fish, coral, cone snails, man-of-war jellyfish}

Discussion:
pages 102–3

➤ **Describe a mollusk.** *{A mollusk has a soft body and a mantle, a special part of the body that sometimes forms a shell.}*

➤ **Do all mollusks have shells?** *{no}*

➤ **What is a difference between a univalve and a bivalve?** *{Univalves have one shell. Bivalves have two shells connected to each other.}*

➤ **How do most bivalves move?** *{They use a muscular foot that extends out of the shell.}*

➤ **What is another way that some bivalves move?** *{Some clams gather water in their shells and squirt it out like a miniature jet propulsion system.}*

➤ **What does *gastropod* mean?** *{stomach footed}*

💡 **What do you think gastropods have in common?** *{They move on their stomachs.}*

➤ **What is a common gastropod that does not have a shell?** *{a slug}*

💡 **Why do you think slugs leave slimy paths?** *{Answers will vary, but elicit that the slime helps protect the slug's body as it slides across surfaces.}*

Refer the student to *Science and Math*.

➤ **What is the chambered nautilus famous for?** *{It is thought to have the most perfectly proportioned shell.}*

➤ **What mathematical pattern is evident in the size of the chambers of a nautilus?** *{the Fibonacci spiral}*

➤ **What are some other organisms in nature that demonstrate this pattern?** *{Possible answers: pine cones, snail shells, sunflowers}*

Mollusks

Mollusks are animals that have soft bodies and mantles, special parts of the body that sometimes form a shell. Snails, oysters, and clams are mollusks that have shells protecting their soft bodies. Some mollusks with shells are bivalves.

Bivalve means that the mollusk has two shells. Clams, mussels, and oysters are bivalves. Many bivalves protect themselves by hiding in the mud or sand. A bivalve usually moves slowly by using a muscular foot that extends out of the shell. A few clams, however, move quickly by gathering water into their shells and then squirting it out like a miniature jet propulsion system. Though most bivalves are small, some are quite large. The giant clam can measure as much as 1.2 m (4 ft) across and weigh 250 kg (550 lb).

bivalve

Other mollusks are gastropods, meaning "stomach footed." Many of these mollusks are univalves. *Univalve* means that the animal has only one shell. A snail is a univalve. Univalves take their shells with them as they move around. Some gastropods have no shells. You may have seen the slimy path left in your garden by one of these mollusks. A slug is a mollusk without a shell. Garden slugs eat leaves and can cause

The chambered nautilus is thought to have one of the most perfectly proportioned shells. As the animal grows, it builds a shell with increasingly larger chambers. The animal always lives in the largest chamber. The nautilus uses the other chambers to help control its ability to float. The size of the chambers of the shell fit closely to a mathematical pattern called the Fibonacci spiral. This pattern is based on a number sequence, the Fibonacci sequence, where each new number is obtained by adding the two previous numbers. For example, the sequence might be 1, 1, 2, 3, 5, 8, 13, and so on. It is a pattern that occurs in living organisms such as pinecones, snail shells, sunflowers, and, of course, the chambered nautilus. What a wonderful example of God's order and design!

102

Heliculture is the farming of snails. From the time of the Roman Empire until the present, man has considered snails an edible delicacy. This treat is still popular in some European countries, such as France. *Escargot* is the French word for snail and the popular name used to identify any dish that contains snails. Snails, like other mollusks, are safe for human consumption if they are properly prepared and cooked. Cooking kills any parasites that the snails may carry. Eating raw snails is dangerous.

considerable damage to flowers and other garden plants. Because slugs do not have shells to protect their bodies, they feed at night and stay hidden during the day. Slugs called *nudibranchs* (NOO di BRAHNKS) can also be found in the ocean. Different nudibranchs use camouflage, nematocysts, and poisons to protect themselves. Many nudibranchs are brightly colored. The colors warn predators that these mollusks are not tasty.

Another group of mollusks, called *cephalopods* (SEF uh luh PODZ), meaning "head-footed," are speedy. A cephalopod moves with a jetlike motion by forcing water through a tube in its body. Squids, octopuses, and cuttlefish are all cephalopods. Even though some cephalopods have shells, their shells are not always on the outside of their bodies. A squid has a thin shell under its mantle, but an octopus is protected only by the special skin of its mantle. The nautilus, though, has a large shell around its body. Each cephalopod is a little different from the others. But most cephalopods have large eyes, arms (tentacles) with suckers around their mouths, and beaks.

Cephalopods are usually small and of little or no danger to man. A few have poison that they use to capture food. The most poisonous cephalopod is the blue-ringed octopus. Its poison paralyzes its prey. Without immediate attention, a person who is poisoned will suffocate.

Many scary stories show the arms of octopuses or squids attacking ships. Though it seems far-fetched, sailors have actually seen giant squids attack ships. Scientists once thought that the largest invertebrate was the giant squid. However, recently they have discovered a squid that is even larger and more dangerous, the colossal squid. Instead of just having suckers, the colossal squid also has hooks on the ends of its tentacles. The discovery of this squid helped explain the cuts and scars that whalers often find on large whales.

1. How do scientists divide animals into two main groups? Which group is bigger?
2. How do nematocysts help some animals get food?
3. How do cephalopods move?

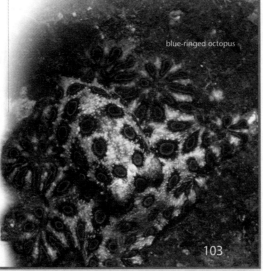

blue-ringed octopus

Discussion:
pages 102–3

➤ What are *nudibranchs*? {slugs that live in the ocean}

💡 How are some nudibranchs similar to sea anemones? {Some nudibranchs have nematocysts.}

➤ What does *cephalopod* mean? {head-footed}

➤ What is unique about the shells of some cephalopods? {The shells are not always on the outside of the animals' bodies.}

➤ How does a cephalopod move? {with a jetlike motion by forcing water through a tube in its body}

💡 Why is an octopus called head-footed? {Its tentacles (arms and legs) come directly from its head.}

➤ Give some examples of cephalopods. {Possible answers: octopus, cuttlefish, nautilus, squid}

Answers

1. Animals are grouped by whether or not they have backbones. Invertebrates make up the largest group of animals.
2. Animals such as jellyfish, sea anemones, and coral use nematocysts to sting and paralyze small prey.
3. A cephalopod moves with a jetlike motion by forcing water through a tube in its body.

Activity Manual
Reinforcement—page 66

Lesson 52

Student Text Page 104
Activity Manual page 67

Objectives

- Construct a terrarium
- Observe land snails
- Record observations

Materials

- fishbowl or large glass jar
- small pebbles or sand
- damp soil
- twigs and dried leaves
- one land snail
- wire screen or lid with holes

Introduction

➤ What is a *terrarium?* {a container of plants and sometimes animals that is made to imitate a specific environment}

➤ Have you ever seen a terrarium?

Purpose for reading:
Student Text page 104
Activity Manual page 67

➤ Why are land snails important for our environment?

➤ To which phylum does the land snail belong?

The student should read all the pages before beginning the exploration.

Discussion:
page 104

➤ Why are snails called univalve mollusks? {because they have only one shell}

➤ Where do land snails like to live? {in cool, dark, shady places}

➤ Why are snails important for the environment? {They help balance the environment by turning decaying matter into materials useful for plants.}

God uses all of His creatures to provide for the needs of His world.

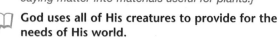

Explorations—page 67

Assessment

Rubric—TE pages A50–A53

Select one of the prepared rubrics, or design a rubric to include your chosen criteria.

Explorations: Snail Terrarium

Land snails are soft-bodied univalves—mollusks that have only one shell. They can be found throughout the world in places where there is moisture and vegetation. Generally, snails live in cool, dark, and shady places. Evening and early morning hours are the best times to find snails under leaves, twigs, and rocks, or in gardens munching on plants.

A large foot underneath the snail makes the snail look like it is dragging its belly while hauling its shell on its back. To make movement easier and less destructive to its soft body, the land snail secretes slime and moves on top of it.

Snails are important for the balance of Earth's ecological system. Millions of snails live in dark, damp forests, shredding decaying leaves and scraping off pieces of fungi for food. Snails help to balance our environment by recycling decaying matter into chemicals that plants can use. Someday scientists may even discover a way to make snail slime useful to man.

What to do

1. Make a terrarium using a fishbowl or a large glass jar. Put one inch of small pebbles or sand on the bottom of the bowl. Add a layer of damp soil and top it with twigs, dried leaves, and a few small stones. Add one land snail. Cover the bowl with a wire screen or the jar with a lid containing nail holes. The snail can climb out of the terrarium if it is left uncovered.

2. Keep plenty of fresh, soft, green leaves or lettuce in the terrarium. Leftover food should be removed regularly. Add fresh water in the lid of a baby food jar each day. The terrarium will need to be cleaned after several days.

 3. Follow the directions and record your findings in your Activity Manual.

The Achatina achatina snail (giant African landsnail) is the largest land snail in the world.

 Aquariums and large glass jars make great terrariums.

Since snails reproduce quickly, start with only one land snail in the terrarium.

A slug may be substituted for a snail.

Student Text pages 105–7
Activity Manual pages 68–70

Lesson 53

Objectives
- Describe three types of worms
- Compare a free-living worm with a parasite
- Explain why worms can be helpful to man
- Explain why worms can be harmful to man

Vocabulary
filter feeder   free-living
parasite   segment

Materials
- several suction cups
- small notepad with plastic cover

crown-of-thorns starfish

Echinoderms

Echinoderms (ih KYE nuh DURMZ) are animals that have *radial symmetry*. When echinoderms are adults, their bodies are shaped like the spokes of a bicycle wheel. Each of the spokes is the same. All echinoderms live in water and move around by using thousands of little *tube feet* located on their undersides. All of these feet have tiny suckers on them that help the animals stick to different surfaces, such as the ground or a rock. Most echinoderms have arms in multiples of five. Sea stars, commonly called starfish, and sea urchins are echinoderms.

Since their mouths are on their undersides, echinoderms have to be on top of prey in order to eat it. Some echinoderms eat whatever comes floating through the water, such as plankton and other tiny life forms. These echinoderms are called **filter feeders**.

Many echinoderms, however, eat mollusks or even other echinoderms.

Unlike mollusks and sponges, most echinoderms have hard skeletons. Perhaps you have found a sand dollar washed up on the seashore. You were looking at the skeleton that was left after the animal had died and the soft body parts had been eaten or rotted away.

Echinoderms cannot move quickly, so they depend on other ways to protect themselves. The spines on their bodies offer one line of defense. They also defend themselves by hiding in cracks or by using camouflage. If a predator grabs the arm of a brittle star, a kind of sea star, the brittle star breaks off its own arm. While the predator eats the arm, the brittle star escapes. The brittle star slowly grows another arm. Some other types of sea stars can also regrow arms that have been lost.

105

Introduction

Place one suction cup on the notebook. Try to lift the notebook with the suction cup. Keep adding suction cups until it is possible to lift the notebook.

Sea stars (starfish) have tiny suction cup-like suckers on their undersides. These suckers help the sea stars move.

▶ What other animals have suckers? *{cephalopods}*

Purpose for reading:
Student Text pages 105–7

▶ What is *radial symmetry*?
▶ How are worms helpful to man?
▶ How are worms harmful to man?

Discussion:
page 105

▶ What does *radial symmetry* mean? *{Equal parts radiate from the center.}*
▶ How do echinoderms move? *{They use thousands of tube feet located on their undersides.}*
▶ Why do echinoderms not slide off the surfaces they move on? *{The tube feet have tiny suckers that help them stick to surfaces.}*
▶ Which term describes echinoderms that eat life forms that float in the water? *{filter feeders}*
▶ What is the outside of an echinoderm like? *{It has a hard skeleton.}*
▶ Which method of defense do echinoderms such as the brittle star have? *{They can lose an arm if caught and can regrow it.}*

SCIENCE BACKGROUND

Radial symmetry—The *radius* of a circle is a line from the center of the circle to the outer edge. *Radial symmetry* follows lines from the center of an object. For example, the petals of a flower show radial symmetry because they each look similar and radiate from the center in a similar way.

How do sea stars get food?—Sea stars may feast on oysters, clams, mussels, and barnacles. They open shells by using their sucker-tipped tube feet. The tube feet are located in rows on the underside of each of their arms. Each tube foot is like a tiny finger with a sucker on the end. The sea star attaches itself to a shell, and its tube feet pull the shell open. One tube foot would not be able to open the shell. Prying apart the shell takes the effort of the legs working together.

 ART Guide the student in making a design that has radial symmetry. Direct the student to draw a large circle. Tell him to cut out the circle and fold it in half several times. Then he should open the circle and decorate each section of the circle so that it shows radial symmetry.

Encourage the student to decorate every other section to look alike rather than make every section the same.

You may choose to hang several circles together to form a mobile.

Chapter 5   Lesson 53 115

Discussion:
pages 106–7

➤ **What is the distinguishing characteristic of a flatworm?** *{It is flat.}*

➤ **What does bilaterally symmetrical mean?** *{Something that is bilaterally symmetrical can be divided down the middle and be the same on both sides.}*

➤ **What is the difference between a free-living worm and a parasitic worm?** *{Free-living worms are independent of other organisms. Parasitic worms live on or in other living organisms.}*

➤ **Why is it important to cook meat thoroughly?** *{Possible answer: Some parasitic worms can get into humans if meat is not properly cooked to kill the parasite's eggs.}*

➤ **What conditions may increase the chances of something getting parasitic worms?** *{unclean, unsanitary conditions}*

Unsanitary conditions occur where trash and animal or human waste are not properly disposed of.

➤ **What are the distinguishing characteristics of roundworms?** *{They are smooth and round on the outside.}*

➤ **What do free-living roundworms do to the soil to help plants?** *{They help decompose dead organisms, which fertilizes the soil and helps plants get the proper food.}*

➤ **What is another name for a segmented worm?** *{annelid}*

➤ **What are the parts of an earthworm made up of?** *{segments}*

➤ **What is inside each segment?** *{some of the same things}*

➤ **What parts of the earthworm are not segmented?** *{the head and the end area}*

➤ **How does a segmented worm move?** *{Tiny hairs called* setae *help the worm to grip the ground and move.}*

➤ **How are sea worms different from land worms?** *{Sea worms have paddles and bristles instead of setae.}*

Flatworms

Flatworms are just that—flat. These worms have *bilateral symmetry,* meaning that they can be divided down the middle and be the same on each side. Flatworms are either parasitic or free-living. **Parasites** live on or in other living organisms, called *hosts.* Parasites depend on their hosts for nourishment. Animals, and even humans, can be hosts for parasitic flatworms. These worms can get into humans when humans eat meat that has not been cooked enough to kill all of the parasite's eggs.

Other flatworms are **free-living,** meaning they are independent of other organisms. Free-living worms are very small. They feed on tiny organisms in the places where they live. Planarians are free-living flatworms that live in soil or fresh water. Usually brown or gray, freshwater flatworms blend in with their environments. Some brightly colored flatworms live in salt water.

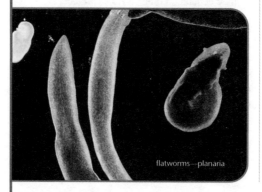

flatworms—planaria

106

Roundworms

Roundworms are smooth and round on the outside. Many of these worms are parasites. Livestock are the most common hosts of roundworms. However, people can become infected by roundworms when they are around areas where infected animals have been. Some roundworms cause serious diseases in humans. Parasitic roundworms are usually found in places where conditions are unsanitary.

Not all roundworms are parasites. Free-living roundworms usually live in soil, water, dead plants, and dead animals. They help decompose dead organisms, and thus they help to fertilize the soil. Thousands of worms may live in the soil in your flower bed, but they are so small that you may not be able to see them.

Segmented Worms

When you think of a worm, you probably think first of a segmented worm, or *annelid.* Annelids, such as earthworms, have soft bodies and are made up of many **segments,** or similar pieces. Only the head and the end area of the worm are not segmented. Each segment has some of the same things

roundworms—ascaris worm

SCIENCE BACKGROUND

Worm phyla
Flatworms—phylum Platyhelminthes
Roundworm—phylum Nematoda
Segmented worm—phylum Annelida

Trichinella spiralis—This flatworm parasite is sometimes carried by pigs and other wild animals and can be transmitted to humans. *Trichinosis* is the name for the human disease. If left untreated, these flatworms can attack the brain, eventually resulting in death.

Flatworms—Flatworms do not die when cut in half. They can grow back the missing parts to form two complete flatworms.

Nicknames—"Intestines of the earth" was an earthworm name used by Aristotle. "Nature's plough" is another frequently used nickname for an earthworm.

Large earthworms—The Giant Gippsland Earthworm of Australia can measure 80 cm long and 2 cm in diameter.

Leech locomotion—Leeches do not have setae like other annelids. They have mouthparts at both ends of their bodies. These mouthparts grip the surface of the earth and move the leech along, similar to the way inchworms and caterpillars move.

inside it. Sometimes a worm can regenerate segments that have been broken off. Many segmented worms move using hairlike structures called *setae* (SEE tee). The worm grips the ground (or whatever it is crawling on) with the setae and pulls itself along.

Most annelids are free-living. Even leeches, unlike flatworms and roundworms, do not get inside their hosts. Leeches suck blood only from the outside. Some annelids live in the sea and eat plankton and algae in the water. Unlike land worms, sea worms have paddles and bristles along their bodies. Some sea worms have long tentacles on their heads. The tubeworms that live in deep-sea vents are annelids.

Earthworms are probably the most familiar annelids. As they get their food from the soil, earthworms also serve an important purpose in the soil. They burrow around and make holes for air to get into the soil. The air helps plants in the

segmented worm—earthworm

soil to grow. Also, like the roundworms, earthworms break down complex plant matter into nutrients that the plants around it can use. The earthworm is so useful that some people buy them to put in their flower gardens.

1. What kinds of food do echinoderms eat?
2. What is the difference between parasites and free-living worms?
3. How are worms helpful? How are worms harmful?

Science and History

In the nineteenth century, doctors often placed leeches on sick people to get the "bad blood" out of them. Eventually, people realized that bad blood was not the cause of illnesses. Doctors gradually stopped using leeches for medicinal purposes. However, leeches are becoming popular again for certain medical treatments. Today doctors may apply leeches to the area around a reattached body part. They have discovered that the leech's saliva has chemicals in it that numb the area of sucking and keep blood from clotting. The leech keeps blood flowing through the veins and arteries around the wound, which helps the wound heal more quickly.

107

Direct a Demonstration
Observe earthworms

Materials: earthworm from garden or pet store, piece of clear glass or plastic, magnifying glass (optional)

The student should keep his hands moist when handling earthworms to prevent the worms from drying out and sticking to his hands.

Direct the student to place the earthworm on a piece of clear glass or plastic.

➤ **Describe an earthworm's body.** *{It is made up of parts, or segments.}*

➤ **How does an earthworm move?** *{It inches along, thrusting forward and pulling the segments along.}*

Direct the student to look at the underside of the earthworm through the glass or plastic to see how the earthworm moves.

➤ **What do you see under the earthworm that helps the earthworm to move?** *{little hairs}*

➤ **How many eyes does the earthworm have on its head?** *{none}*

Explain to the student that the earthworm has no eyes but that it has cells that are sensitive to light. Earthworms do not have lungs to help them breathe. They breathe through their skin. To take in oxygen and let out carbon dioxide, their skin must stay moist.

Discussion:
pages 106–7

➤ **How do leeches get their food?** *{They suck blood from outside the host.}*

➤ **How do earthworms get their food?** *{from the soil}*

➤ **Why do some gardeners buy earthworms?** *{Earthworms make holes in the ground so that air can get into the soil. Worms break down plant matter into useful nutrients.}*

📖 Farmers can rely on the fact that earthworms live and work in the soil, just as God designed them to. God desires that Christians be faithful workers that others can depend on. [BATs: 2c Faithfulness; 2e Work]

Refer the student to *Science and History*.

➤ **Why did doctors in the nineteenth century use leeches on people?** *{They were trying to get the bad blood out of them.}*

💡 **Why might a doctor use leeches today?** *{Doctors may use leeches on reattached body parts. A leech's saliva contains a numbing chemical, which keeps the blood from clotting and speeds up the healing process.}*

Answers

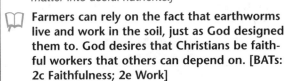

1. Some echinoderms eat mollusks or other echinoderms. Other echinoderms are filter feeders, eating whatever comes floating through the water.
2. Free-living worms are independent of other organisms. Parasitic worms live on or in other living organisms.
3. Roundworms and segmented worms help break down plant matter in the soil and burrow holes in the soil, thus increasing air circulation needed by plants. Some worms are parasites and can cause diseases.

Activity Manual
Bible Integration—page 68
This page compares metamorphosis with the process of sanctification. The page may be used with Lesson 54, which talks about metamorphosis.

Study Guide—pages 69–70
These pages review Lessons 51 and 53.

Lesson 54

Student Text pages 108–11
Activity Manual pages 71–72

Objectives

- Identify crustaceans, arachnids, centipedes, millipedes, and insects as arthropods
- Describe basic characteristics of each kind of arthropod

Vocabulary

arthropod
exoskeleton
molt
crustacean
arachnid
insect
metamorphosis
incomplete metamorphosis
complete metamorphosis

Materials

- real grasshopper in a jar or a picture of a grasshopper
- magnifying glass

Introduction

Display the grasshopper.

➤ **How many body segments do you see?** {3}

➤ **To which group does the grasshopper belong?** {Possible answers: arthropod, insect}

➤ **What other body parts do you see?** {Possible answers: six legs, thorax, abdomen, antennae}

➤ **Have you ever held a grasshopper and watched it spit brown liquid? Why do you think it does this?** {Possible answer: The liquid scares off predators.}

Give the student an opportunity to view the grasshopper with a magnifying glass.

Purpose for reading:
Student Text pages 108–11

➤ **What is the difference between crustaceans and arachnids?**

➤ **Why is a spider not an insect?**

Arthropods

Arthropods (AHR thruh pahdz) are the most numerous animals on the earth. Scientists estimate that at least half of the known animal species are arthropods. **Arthropod** means "jointed foot." Arthropods have jointed legs and segmented bodies. But the segments of an arthropod are not like those of an annelid. Each segment of an arthropod has a specific purpose.

Another characteristic of an arthropod is its exoskeleton. An **exoskeleton** (EK so SKEL ih tuhn) is a hard covering that acts like a knight's armor, protecting the arthropod's body. In order to grow, arthropods must **molt**, or shed this exoskeleton and grow a new one. Sometimes you can find the old exoskeletons of locusts on trees. Arthropods also have antennae. All arthropods have some of the same characteristics, but scientists divide them further by some of their unique features.

Crustaceans

If you go to a seafood restaurant, you will probably see several crustaceans, such as shrimp, lobsters, and crabs. **Crustaceans** (kruh STAY shunz) have at least five pairs of jointed legs and two pairs of antennae. They breathe through gills, and most crustaceans have some sort of claw.

The blue crab is a very common North American crustacean. It has five pairs of legs. The middle three pairs are used for walking, the back pair is used for swimming, and the front pair has claws for getting food. The blue crab's claws can hold on to food or a predator with a vise-like grip.

One crustacean that does not look like a typical crustacean is the barnacle. Barnacles look more like mollusks because they have shells. Barnacles attach themselves to surfaces by their sticky antennae and then build themselves little caves to live in. This means that a barnacle is always standing on its head! Inside its mollusklike shell, a barnacle has a segmented body and six pairs of legs. Barnacles use their twelve legs to filter food out of the water. Barnacles can live in tide pool areas where they are not covered by water all day, but many live fully submerged all the time. Although they do not look like other crustaceans, barnacles have the antennae, segmented bodies, and legs necessary to be classified as crustaceans.

blue crab molting

108

SCIENCE BACKGROUND

Rocky Mountain spotted fever—The rickettsia pathogen causes this disease. The wood tick, dog tick, and lone star tick can transmit it.

Lyme disease—This is a bacterial disease spread most often by tiny deer ticks. One of the first symptoms of this disease is inflammation in the shape of a bull's eye surrounding the bite site. As the disease progresses, it affects the muscles and other systems. In 1975 the disease was named after Lyme, Connecticut, the community where it was first discovered. In 1982 Willy Burgdorfer identified the specific bacteria that causes Lyme disease.

Arachnids

Arachnids (uh RAK nihdz), such as spiders, scorpions, ticks, and mites, have eight legs and two body segments. Most arachnids are not harmful to humans. But a few can be both painful and dangerous. A black widow spider's bite can cause severe pain and muscle spasms. A provoked scorpion can deliver a painful sting. But tiny ticks probably cause the most serious human health problems. Ticks are parasitic and use both animals and humans as hosts. Though their bites are not often painful, ticks can spread diseases such as Rocky Mountain spotted fever and Lyme disease.

Spiders are the most familiar arachnids. Most spiders have eight eyes. However, this many eyes does not necessarily give a spider good eyesight. Only spiders that actively seek their prey have good eyesight. Other spiders, which just sit in their webs and wait for food to come to them, usually detect prey from the vibrations of the web.

Spider webs are some of the most beautiful and complex creations of any arthropod. The silk is secreted from *spinnerets*, silk-spinning organs in the back of the spider. The spider weaves a web that snares passing insects and other prey. But not all webs are the same. Some webs are like lace, and others are like tunnels. Still others look as if they are just random strands of silk. But each web fits its spider's needs perfectly.

Centipedes and millipedes

Centipedes and millipedes are also arthropods. They have many body segments, though the number depends on the creature. A centipede has one pair of legs on each segment, and a millipede has two pairs on each segment. Although *centi-* means "one hundred," centipedes may have as few as 30 or as many as 274 legs. And millipedes, though they may have hundreds of legs, certainly do not have one thousand! Centipedes' front pair of legs have poisonous fangs. They pounce on their prey and inject poison from those fangs. Millipedes have no poisonous fangs and are quite harmless. However, they do secrete a fluid that smells foul and can cause allergic reactions.

centipede

millipede

According to Greek mythology, Arachne, whose name means "spider," was a mortal weaver. She challenged the Greek goddess Athena to a weaving contest to see who really was the goddess of the loom. Arachne won the contest with a tapestry that mocked the gods. The tapestry depicted both the gods' successes and errors. Athena, though very angry and jealous, mercifully spared Arachne's life but turned her into a spider.

This fictitious story is part of Ovid's *Metamorphoses*. *Arachnid*, the name for the class of spiders, scorpions, ticks, and mites, probably came from this piece of Greek mythology.

Discussion:
pages 108–9

➤ **What does the word *arthropod* mean?** {jointed leg}

➤ **Describe the legs and body of an arthropod.** {Arthropods have jointed legs and segmented bodies.}

➤ **How are the segments of an arthropod different from the segments of an earthworm?** {The segments of an earthworm are alike. The segments of an arthropod are different and have different functions.}

➤ **What is an *exoskeleton*?** {a hard covering that protects the body of an arthropod}

➤ **Why must an arthropod molt?** {As the arthropod grows, its exoskeleton becomes too small. It sheds the small exoskeleton and grows a new one that fits.}

➤ **What are *crustaceans*?** {Crustaceans are arthropods with at least five pairs of jointed legs and two pairs of antennae. They breathe with gills, and most have some sort of claw.}

➤ **What are some examples of crustaceans?** {Possible answers: shrimp, lobsters, crabs, barnacles}

➤ **How is a barnacle different from other crustaceans?** {It attaches itself to a surface and builds a shell around its body.}

➤ **Describe the characteristics of an arachnid.** {Arachnids are arthropods with eight legs and two body segments.}

➤ **Give two examples of arachnids.** {Possible answers: spiders, scorpions, ticks, mites}

➤ **What dangers do some arachnids pose to humans?** {They can deliver painful bites and stings. Some can transmit diseases to humans.}

➤ **What are some diseases that ticks can transmit to humans?** {Possible answers: Lyme disease and Rocky Mountain spotted fever}

➤ **Which body part does a spider use to spin its web?** {spinnerets}

➤ **Why are all spider webs not the same?** {They are designed to suit each spider's method of obtaining food.}

➤ **What is the difference between a centipede and a millipede?** {Centipedes have one pair of legs on each segment, and millipedes have two pairs of legs on each segment. Centipedes have poisonous fangs, but millipedes are harmless.}

Discussion:
pages 110–11

➤ **What is the most common type of arthropod?** *{the insect}*

➤ **What percentage of arthropods are insects?** *{about 90 percent}*

Refer the student to the circle graph on Student Text page 110. Ask questions about the graph.

➤ **Describe the body of an insect.** *{Insects have three body segments: head, thorax, and abdomen. Three pairs of legs are on the thorax. Most insects have two pairs of wings.}*

💡 **Why are spiders not classified as insects?** *{Spiders have two body parts, and insects have three. Spiders have eight legs, and insects have six legs.}*

➤ **What type of insect has chewing mouthparts?** *{a beetle}*

💡 **Why do mosquitoes need piercing and sucking mouthparts?** *{They need piercing mouthparts to break the skin or hide of their victims and sucking mouthparts to suck up the blood.}*

💡 **Why do butterflies need siphoning mouthparts rather than chewing mouthparts?** *{They need a way to get nectar from flowers.}*

➤ **What is *metamorphosis*?** *{the process of an insect becoming an adult}*

➤ **What are the stages of incomplete metamorphosis?** *{egg, nymph, and adult}*

➤ **What does a nymph resemble?** *{an adult}*

➤ **Although a nymph looks like an adult, how is it different?** *{Its wings are not fully functional.}*

➤ **What happens to a nymph as it changes to an adult?** *{It grows and molts.}*

➤ **Why does a nymph need to molt as it grows?** *{Its exoskeleton does not grow.}*

Insects

Insects are one of the largest and most diverse groups of animals. In fact, scientists estimate that 90 percent of all arthropods are insects. Although there are many kinds of insects, all insects have certain specific characteristics. **Insects** have three body segments: the *head*, the *thorax*, and the *abdomen*. They have three pairs of legs on the thorax, and most also have two pairs of wings.

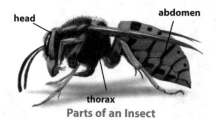

Parts of an Insect

Insects have different mouthparts depending on what the insects eat. Beetles have *chewing* mouthparts because they chew the things they eat. But insects such as mosquitoes have *piercing* and *sucking* mouthparts because they suck blood. Butterflies and moths have *siphoning* mouthparts for getting nectar out of flowers.

An insect becomes an adult through the process of **metamorphosis** (MET uh MOR fuh sis). There are two different types of metamorphosis: incomplete and complete. **Incomplete metamorphosis** has three stages and takes

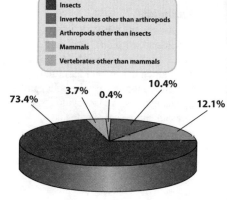

place in dragonflies, grasshoppers, and other similar insects. When an egg hatches, the immature insect looks much like the adult; however, its wings are not fully functional. This immature insect is called a *nymph*. The nymph molts many times as it grows and becomes an adult. During this time of molting, the insect's wings become functional.

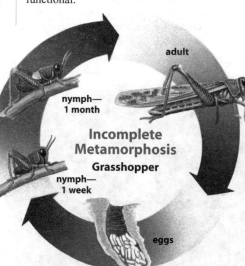

Incomplete Metamorphosis Grasshopper

SCIENCE BACKGROUND

Siphon—a tube used to take in or expel a liquid through the use of pressure

Transition—the process of changing from one form or stage to another; The pupal stage is called a transition because the animal changes significantly from its previous form.

Complete metamorphosis has four stages. The egg hatches into what is called the *larva*. The larva does not look like the adult insect; in fact, you would not think it was even related to the adult. Caterpillars and grubs are larvae of different-looking adult insects. The larva eats as much as possible in order to be ready for the next stage of its growth—the *pupa*. During the pupal stage the insect is in transition. It may be covered with a chrysalis or cocoon. The insect does not eat while it is in the pupal stage. It grows wings and body segments during this time. Finally, the insect emerges from its covering to begin its adult stage. When an adult female lays eggs, the cycle begins all over again.

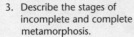

Complete Metamorphosis
European swallowtail butterfly

- pupa (chrysalis)
- larva (caterpillar)
- adult
- egg

QUICK CHECK
1. Describe two characteristics of arthropods.
2. Why is a spider not an insect?
3. Describe the stages of incomplete and complete metamorphosis.

Science and the BIBLE

The Bible mentions several instances in which God used arthropods. God sent locusts and flies to plague Pharaoh when the children of Israel were in Egypt. John the Baptist ate locusts and honey during his wandering in the wilderness.

God also uses arthropods to teach us lessons. Proverbs 30:25–28 says of the ants, "The ants are a people not strong, yet they prepare their meat in the summer. . . . The locusts have no king, yet go they forth all of them by bands; The spider taketh hold with her hands, and is in kings' palaces." Even God's animals behave in a manner that glorifies Him. Should not we as humans who have the choice of behavior honor God in how we act as well?

111

 Encourage the students to do a Bible search to find other places in the Bible where arthropods are mentioned. Ants, locusts, spiders, grasshoppers, flies, beetles, bees, lice, hornets, and moths are some of the arthropods mentioned in the Bible.

 You may choose to use Activity Manual page 68 with this lesson. The page compares metamorphosis with the process of sanctification.

Discussion:
pages 110–11

Refer the student to the diagram on Student Text page 111.

➤ What is the insect called at each stage of complete metamorphosis? {egg, larva, pupa, and adult}

➤ What is another name for a caterpillar? {larva}

➤ Which stage of metamorphosis usually has a chrysalis or cocoon? {pupal}

➤ Why does the larva of an insect eat as much as possible in the larva stage? {It needs to store energy for the pupal stage, when it does not eat.}

➤ In which stage is the insect when it emerges from its cocoon? {adult}

How does metamorphosis demonstrate one of the characteristics of living things? {It shows a life cycle.}

Refer the student to *Science and the Bible*.

➤ Which two insects were used by God to plague Egypt? {locusts and flies}

➤ God's creatures always give glory to Him. What should that teach us? {that we should give God glory as well}

Answers

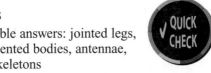

1. Possible answers: jointed legs, segmented bodies, antennae, exoskeletons

2. A spider has eight legs and only two body parts. An insect has six legs and three body parts.

3. An insect such as a grasshopper goes through incomplete metamorphosis, developing from an egg to a nymph to an adult. Complete metamorphosis has four stages. For example, a butterfly grows from an egg to a larva (a caterpillar) to a pupa (a chrysalis) to an adult (a butterfly).

Activity Manual
Study Guide—pages 71–72
These pages review Lesson 54.

Assessment
Test Packet—Quiz 5-A
The quiz may be given any time after completion of this lesson.

Lesson 55

Student Text pages 112–13
Activity Manual pages 73–74

Objectives

- Observe the larval stage of complete metamorphosis
- Observe the pupal stage of compete metamorphosis
- Collect and record observation data

Materials

- See Student Text page
- copies of TE page A25, *Observation Record*

Introduction

➤ Why are observations important for learning? *{Possible answers: to utilize all five senses when gathering information; to understand things better; to note the effects of changed variables}*

➤ Have you ever seen a beetle or a mealworm? Where?

Mealworms and beetles infest grains and their products, such as flours and cereals.

Purpose for reading:
Student Text pages 112–13
Activity Manual pages 73–74

➤ What kinds of mealworm reactions will you test?

The student should read all the pages before beginning the activity.

Discussion:
pages 112–13

➤ What are the four stages of complete metamorphosis? *{egg, larval, pupal, and adult}*

➤ At what growth stage is the mealworm? *{larval}*

➤ Why is a mealworm not a worm? *{because in the adult stage it has all the characteristics of an insect—three body parts, six legs, and two antennae}*

➤ What kind of activity can you expect during the pupal stage? *{no movement and no eating}*

➤ What does the mealworm become by the time it reaches the adult stage? *{a beetle}*

ACTIVITY: Mealworm Movement

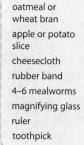

Process Skills
- Experimenting
- Observing
- Identifying and controlling variables
- Recording data

The grain beetle goes through the four stages of complete metamorphosis. The egg is only about 12 mm long and is very difficult to see. The larva of a grain beetle is called a mealworm. Observation of the body parts of a mealworm shows that it is an insect and should not be confused with members of the worm family. Mealworms can be found feeding on grain or grain products. During the pupal stage, the mealworm does not move or eat, and it remains in a small, firm form. After a two- to three-week pupa, the adult beetle appears.

Problem
How does a mealworm respond to different stimuli?

Procedure

1. Prepare a habitat for your mealworms. Place 100 mL of oatmeal and a slice of apple in the jar.

2. Observe the characteristics of your mealworms by placing them on a sheet of paper. Use a toothpick to gently move the mealworms while you observe them with the magnifying glass. Count the number of legs and measure the length and width of each mealworm. Record your observations in your Activity Manual.

3. Test the reaction of a mealworm to light and dark. Tape together a piece of white paper and a piece of black paper. Place a mealworm in the center. Record the color the mealworm crawls toward.

Materials:
- large-mouth glass jar
- oatmeal or wheat bran
- apple or potato slice
- cheesecloth
- rubber band
- 4–6 mealworms
- magnifying glass
- ruler
- toothpick
- white and black paper
- ice
- lamp
- cotton swabs
- ammonia or vinegar
- blocks or other materials to make a maze
- observation log
- Activity Manual

larva

pupa

112

SCIENCE BACKGROUND

Mealworm development—The egg of a mealworm is very difficult, if not impossible, to see.

The larval stage of a mealworm lasts about ten weeks. As the larva grows, it sheds its exoskeleton 9–20 times. The temperature and amount of moisture have some effect on the number of times shedding takes place. Also the larva turns into a pupa more quickly in warmer temperatures.

The pupal stage may last only 1–2 weeks.

Complete metamorphosis from egg to adult may take 6–9 months.

The types of beetles that develop from mealworms vary. Some common names are grain beetle and darkling beetle.

Teacher Helps

Most pet or bait stores sell mealworms.

Glass or metal containers are the best to use.

Apple or potato slices provide moisture. Check regularly for mold, and replace the slices as needed.

After the adult develops from the pupal stage, you may allow the beetles to continue to multiply. Because of waste residue in the bottom of the habitat, eggs are nearly impossible to see. If eggs are laid and hatch, small mealworms may become visible about a month after the beetles develop.

4. Test the reaction of a mealworm to warmth and coldness. Place the mealworm in a plastic cup. Observe its reactions when the cup is placed over a lit light bulb and when it is placed over an ice cube. Record your observations.

adult beetle

5. Test the reaction of a mealworm to a smell. Place a mealworm on a sheet of paper. Dip a cotton swab in water and hold it near the head of the mealworm. Dip a cotton swab in the ammonia or vinegar and hold it near the head of the mealworm. Record your observations.

6. Test to find out if a mealworm repeatedly goes the same direction. Make a maze with blocks similar to illustration A in your Activity Manual. Place a mealworm at the start of the maze. Record your observations. Make a maze with blocks similar to illustration B. Place a mealworm at the start of the maze. Record your observations. Compare the results of each test.

7. Place your mealworms in their new home. Cover the opening of the jar with a piece of cheesecloth and secure it with a rubber band.

8. After testing, observe your mealworms two to three times a week. Record your observations on the observation log that your teacher gives you.

Conclusions
- How did the mealworms respond to the different stimuli?
- Did any of your mealworms develop into other stages?

Follow-up
- Perform the same tests on adult beetles and compare the results.

SCIENCE PROCESS SKILLS

Observing

➤ What information can a person gain by making observations and collecting and recording data? *{growth records; responses to stimuli such as light, cold, and dampness}*

➤ Why are observations important to scientific research and thinking? *{to increase knowledge}*

➤ Why is recording data important in scientific research? *{to assure accuracy}*

God desires that Christians grow to know and love Him through Bible reading and prayer. Many Christians find it helpful to keep a written record of answers to prayer, along with other truths that God teaches. [BATs: 6a Bible study; 6b Prayer]

Procedure
Examine the mealworms according to the given criteria before placing them in the prepared habitat.

Observe the mealworms two or three days each week.

Use TE page A25, *Observation Record*, to record the observations.

The Conclusions and Follow-up sections of the pages should be completed at the end of the observation period.

Conclusions

➤ **Describe the larval stage of metamorphosis in a grain beetle.** *{wormlike, six legs, two antennae, hard exoskeleton or covering}*

➤ **What kind of response did the mealworm have to the light bulb, the ice cube, and the ammonia or vinegar?** *{It pulled away from the light and the ice cube and stayed on the damp towel.}*

➤ **Describe the pupal stage of complete metamorphosis of the grain beetle.** *{small, hard, motionless, does not eat}*

➤ **Describe the adult grain beetle.** *{hard outer covering, dark, six legs, three body parts}*

➤ **How did the adult grain beetle respond to the light bulb, ice cube, and ammonia or vinegar?** *{The grain beetle responded in the same manner as the mealworm.}*

➤ **How can these observations be helpful to man?** *{Grain storage businesses might need to have moisture and temperature controls to discourage the presence of mealworms and grain beetles.}*

Activity Manual
Activity—pages 73–74

Assessment
Rubric—TE pages A50–A53

Select one of the prepared rubrics, or design a rubric to include your chosen criteria.

Lesson 56

Student Text pages 114–17

Objectives
- Identify the characteristics of fish
- Identify the characteristics of amphibians
- Recognize how fish and amphibians are alike
- Describe the life cycle of most amphibians

Vocabulary
cold-blooded amphibian
cartilage

Introduction
➤ Have you ever had a pet fish?
➤ Did you have only one kind of fish or many different kinds?
➤ Did the fish have bright colors or dull colors?
➤ Why do you think fish would be either brightly colored or dull? *{Its color depends on its natural environment.}*

Purpose for reading:
Student Text pages 114–17
➤ What are two of the ways we group fish?
➤ Why is the name *amphibian* fitting for animals in that group?

Vertebrates

Though invertebrates make up most of the species of the animal kingdom, vertebrates make up most of its size. The vertebrate's backbone gives support for its greater weight. If vertebrates did not have this support, they would collapse under their own weight.

Just as scientists group invertebrates by common characteristics, scientists also group vertebrates by common traits. Five of the groups that scientists use to classify vertebrates are: fish, amphibians, reptiles, birds, and mammals.

Fish

Fish come in many different shapes and sizes. Certain characteristics, however, identify an animal as a fish. All fish breathe through gills and are **cold-blooded**. Cold-blooded animals must find warmth or coolness from their environments. Their blood does not maintain a constant temperature.

One way that scientists group fish is based on what their skeletons are made of—cartilage or bone.

Cartilage fish

Sharks, rays, and skates have skeletons made completely of cartilage. **Cartilage** (KAR tl ij) is a bonelike substance, but it is softer and more bendable than bone. Your nose and the outside of your ear are made out of cartilage rather than bone.

Rays and skates look very similar. Both have "wings" of skin that make them look a little like stealth bombers. But rays have whiplike tails that may have painful stingers, while skates are harmless. Rays and skates, as well as sharks, have mouths on the bottoms of their bodies, so whatever they eat must be below them.

Sharks have a reputation for being human killers. Although sharks do bite humans every year, a person is more likely to be struck by lightning than to be bitten by a shark. Sharks that eat sea lions and small whales may be big enough to bite people, but they usually shy away from humans. Most sharks eat only small fish and plankton. Many of

manta ray

114

SCIENCE BACKGROUND

Buoyancy—Buoyancy is the upward thrust of a liquid, such as water. The swim bladder of a fish acts much like a balloon. A fish increases in size as its bladder fills with air. As a result, the water molecules under the fish have a greater surface area to push the fish upward.

Fish classification—In addition to cartilaginous and bony fish, there is a small group of fish without jaws that is classified separately because of the unique characteristics of these fish. This group includes fish such as lampreys.

Fish eyes—Fish never close their eyes, because they have no eyelids. Some fish, such as halibut and flounder, lie on the same side all the time. Their eyes are both on the same side of their heads. Usually these flat fish live on the ocean bottom. God designed this position of their eyes to give them the ability to see greater areas.

butterfly fish

the fish appear larger and scares predators away. Colors also warn predators of the danger of eating certain fish.

The lionfish has brightly colored but very poisonous spines that warn away other animals. The anglerfish uses its body to attract food. This fish has a light on a long string of skin on the front of its head. The light attracts other fish in the deep sea where the anglerfish lives. When a fish comes and tries to bite the light, the anglerfish bites the fish instead.

Some fish do not look like fish at all. The seahorse is a fish that uses its shape to blend in with seaweed. The eel looks more like a snake, but it is a bony fish. Some fish, like the appropriately named rockfish, look like rocks.

Whatever the color, shape, or size of the fish, God has given each species exactly what it needs to survive in its habitat.

the largest sharks, such as the whale shark and the basking shark, are filter feeders, filtering food through their gills.

Bony fish

Most fish are bony fish. Bony fish have skeletons that are stronger and harder than those of cartilage fish. Bony fish live in both fresh water and salt water. Freshwater bony fish, such as bluegill, bass, and trout, are usually brown or gray, so they can blend in with the mud and water in their freshwater habitats. Most fish have scales, but some, like catfish, have only skin to cover their skeletons.

Saltwater fish are often brightly colored. They use their unique coloring for camouflage, for warning, and sometimes for attracting food. The flounder changes its color while it sits on the sea floor in order to avoid detection. The butterfly fish has a huge spot on its side that looks like an eye. This "eye" makes

seahorse

WRITING Provide pictures of some distinctive and unusual fish. Allow the student to choose a picture and write a tall tale explaining the fish's distinctive features.

Discussion:
pages 114–15

▶ Why can a vertebrate grow larger than an invertebrate? {*A vertebrate has a backbone that supports its weight.*}

▶ Name two characteristics of all fish. {*All fish breathe with gills and are cold-blooded.*}

▶ What does *cold-blooded* mean? {*An animal must rely on its environment for warmth.*}

💡 List two other groups of vertebrates that are cold-blooded. {*amphibians and reptiles*}

▶ What is one way scientists group fish? {*by whether the fish has a skeleton made of cartilage or a backbone made of bone*}

▶ What is *cartilage*? {*a material similar to bone but pliable and not as hard as bone*}

▶ Which parts of the human body are made of cartilage? {*the nose, the outside of the ear*}

▶ Describe a ray. {*a fish that has wings of skin and a whiplike tail*}

💡 Some sharks are filter feeders. What other animals are sometimes filter feeders? {*echinoderms*}

💡 Why do you think sharks have such a bad reputation? {*Possible answers: They look scary. Some have attacked humans. Movies and books often portray sharks as vicious.*}

▶ Why do freshwater fish usually have dull colors? {*to blend in with surroundings*}

💡 What word means "to blend in"? {*camouflage*}

▶ Why are many saltwater fish brightly colored? {*for camouflage, for warning, to attract food*}

💡 How could being brightly colored provide camouflage? {*Answers may vary, but elicit that many sea animals, such as coral, are brightly colored, and brightly colored fish blend in well with them.*}

▶ What are some fish that do not look like fish at all? {*Possible answers: sea horse, eel, rockfish*}

Discussion:
pages 116–17

➤ What does *amphibian* mean? *{double life}*

➤ Why is *amphibian* a good name for this group of animals? *{Amphibians lead double lives, living part of their lives in the water and the other part on land.}*

➤ How are amphibians like fish? *{They are cold-blooded.}*

➤ Describe the common life cycle of an amphibian. *{Most amphibians lay eggs in the water. The eggs hatch into the larval stage, or tadpoles. Most tadpoles then lose their gills, gain legs, and move onto land as adults.}*

💡 What other animals have a larval stage? *{insects that go through complete metamorphosis}*

➤ Why do many tadpoles never reach adulthood? *{They are eaten by other aquatic creatures.}*

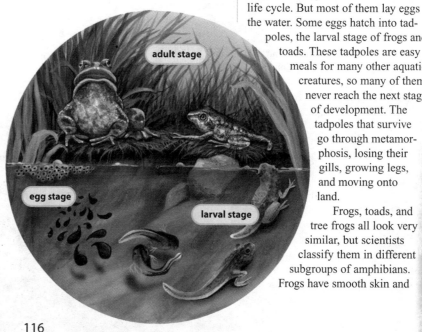

Amphibians

Perhaps you have been on a camping trip to a lake. The quietness of the night was rudely interrupted by a huge blast of sound. You were ready to take cover until someone laughingly told you that a male bullfrog made the sound. The background chirping you heard might have been tree frogs. Frogs, along with toads and salamanders, are amphibians. The term **amphibian** means "double life." And amphibians do lead double lives—part in the water and part on land. They are cold-blooded like fish. As adults, amphibians have lungs, but they also use their thin skin to help them breathe.

Not all amphibians have the same life cycle. But most of them lay eggs in the water. Some eggs hatch into tadpoles, the larval stage of frogs and toads. These tadpoles are easy meals for many other aquatic creatures, so many of them never reach the next stage of development. The tadpoles that survive go through metamorphosis, losing their gills, growing legs, and moving onto land.

Frogs, toads, and tree frogs all look very similar, but scientists classify them in different subgroups of amphibians. Frogs have smooth skin and

116

SCIENCE BACKGROUND

Marbled salamander—Sometimes known as *mud puppies, grampuses,* or *spring lizards,* marbled salamanders have very distinct markings. This salamander moves slowly rather than quickly like most salamanders.

Red-spotted newt—The red-spotted newt is usually green with red spots, but during its immature stage, called an eft stage, it is bright reddish-orange with red spots.

126 Lesson 56

always live near the water. They have large, powerful hind legs for jumping. Toads have short legs because they hop only short distances at a time. They have nubby skin that makes them look like they have warts. Toads lay their eggs in long chains, but frogs lay theirs in clusters.

Although they look like regular frogs, tree frogs belong to a different group of amphibians. Most tree frogs are brown, gray, or green in order to blend in with their environments. But some, like the poison dart frogs of South and Central America, come in an array of beautiful color patterns. Like some fish, these bright colors warn predators that the frogs are poisonous. Many of these frogs skip the tadpole stage of life and hatch as immature adults.

Frogs and toads are not the only amphibians. Salamanders and newts

marbled salamander

are tailed amphibians. They are often brightly colored and live in moist areas under rocks and logs.

Amphibians are useful to humans because they eat insects, such as flies and mosquitoes, that humans consider pests. God used amphibians, too. He judged Pharaoh with frogs during the ten plagues. "And Aaron stretched out his hand over the waters of Egypt; and the frogs came up, and covered the land of Egypt." (Exod. 8:6).

1. How are fish and amphibians alike?
2. Why is amphibian an appropriate name for this class of animals?
3. Describe the life cycle of many amphibians.

red-spotted newt

 Amphibious vehicles can operate in the water and on land. The military has special ships that have well decks for transporting amphibious vehicles.

Discussion:
pages 116–17

➤ **What are some examples of amphibians?** *{frogs, toads, and salamanders}*

➤ **Which three types of amphibians look similar but are actually in different subgroups?** *{frogs, toads, and tree frogs}*

➤ **How are toads and frogs different?** *{Frogs have smooth skin and live near water. They have powerful jumping legs. Frogs lay their eggs in clusters. Toads have nubby skin and short legs. They lay eggs in long chains.}*

➤ **Why are most tree frogs colored green or gray?** *{to blend in with the trees they live on}*

💡 **Why do you think many poisonous frogs have dart as part of their names?** *{Their poison is often used to coat the tips of blow darts or other weapons.}*

💡 **Why do you think tree frogs in Central and South America are sometimes brightly colored?** *{Possible answers: The color serves as a warning. The foliage and flowers on tropical trees are often brighter than they are on other kinds of trees.}*

➤ **Where do most salamanders and newts live?** *{in moist areas under rocks and logs}*

💡 **Why do people find some amphibians useful?** *{Amphibians eat pesky insects like flies and mosquitoes.}*

📖 **What amphibian did God use as a plague to judge Egypt?** *{frogs}*

Answers

1. Fish and amphibians are cold-blooded, and most amphibians breathe with gills until they become adults.

2. Amphibian means "double life." Most amphibians live part of their lives in the water and part of their lives on land.

3. Most amphibians hatch from eggs in the water, develop into the larval stage, and mature into an adult. Usually they lose their gills and gain legs before leaving the larval stage and moving to land.

Lesson 57

Student Text pages 118–21
Activity Manual page 75

Objectives

- Identify two characteristics of reptiles
- Identify two characteristics of birds
- Explain how birds and reptiles are similar and different

Vocabulary

omnivore carnivore
herbivore warm-blooded

Materials

- world map

Introduction

Note: This lesson covers two different groups of animals. You may choose to read and discuss pages 118–19 first and then read and discuss pages 120–21.

➤ As I read the poem "A Narrow Fellow in the Grass" by Emily Dickinson, listen for the phrases that describe a snake.

The poem is in the *Language Link*.

➤ Listen to the first verse of the poem. What might make you suddenly notice a snake? *{Possible answers: a rattle, a hiss, a sudden swish into the bushes or grass}*

➤ Listen to the next two verses. What other characteristics of a snake do you notice? *{Possible answers: likes cool places, lays in the sun, "wrinkles" or slithers}*

➤ Listen to the last two verses. What emotion do you think the author is feeling when she says she has "tighter breathing and Zero at the Bone"? *{fear}*

➤ What other words might you use for "Zero at the Bone"? *{Possible answers: frozen, chilled to the bone, scared stiff}*

Purpose for reading:
Student Text pages 118–21

➤ What are two characteristics of reptiles?
➤ What are two characteristics of birds?

Reptiles

Reptiles exhibit some characteristics similar to amphibians; however, they are a different group. Like fish and amphibians, reptiles are cold-blooded. Most reptiles lay eggs, but unlike amphibians, they lay them on the land instead of in water. Reptiles have scaly skin that allows them to live in areas away from water. Scientists have divided reptiles into three major groups: turtles, lizards and snakes, and crocodilians.

The Galápagos tortoise is the largest known land turtle. It weighs about 225 kg (500 lb).

Turtles

The unique skeletal structure of a turtle clearly identifies this reptile. Most turtles have a layer of hard, bony plates on their backs that provide protection. Some turtles, such as the common box turtle, can completely enclose their heads and legs inside their shells.

Turtles live in a wide variety of locations, including deserts and oceans. Turtles that live in or around water usually have streamlined shells and webbed feet for better swimming.

Many of these turtles are **omnivores** (AHM nuh VORZ), eating both plants and animals. Some turtles are called *tortoises*. Tortoises are usually land-dwellers with high, domed shells. These land tortoises often have thick legs and feet that can support their heavy shells. Tortoises are often **herbivores** (HUR buh VORZ), eating only plants. Some small turtles are sometimes called *terrapins*.

Several large turtles, such as the leatherback turtle and the Galápagos tortoise are endangered species. The leatherback is a marine turtle. The Galápagos tortoise lives in the Galápagos Islands near South America. Known not only for weight but also for age, a Galápagos tortoise can live for up to two hundred years.

Lizards and snakes

Some lizards look much like salamanders, which are amphibians. Lizards, though, belong to the reptile group. Lizards have scaly skins and can live almost anywhere. There is such a variety of lizards that even a small ecosystem can support many different kinds.

Most lizards are small and harmless. Only a few, such as the Gila monsters of the American Southwest, are poisonous. One lizard, however, is

Komodo dragons can grow to 3 m (10 ft) and can weigh up to 165 kg (364 lb).

118

 A Narrow Fellow in the Grass
by Emily Dickinson

A narrow Fellow in the Grass
Occasionally rides—
You may have met Him—did you not
His notice sudden is—

The Grass divides as with a Comb—
A spotted shaft is seen—
And then it closes at your feet
And opens further on—

He likes a Boggy Acre
A Floor too cool for Corn—
Yet when a Boy, and Barefoot—
I more than once at Noon
Have passed, I thought, a Whip lash
Unbraiding in the Sun
When stooping to secure it
It wrinkled, and was gone—

Several of Nature's People
I know, and they know me—
I feel for them a transport
Of cordiality—

But never met this Fellow
Attended, or alone
Without a tighter breathing
And Zero at the Bone—

Reprinted by permission of the publishers and the Trustees of Amherst College from THE POEMS OF EMILY DICKINSON, Thomas H. Johnson, ed., Cambridge, Mass.: The Belknap Press of Harvard University Press, Copyright © 1951, 1955, 1979 by the President and Fellows of Harvard College.

neither small nor harmless. Komodo dragons grow to enormous sizes and are very fierce when disturbed.

Like lizards, snakes live in almost every area of the world. A snake's most obvious feature is its long, legless body. Without legs, the snake must slither along on its belly. Snakes have other characteristic traits as well. One of these is a clear scale that covers each eye. Snakes do not have moveable eyelids. They live their entire lives without blinking even once!

All snakes are **carnivores** (KAR nuh VORZ), meaning they eat only animals. However, they cannot tear or chew their prey, so they must be able to swallow their meals whole. God has specially designed their jaws and bodies to accommodate this need. Unlike most animals, a snake has upper and lower jaws that are not tightly attached. Instead, a strong ligament allows the jaws to separate widely. Snakes can swallow prey considerably larger than the diameter of their own bodies. So whether the snake squeezes its prey to death like a python, poisons it like a cobra, or just quickly catches it like a garden snake, it can swallow its prey.

People often fear snakes and other reptiles without real cause. Snakes are actually beneficial to humans. Many snakes are predators of mice and rats. Others eat insects and slugs that can destroy gardens. God created snakes to serve an important role in helping humans.

Crocodilians

Crocodilians, such as alligators, caimans, and crocodiles, are often thought of as fierce predators. These animals are excellent hunters, especially in the water. Their primary food is fish, but anything that comes near their water habitat, including humans, may become prey.

All crocodilians look similar. They have scaly skin, large bodies, and short legs. The biggest difference in their appearance is the width of their snouts. The crocodilians are the largest reptiles, but they vary greatly in size. Caimans are the smallest crocodilians, approximately 2–3 m (about 7 ft) in length. The Indo-Pacific crocodile, on the other hand, is the largest crocodilian. It can measure 7 m (23 ft) and can weigh more than 1000 kg (over 1 ton).

saltwater crocodile

SCIENCE BACKGROUND

Eating habits—Alligators and crocodiles swallow their food whole. In addition to food, they also swallow stones. These stones remain in their stomachs to help grind the food they swallow.

A snake can separate its upper and lower jaws so that it can devour and swallow its prey whole.

 In the Garden of Eden, snakes may have had legs, but God cursed them to crawl on their bellies. The anatomy of the snake is a reminder of God's triumph over Satan (Gen. 3:15).

 Endangered species lists—Before an organism can receive protection under the Endangered Species Act, it must have met the specific criteria needed to be placed on the list of endangered and threatened wildlife and plants.

Endangered Species Act—The U.S. Congress passed the Endangered Species Act in 1973. Its purpose is to protect and preserve endangered species and their habitats.

Discussion:
pages 118–19

▶ **What are three characteristics of reptiles?** *{cold-blooded, lay eggs, scaly skin}*

▶ **How are reptiles like amphibians?** *{Possible answers: They are cold-blooded. They lay eggs.}*

💡 **What is the difference between a reptile and an amphibian?** *{Reptiles have scaly skin, can live away from the water, and lay eggs on land. Amphibians have smooth skin, live in or near water, and lay eggs in water.}*

▶ **Into what three major groups do scientists divide reptiles?** *{turtles, lizards and snakes, crocodilians}*

▶ **What unique skeletal feature identifies a turtle?** *{Possible answers: the hard, bony plates on its back; the shell}*

💡 **What does *streamlined* mean?** *{Answers will vary, but elicit that it means smooth and straight.}*

▶ **Why would having a streamlined shell and webbed feet help turtles that live near the water?** *{They make swimming easier.}*

▶ **Explain the difference between omnivores and herbivores.** *{Omnivores eat plants and animals. Herbivores eat plants only.}*

💡 **What does *endangered species* mean?** *{The species exists in limited numbers in its natural habitat and is dangerously close to extinction.}*

🗺 Point out the Galapagos Islands on the world map.

▶ **Most lizards are small and harmless. What are some examples of lizards that are dangerous?** *{the Gila monster of the southwestern United States and the Komodo dragon}*

▶ **What are two characteristics of snakes?** *{Possible answers: legless bodies, clear scales over each eye, no moveable eyelids}*

▶ **What is a *carnivore*?** *{a meat eater}*

▶ **How do snakes swallow prey larger than the diameter of their bodies?** *{Snakes' jaws are not attached, but they have strong ligaments that allow the jaws to open widely.}*

▶ **Why do snakes need such jaws?** *{They cannot tear or chew their prey.}*

▶ **How are snakes useful to man?** *{They kill unwanted pests, such as rats, insects, and slugs.}*

▶ **What are the largest reptiles?** *{the crocodilians}*

▶ **Which kind of crocodilian is the largest?** *{the Indo-Pacific crocodile}*

▶ **What is the primary food of crocodilians?** *{fish}*

Chapter 5   Lesson 57

Discussion:
pages 120–21

➤ Why are birds classified as birds? *{They have feathers.}*

➤ What are some purposes for feathers? *{Possible answers: assist in flight, protect from water, warmth}*

➤ Why is the ability to fly not a unique characteristic of birds? *{Not all birds can fly, and other animals, such as bats and insects, can also fly.}*

➤ Describe the unique design of a bird's skeleton and how it helps the bird to fly. *{The bones are very strong, but they contain hollow, air-filled cavities. The skeleton is strong but light.}*

➤ Not all birds fly. Name some flightless birds and tell how they move. *{Possible answers: Penguins swim. Emus and ostriches run.}*

Refer the student to *Creation Corner*.

➤ What are the largest flying birds? *{albatrosses}*

💡 How do albatrosses fly such long distances without tiring? *{They don't need to move their wings very much. Their huge wingspans help them float on wind currents.}*

📖 Why do birds have differently shaped beaks? *{God has designed the beak of each bird for the type of food it eats.}*

God has also designed every person with special abilities and talents. [BAT: 3a Self-concept]

➤ What is a strong, thick beak typically used for? *{cracking seeds and nuts and pulling the nut meat out of the seed's outer covering}*

Refer the student to the pictures of beaks.

💡 A parrot eats nuts and berries. Which picture do you think shows a parrot beak? *{the left one}* Why? *{It is thick and looks strong.}*

💡 Which beak would be used for sucking nectar? *{the one on the right; hummingbird}*

💡 Which beak would be used for tearing flesh? *{the middle beak}* (This picture is of an eagle's beak.)

💡 What do we call animals that eat other animals? *{carnivores}*

💡 What kinds of birds are carnivores? *{Possible answers: eagles, hawks, falcons}*

emu

Birds

What makes a bird a bird? Is it the wings, the ability to fly, or the ability to lay eggs? No, insects and certain mammals have wings. Not all birds can fly, and yet bats and many insects can fly. And reptiles, amphibians, fish, and arthropods all lay eggs.

Birds are birds because they have feathers. Feathers serve many purposes for birds. For some birds, feathers assist in flight. For other birds, feathers protect them from the water they swim in. Feathers also provide needed warmth. God has designed feathers perfectly so that birds can use them to the fullest.

Birds that fly have very lightweight skeletons. Their bones are very hard, but they contain hollow, air-filled cavities. Thus, the skeleton is strong yet light. This unique skeleton enables most birds to fly. However, some birds, such as the kiwi, have wings that are not meant to fly. The penguin cannot fly, but it can swim underwater better than any other bird. The ostrich and emu cannot fly because of their body

Creation CORNER

Albatrosses, the largest flying birds, are also some of the hardest-working birds. They nest on the shore, but they get food from the open sea. Sometimes adult albatrosses have to fly thousands of kilometers to get food for their young. God has given them the ability to fly long distances without ever moving their wings. They use wind currents to float along at high speeds. Albatrosses have huge wingspans to help them "catch the wind." Your outstretched arms measure approximately 1.3 m (about 4 ft). An average professional basketball player's arms would measure about 2.1 m (about 6.8 ft). However, some albatrosses have wingspans of over 3 m (about 9.84 ft)! Albatrosses are some of the most amazing birds in the animal kingdom.

120

Teacher Helps

Since both the parrot and the eagle beaks have a "hook," emphasize the thick beak of the parrot as the deciding factor of whether the beak is used to eat meat or hard materials, such as nuts.

SCIENCE BACKGROUND

Bird eggs—Birds' eggs are covered with tiny pores. These pores are so tiny that they are visible only with the aid of a microscope. The pores allow the embryo to take in oxygen and give off carbon dioxide and water. Each bird egg can have as many as 10,000 pores.

Penguins—Closely packed feathers and blubber help to maintain the body temperature of the penguin in arctic waters.

Name That Beak!

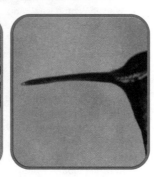

weight, but they have been equipped with long, powerful legs. These birds are excellent runners.

God has also given each bird the perfect beak for the food it eats. A bird that eats seeds has a strong, thick beak to crack seeds and nuts and to pull the meat out of the seed's outer covering. But a bird that eats other animals has a sharp hook to tear flesh off of the bones. A hummingbird has a thin, strawlike beak to suck nectar out of flowers.

Birds are **warm-blooded**, having body temperatures that stay basically the same, regardless of where the birds are. Reptiles, fish, and amphibians are cold-blooded.

All birds lay their eggs in nests. The variety of nest styles is just as amazing as the variety of birds. Some birds take a long time to build their nests, and they make them very elaborate. Others, like the cowbird and the European cuckoo, lay their eggs in other birds' nests. The cuckoo often lays one of her eggs in a magpie's nest. The magpie usually takes care of the cuckoo egg and the baby bird as if it were her own.

1. What is the difference between an omnivore and an herbivore?
2. What is unique about a snake's jaw?
3. How do the beaks of different birds reflect God's perfect design?

Discussion:
pages 120–21

➤ **What does *warm-blooded* mean?** {The body temperature of the animal stays basically the same no matter where the animal is.}

➤ **Where might cowbirds and the European cuckoo lay their eggs?** {in other bird's nests}

➤ **What happens when a cuckoo's egg hatches in a magpie's nest?** {The magpie takes care of the baby bird as if it were her own.}

💡 **What features of birds has man copied in his inventions?** {Possible answers: how their wings work; their lightweight skeletons; the waterproofing of their feathers}

📖 Man can learn from God's creation. Our gracious God has provided many examples and patterns from which man can learn.

Answers

1. An omnivore eats both plants and animals. An herbivore eats plants only.
2. The upper and lower jaws of a snake are not tightly attached. A strong ligament allows the jaws to separate widely. The snake can then swallow its prey whole.
3. The shape of a bird's beak matches the kind of food it eats.

Activity Manual
Study Guide—page 75
This page reviews Lessons 56–57.

Assessment
Test Packet—Quiz 5-B
The quiz may be given any time after completion of this lesson.

Direct a DEMONSTRATION
Demonstrate the wingspan of an albatross

Allow a student to measure a distance of 3 meters to show how large the albatross wingspan is.

 MATH Measure the outstretched arms of several students. Average the measurements. Determine how much longer the wingspan of an albatross is than the average arm span of a sixth grader.

Chapter 5 — Lesson 57

Lessons 58–59

Student Text pages 122–27
Activity Manual page 76

Objectives

- Identify four characteristics of mammals
- Explain how marsupials are different from other mammals
- Recognize how humans are different from mammals

Vocabulary

insectivore
nocturnal
echolocation
pride
blubber
pod

Materials

- pictures of a whale, rhinoceros, and porcupine
- world map

Introduction

Note: This material has been allotted two lessons. Divide the material where it is most convenient for your class.

Show the pictures of the whale, rhinoceros, and porcupine.

➤ **All mammals have hair. Describe the hair on each of the animals pictured.** *{Answers will vary.}*

Hair does not always look like a lion's mane or like the hair on our heads. A whale has very little hair. The horns of a rhinoceros are actually made of hair packed together tightly. The quills of a porcupine are thick hairs.

Purpose for reading:
Student Text pages 122–27

➤ Which characteristics do all mammals have in common?
➤ Why do the platypus and the echidna appear to belong to another group?

Mammals

Though fewer in number, mammals probably have the widest diversity of any group of animals. These vertebrates range in size from a tiny mouse to an enormous blue whale. Scientists have determined certain characteristics that categorize an animal as a mammal, regardless of its size or uniqueness. All mammals, even the aquatic mammals, have hair or fur. Mammals are warm-blooded, and the fur or hair helps land mammals maintain their internal body temperatures. Most mammals bear live young, and all mammals, including the egg-layers, feed their young with milk from the mother's body.

Other characteristics of mammals are less obvious. Unlike most other animals, every mammal has a four-chambered heart. Mammals also have three ear bones. Even whales, which have no outer ears, have middle ears with bones very similar to the ones in your ear. The whale receives sound vibrations through the tissues in its head.

Take a deep breath. You should be able to feel the muscle in your abdomen move. Most mammals use this muscle, the diaphragm, for moving air in and out of the lungs. All mammals breathe using lungs.

Monotremes

Monotremes are a unique kind of mammal. The Australian platypus (PLAT ih puhs) and echidna (ih KIHD nuh)

echidna

both lay eggs. No other mammal does that. But once the eggs hatch, the babies drink milk and have hair, just like other mammals. The platypus has a ducklike bill and poisonous spurs on its back legs. Though these features may look like those of reptiles or birds, the platypus still has all the characteristics of mammals. Some mammals, such as echidnas, do not have teeth. An echidna sticks out its tongue on top of an anthill or termite mound and waits for the insects to crawl onto it.

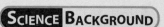
platypus

SCIENCE BACKGROUND

Marsupials—The majority of marsupials live in Australia. Some of these marsupials are the Tasmanian devil, numbat, bandicoot, wombat, mouse opossum, and cuscus.

North American marsupial—The only marsupial native to North America is the Virginia, or common, opossum. This opossum is known for "playing possum," or pretending to be dead when frightened or attacked.

South American marsupials—South America has two other marsupials. Each is in a separate class of marsupial. The *shrew opossum* is not an opossum except in name. It lives in western South America. The *monito del monte* is another South American marsupial that lives in the Andes mountains.

Monotremes—Australia is the home to all known monotremes. The echidna is also known as the spiny anteater.

kangaroos

Marsupials

These unusual mammals have pouches outside their bodies. In these pouches their underdeveloped babies grow big enough to function. When the baby comes out of the pouch, it is like other mammal babies. Many marsupials, such as kangaroos, koalas, and wallabies, live in Australia. All but one species of marsupials living in North and South America are called opossums.

Rodents, rabbits, and moles

Forty percent of all mammals are rodents. Mice, rats, and squirrels are common rodents, but beavers and porcupines also belong to this group. Rodents have large front teeth that never stop growing. The largest rodent, the capybara (KAP uh BAHR uh), lives in South America and averages 50 kg (110 lb) in weight. Rodents are found on every continent except Antarctica. They are often considered pests because they eat crops and gardens and get into people's houses. But some rodents are beneficial to humans because they eat harmful insects.

Rabbits share some characteristics with rodents, but a rabbit's teeth form differently than the teeth of a rodent. Moles and shrews are also similar to rodents, but they are **insectivores** (in SEK tuh VORZ). They eat insects as their primary food.

capybaras

SCIENCE MISCONCEPTIONS

Flying squirrels—Some students may suggest that a flying squirrel is another mammal that flies. Flying squirrels do not really fly. Flaps of skin between their front and back legs help them glide from high to low points.

Discussion:
pages 122–23

➤ **What characteristics do all mammals have?** *{Possible answers: have hair or fur, are warm-blooded, bear live young, feed babies milk from the mother's body, have a four-chambered heart, have three middle ear bones, breathe with lungs}*

➤ **Why are monotremes unusual mammals?** *{They lay eggs.}*

➤ **Give two examples of monotremes.** *{echidna, platypus}*

➤ **Where do these animals live?** *{Australia}*

➤ **Why are platypuses and echidnas not classified as reptiles or birds?** *{They have all the requirements to be mammals. They do not have scales or feathers.}*

➤ **What distinctive feature do marsupials have?** *{They have pouches outside their bodies, where their babies finish developing.}*

➤ **What are some examples of marsupials?** *{Possible answers: kangaroos, koalas, wallabies, opossums}*

➤ **What is the largest group of mammals?** *{rodents}*

➤ **What is a distinctive feature of rodents?** *{Rodents have large front teeth that never stop growing.}*

💡 **Beavers are rodents. Why would having continuously growing teeth be useful to a beaver?** *{Possible answer: Beavers wear down their teeth as they gnaw trees, but the teeth continue to grow, so the teeth do not wear down.}*

➤ **What is the largest rodent?** *{capybara}*

➤ **How do some rodents help man?** *{by eating insects}*

➤ **What other groups of animals are similar to rodents?** *{rabbits, moles, and shrews}*

➤ **What is an *insectivore*?** *{an animal that eats insects only}*

Discussion:
pages 124–25

➤ What are the only flying mammals? {bats}

➤ What does *nocturnal* mean? {comes out at night}

➤ What is *echolocation?* {a technique used by some animals to determine the location of objects; The animal produces high-frequency clicking sounds that bounce off objects. The length of time it takes for the sound to bounce back off an object and reach the animal tells the animal how far away the object is.}

➤ What man-made technology uses the technique of echolocation? {sonar}

➤ How do scientists group hoofed mammals? {by the number of toes on each hoof—an odd number or an even number}

➤ What are some common odd-toed mammals? {Possible answers: horses, zebras, burros, and mules}

➤ What is another name for an even-toed animal? {cloven-foot}

💡 Give an example of an even-toed animal that lives in each of these places. (Possible answers given.) In the plains of Africa. {giraffe} In the forest. {deer} In the desert. {camel} On a farm. {cow, sheep, pig}

Bats

The next largest group of mammals after rodents is bats. Bats are the only mammals that can fly. They are usually **nocturnal**, meaning they come out at night. Bats can see, but just like you, they cannot see well in the dark. In order for them to know where they are flying, bats use a technique called **echolocation.** They make high-frequency clicks that bounce off objects. The bats judge the distance to the object by the time it takes the sound to return. Bats hear and use sounds that are too high for humans to hear. Humans use an artificial form of echolocation. We call it sonar.

Sometimes thousands of bats live in a single place. Bats eat mainly insects and fruit, and some bats help pollinate trees as they search for food.

Hoofed mammals

Scientists divide hoofed mammals into two groups: those with an odd number of toes on each hoof and those with an even number. Odd-toed animals include horses, zebras, burros, and mules. Tapirs and rhinoceroses are also odd-toed hoofed mammals.

Even-toed mammals are also called "cloven hoof." Deer, giraffes, camels, cattle, and sheep are all examples of this group of mammals. Some even-toed mammals live on the plains, some in forests, and others on farms. Antelopes, some of the fastest mammals, and hippopotamuses, some of the slowest, belong to this group.

even-toed— camel

odd-toed— zebra

124

SCIENCE BACKGROUND

Tapir—A tapir looks like a large pig with a short trunk instead of a snout. It lives in Central America, South America, and Southeast Asia.

Vampire bats—Most bats eat insects, but the vampire bat eats the blood of large birds and animals. Vampire bats do not suck blood, but rather lick the blood after making a small bite. The bat can lick up as much blood as its own body weighs. Much like the leech, the vampire bat's saliva contains chemicals that numb the victim's skin and keep the blood from clotting.

SCIENCE MISCONCEPTIONS

Blind bats—Bats are not blind. They just have poor night vision.

Carnivores

Many kinds of mammals are omnivores, but only a few are exclusively carnivores, or meat-eaters. Cats are carnivorous mammals. Although many people have domestic cats as pets, most of the cats in the world would not make good pets. They need space to roam. Most cats have *retractable claws*. A cat can make its claws disappear into its paw when it does not need them for hunting or climbing. This tool enables cats to move quickly and quietly, thus allowing them to stalk their prey.

Cats can live nearly anywhere and are native to every continent except Australia and Antarctica. With the exception of lions, most cats are solitary. Lions live in groups called **prides.** The prides establish very clear territory. Solitary cats also establish territories. Lions, cougars, tigers, and jaguars are just a few of the cats that roam the earth.

Dogs have been domesticated since Bible times to herd sheep, guard houses, and do other things. However, wolves, jackals, and dingoes are also dogs. Many kinds of untamed dogs hunt in packs. Every pack has a leader, and the other dogs obey him. Dogs have a very good sense of smell that helps them locate prey, and they can pursue prey a long distance. Dogs communicate by barks and howls. One of the eeriest nighttime sounds is the mournful howl of the coyote "talking" to his pack.

Another group of carnivores is called *pinnipeds* (PIN uh PEDZ). This group includes seals, walruses, and sea lions. Pinnipeds' primary food is fish, though they will eat mollusks, crustaceans, and even careless penguins.

African wild dog

sea lion

cougar

Discussion:
pages 124–25

▶ **What is a *retractable claw?*** {a claw that disappears into the paw when not needed}

▶ **Why are retractable claws important to cats?** {They allows cats to move quietly when stalking prey.}

▶ **Which cat lives in a group?** {lion} **What is the group called?** {a pride}

▶ **What are some examples of wild cats?** {Possible answers: lions, cougars, tigers, jaguars}

💡 **What does *domesticated* mean?** {tamed for man's use}

▶ **How are domesticated dogs used?** {Possible answers: to herd sheep; to guard houses}

▶ **Name some kinds of nondomesticated dogs.** {Possible answers: wolves, jackals, African wild dogs, dingoes}

▶ **How do most wild dogs hunt?** {in packs}

▶ **Why are dogs used to track other animals or to find people?** {Dogs have a good sense of smell.}

▶ **What is another group of carnivores?** {pinnipeds}

▶ **What is distinctive about these mammals?** {They live in and around water.}

▶ **Name two examples of pinnipeds.** {Possible answers: seals, walruses, sea lions}

▶ **What is a pinniped's primary food?** {fish}

LANGUAGE

The term *pinniped* comes from two Latin words: *pinna* meaning "feather," and *ped* meaning "foot." This group includes carnivorous water mammals that use flippers for moving.

Echolocation—*Echolocation* is a compound word.

▶ **What two words do you see in the term *echolocation?*** {echo and location}

▶ **How does knowing this is a compound word help you understand its meaning?** {Answers will vary, but elicit that echo refers to sound that bounces back and location refers to a specific place. The words together define a technique that shows where something is by how sounds bounce back.}

Discussion:
pages 126–27

- How are pinnipeds different from marine mammals? {Pinnipeds can live on land, but marine mammals live only in the ocean.}
- Why do scientists identify whales as mammals instead of fish? {Possible answers: They have the characteristics of mammals—hair, ear bones, and feeding their young with milk.}
- 💡 Is a marine mammal able to breathe underwater? Why? {No; because as a mammal it has lungs, which require air.}
- What fatty substance insulates whales against the cold? {blubber}
- What do baleen whales have instead of teeth? {baleen plates}
- What do baleen whales eat? {plankton and tiny crustaceans called krill}
- What is a *pod*? {a group of whales}
- Why are whales considered social animals? {They travel in pods and communicate with each other by making noises.}
- What are some examples of toothed whales? {Possible answers: dolphins, porpoises, orcas, and sperm whales}
- How are toothed whales similar to bats? {They use echolocation to navigate the waters where they travel.}
- What are some characteristics of primates? {good eyesight; "hands" that grasp}
- What are two groups of primates? {lemurs and monkeys}

 Find Madagascar on the map. Like Australia, this island nation is home to animals that are found in only a few other places.

- Which primates are found in Madagascar? {lemurs and related primates}
- What are some characteristics of New World monkeys? {broad noses; long, useful tails}
- 💡 Why are these monkeys called New World monkeys? {They are found in the New World.}
- 💡 What part of the world is considered the New World? {North, South, and Central America}
- Which areas of the New World have monkeys? {Central America and South America}

 Find Central and South America on the map.

humpback whales

Marine mammals

Pinnipeds spend much time in the ocean, but they can also live on land. However, some mammals live only in the ocean. Marine mammals may seem to lack the requirements for being mammals, but they too have hair, ear bones, and milk to feed their young. These mammals belong to the whale family. Whales have a fatty substance called **blubber** that insulates them against cold. Blubber is so rich in oil that people used to hunt whales to get this oil.

Some whales, called baleen whales, strain their food out of the water. Instead of teeth, these whales have giant plates, called *baleen plates,* in their mouths. Baleen plates help the whales gather plankton and tiny crustaceans called *krill*. The blue whale, the largest whale, can eat around 3500 kg (about 7,700 lb) of krill per day!

Whales are social creatures. They travel in groups called **pods.** Whales communicate to their pods with sounds.

The most famous noise is the male humpback whale's "song." Scientists do not know exactly why the humpback whale sings, but the song is one of the most interesting sounds of the ocean.

Toothed whales are usually smaller than the baleen whales. The smallest toothed whale is about 1.3 m (4 ft) long, and the largest is about 18 m (59 ft) long. Toothed whales can bite into their food. Dolphins, porpoises, orcas, and sperm whales all belong to this group. Toothed whales use echolocation in a manner similar to that of bats. Dolphins are well known for the clicking sounds that they use to navigate the ocean.

dolphin

126

SCIENCE BACKGROUND

Old World monkeys—Scientists differ on which families of primates to include in this group.

SCIENCE MISCONCEPTIONS

Students will generally hear mankind referred to as a mammal. In fact, in the classification system, man is classified as a mammal and as a primate. Although man's physical characteristics are similar to those of mammals, students should always be taught that man is a separate and special creation of God. Although all of God's creation brings glory to Him, only man can glorify God because he was made in God's image.

Primates

Many primates are tree-dwelling mammals. They typically have good eyesight and have "hands" that can grasp. Primates are divided into two groups, lemurs and monkeys. Most lemurs have long snouts, similar to those of dogs. Many lemurs and related primates live in Madagascar, an island off the east coast of Africa.

Scientists usually divide monkeys into two groups, New World monkeys and Old World monkeys. New World monkeys are found in Central and South America. They have broad noses, and most have tails that can be used almost like other arms.

Old World monkeys include baboons and several monkeys without tails, such as apes, gorillas, and chimpanzees. Many of these monkeys live in groups and appear to have a kind of social order within the group. Unlike the New World monkeys, many of these monkeys spend much of their time on the ground instead of in trees. Old World monkeys are found primarily in Africa and in both South and East Asia.

gorilla

Humans

Humans also have all the physical characteristics necessary to be mammals. But humans are not animals. Some scientists say man is different from animals only because he is rational—he can think. But Christians know that one difference between man and animals is that God gave man a soul. Man was created separately from the rest of creation and was formed in the likeness and image of God.

QUICK CHECK
1. Name four characteristics of all mammals.
2. How do bats and toothed whales use echolocation?
3. How are marsupials different from other mammals?

lemurs

127

Discussion:
pages 126–27

➤ **Where are Old World monkeys found?** {Africa, South Asia, and East Asia}

Find Africa, South Asia, and East Asia on the map.

➤ **What are some characteristics of Old World monkeys?** {tend to live in groups with social orders; small or no tail; live mainly on the ground}

➤ **Give some examples of Old World monkeys.** {baboons, apes, gorillas, and chimpanzees}

➤ **Why do some scientists classify man as a mammal?** {because man has all the physical characteristics of mammals}

➤ **Why is a man not an animal?** {Men have souls.}

Read aloud Genesis 1:24–27 and 2:7.

➤ **Is man's special glory the fact that he is rational?** {no}

➤ **Was man created as one of the animal groups? Explain.** {No; God created man separately.}

➤ **How was man made?** {in the likeness of God}

Answers

1. Possible answers: have hair or fur, are warm-blooded, bear live young, feed babies milk from the mother's body, have a four-chambered heart, have three middle ear bones, breathe with lungs

2. Bats and toothed whales bounce sounds off objects. The length of time needed for the sound to return tells bats and whales their distance from the objects.

3. Marsupials give birth to underdeveloped babies that then live in their mothers' pouches until they grow big enough to function. Most other mammals give birth to developed young.

Activity Manual
Reinforcement—page 76

Assessment
Test Packet—Quiz 5-C
The quiz may be given any time after completion of this lesson.

Direct a DEMONSTRATION
Demonstrate water vapor condensing

Materials: bowl of ice, mirror or window

The water seen spouting from a whale is actually the water vapor of the whale's exhaled air condensing as it hits air of a different temperature. To illustrate this concept, place the bowl of ice against the mirror or against the window. Forcefully exhale over the ice so that the air hits the mirror or window. As the air blows over the cooler ice and hits the mirror or window, it condenses as water droplets.

HISTORY The United States Navy has trained bottle-nosed Atlantic dolphins to search for explosive mines planted by enemy troops. The dolphins use echolocation to find the mines so that the mines can be removed or disarmed safely.

Chapter 5

Lessons 58–59   137

Lesson 60

Student Text pages 128–29
Activity Manual pages 77–78

Objectives

- Write a hypothesis
- Record temperatures and observations
- Relate the effectiveness of shortening or lard as an insulator to the effectiveness of animal blubber

Materials

- See Student Text page
- buckets or large containers to transport and dispose of the water and ice

Introduction

Review the usefulness of blubber in whales.

➤ **What is *blubber*?** {a fatty substance}

➤ **How is blubber helpful to whales?** {Blubber insulates their bodies against the arctic cold.}

➤ **What other animals have blubber to help keep them warm?** {Possible answers: polar bears, seals}

Purpose for reading:
Student Text pages 128–29
Activity Manual pages 77–78

Display shortening, batting, and plastic bags.

➤ **How do you think these materials could be used to show how walrus blubber insulates animals from cold temperatures and keeps body heat from escaping?** {Answers will vary.}

The student should read all the pages before beginning the activity.

ACTIVITY: Blubber Mitts

Walruses live in arctic conditions. They rest and bear their young on snow-covered moving ice called *ice floes*, where the air temperature may be as low as –50°C (–58°F). Walruses can dive deep in the icy, arctic waters. To survive the frigid arctic conditions, walruses have thick, tough skin, much like that of a rhinoceros. Under the skin is a thick yellow layer of blubber. Why has God given walruses this layer of blubber? Experiment to find out how effective different materials are at insulating against the cold.

Problem

What materials best insulate against cold?

Preparation

1. Make the **non-insulated mitt** with two plastic bags. Carefully turn one plastic bag inside out. Place your hand inside this bag and push it into the bag that is right-side-out. Zip together the two bags so you can still insert your hand.

2. Make the **batting-insulated mitt** with two plastic bags and the piece of quilt batting. Repeat step 1, but fold the batting in half and insert it into the right-side-out bag before inserting the inside-out bag.

3. Make the **blubber-insulated mitt** with two plastic bags and the shortening or lard. Repeat step 1, but place 500 mL of shortening in the right-side-out bag before inserting the inside-out bag. Tape the top edge as needed for a better seal. Squish the shortening until it makes a layer about as thick as the batting in the batting-insulated mitt.

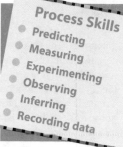

Process Skills
- Predicting
- Measuring
- Experimenting
- Observing
- Inferring
- Recording data

Materials:
resealable bags
15 cm × 30 cm piece of quilt batting, 1 cm thick
solid shortening or lard
metric measuring cup
rubber spatula or scraper
heavy tape (optional)
deep dishpan or wide bucket
ice cubes to fill dishpan half full
water
4 thermometers
clock or timer
Activity Manual

128

 Teacher Helps

This activity may be done in one or two days. If continuing the activity a second day, allow the students to bring other insulating materials to test.

Prepare ice cubes the day before. Keep them in an ice chest.

Self-sealing plastic bags with sliding zipper closures will not work for this activity.

To reduce the mess when measuring the shortening, place the plastic bag inside the measuring cup and flatten the sides of the bag around the edges of the measuring cup. Measure the correct amount of shortening directly into the bag.

If you would rather not have the student measure the shortening, prepare the bags ahead of time.

Science Background

Desert mammals—Camels, alpacas, and llamas have fur, which helps to insulate their bodies from the desert heat.

Body insulators—Insulators such as blubber not only protect the body from the cold temperatures on the outside, but they also decrease the loss of body heat. The thick skin of the walrus protects it against the cold, and the food eaten by the walrus provides energy that is transformed into heat. As an animal eats, it replenishes lost heat.

Procedures

1. Write your hypothesis in the Activity Manual, stating which mitt you think will insulate the best against the cold.

2. Fill the dishpan half full with ice. Add water, filling to about 5 cm from the top of the dishpan.

3. In *Column A* on the chart record the temperature of each thermometer at room temperature.

4. Two or three people are needed. Insert a hand and a thermometer into each mitt. Keep the hands inside the mitts until step 8 is finished. Hold the fourth thermometer in the water. (Do not place the mitts into the water yet.)

5. After 2 minutes, check and record in *Column B* the temperatures of each mitt and the water. Then replace each thermometer.

6. Place each mitt into the ice water. Keep the fourth thermometer in the water.

7. After 2 minutes, check and record in *Column C* the temperatures of each mitt and the water.

8. Return the thermometers. After 2 more minutes check and record in *Column D* the temperatures of each mitt and the water. Repeat again after 2 more minutes, recording your temperatures in *Column E*.

Conclusions

- Were your predictions correct?
- How can you apply what you have learned to other areas, such as special clothing for arctic explorations and mountain rescue teams?

Follow-up

- Try the activity using warm water instead of ice water. Will the results be the same?

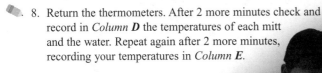

SCIENCE PROCESS SKILLS

Collecting and Recording Data

Discuss after the activity.

➤ Why is it necessary to record the temperature before putting the mitts in the ice water? *{To know how much the temperature changes, you must know what the starting temperature is.}*

➤ Why is the data gathered at two minute intervals rather than at five minute intervals? *{Elicit that at five minute intervals you may not be able to see the progression of temperature changes.}*

Gathering data at shorter intervals of time enables the observer to see the progression of changes in the experiment. If the observer gathered data at longer intervals of time, he would see the end result but not the progression.

➤ Why is recording the data important? *{With four different temperatures to check each time, it would be easy to forget or mix up the temperatures.}*

Discussion:
pages 128–29

💡 What do we mean when we say that walruses live in arctic conditions? *{They live where it is very cold.}*

➤ On what do walruses often rest and bear their young? *{ice floes}*

Procedure

Guide the students in preparing the different mitts. Make sure the narrow edges of the insulated bags have as much insulation as the flat sides.

Review how to read a thermometer accurately.

Conclusions

➤ Which material was the best insulator against the cold water? *{shortening or lard}*

➤ What do you think would happen if you decreased the amount of shortening in the blubber mitt? *{The mitt probably would not have insulated as well.}*

💡 What other materials would make good insulators? *{Possible answers: feathers, foam bits, wool, fake or real fur, sawdust, scrap fabric}*

➤ How can you apply what you learned about insulation and put it to practical use? *{Answers will vary but may include a discussion of clothing and building materials.}*

Activity Manual
Activity—pages 77–78

Assessment

Rubric—TE pages A50–A53

Select one of the prepared rubrics, or design a rubric to include your chosen criteria.

Chapter 5

Lesson 60

Lessons 61–62

Student Text pages 130–31
Activity Manual page 79

Explorations: Animal Robotics

Objectives
- Associate animal parts with mechanical tools
- Research to design a robotic animal
- Prepare a drawing and description of the robotic animal

Materials
- picture of a hang glider
- reference books, such as encyclopedias and library books
- unlined paper or poster board
- colored pencils, felt-tip pens, or crayons

Introduction

Note: Two lesson days are allotted for this exploration. On the first day, introduce the activity, set guidelines and a due date for the drawings and reports to be completed, and begin planning. The second lesson day may occur later after the drawings and reports are finished.

➤ **Which man-made machine models a bat's or dolphin's echolocation?** {sonar}

We can learn and apply many concepts of movement and operation by watching the world around us.

Show the picture of the hang glider.

➤ **Of what animal does the hang glider remind you?** {Possible answers: any large soaring bird, such as an eagle, a hawk, or an albatross}

➤ **How does a hang glider remind you of the wingspan of an eagle?** {The shape and support rods on the glider are like the shape and bone structure of an eagle's wings.}

Purpose for reading:
Student Text pages 130–31
Activity Manual page 79

➤ **How does the mantis shrimp use its hard "elbow" to get food?**

➤ **How are the mantis shrimp's eyes shaped?**

The student should read all the pages before beginning the exploration.

Many of the tools and machines that man designs are based on the superior design of God's creation. For example, man designed sonar, but many of God's creatures, such as bats and dolphins, have built-in sonar. Man uses scuba flippers to move around underwater, but God has equipped His aquatic creatures with ready-made flippers. God has created many complex creatures with bodies perfectly suited to their needs.

Perhaps you have seen someone with a mechanical part to replace a joint or missing limb. That mechanical part is designed to function as closely as possible to what it replaces. Suppose you were to replace some parts of an animal with mechanical parts. What machines or tools would you use? Of course what you use would depend on the animal you are making. Let's use a mantis shrimp as an example.

Mantis shrimp are crustaceans that pack a powerful punch for their size. The front appendages of the mantis shrimp fold under its head like those of a praying mantis. When it sees prey, the shrimp unfolds its appendages with the speed of a bullet. Some mantis shrimp are called spearers because the appendages have sharp spines on them that stab the victim. Others are called smashers because each appendage has a hard "elbow" that the shrimp uses to smash mollusks and other crustaceans. The mantis shrimp has been known to break divers' fingers and aquarium glass.

Good vision is also important for catching prey with lightning speed. The mantis shrimp has compound eyes that can see in every direction. These eyes rest on the top of short stalks. Each compound eye is radar-dish-shaped and has three pupils. Divers often see only the eyes of the mantis shrimp peering out from a sand burrow.

Use the picture of the robotic mantis shrimp in the Activity Manual to help you design your own robotic animal.

What to do
1. Choose an animal to use as your robotic animal. Pay close attention to unusual body parts that might be replaced by mechanical devices. You may use one of the following animals

130

Teacher Helps: Students may work in groups to research and design the robotic animal. Each student should write his own description.

Encourage students who are not confident drawing their animal to trace the animal picture.

You may choose to have the students make models of their animals.

or choose one of your own: hummingbird, beaver, bombardier beetle, Komodo dragon, spade-foot toad, stalk-eyed mud crab.

2. Gather information about the animal you selected. How does your animal use its body parts? What parts remind you of other tools or equipment that are used around us? Where does this animal live, and how does its environment affect the way it looks and uses its body?

3. Select at least four body parts of the animal to make robotic. Label the tool, machine part, or piece of equipment that is used to replace an animal body part. Label its function. In parentheses, label the actual name of the animal's body part. For example, the paddles on the mantis shrimp are used for movement. The actual name for these paddles is swimmerets.

4. Write a brief description of where the actual animal lives, what it eats, and any other information that you would like to include.

5. Display and explain your drawing. Read the description about your animal.

131

Direct an Activity

Materials: Pictures of animals, such as a swan or duck, woodpecker, flying gecko or flying squirrel, mole

Choose several animal parts. Discuss what function each part has. Allow a student to draw a mechanical device that corresponds to an animal part.

Suggestions:

swan's or duck's feet—paddles

woodpecker beak—drill

gecko frill or flying squirrel—hang glider

mole's front paws—shovel

Discussion:
pages 130–31

God has made man to have dominion over the earth. Part of that dominion is man's curiosity and creativity in producing new things. Often God's creation provides the inspiration and pattern for these new things.

Stimulate student interest and imagination by discussing the features of the mantis shrimp. Relate each part of the shrimp to a mechanical device or tool.

➤ **What is an *appendage*?** *{Answers may vary. Elicit that it is a claw- or armlike body part.}*

Look at the pictures on Activity Manual page 79 as you discuss each feature of the mantis shrimp.

💡 **Why can a small appendage pack such a powerful punch?** *{Answers may vary, but elicit that the speed increases the force the appendage produces.}*

➤ **What does the shrimp use to smash mollusks and other crustaceans?** *{its hard "elbow"}*

💡 **Do you think the shrimp actually moves as fast as a bullet or as fast as lightning? Why do we use these phrases?** *{no; Accept any answer, but elicit that those descriptive phrases help us visualize that the shrimp is very fast.}*

➤ **Why would having eyes on stalks allow an animal to see in more directions than having eyes directly on its head would?** *{Answers will vary, but elicit that the head interferes with the field of vision.}*

➤ **How would periscope-mounted eyes be helpful to the mantis shrimp waiting in a sand burrow or crevice in a rock.** *{Answers will vary but may include that the shrimp can see prey from all angles and can be ready to attack without leaving the security of its hiding place.}*

Procedure

Guide the student in choosing and researching his robotic animal.

Other possible animals include a pelican, armadillo, rhinoceros beetle, and hognose snake.

Activity Manual
Explorations—page 79

Assessment
Rubric—TE pages A50–A53

Select one of the prepared rubrics, or design a rubric to include your chosen criteria.

Lesson 63

Student Text page 132
Activity Manual pages 80–82

Objectives

- Recall concepts and terms from Chapter 5
- Apply knowledge to everyday situations

Introduction

Material for the Chapter 5 Test will be taken from Activity Manual pages 69–72, 75, and 80–82. You may review any or all of the material during this lesson.

You may choose to review Chapter 5 by playing "Vertebrate vs. Invertebrate" or a game from the *Game Bank* on TE pages A56–A57.

Diving Deep into Science

Questions similar to these may appear on the test.

Answer the Questions

Information relating to the questions can be found on the following Student Text pages:

1. page 114
2. pages 120, 126
3. page 115

Solve the Problem

In order to solve the problem, the student must apply material he has learned. The answer for this Solve the Problem is based on the material on Student Text pages 116–19. The student should attempt the problem independently. Answers will vary and may be discussed.

Activity Manual

Study Guide—pages 80–82

These pages review some of the information from Chapter 5.

Assessment

Test Packet—Chapter 5 Test

The introductory lesson for Chapter 6 has been shortened so that it may be taught following the Chapter 5 test.

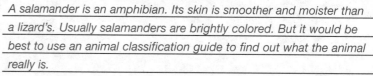

Answer the Questions

1. Why would cold-blooded animals have a difficult time surviving in areas such as the polar regions?
   Because cold-blooded animals depend on the environment for their warmth or coolness, they cannot keep enough body warmth to survive in consistently cold conditions.

2. Penguins have blubber, solid bones, and small wings. How does God's design for penguins help them to function in their natural habitat of Antarctica?
   Answers will vary, but elicit that blubber helps the penguins stay warm. Solid bones help the penguin sink deeper, and small wings are not good for flying but are good for swimming.

3. Many ocean animals are brightly colored. Why is this a better camouflage in the ocean than it would be on the land?
   Answers will vary, but elicit that many areas of the oceans abound with brightly colored plants and animals, so brightly colored animals can blend in easily. Areas on land are more often predominantly green or brown.

Solve the Problem

Your little cousin brings a four-legged animal into the house. Your older brother is positive that it is a lizard and firmly declares it is a reptile. But you are not so sure. You think it might be a salamander. What animal group does a salamander belong to? What could you do to find out whether you or your brother is correctly classifying the animal?
A salamander is an amphibian. Its skin is smoother and moister than a lizard's. Usually salamanders are brightly colored. But it would be best to use an animal classification guide to find out what the animal really is.

Review Game

Vertebrate vs. Invertebrate

Divide the students into two teams. Prepare a tic-tac-toe grid. The team using *X* as its symbol should receive questions about invertebrates. The team using *O* as its symbol should receive questions about vertebrates. After each game, or halfway through the review time, switch the question designation so that each team answers questions on all of the material.

6 Plant Classification

In the frozen tundra grows an unusual "plant." Reindeer moss is an important source of food for caribou, moose, and reindeer. But actually, it is not a moss at all. In fact, it is not even a plant. Reindeer moss is a lichen. Lichens are combinations of two organisms, algae and fungi. The algae provide food through photosynthesis, and the fungi provide water and protection. The organisms live together in a symbiotic relationship—each benefits from the other. Lichens are important not only for animals, but also for breaking down rocks to help produce soil. Today, reindeer moss is often more correctly called reindeer lichen. Man has to continually re-evaluate his understanding of creation. But God never has to update His knowledge. He knows everything about all the organisms that He created.

Student Text page 133
Activity Manual page 83

Lesson 64

Objectives
- Demonstrate knowledge of concepts taught in Chapter 5
- Recognize that man's knowledge must continually be re-evaluated

Introduction

Note: This introductory lesson has been shortened so that it may be taught following the Chapter 5 test.

Man tries to organize and name living things to the best of his knowledge and abilities. But sometimes his knowledge is incomplete. As God allows him to learn more, man may need to change or reorganize living things that have already been classified. [Bible Promise: I. God as Master]

The lichen is one example of an organism that was reclassified because man developed better methods to study it.

➤ **Why would men have named the lichen "reindeer moss"?** *{Answers may vary, but suggest that it looks like moss and is food for reindeer.}*

➤ **To which kingdoms does lichen belong?** *{protista and fungi}*

➤ **What is a *symbiotic relationship*?** *{organisms living together and benefiting each other}*

Chapter Outline
I. Nonvascular Plants
   A. Mosses
   B. Liverworts
II. Vascular Plants
   A. Seedless Vascular Plants
   B. Seed-Bearing Vascular Plants
   C. Plant Parts

Activity Manual
Preview—page 83

The purpose of this Preview page is to generate student interest and determine prior student knowledge. The concepts are not intended to be taught in this lesson.

Lesson 65

Student Text pages 134–37
Activity Manual page 84

Objectives

- Explain the difference between vascular and nonvascular plants
- Recognize that vascular plants can be classified as seed-bearing plants and seedless plants
- Identify ferns, horsetails, and club mosses as seedless vascular plants
- Identify the parts of a fern: rhizome, fronds, fiddleheads

Vocabulary

vascular plant   rhizome
nonvascular plant   frond
rhizoid   fiddlehead

Materials

- pictures of common plants that students would easily recognize: trees, mosses, ferns, flowering plants, grass, and shrubs or bushes

Introduction

Display the pictures of common plants.

➤ **What do these pictures have in common?** *{Each picture shows plants.}*

➤ **How would you identify each picture?** *{Answers will vary.}*

➤ Today we will learn how scientists classify plants.

Purpose for Reading:
Student Text pages 134–37

➤ How do scientists classify plants?
➤ How does a vascular system benefit a plant?

Discussion:
pages 134–35

 Refer to the classification diagram on Student Text page 134 as each new plant group is discussed.

💡 What are some common characteristics of most plants? *{Possible answers: roots, stems, leaves, ability to make their own food}*

📖 Plants show God's love of beauty. God could have made all plants exactly the same, but He chose to make them fulfill their purpose through variety and beauty.

What do you think of when you think of plants? You might think of flowers in a yard, vegetables in a garden, or a forest of tall trees. The Bible tells us in Genesis 1:11–13 that God created plants on day three of Creation. The plants that God created come in all shapes and sizes. Most of them have roots, stems, and leaves and can make their own food through photosynthesis.

Plants can be classified in many different ways—by how they reproduce, their growing habits, their seed structures, or even their height. However, when scientists separate plants into two big categories, they usually classify them by how each plant transports water. Most plants have tubelike structures that transport water from the roots to the stems and leaves. These plants are called **vascular** (VAS kyuh lur) **plants**. Plants that do not have these structures are called **nonvascular plants**.

Nonvascular Plants

All plants need water to survive. Nonvascular plants usually grow in moist places. Because they do not have vascular systems, these plants absorb water and nutrients through their leaves. As a result, most nonvascular plants do not grow very big or tall.

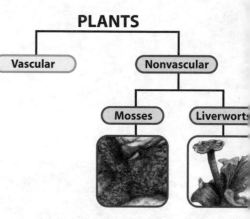

Mosses

Scientists have identified more than ten thousand species of moss. Moss is a nonvascular plant found in most places around the world. Most species of moss are only a few millimeters tall, but some tropical mosses can be as tall as 70 cm (27½ in.). Moss plants usually grow in groups and spread out over large areas. They can grow on rocks, in soil, and sometimes even on other plants, such as tree trunks.

SCIENCE BACKGROUND

Mosses—Mosses can be divided into three categories: peat mosses, granite mosses, and true mosses. Peat mosses, especially sphagnum, often grow in bogs. The acidity in bog water prevents plants from decaying, so dead moss compresses into layers, forming peat. Some people use peat as a fuel for heating and cooking, as well as for medicine. Peat is also used in gardening.

Hornworts—Hornworts, another nonvascular plant, are similar to liverworts, but they are seldom found on rocks and tree trunks. They usually live in damp soil among grasses. Hornworts produce spores inside horn-shaped stalks.

Reproduction—All of the plants mentioned in this lesson reproduce by spores. Unlike seeds, spores have no stored food available for the new plant. Spores are discussed in Chapter 12, *Reproduction*.

SCIENCE MISCONCEPTIONS

Not everything named "moss" is a moss. Spanish moss, a member of the pineapple family, is a flowering plant. Reindeer moss is actually a lichen, and club mosses are vascular plants.

---

144   Lesson 65

Mosses do not have true roots. Instead, they have thin, rootlike structures called **rhizoids** (RY zoydz). Mosses also have tiny leaves that grow from stemlike structures. These leaves are usually only one cell thick.

Even though they are small, mosses are very beneficial. They are often one of the first plants to grow in areas destroyed by volcanoes or fires. They also help prevent erosion by holding dirt in place. Other mosses, such as peat mosses, have been used for heating, cooking, and medicines.

leafy liverwort

Moss Plant

rhizoids

Liverworts

Liverworts can look similar to mosses, but their leaves are arranged differently. More than eight thousand species of liverworts have been identified. Liverworts got their name because people thought that their leaves resembled the shape of a liver.

Like mosses, liverworts usually grow in moist places, especially on rocks near streams and waterfalls. Leafy liverworts are often mistaken for mosses. Their leaves are on stemlike structures similar to those of mosses. Other liverworts, though, do not have a stem structure. Their leaves are usually flat like plates.

LANGUAGE

The name *liverwort* comes from an old English name meaning "liver plant" (*lifer* = "liver" and *wyrt* = "plant"). People used to think that the shape of a plant indicated its medicinal value, so they believed that liverworts could help liver problems. The modern scientific name for a liverwort is *Hepaticae*, which is Latin for *liver*.

Discussion:
pages 134–35

➤ **What are some ways that scientists classify plants?** *{Possible answers: how they reproduce, growing habits, seed structure, height}*

➤ **How do scientists separate plants into two main categories?** *{by how each plant transports water}*

➤ **What is the difference between vascular plants and nonvascular plants?** *{Vascular plants have tubelike structures that can carry water to all parts of the plant. Nonvascular plants do not have these structures.}*

➤ **What do all plants need to survive?** *{water}*

➤ **Why are nonvascular plants usually small?** *{Since they do not have vascular systems, the leaves absorb water and nutrients.}*

💡 **Where have you seen mosses grow?** *{Answers will vary but may include along the rocks of a stream or on live and dead trees.}*

💡 **Why might tropical mosses be able to grow so much larger than mosses in other parts of the world?** *{Tropical areas, such as rain forests, often receive more moisture. The high humidity allows mosses to get water even away from the ground.}*

➤ **What are *rhizoids*?** *{thin rootlike structures of nonvascular plants}*

➤ **How thick are most tiny moss leaves?** *{one cell thick}*

➤ **What are some benefits of mosses?** *{Possible answers: They are the first plants to grow in areas burnt by volcano eruptions or fires. They help prevent erosion. They are used in cooking, heating, and medicines.}*

💡 **How are mosses related to weathering and erosion?** *{They help prevent erosion, but they also secrete acids that help break down rocks into soil.}*

➤ **How did liverworts get their name?** *{People thought their leaves resembled the shape of a liver.}*

➤ **Where do liverworts grow?** *{in moist places, especially near streams and waterfalls}*

➤ **Which type of liverworts are often mistaken for mosses?** *{leafy liverworts}*

➤ **How are leafy liverworts similar to mosses?** *{Their leaves are on stemlike structures, similar to the leaves of mosses.}*

➤ **Although leafy liverworts have a stem and leaf structure similar to that of mosses, what other leaf shape can liverworts have?** *{flat, like plates}*

Discussion:
pages 136–37

▶ Why can vascular plants grow larger than nonvascular plants? *{The vascular system transports water and food and helps to strengthen and support the plant.}*

▶ How are ferns, horsetails, and club mosses different from other vascular plants? *{They do not have seeds.}*

▶ Horsetails, club mosses, and ferns all grow from rhizomes. How are rhizomes different from rhizoids? *{A rhizome is an underground stem. Rhizoids are the rootlike structures of nonvascular plants.}*

▶ What kind of stems do horsetails have? *{tall, hollow, jointed stems}*

▶ What is unusual about the leaves of a horsetail? *{Possible answers: They are often colorless. They look like scales. They clasp the stem close to each joint.}*

💡 Why do you think horsetails produce food in their stems and branches instead of in their leaves? *{Answers will vary, but elicit that their brown or colorless leaves do not contain chlorophyll, so they cannot make food in their leaves.}*

▶ Where are horsetails usually found? *{Possible answers: riverbanks, marshes, ditches, and meadows}*

Refer the student to *Science and History*.

▶ Why did people in Colonial times call horsetails "scouring rushes"? *{They used horsetails to scour and scrub their pots.}*

▶ What other names did horsetails have? *{pewterworts, shavegrass}*

146 Lesson 65

Vascular Plants

Plants with vascular systems are able to grow larger than mosses and liverworts. This is because the vascular system strengthens and supports the plant and also transports water and food to all parts of the plant.

Seedless Vascular Plants
Horsetails

Most vascular plants have seeds. However, horsetails, club mosses, and ferns are three types of vascular plants that do not have any seeds. Horsetails grow from underground stems called **rhizomes** (RYE zohmz). These plants are mostly tall, hollow, jointed stems. Some horsetails have needlelike branches that grow in a circle around each joint, but other horsetails do not have any branches. The leaves are very small and are often brown or colorless.

horsetails with branches

136

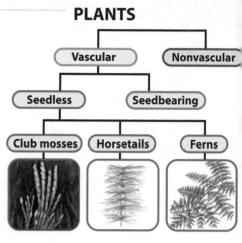

They look like tiny scales that clasp the stem close to each joint. Horsetails usually produce their food in the stems and branches instead of in the leaves. Even though only about thirty species of horsetails exist, they can be found in many different locations. Horsetails thrive in moist places such as riverbanks, marshes, ditches, and meadows.

Science and History

Horsetails have many different names and many different uses. In Colonial times, horsetails were often called scouring rushes. Early Americans used the horsetails to scour and scrub their pots. They were also called pewterworts, since they could be used to polish metal and pewter. Another name was shavegrass because they could even be used to sand wood.

horsetails without branches

SCIENCE BACKGROUND

Spike mosses and quillworts—Two other types of plants closely related to club mosses are spike mosses and quillworts. A well-known spike moss is the resurrection plant. This plant curls up when dry and unfurls when moistened. It can survive several years in dry conditions. Spike mosses are usually found in the tropics. Quillworts grow in water or in areas where they can be partially underwater for part of the year. Quillworts are sometimes called Merlin's grass.

SCIENCE MISCONCEPTIONS

The asparagus fern is not really a fern. It is actually an asparagus, a member of the lily family. This fern is a flowering plant and does not produce spores.

club moss

Club mosses

Although club mosses may look similar to mosses, they really have little in common with them. Club mosses do not usually grow taller than 30 cm (almost 12 in.). They can resemble small evergreen trees and are often called ground pines. Club mosses are frequently found in forests near streams or other moist places. In tropical climates, however, some species of club mosses live on trees.

Ferns

Scientists have identified at least twelve thousand species of ferns. Some ferns, such as rock ferns, can tolerate heat and drought. But most ferns need moisture. Ferns are often found by streams and waterfalls or in wooded areas and pastures.

Like horsetails, ferns grow from rhizomes. Rhizomes can grow to be quite long, and many **fronds,** the leafy branches of a fern, can grow from just one rhizome. Tree ferns found in the tropics sometimes have fronds that are almost 4 m (13 ft) long. However, other ferns are small, having fronds that are only about 1.5 cm (0.6 in.) long.

When a fern is just beginning to grow, its fronds are coiled up tightly. The coiled-up frond is called a **fiddlehead** because it resembles the top of a violin. Some fiddleheads, such as those from an ostrich fern, are edible and are sometimes used in salads or vegetable dishes. Others, though, are quite poisonous.

One of the most common ferns is the bracken fern. The fronds of this fern may grow to be as long as 2 m (6½ ft). They are shaped like triangles, and each frond has three leaflets. Bracken ferns are often found in wooded areas, especially near oak, pine, and maple trees.

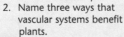

1. What are the two main classifications of plants?
2. Name three ways that vascular systems benefit plants.

Bracken fern

Fern — frond, rhizomes

Spores from club mosses have been used in fireworks, dusting powder, and flash powder used in early photography.

Direct a Demonstration
Look at the parts of a fern

Materials: a fern

Point out the fronds and any fiddleheads that may be growing. If possible, show the underground rhizomes.

Discussion:
pages 136–37

➤ Why are club mosses often called ground pines? {because they resemble small evergreen trees}

➤ Where are club mosses typically found? {near streams or other moist places}

➤ Where can ferns usually be found? {Possible answers: in moist places, wooded areas, or pastures}

➤ Which part of a fern is the frond? {the leafy branch}

➤ Where are tree ferns usually found? {in tropical areas}

💡 Why do you think the fronds of tree ferns are so much longer than the fronds of other ferns? {because tropical areas usually receive more water}

➤ What is a *fiddlehead*? {a tightly-coiled, developing frond}

➤ Why is it called a fiddlehead? {because it resembles the top of a violin}

➤ Why are fiddleheads sometimes dangerous? {Although some fiddleheads can be eaten, others are poisonous.}

➤ Which fern is sometimes used in salads or vegetable dishes? {ostrich fern}

➤ What is one of the most common ferns? {bracken fern}

➤ How large are bracken ferns? {sometimes as long as 2 m (6½ ft)}

➤ Where are bracken ferns often found? {in wooded areas}

Answers
1. vascular and nonvascular
2. They strengthen and support the plant, transport water and food to all parts of the plant, and enable the plant to grow larger.

Activity Manual
Reinforcement—page 84

Lesson 66

Student Text pages 138–41
Activity Manual pages 85–86

Objectives

- Recognize that seed-producing plants can be classified as gymnosperms and angiosperms
- Identify four kinds of gymnosperms
- Identify two kinds of conifers
- Describe several ways that man uses conifers

Vocabulary

angiosperm   gymnosperm

Materials

- box of tea containing gingko
- a picture of the Statue of Liberty

Introduction

➤ How many of you have at least one pet at your house?

➤ We can divide the class into two groups: pet owners and non–pet-owners.

➤ How can we divide the pet owner group into categories? *{Possible answer: Separate by types of pets.}*

➤ Name the two main categories of plants we studied in the previous lesson. *{vascular and nonvascular}*

➤ Just as we can separate the group of pet owners into smaller categories, scientists can separate vascular plants with seeds into smaller categories.

Purpose for reading:
Student Text pages 138–41

➤ How are gymnosperms and angiosperms different?
➤ Which is the largest group of gymnosperms?
➤ Why are conifers suited to colder climates?

Discussion:
pages 138–39

➤ What two things do seeds contain? *{the embryo of a new plant and food reserves}*
➤ How do scientists further separate seed-producing vascular plants? *{by the way the seeds are produced}*
➤ What are two main differences between angiosperms and gymnosperms? *{Angiosperms have flowers and seeds protected by a fruit. Gymnosperms do not have flowers. Their seeds usually develop in cones and do not have any protective covering other than a seed coat.}*

cycad

Seed-Bearing Vascular Plants

Seeds can be smaller than a flake of oatmeal or bigger than a hand. However, no matter the size, each seed contains the embryo of a new plant and has food reserves stored for that plant. Vascular plants that have seeds can be classified by how those seeds are produced.

Angiosperms (AN jee uh SPURMZ) are vascular plants that have flowers, and their seeds are protected inside a fruit. **Gymnosperms** (JIM nuh SPURMZ) do not have flowers, and their seeds are usually produced inside cones. The seed coat provides the only protection for these seeds.

Gymnosperms

Scientists divide gymnosperms into four smaller groups: cycads (SY kadz), ginkgoes (GING kohz), gnetophytes (NEE tuh FYTES), and conifers (KAHN ih furz). Cycads often are mistaken for palm trees because they look like tree ferns or palms. However, they produce pollen in a cone that can grow to be quite large. The trunk of a cycad can be above the ground or below the ground. Cycads are often used in landscaping as ornamental plants.

The ginkgo tree has flat, fan-shaped leaves that turn yellow in autumn

ginkgo tree

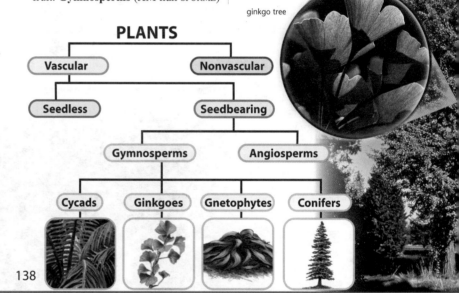

138

SCIENCE BACKGROUND

Gymnosperm seeds—These seeds usually develop in cones, but they are sometimes protected by fleshy seed coats. Even though yews and junipers have seeds that look like berries, the "berry" is actually the seed coat.

Grouping—The families of conifers include the pines (in both the northern and southern hemispheres), yellowwoods, yews, plum yews, junipers, and cypresses, as well as redwoods and sequoias. Spruces, firs, larches, cedars, and pines are some members of the pine family.

SCIENCE MISCONCEPTIONS

Be sure that students realize that conifers are not the only gymnosperms. Conifers are the most numerous and well-known gymnosperms.

BIBLE

Just as evergreens produce food all year long, Christians should be continually praising and glorifying God. The Bible compares a person who delights in the Lord to a tree whose leaves will not wither. Such a person will grow and prosper spiritually (Ps. 1:3). [BATs: 7b Exaltation of Christ; 7c Praise]

148 Lesson 66

gnetophyte

Fantastic FACTS

The tallest conifer is a redwood tree in California. It was measured in 1998 and was found to be a little over 112 m (367 ft) tall—about 19 m (62 ft) taller than the Statue of Liberty. The largest conifer is a sequoia named General Sherman. Its largest branch is a little more than 2 m (7 ft) in diameter. It is 31 m (103 ft) wide at its base and has bark that is almost 1 m (3 ft) thick.

before they fall to the ground. Ginkgoes are often planted in cities because they can tolerate air pollution. They also are very resistant to pests and diseases. Their dried leaves have been used in herbal medicines and teas for many years.

Gnetophytes are usually found in hot, dry deserts or in tropical rain forests. They can be trees, shrubs, or vines. The cones of some gnetophytes can resemble flowers, and their vascular systems are somewhat similar to those of angiosperms. However, gnetophytes are gymnosperms.

The majority of gymnosperms are *conifers*. Most conifers are tall, straight trees, but some are woody shrubs. Their leaves are often needlelike or scalelike. Almost all conifers are evergreen and are able to make food year-round through the process of photosynthesis. This food making allows conifers to live in colder climates than deciduous (dih SIJ oo uhs) trees, which lose their leaves during winter. The needle-shaped leaves also shed snow well, which helps keep branches from breaking. The smaller surface area of the leaves also helps the trees lose less water, making them more able to resist droughts and dry conditions in many climates.

redwoods—a type of conifer

139

Discussion:
pages 138–39

▶ How many groups are gymnosperms divided into? {four}

▶ What are these groups? {cycads, ginkgoes, gnetophytes, and conifers}

▶ Which type of plants are often mistaken for cycads? {palm trees}

▶ Why are ginkgoes often planted in cities? {because they can tolerate air pollution}

▶ What other uses do gingkoes have? {Their crushed leaves are sometimes used for herbal medicines and teas.}

Display the tea box that lists ginkgo as an ingredient.

💡 Which of the gymnosperm groups are you least likely to see on a daily basis? {gnetophytes} Why? {They are usually found only in deserts or rain forests.}

💡 Why do you think conifers are probably the most well-known tree group? {because there are more conifers than other gymnosperms}

▶ Why can conifers survive in the cold better than deciduous trees can? {Possible answers: They can produce food year round. Their needles shed snow more easily.}

💡 What are deciduous trees? {trees that shed their leaves}

💡 Why would being able to shed snow well be more important to evergreen trees than deciduous trees? {Possible answers: Evergreen trees live in colder climates where there is more snow. During the winter, deciduous trees do not have leaves to hold snow, but the weight of snow on evergreens might cause branches to break.}

▶ How does the shape of a conifer's leaves help the conifer to resist droughts? {Needle-shaped leaves lose less water, so they keep more water available for the tree.}

Refer the student to *Fantastic Facts*.

▶ What kind of tree is the tallest conifer? {redwood}

Show the student the picture of the Statue of Liberty

▶ Which is taller, the tallest redwood or the Statue of Liberty? {the tallest redwood}

▶ What is the name of the largest conifer? {General Sherman}

Direct a DEMONSTRATION
Demonstrate how conifer leaves retain moisture

Materials: two paper towels

Direct the student to wet each paper towel. The student should lay one paper towel flat to dry and roll the other into a tight tube. Let the paper towels sit for several hours.

▶ Which paper towel dried first? {the flat one}

▶ How does this show why conifer leaves retain moisture better than flat leaves? {Answers will vary, but elicit that leaves with a greater surface area dry faster. Conifer leaves have less surface area, so they retain water better.}

GEOGRAPHY

The world's largest plantation of ginkgo trees is located in Sumter, South Carolina. The trees are kept shrub-sized, and the leaves are gathered and shipped to Europe. There the dried leaves are used to produce ginkgo extracts.

Find Sumter, South Carolina, on a map. Encourage the student to research the uses of ginkgo leaves.

MATH

Allow students to measure some of the sizes discussed in *Fantastic Facts*.

Chapter 6

Lesson 66 149

Discussion:
pages 140–41

▶ **What is the easiest way to identify a conifer?** {looking at its leaves}

▶ **Why is it difficult to determine a standard color or length for a conifer's needles and cones?** {Climate and growing conditions affect the growth and development of needles and cones. Growth and development can vary from tree to tree even within the same species.}

▶ **Which conifers are probably the most familiar?** {pine trees}

▶ **What is one way to identify different kinds of pine trees?** {by the number of needles that are bound together}

▶ **In what ways do people use pine trees?** {Possible answers: lumber, cabinets, fence posts, telephone poles, paper}

▶ **How is resin beneficial to conifers?** {It seals off wounds on the tree and protects the tree from diseases and insects while it heals.}

💡 **Compare what happens when a pine tree is injured with the healing process when your skin is cut.** {Possible answers: The resin seals and protects the tree's wound in a way similar to how a scab forms a seal and protects you from infection.}

▶ **What are some uses for resin?** {Possible answers: tar, turpentine, inks, paints, adhesives}

▶ **What is *amber*?** {hardened resin}

Point out the different kinds of pines pictured at the top of the page.

▶ **Which two conifers are often mistaken for one another?** {firs and spruces}

▶ **How are the needles of firs and spruces different from the needles of other conifers?** {Each fir and spruce needle is directly attached to the branch.}

▶ **How are spruce needles different from fir needles?** {Spruce needles are attached to the branches on little woody pegs. They are also stiff, prickly, and four-sided. Fir needles are flat and flexible.}

white pine

lodgepole pine

Ponderosa pine

The easiest way to identify a conifer is to look at its leaves. Depending on the climate and growing conditions, the color, length, and texture of the needles and cones may vary from tree to tree. However, each family of conifers has certain characteristics that are unique to its group.

Pine trees

Pine trees are probably the most well-known conifers. They have woody cones that are often egg shaped and needles that are bound in bundles. Lodgepole pines have needles bound in groups of two, while ponderosa pines have needles in bundles of three. White pines and sugar pines have needles in bunches of five. People use pine trees for many different things, including lumber, cabinets, fence posts, telephone poles, and paper.

When a pine tree is cut or its branches are broken, sticky resin (REZ in) seeps out to clog up the wound. This protects the tree from diseases and insects while it heals. Resin from pine trees and other conifers can be used for a large variety of products, such as tar, turpentine, inks, and paints. Resin is also used for adhesives, such as the sticky part of a bandage. Musicians use a form of resin called rosin (RAHZ in) on the bows of stringed instruments such as violins and cellos. Sometimes fossils are found in hardened resin called amber.

Firs and spruces

Firs and spruces are often mistaken for each other. Unlike those of other conifers, each of their needles is attached directly to their branches. Fir needles are flat and flexible. Spruce needles, however, are stiff and prickly and are attached to the branches on little woody pegs. Most spruce needles are four sided.

spruce

SCIENCE BACKGROUND

Douglas fir—The Douglas fir is not really a fir. It was named by Scottish botanist David Douglas and is similar to both the fir and the spruce. In 1867 it was separated from the firs and spruces and placed into a different group.

The cones of a Douglas fir hang down and have three-pointed bracts. Its needles are soft and flat with round tips. Douglas firs have been used for railroad ties, shipbuilding, flooring, boxes, and ladders. The Douglas fir is also the state tree of Oregon.

Science and the Bible

cedar of Lebanon

When King Solomon built the temple and his palace, he used cedars and firs from Lebanon. Hiram, king of Tyre, provided as many firs and cedars as Solomon needed (I Kings 5). The cedar of Lebanon has a deep-red-colored wood that has a pleasant aroma. This tree can grow to be over 30 m (100 ft) tall.

Another way to distinguish between firs and spruces is to look at their cones. Fir cones stand upright on the branches and are sometimes violet colored when young. In autumn, the cone scales fall off as the seeds are dispersed. The small stem of the cone stays attached to the branch. Spruce cones hang down from the branches and can stay on the tree for several years. Therefore, a spruce tree can have both old and new cones on the tree at the same time.

Firs are used for pencils and plywood, as well as in construction and landscaping.

Douglas fir

Spruce trees are used for canoe paddles, furniture, paper, and musical instruments.

QUICK CHECK
1. What are the four groups of gymnosperms?
2. Which group is the largest?
3. What is the easiest way to identify a conifer?

141

Discussion:
pages 140–41

➤ **How can the cones on a tree identify whether the tree is a fir or a spruce?** {Fir cones stand upright on the tree. Spruce cones hang down.}

➤ **How can spruce trees have old and new cones on the tree at the same time?** {because their cones can stay on the tree for several years while the seeds develop}

Point out the photo of the Douglas fir. A Douglas fir has characteristics of both a spruce and a fir, so scientists eventually classified it in a completely different group.

➤ **What are some ways that fir trees are used?** {Possible answers: pencils, plywood, construction, and landscaping}

➤ **What are some ways that spruce trees are used?** {Possible answers: canoe paddles, furniture, paper, and musical instruments}

💡 **What are some other common uses for firs, spruces, pine trees, and other conifers?** {Possible answers: Christmas trees and decorations}

📖 Refer the student to *Science and the Bible*.

➤ **Which conifers did Solomon use extensively in the building of the temple?** {the firs and cedars of Lebanon}

➤ **Who provided the firs and cedars?** {Hiram, King of Tyre}

💡 **Why do you think cedar would have been an appropriate wood for Solomon's temple?** {Answers will vary, but elicit that cedar is both beautiful and aromatic. The temple was built to exalt God, therefore requiring the best.} [BAT: 7b Exaltation of Christ]

Answers
1. cycads, ginkgoes, gnetophytes, and conifers
2. conifers
3. Look at the leaves.

Activity Manual
Reinforcement—page 85

Study Guide—page 86
This page reviews Lessons 65–66.

Assessment
Test Packet—Quiz 6-A
The quiz may be given any time after completion of this lesson.

Chapter 6 — Lesson 66

Lesson 67

Student Text pages 142–45
Activity Manual pages 87–88

Objectives

- Recognize that angiosperms include trees, shrubs, and flowering plants
- Distinguish among annuals, biennials, and perennials
- Name some ways that angiosperms are used
- Compare monocotyledons and dicotyledons

Vocabulary

annual
biennial
perennial
cotyledon
dicotyledon
monocotyledon

Introduction

Direct the students who own pets to stand. Separate the students into groups according to their pets (dogs, cats, fish, turtles, etc.).

➤ **How can we separate the dog owners group into smaller groups?** {Divide into groups according to the type of dog owned, the age of the dog, the color of the dog, or the number of dogs owned.}

Allow the students who own dogs to separate into smaller groups.

➤ Today we will be studying how scientists separate flowering plants into groups.

Purpose for reading:
Student Text pages 142–45

➤ How are perennials different from annuals?
➤ How are monocotyledons different from dicotyledons?

Angiosperms

What do soap, medicines, food, and clothing have in common? Each one of these can be made from some part of an angiosperm. Angiosperms can be as tall as an oak tree or as small as a blade of grass. They include peppers and tomatoes as well as roses and daisies. Any vascular plant that produces flowers and fruit is classified as an angiosperm. Although most angiosperms have flexible, green stems, angiosperms also include many woody shrubs and trees.

All angiosperms have flowers. When pollinated, the flowers produce seeds protected by an outer covering, or fruit. In some species, the flowers are small and inconspicuous. Other species, such as corn, may have flowers without petals. The leaves of angiosperms are quite different from the leaves of conifers and other gymnosperms. Angiosperm trees usually have leaves that are wide and flat. They are often called broad-leaved trees to distinguish them from conifers.

Scientists have identified more than 250,000 species of angiosperms. Though most angiosperms live on land, some are aquatic plants. Aquatic angiosperms are found in both salt water and fresh water. Some of the plants, such as sea grasses, are completely submerged. Others, such as water lilies, float on top of the water. Still others, such as cattails and bulrushes, grow along the edges of bodies of water.

PLANTS
- Vascular
  - Seedless
  - Seedbearing
    - Gymnosperms
    - Angiosperms
- Nonvascular

142

SCIENCE BACKGROUND

Angiosperms—In an angiosperm, the ovary of the flower forms the fruit, or protective covering for the seed. This is mentioned again in Chapter 12, Lesson 126. Sometimes angiosperms are called *anthophytes*.

Grouping plants—Plants do not always fit neatly into the categories of annual, biennial, and perennial. Their grouping often depends on where, when, and how they grow. Short-lived perennials, such as hollyhocks, often get mistaken for biennials. Biennials that begin growing indoors can sometimes act as perennials. Also, many biennials are sold as one-year-old plants that are ready to flower.

SCIENCE MISCONCEPTIONS

Make sure that students understand that not all flowers have petals. The parts of a flower are discussed in Chapter 12, *Reproduction*.

152   Lesson 67

An angiosperm that lives for only one growing season is called an **annual** (AN yoo uhl). Annuals grow, flower, produce seeds, and then die all in the same growing season. Marigolds, tomatoes, and sunflowers are some examples of annuals. A **biennial** (by EN ee uhl) needs two growing seasons to fully develop. In the first season, the plant produces leaves. It rests during the winter and then flowers, produces seeds, and dies in the second year. Parsley, carrots, cabbage, onions, and foxgloves are biennials.

A **perennial** (puh REN ee uhl) can live for three or more years. It grows, flowers, and produces new seeds year after year. Perennials include trees and bushes, such as oak trees and roses. However, plants such as spearmint and carnations are also perennials.

cacao tree and chocolate

We use angiosperms for many things. The most obvious use is for food, but many beverages are also from angiosperms. Trees in the Amazon rain forest provide some of the ingredients necessary for coffee, hot chocolate, and cola. Other angiosperms are used for medicines. The purple foxglove is used to make digitalis, a heart medicine. Aspirin, a medicine used to treat fever and inflammation, was once prepared from angiosperms. Even products such as rubber, cork, rope, and chewing gum are made from angiosperms.

foxglove and digitalis pills

143

Discussion:
pages 142–43

➤ Which two features identify a plant as an angiosperm? {produces flowers and fruit}

➤ Which three types of plants are in the angiosperm category? {Possible answers: trees, shrubs, and plants with green, flexible stems}

➤ How are the seeds of angiosperms protected? {by an outer covering, or fruit}

💡 Why are angiosperm trees called broad-leaved trees? {to distinguish them from conifers; Broad-leaved trees have leaves that are wide and flat, and conifers have needle- or scale-like leaves.}

➤ What are some aquatic angiosperms? {Possible answers: sea grasses, water lilies, cattails, and bulrushes}

📖 Can you think of a Bible character who was found hidden among some aquatic angiosperms? {Moses; His mother hid him in the bulrushes.}

➤ How are annuals and biennials different? {Annuals grow, flower, and die in one growing season. Biennials need two growing seasons to fully develop.}

💡 When do carrot flowers form? {the second year} Why? {because carrots are biennials}

💡 What usually prevents us from seeing carrot flowers? {Most garden carrots are pulled and eaten before the second year.}

💡 Why are flowers necessary? {They produce seeds for new plants.}

➤ What name is given to plants that grow, flower, and produce seed year after year? {perennials}

➤ What are some common perennials? {Possible answers: trees, roses, spearmint, carnations}

➤ What are some ways that angiosperms are used? {Possible answers: food, beverages, medicine, rubber, cork, rope, chewing gum}

📖 Why should man use plants to produce food and materials? {God gave man dominion over the earth to make the best use of what He has given to man.}

Direct an ACTIVITY

Materials: seed packets, seed and bulb catalogs

Instruct the student to use seed packets and catalogs to plan a flower garden with twelve different varieties of flowers. One-fourth of the flowers should be annuals, one-fourth biennials, and the other one-half should be perennials.

➤ Why would someone choose a perennial or biennial instead of an annual? {Possible answer: Perennials and biennials do not need to be planted each year.}

Discussion:
pages 144–45

 Point out the addition of monocots and dicots to the classification diagram.

▶ How are angiosperms usually classified? *{by their seed structure}*

▶ What are *cotyledons*? *{the seed leaves that contain stored food for the new plant}*

▶ How many cotyledons do dicots have? *{two}*

▶ How many cotyledons do monocots have? *{one}*

▶ How are the tubelike structures arranged in the stem of a dicot? *{in a ring shape}*

▶ How does the arrangement of vascular bundles affect the size of a dicot, such as an oak tree? *{Because the vascular bundles are in a ring shape, the stem (or trunk) of a dicot grows a little thicker each year.}*

▶ What kind of leaves do dicots have? *{Dicot leaves are broad and have networks of veins.}*

▶ What kind of roots does a dicot have? *{one long taproot with secondary roots branching off of it}*

▶ How can you use an angiosperm's flower to identify the plant as a monocot or dicot? *{The flowers of dicots have petals in multiples of four or five. Monocots have petals in multiples of three.}*

▶ Name some common dicotyledons. *{Possible answers: broad-leafed trees, roses, dandelions, cactuses, beans}*

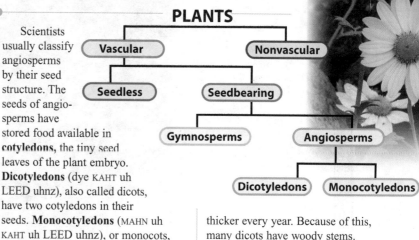

PLANTS

Scientists usually classify angiosperms by their seed structure. The seeds of angiosperms have stored food available in **cotyledons**, the tiny seed leaves of the plant embryo. **Dicotyledons** (dye KAHT uh LEED uhnz), also called dicots, have two cotyledons in their seeds. **Monocotyledons** (MAHN uh KAHT uh LEED uhnz), or monocots, have only one cotyledon in their seeds.

Dicotyledons

Most angiosperms are dicots. In addition to their seed structure, dicots also have other similar characteristics. Their tubelike structures are arranged in the shape of a ring. This arrangement allows the stems of dicots to grow thicker every year. Because of this, many dicots have woody stems.

Dicots' leaves are usually broad and flat and have a network of veins. As a dicot grows, its first root lengthens and branches out into smaller secondary roots. An easy way to identify dicots is to observe their flowers. The flower of a dicot will have either four or five petals or petals in multiples of four or five.

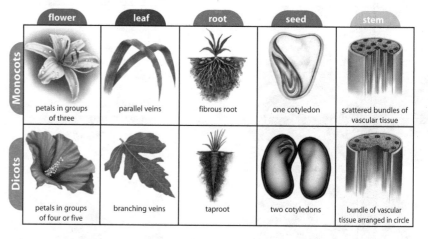

144

SCIENCE BACKGROUND

Cotyledon classification—Angiosperms are classified into monocots and dicots, depending on the number of cotyledons in the seed. Gymnosperms, however, are called *polycots* because their seeds have multiple cotyledons. The vascular structure of gymnosperms is also slightly different from that of angiosperms.

Types of roots—The types of roots, taproots and fibrous roots, will be discussed in Lesson 70.

Longest leaf—A monocot holds the record for having the longest leaf of any flowering plant. The leaves of the raffia palm can be as long as 20 meters (65 feet). The raffia palm is found in Madagascar (a country in Africa). Its leaves are often used to make woven hats and baskets.

LANGUAGE

A prefix added to the beginning of a word changes the meaning of the word.

The prefix *mono-* means "one." So *monocotyledon* means "one cotyledon."

▶ Can you think of other words that have the prefix *mono-*? *{Possible answers: monotone, monocle, monochromatic, monopoly}*

The prefixes *di-* and *bi-* mean "two."

▶ Which words from this lesson begin with these prefixes? *{dicot, biennial}*

▶ Can you think of other words that have the prefix *bi-*? *{Possible answers: bicycle, biped, binary}*

154 Lesson 67

Dicotyledons can be large or small. All broad-leafed trees are dicots, as are roses, dandelions, cactuses, and beans. Dicots are the largest group of angiosperms.

Monocotyledons

Monocots also have similar characteristics in addition to their seed structure. Because their tubelike structures are not arranged in any particular order, monocot stems do not become thicker each year. Most monocots are small, with soft, green stems.

Monocots have long and narrow leaves with parallel veins. Many of them, such as bananas and pineapples, are used for food. Monocots also include about eight thousand species of grasses. Some of these grasses are cereal grains that we eat, such as rice, wheat, corn, and oats. Some monocots have many thin roots instead of secondary roots branching off the first root.

Monocot flowers may be large and attractive, such as daffodils and orchids. They can also be hardly noticeable, such as the flowers on grasses. If the flower has petals, it will usually have three petals or be arranged in multiples of three.

 QUICK CHECK
1. What are the three classifications of angiosperms according to their growing seasons?
2. How are the seeds of monocots and dicots different?

wheat

oats

Fantastic FACTS
Have you ever thought much about grass? The grass, or turf, found on golf courses and sports fields is often a product of turf management. People in turf management study grasses, soils, fertilizers, insects, and landscaping. They maintain and manage athletic fields and parks as well as commercial and residential lawns.

Direct an ACTIVITY
Provide the student with pictures of many different flowers that have petals, stems, and leaves visible, such as a daisy, daffodil, tulip, lily, geranium, snapdragon, morning glory, apple blossom, buttercup, iris, grass, or mint.

Allow the student to classify the pictures of the plants as either dicots or monocots. The student should be able to state a reason for each classification.

Discussion:
pages 144–45

➤ **How are the tubelike structures in a monocot arranged?** *{not arranged in any particular order; scattered}*

➤ **Why do you think most monocots are smaller than dicots?** *{Their stems do not grow thicker each year because their vascular bundles are scattered throughout the stem.}*

➤ **What kind of stems do most monocots have?** *{soft, green stems}*

➤ **What kind of leaves do monocots usually have?** *{Monocots have long and narrow leaves with parallel veins.}*

➤ **What kind of roots do most monocots have?** *{many thin roots; fibrous roots}*

➤ **What are some common monocots?** *{Possible answers: bananas, pineapples, grass, rice, wheat, corn, oats, daffodils, and orchids}*

➤ **Are most angiosperms monocotyledons or dicotyledons?** *{dicotyledons}*

Point out the chart on Student Text page 144. Review the characteristics of monocots and dicots.

Refer the student to *Fantastic Facts*.

➤ **What is another name for the grass found on golf courses?** *{turf}*

➤ **What kinds of subjects does someone in turf management study?** *{grasses, soils, fertilizers, insects, and landscaping}*

➤ **Where are some places that turf management is used?** *{Possible answers: athletic fields, golf courses, commercial and residential lawns}*

Answers
1. annual, biennial, perennial
2. Monocot seeds have one cotyledon. Dicot seeds have two cotyledons.

QUICK CHECK

Activity Manual
Study Guide—page 87
This page reviews Lesson 67.

Bible Integration—page 88
This page explores some of the plants recorded in the Bible.

Lesson 68

Student Text page 146
Activity Manual pages 89–90

Objective
- Plan a visual to demonstrate how plants are classified

Vocabulary
taxonomy

Materials
- See Student Text page

Introduction

Review the main classification categories of plants.

➤ **What are the two main categories into which scientists classify plants?** {vascular and nonvascular}

➤ **Into what two groups are vascular plants divided?** {seed-bearing and seedless}

➤ **Into what two groups are seed-bearing vascular plants divided?** {gymnosperms and angiosperms}

➤ **What are the four types of gymnosperms?** {cycads, gnetophytes, ginkgoes, and conifers}

➤ **What are the two main groups of angiosperms?** {monocotyledons and dicotyledons}

Purpose for reading:
Student Text page 146
Activity Manual pages 89–90

The student should read all the pages before beginning the activity.

Procedure

Guide the student in deciding how to represent the categories of plant classification.

Remind the student to complete the Activity Manual pages. This will help him remember to include necessary categories and pictures.

Activity Manual
Activity—pages 89–90

Assessment
Rubric—TE pages A50–A53
Select one of the prepared rubrics, or design a rubric to include your chosen criteria.

ACTIVITY: Classification Check

Process Skills
- Observing
- Classifying
- Communicating

Taxonomy is the branch of science that deals with classifying organisms. A scientist who specializes in this branch of science is called a taxonomist. For this activity, you are the taxonomist. You must prepare a visual aid to show how scientists classify plants. It could be a mobile, a chart, a concept web, or any other method that you choose.

Procedure

1. Plan a method to show the plant classification. Describe and draw your plan in your Activity Manual.
2. Make a list of the materials that you will need, and then gather the materials.
3. Your visual aid should include these categories: vascular, nonvascular, seedless vascular, seed-bearing vascular, angiosperms, and gymnosperms.
4. Find a picture for each subcategory (mosses, liverworts, ferns, horsetails, club mosses, cycads, ginkgoes, gnetophytes, conifers, dicotyledons, and monocotyledons).
5. Use the field guide and/or the encyclopedia to specifically identify each plant pictured. For example, you may find a picture of a pine tree to represent the conifers. Try to identify the species of the pine tree. Is it a white pine, lodgepole pine, or another species of pine?
6. Construct your visual aid.

Materials:
- pictures of various plants (from magazines, seed catalogs, Internet sources, etc.)
- chosen materials for your visual aid
- field guide or encyclopedia
- Activity Manual

Follow-up
- Add the scientific names, along with the common names, of the plants that you identify.

146

 Completed projects make excellent classroom displays.

 The Greek word *taxis* means "arrangement." Taxonomy is the classification, or arrangement, of organisms according to specific criteria.

SCIENCE PROCESS SKILLS

Classifying

➤ **What would be the easiest characteristics to identify?** {Answers will vary.}

➤ **Why would it be difficult to accurately classify some plants?** {Answers will vary, but elicit that they may look similar even though they actually are not.}

➤ **Do you think you could identify plants by looking only at the outside of them? Why?** {Answers will vary, but elicit that in order to see some of the characteristics used to classify plants, you would need to split the plants open.}

Explorations: Plant Products

Have you ever thought about the many different products that can be made from plants? Some products, such as furniture or food, might be obvious. But did you realize that lipstick, glue, fabric, and hair spray can also be made from plants and plant products? Scientists may have found other ways to make some products, such as aspirin and marshmallows, that were formerly made from plants. However, many medicines and other products that we use every day would never have been discovered without plants.

What to do

1. Listed below are some plants that are used in a variety of ways. Choose one of the plants.
2. Make a collage or display that shows at least ten different uses for that plant. Five products should be non-food-related. Try to find some very unusual uses for your plant.
3. Present your project to the class and be able to explain which part of the plant is used in each product.

gumballs

olive	onion
soybean	sunflower
cotton	flax
stinging nettle	carnauba tree
douglas fir	hazel tree
kapok tree	spruce
western red cedar	white pine

147

 Vikings used plants for many different things. Club mosses were made into powder or lotion to treat skin diseases. Spagnum moss contains penicillin and was used at one time to treat wounds. The iris flower had many medicinal uses, such as for coughs, poisoning, toothaches, and swelling.

 Displays can be assembled at home or put together in class.

Student Text page 147

Lesson 69

Objectives
- Research products made from a given plant
- Prepare a display to demonstrate research results

Introduction

➤ What do lipstick, the waxy shine of some apples, and some car waxes have in common? *{All of them can be made from the same plant.}*

➤ Diapers, hair spray, and lotion are just a few products that also can be made from plants. In this exploration, you will choose one plant and find ways in which that plant is used.

Purpose for reading:
Student Text page 147
The student should read the page before beginning the exploration.

Procedure
Provide encyclopedias or Internet access as needed for the student to research his plant. Remind him to document his sources as he lists products made from his plant. When possible, the student should also note which part of the plant is used to make each product.

Guide the student in deciding how to display the many different uses for the plant that he chose.

Assessment
Rubric—TE pages A50–A53
Select one of the prepared rubrics, or design a rubric to include your chosen criteria.

Chapter 6   Lesson 69   157

Lesson 70

Student Text pages 148–51
Activity Manual pages 91–92

Objectives

- Identify the two kinds of vascular tissue and their functions
- List three main functions of the stem of a plant
- Describe the difference between herbaceous and woody stems
- List three main functions of root systems
- Describe the differences between taproots and fibrous roots

Vocabulary

xylem
phloem
vascular bundle
cambium
tuber

herbaceous
primary root
taproot
fibrous root

Introduction

Discuss some occupations, such as sanitation engineer, policeman, fireman, emergency medical technician, etc.

➤ Why are these community occupations important? *{to keep communities clean; to help things run smoothly}*

➤ What happens if someone does not do his assigned job? *{The job does not get done. Things do not go as smoothly. Someone else has to take the time to do the job.}*

➤ Just as different people in our community do specific jobs to help us out, different parts of plants have specific jobs, or functions. Today we will be studying some of these plant parts.

Purpose for reading:
Student Text pages 148–51

➤ What three types of cells are found in vascular bundles?

➤ What are three kinds of roots?

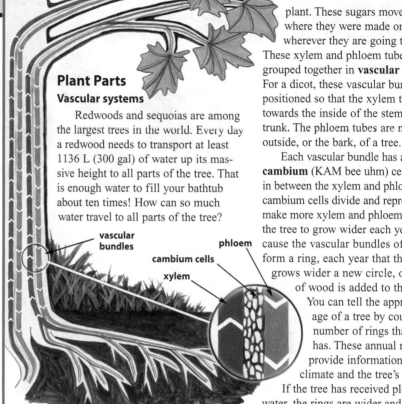

Plant Parts
Vascular systems

Redwoods and sequoias are among the largest trees in the world. Every day a redwood needs to transport at least 1136 L (300 gal) of water up its massive height to all parts of the tree. That is enough water to fill your bathtub about ten times! How can so much water travel to all parts of the tree?

As we have already seen, most plants, including trees, have vascular systems. This system of tubes transports water, food, and nutrients throughout the plant.

Xylem (ZY luhm) tubes carry water and minerals from the roots to the top of the plant. **Phloem** (FLO uhm) tubes carry sugars and food throughout the plant. These sugars move from where they were made or stored to wherever they are going to be used. These xylem and phloem tubes are grouped together in **vascular bundles**. For a dicot, these vascular bundles are positioned so that the xylem tubes are towards the inside of the stem or tree trunk. The phloem tubes are nearer the outside, or the bark, of a tree.

Each vascular bundle has a layer of **cambium** (KAM bee uhm) cells tucked in between the xylem and phloem. These cambium cells divide and reproduce to make more xylem and phloem, allowing the tree to grow wider each year. Because the vascular bundles of dicots form a ring, each year that the tree grows wider a new circle, or ring, of wood is added to the tree.

You can tell the approximate age of a tree by counting the number of rings that the tree has. These annual rings also provide information about the climate and the tree's health. If the tree has received plenty of water, the rings are wider and farther apart. However, narrow rings form during times when the tree has less water.

148

SCIENCE BACKGROUND

Stem growth—Stems show both primary and secondary growth. Changes in the height and length of a stem are primary growth. Secondary growth refers to the increase in the width of the stem.

Tree trunks—The cambium cells are the only living cells in a tree trunk. Even the wood in a living tree is made up of dead cells. In the center of the trunk is the pith, a food storage area. The pith is surrounded by the heartwood, which consists of old layers of xylem that no longer transport water and nutrients. The heartwood adds additional strength and support to the tree.

SCIENCE MISCONCEPTIONS

Even though both the sweet potato and the white potato are referred to as tubers, they come from different parts of the plants. A white potato is actually an underground stem, but a sweet potato is part of the plant's root.

158 Lesson 70

flower with herbaceous stem

Stems

Scientists define a stem as any part of the plant that will grow leaves, shoots, or buds rather than roots. Even though stems come in many different widths and textures, every stem has the same two important jobs. Stems provide support to hold the plants upright, and they also provide for the transportation of food, water, and minerals. Some stems even help store food. The baked or mashed potatoes that you may have eaten were actually the **tuber**, or food storage stem, of the plant.

Stems can be above or below ground. They do not always look the same as a typical green stem or a brown tree trunk. Most stems are either herbaceous or woody. **Herbaceous** (hur BAY shuhs) stems are soft and green, like the stems of most flowers and vegetables. Most herbaceous stems belong to annual plants and live for only one growing season. However, some perennials, such as tulips and daffodils,

TRY IT Yourself

You can make a "leaf skeleton" that shows the tubes in a leaf. Choose a leaf and cover it with water in a bucket or pan. The water will speed up the rotting process of the leaf. Change the water every few days. Be careful not to tear your leaf. After a few weeks the green part of the leaf will rot away, leaving only the veins, or tubes, in the leaf. Then you will have a "leaf skeleton."

also have herbaceous stems. When cold weather comes, these stems die and then grow from the roots again in the spring.

Woody stems are usually found in plants that have been growing for at least two years. The stems are stiff because a type of bark or cork forms a layer on the outside of the stem. This protective layer helps the plant resist diseases, insects, and extreme temperatures. A layer of cambium cells just underneath the bark keeps producing new layers of bark. Old outer bark cells are shed as new cells take their place.

flower with woody stem

Direct a Demonstration
Demonstrate how water travels through a stem

Materials: fresh white carnation, water, food coloring, one or two clear plastic cups

Fill the cup(s) with five centimeters or so of colored water. If desired, split the stem of the carnation for a few centimeters so that it can be placed in two different containers of colored water at the same time. Place the stem of the carnation in the colored water. Trim the stem as needed so that it will absorb the water. Direct the student to observe the flower as the colored water moves up its stem.

Finding the age of a tree— Scientists do not have to cut a tree down to determine its age. With a borer, a scientist can drill into a tree and bring out a small sample of wood, about 4 mm in diameter. The hole is then sealed to protect the health of the tree. The sample shows the rings and other features the scientist needs. Scientists use computers to help track and record data as well as to detect possible patterns in the weather and environment.

Discussion:
pages 148–49

▶ **What is the purpose of a vascular system?** {to transport food, water, and nutrients to all parts of a plant}

 Refer the student to the diagram of the parts of a plant.

▶ **What three types of cells are found in vascular bundles?** {xylem, phloem, and cambium}

▶ **Which part of the vascular system carries water?** {xylem} **In which direction do the xylem tubes usually transport water?** {up from the roots to the top of the plant}

▶ **What does the phloem transport?** {sugars and food}

▶ **How are the xylem and phloem grouped in a plant?** {in vascular bundles}

▶ **In a dicot, how are these vascular bundles positioned?** {The xylem is closer to the center of the stem, and the phloem is toward the outer part of the stem.}

▶ **What is between the xylem and phloem cells?** {cambium}

▶ **What is the purpose of the layer of cambium cells?** {It divides and reproduces to make more xylem and phloem tubes.}

▶ **What is different about tree rings formed during periods of drought and those formed when rainfall is plentiful?** {Narrow rings form when water is scarce. When the tree receives plenty of water, the rings are wider and farther apart.}

▶ **How do scientists define a *stem*?** {any part of the plant that will grow leaves, shoots, or buds rather than roots}

▶ **What are the two main jobs of a stem?** {to support the plant and to transport food, water, and minerals}

▶ **What is another job that some stems do?** {store food}

▶ **What are underground stems that store food called?** {tubers}

▶ **How are herbaceous stems different from woody stems?** {Herbaceous stems are soft and green. Woody stems have a protective layer of bark or cork on the outside of the stem.}

▶ **Do most herbaceous stems belong to annuals or perennials?** {annuals}

▶ **What is the purpose of the bark on a woody stem?** {It helps protect the plant from diseases, insects, and extreme temperatures.}

Chapter 6 · Lesson 70

Discussion:
pages 150–51

- What are four jobs of a plant's roots? {to anchor the plant in the soil, support the stem, absorb water and nutrients, and help transport water and food}
- What do some roots store for the plant? {starches and sugars}
- What are some common roots that we eat? {Possible answers: beets, sweet potatoes, radishes, and carrots}
- What will happen to a plant if its roots are diseased or injured? {The entire plant will suffer.}
- What is a *taproot*? {a long root that grows straight down into the soil}
- What is the first root that emerges from a seed? {primary root}
- Why is it called a *primary root*? {Answers may vary but should include that primary means first.}
- What is the purpose of root hairs? {They allow the root to touch more soil and absorb more moisture and nutrients. They also help to anchor the plant.}
- What are three kinds of roots? {taproots, fibrous roots, aerial roots}
- Both taproots and fibrous roots develop from a primary root. How are they different? {A taproot is the primary root that has continued to grow and enlarge. After the primary root stops growing, a fibrous root has many thinner roots grow out from it in all directions.}
- What does Paul mean in Colossians 2:7 when he says that Christians are "rooted and built up in him [Christ]"? {Answers will vary, but elicit that Christians are "rooted" in Christ because their salvation and spiritual lives are in Him. A Christian is "built up" in Christ as he grows spiritually.} [Bible Promise: E. Christ as Sacrifice]
- Aerial roots do not touch the soil. How do the roots of these plants get moisture? {They absorb moisture from the air.}
- What are some examples of plants that have aerial roots? {Possible answers: orchids, Spanish moss}
- Why are plants with aerial roots not considered parasites? {They do not use the supporting plant for food.}

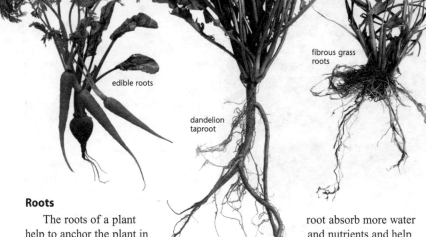

edible roots
dandelion taproot
fibrous grass roots

Roots

The roots of a plant help to anchor the plant in the soil and support the stem. They also absorb water and nutrients and are involved in transporting water and food. Some roots store starches and sugars for the plant. Beets, sweet potatoes, and radishes are some roots that people often eat.

Roots also affect the size and health of the plant. If the roots of a plant are diseased, the entire plant suffers. The root is the largest part of many plants. Since the roots can stretch out underground, some plants may have roots that spread out for thousands of square meters.

The first root that emerges from a seed is called the **primary root.** For most dicots, this root continues to grow and enlarges into the taproot. The **taproot** continues to grow straight down into the soil and may have secondary roots branching off it. Small root hairs cover the root, allowing the root to touch more soil. The root hairs help the root absorb more water and nutrients and help to anchor the plant. New root hairs continually replace old root hairs as the roots grow.

Most monocots and some gymnosperms, however, have a **fibrous root** system. The primary root stops growing, and the plant develops many thin roots that spread out in all directions. These thin roots branch and divide again and again. Fibrous roots are usually found near the top of the soil, but some sections branch downward to absorb water and minerals. Roots like these help to anchor soil in place and prevent erosion. The Bible uses roots to teach us about a Christian's relationship to Christ. In Colossians 2:7, Paul says that Christians are "rooted and built up in him [Christ]."

150

SCIENCE BACKGROUND

Taproot or fibrous roots—If a plant that normally has a taproot suffers an injury to its root early in its development, that plant may develop fibrous roots instead of a taproot. The type of soil and amount of moisture in the soil also affects root development. Some plant nurseries sell trees whose taproots have been changed to fibrous roots. The fibrous roots are easier to transplant than a taproot would be.

Branches of botany—Botany is the study of plants. Many branches of science deal with botany. *Dendrochronology* includes studying tree ring patterns to identify periods of drought and changes in climate. *Agriculture* deals with the study of crops related to food. *Agronomy* includes the study of soil as well as crop management. *Taxonomy* is the branch of science that deals with the classification and naming of organisms.

Not all plants have taproots or fibrous roots, though. Some plants have *aerial* roots that never touch the soil. These plants are most often found in rain forests or in other areas of high humidity. Orchids and Spanish moss are examples of plants with aerial roots. Their roots absorb moisture from the air. These two plants use other plants and objects for support and height. Since they do not use the supporting plant for food, they are not considered parasites. Mistletoe, however, is a parasitic plant. Mistletoe's roots grow into the branches of the tree supporting it. It absorbs moisture from the tree's vascular system.

Many evolutionists believe that one type of plant evolved into another more sophisticated type. For example, an evolutionist might claim that seedless plants, such as ferns, evolved into gymnosperms and that gymnosperms evolved into angiosperms. However, we know from God's Word that this did not happen. All of the different plants that exist simply showcase God's magnificent designs. God created each plant with exactly what it needs for survival. Our God, who cares and provides for plants, also provides all things needful for humans (Matt. 6:28–34).

1. What two kinds of vascular tubes are bundled together?
2. How do scientists define a stem?
3. What is the difference between a taproot and a fibrous root?

epiphyte

Fibrous roots can be quite large! One rye plant that was about 51 cm (20 in.) tall was found to have about 611 km (380 mi) of roots. Scientists counted at least 14 billion root hairs on these roots. If these root hairs could have been spread out flat, they would have probably covered an area the size of two or three houses.

The bark of sequoias can be as much as 79 centimeters (31 inches) thick, and the trees can have branches as large as 2.4 meters (8 feet) in diameter.

Discussion:
pages 150–51

- How was *parasite* defined in Chapter 5? {*an organism that lives and feeds off another organism*}
- Why is mistletoe considered a parasite? {*It grows into the branches of the tree it is on and absorbs moisture from the tree's vascular system.*}
- Many evolutionists believe that plants evolved in stages over billions of years. How do we know that this could not have happened? {*God's Word tells us that God created all plants on Day 3 of Creation.*}

Creation scientists have discovered fossil evidence that proves that "higher order plants," such as angiosperms, existed at the same time as nonvascular and seedless vascular plants. God provides exactly what each plant needs to grow. Read Matthew 6:28–34 and discuss how God also cares for humans. [Bible Promise: H. God as Father]

Refer the student to *Fantastic Facts*.

- How many root hairs did scientists think the rye plant had? {*14 billion*}
- Do you think they actually counted each one? {*no*}
- How do you think scientists arrived at that number? {*Possible answer: They probably divided the root hairs into sections and counted the number of hairs in one section, then multiplied the number of hairs by the number of sections.*}

Answers
1. xylem and phloem
2. A stem is any part of the plant that will grow leaves, shoots, or buds rather than roots.
3. A taproot grows straight down in the ground and has secondary roots branching off it. For fibrous roots, the primary root stops growing, and many thin roots grow out in all directions.

Activity Manual
Reinforcement—page 91

Study Guide—page 92
This page reviews Lessons 67 and 70.

Assessment
Test Packet—Quiz 6-B
The quiz may be given any time after completion of this lesson.

Lesson 71

Student Text pages 152–53
Activity Manual pages 93–94

Objectives

- Measure the circumference, height, and crown of a tree and calculate the tree's point value
- Produce and read a graph to show relationships

Materials

- See Student Text page

Introduction

Note: This activity uses the English measuring system because the point value system for trees is based on the English system.

➤ **How would you measure the height of a person?** {Possible answers: Using a ruler, start at the person's foot and measure up to the top of his head. Have the person stand against a wall or tape, and make a mark on the wall at the top of his head. Then use a ruler to measure from the floor to the mark.}

➤ **Rulers cannot be used to measure every distance. Sometimes you may need to measure something that is not flat. At other times it is not possible to get from one end to the other end of an object to measure it. In today's activity, we will learn one way to calculate the size of a tree.**

Purpose for reading:
Student Text pages 152–53
Activity Manual pages 93–94

The student should read all the pages before beginning the activity.

How Big Is My Tree?

Process Skills
- Measuring
- Observing
- Inferring
- Communicating
- Collecting, recording, and interpreting data

Is your tree the biggest in your state? Many states have a registry of the biggest trees that grow there naturally. These trees are listed by species and ranked according to a point system. Measurements of a tree's circumference, height, and crown are needed to calculate the point value for that tree. How can you measure the height of a tree if you can't reach the top? How can you measure the crown, or upper part of the tree where the branches and leaves grow? In this activity you will measure the size of a tree while keeping both feet firmly on the ground.

Problem

How can you measure the circumference, height, and crown of a tree?

Procedure

Note: This activity uses English rather than metric measurements.

1. Choose a tree that you would like to measure. Identify it as a gymnosperm or an angiosperm. Record your classification in your Activity Manual.

2. Measure the circumference of your tree. Use string to measure the circumference of your tree in inches at 4½ feet above the ground. Record your measurement.

3. Measure the height of your tree. Have a partner stand at the base of the tree. With your arm straight out in front of you, hold one end of the ruler so the other end points up. Line up the point where the top of your hand is on the ruler so that it is even with the base of the tree. Back away from the tree until the top of the tree appears to be even with the top of the ruler. The top of your hand should be even with the base of the tree.

Materials:
string 15–20 ft long
12 in. ruler
yardstick or tape measure
4 short sticks or pencils
calculator
tree field guide or encyclopedia (optional)
Activity Manual

152

If you have a small class, you may want the students to measure more than one tree. The more trees that the students measure, the more data they will have to analyze.

Encourage groups to measure different sizes and kinds of trees to get the best information for their comparisons. If possible, at least two groups should measure trees of different sizes, but of the same species. At least one person in the group should record the measurements for circumference, height, and crown as they are measured.

There are many ways to measure the height of very tall objects. The student could research and try some of these methods then compare the results.

162 Lesson 71

Rotate your hand until the ruler lies horizontally. Your hand should still be at the base of the tree. Have your partner walk to the place that you see at the end of the ruler. Measure the distance in feet from the tree to your partner. Record your measurement.

 4. Measure the crown of your tree. Use the sticks to mark the places on the ground where the ends of the branches reach overhead. Mark the widest and narrowest spread of branches. Measure in feet both distances. Add both measurements and divide by two. Record this number as the average measurement of the crown.

 5. Calculate and record the point value for your tree.

 6. Compare the circumference, height, and crown of your tree to the measurements of other trees. Record and graph your information.

Conclusions
- Do you think that thinner trees are younger or older than thicker trees of the same species? Why?

Follow-up
- In an encyclopedia, field guide, or tree book, look up the annual growth and mature size of your tree or its species.
- Compare the point value of your tree with other big trees in your state or around the country.

Procedure
After measurements are recorded, provide time for each group to calculate the point value for their tree. Record each group's data on a chart. Allow time for each student to copy the chart onto his Activity Manual page. Instruct each student to graph the information using the three graphs provided in his Activity Manual.

Conclusions
➤ Do you think thinner trees are younger or older than thicker trees of the same species? {younger} **Why?** {The circumference of a tree usually increases as the tree ages.}

➤ Rank measured trees of the same species according to their point values. Would it be a valid comparison to compare point values of all the trees measured? {probably not} **Why?** {because they probably are not all of the same species}

Activity Manual
Activity—pages 93–94

Assessment
Rubric—TE pages A50–A53
Select one of the prepared rubrics, or design a rubric to include your chosen criteria.

153

SCIENCE PROCESS SKILLS
Measuring

➤ How was the measuring done for this activity different from some other measuring? {Answers will vary, but lead to the idea that it was less exact because the student could not physically measure the tree.}

➤ Do you think that different people would get different measurements of the same tree? {yes} **Why?** {because they may measure differently}

➤ How could you minimize error? {Possible answer: Measure the tree several times and find the average measurement of the tree.}

Chapter 6 Lesson 71 163

Lesson 72

Student Text page 154
Activity Manual pages 95–96

Objectives

- Recall concepts and terms from Chapter 6
- Apply knowledge to everyday situations

Introduction

Material for the Chapter 6 Test will be taken from Activity Manual pages 86–87, 92, and 95–96. You may review any or all of the material during the lesson.

You may choose to review Chapter 6 by playing "Build a Flower" or a game from the *Game Bank* on TE pages A56–A57.

Diving Deep into Science

Questions similar to these may appear on the test.

Answer the Questions

Information relating to the questions can be found on the following Student Text pages:

1. page 148
2. pages 144–45
3. pages 134, 137

Solve the Problem

In order to solve the problem, the student must apply material he has learned. The answer for this Solve the Problem is based on the material on Student Text pages 150–51. The student should attempt the problem independently. Answers will vary and may be discussed.

Activity Manual

Study Guide—pages 95–96

These pages review some of the information from Chapter 6.

Assessment

Test Packet—Chapter 6 Test

The introductory lesson for Chapter 7 has been shortened so that it may be taught following the Chapter 6 test.

Diving Deep into Science

Answer the Questions

1. The spring of 2003 was very wet in the Southeast United States. What would you expect to be true of the tree rings in the Southeast for this year?
   They would probably be wider than usual.

2. Why are most dicots larger than monocots?
   Answers will vary but should include understanding that the woody stems of many dicots provide the support for them to grow larger than most monocots.

3. Why would plants such as mosses and ferns in tropical rainforests be able to grow larger than normal?
   Answers will vary, but elicit that these plants are very dependent on easily accessible water. The tropical rainforest would have an abundance of water available.

Solve the Problem

Tanya's father said he would pay her two dollars for every bucket of dandelion weeds that she removed from the yard. So Tanya went to work. She pulled up each plant, including its broad leaves and a few thin roots. There wasn't a dandelion left in the yard. A week later, however, there were dandelions growing everywhere Tanya had pulled them. Tanya's father asked, "Tanya, did you dig up the dandelions or just pull them up?" Why did Tanya's method not work?

Dandelions have long taproots. If Tanya removed only what was above the ground and a few secondary roots, the taproot would remain, and the plant would regrow.

Review Game

Build a Flower

Divide the class into two teams. Draw a completed flower as an example. Ask each team questions. If the team answers correctly, a member of the team may draw a portion of the flower. The drawing is completed with these steps: root, stem, one leaf, add veins, second leaf, add veins, circle for flower head, add a petal until the flower has five petals. The first team to draw a complete flower wins.

**Student Text pages 155–57
Activity Manual page 97**

Lesson 73

Objectives

- Demonstrate knowledge of concepts taught in Chapter 6
- Understand the interrelationship of science concepts
- Recognize that man's inferences are sometimes inaccurate

Unit Introduction

Note: This introductory lesson has been shortened so that it may be taught following the Chapter 6 test.

Someone once said that all science is physics. In Unit 3, students will learn about the most basic unit of matter and energy—the atom. Chapter 7 discusses the structure of the atom. Chapter 8 relates the atom to the flow of electrons that produce the electricity.

Chapter 9 explains how chemical energy (Chapter 7) and electrical energy (Chapter 8) can be changed into mechanical energy.

➤ **Look through Unit 3. What are the topics of the chapters we will be studying in this unit?** *{atoms, molecules, electricity, magnetism, motion, machines}*

➤ **Why do you think these chapters are organized into the same unit?** *{Answers will vary, but elicit that they all have to do with matter and energy.}*

📖 **God's creation is not limited to what we can see. The forces, such as gravity and magnetism, that cause objects to behave in certain ways are also God's creation. This unit will help us see the wonderful way God holds our universe together.**

Energy in Motion

155

 Additional information is available on the BJU Press website.
www.bjup.com/resources

Unit 3   Lesson 73   165

Project Idea

The project idea presented at the beginning of each unit is designed to incorporate elements of each chapter as well as information gathered from other resources. You may choose to use the project as a culminating activity at the end of the unit or as an ongoing activity while the chapters are taught.

Unit 3—Build a Better Mousetrap

Design a mousetrap that uses at least four simple machines. The mousetrap must also include either a chemical reaction or an electrical circuit as a force that causes parts of the mousetrap to move.

Technology Lesson

A correlated lesson for Unit 2 is provided on TE pages A8–A9. This lesson may be taught with this unit or with the other Technology Lessons at the end of the course.

Bulletin Board

Note: This suggested bulletin board idea may be modified for use in a learning center.

Direct the student to collect ingredient labels from household products. Highlight ingredients that are elements.

Attach a copy of the periodic table of the elements to the bulletin board. Display the labels around the periodic table. Record some of the most common elements on a tally chart. Allow the student to add tally marks for the number of times those elements occur on the labels.

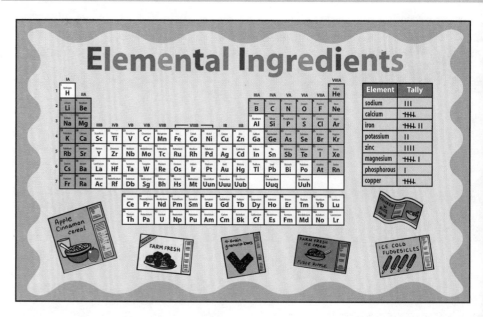

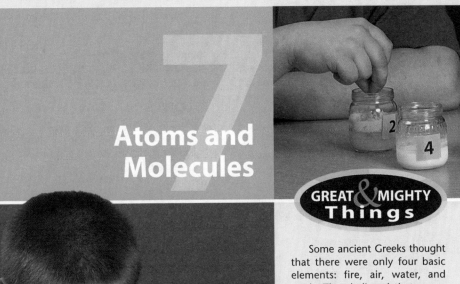

Atoms and Molecules

GREAT & MIGHTY Things

Some ancient Greeks thought that there were only four basic elements: fire, air, water, and earth. They believed that everything was made from different combinations of these elements. These scientists also thought that the four elements could change from one form to another. For example, they noticed that water disappeared when it was boiled. Since they did not know that water changes into vapor, they thought the water had turned into air. Scientists now know that there are over 100 basic elements that are made up of very tiny atoms. Man may have to change his ideas about what things are made of, but none of these elements are unknown to God. He created them long ago. With each new discovery, God's glory is more fully revealed to man. As Colossians 1:16 says, "All things were created by him, and for him."

157

SCIENCE BACKGROUND

The four elements—Aristotle was a strong proponent of the four elements: fire, earth, air, and water. During his lifetime there was much debate as to which element was the main one. Greek philosophers argued for different elements based on their inferences of their observations. Finally, a philosopher named Empedocles convinced others that each element was equally important.

Introduction

Man's understanding of science is always limited by what he can observe. Sometimes his observations are correct, but his conclusions are flawed.

Science is changeable because it is based on man's knowledge. However, God's knowledge is infinite. Nothing is new to God.

Chapter Outline
I. Atoms
   A. Parts of an Atom
   B. Atomic Number
   C. Models of Atoms
   D. Even Smaller Parts
II. Elements
   A. Classifying Elements
   B. Periodic Table of the Elements
III. Compounds
   A. Common Compounds
   B. Chemical Formulas
   C. Chemical Reactions
   D. Atomic Bonding
IV. Acids and Bases
   A. Properties
   B. pH Scale
   C. Indicators
   D. Neutralizing Acids and Bases

Activity Manual
Preview—page 97

The purpose of this Preview page is to generate student interest and determine prior student knowledge. The concepts are not intended to be taught in this lesson.

Lesson 74

Student Text pages 158–61
Activity Manual page 98

Objectives

- Describe and label the size, charge, and location of each part of an atom
- Recognize that an element is made of only one kind of atom
- Differentiate between atomic mass and atomic number

Vocabulary

chemistry	atomic mass
atom	electron
element	shell
nucleus	atomic number
proton	atomic theory
neutron	particle accelerator

Materials

- a piece of paper for each student

Introduction

Give each student a piece of paper. Instruct him to fold the paper in half and then to tear it in half. Tell each student to continue folding and tearing one section of the paper in half until it is too small to continue tearing.

➤ Can you still see the smallest piece?
➤ Is there a way you could continue tearing the very small piece in half?
➤ Is the small piece still paper?
➤ Today you will be studying things that are many times smaller than your tiny piece of paper.

Purpose for reading:
Student Text pages 158–61

➤ Which parts of an atom are found in the nucleus?
➤ How is an atom's atomic number different from its atomic mass?

When you think of chemistry, perhaps you imagine a messy laboratory with lots of test tubes and billows of vapors rising from containers. A slightly unkempt person with a sinister smile and a white lab coat pours material from one test tube into another. Such images are more fiction than fact.

For centuries chemistry was considered somewhat of a mystical science. Some men, called *alchemists,* tried to find a way to turn common metals, such as iron or lead, into gold. Others looked for a potion that would help them live forever. However, not all alchemists were tricksters. Some studied and experimented to find out how one substance changed to another. Their careful investigation and accurate recording formed the basis of modern **chemistry,** the study of matter—what it is made of, what its usual characteristics are, and how it reacts with other matter.

Atoms

To understand matter—anything that takes up space and has mass—you must first start with the smallest part of it. Imagine that someone has just given you a kilogram of gold. Your job is to keep dividing the gold in half until you have the smallest piece possible. Once you divide the gold into the tiniest piece that you can see, are you finished? No! If you put the gold under a microscope, you could continue to divide it into smaller pieces. Eventually the gold would become so small that you could not see it even with a microscope. But the gold could still be divided more. Finally, you would get to a piece that could no longer be divided and still be gold. That tiny piece of gold would be called an atom. An **atom** is the smallest piece of an element, such as gold, that can be recognized as that element. Substances containing only one kind of atom are called **elements.** Some of the elements that you probably know about are gold, silver, iron, oxygen, and copper.

158

SCIENCE BACKGROUND

Concept of atoms—Most people attribute the concept of the atom to the ancient Greeks, especially Leucippe and his disciple Democritus. There is evidence to suggest that at about the same time, philosophers in India were developing some of the same ideas.

Alchemy—One of alchemy's great benefits to scientific advancement was the use of scientific processes, such as observation, experimentation, measurement, and recording. These same skills are used today for scientific study.

SCIENCE MISCONCEPTIONS

Bohr's model of the atom is not the only model, and it is not the most accurate model. Electrons do not actually orbit in fixed paths the way they appear in a Bohr model.

Parts of an Atom

All atoms have the same basic structure. They are made up of three main parts: protons, neutrons, and electrons. Two of these parts, the protons and the neutrons, make up the center section of the atom, called the **nucleus**.

Protons have a positive charge (+) and **neutrons** have no charge (N). Protons and neutrons are by far the heaviest parts of an atom. Adding the number of protons and neutrons together results in the approximate **atomic mass** of an atom. For example, fluorine has nine protons and ten neutrons, so its atomic mass is approximately 19.

The third basic part of the atom is not part of the nucleus. The negatively charged **electrons** (–) travel around the nucleus. They are so light that they contribute almost nothing to the mass of an atom.

In a normal atom there are equal numbers of protons and electrons. Normal atoms have no overall electrical charges because the total negative charge of the electrons balances the total positive charge of the protons.

Though very small, electrons are very important to the atom. Scientists think that electrons constantly move and spin around the nucleus. The electrons move randomly in all directions, much like a swarm of bees around a hive. As they move, the electrons stay within a limited cloudlike space surrounding the nucleus. This space, called a **shell**, represents the average distance of the electrons from the nucleus. The first shell around the nucleus can hold a maximum of two electrons. The second shell can hold a maximum of eight electrons. Each shell farther away from the nucleus can hold more electrons. Because electrons move freely within their shells around the nucleus, they provide the means for atoms to combine with each other to form other substances.

Structure of an Atom

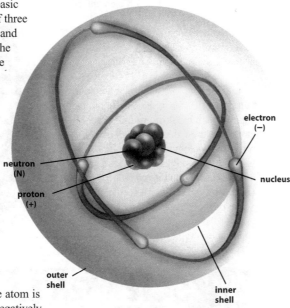

electron (–)
neutron (N)
proton (+)
nucleus
outer shell
inner shell

 The word *atom* comes from the Greek word *atomos*, which means "indivisible." The ancient Greeks considered the particles that make up all matter as being the smallest bit of matter possible.

Discussion:
pages 158–59

➤ What is **chemistry**? *{the study of matter—what it is made of, its usual characteristics, and how it reacts with other matter}*

➤ What is the smallest piece of an element that still can be recognized as that element? *{an atom}*

➤ What do we call substances that are made of only one kind of atom? *{elements}*

 Can an element have more than one atom? Explain. *{Yes, but all the atoms are the same kind.}*

➤ What are the three main parts of an atom? *{protons, neutrons, and electrons}*

➤ What is the center section of an atom called? *{nucleus}*

➤ Which parts of the atom are contained in the nucleus? *{protons and neutrons}*

➤ Protons have a positive charge. What kind of charge do neutrons have? *{no charge}*

➤ How do scientists calculate the approximate atomic mass of an atom? *{by adding together the number of protons and neutrons}*

Refer the student to the illustration *Structure of an Atom* on Student Text page 159.

➤ Which symbols are used to represent a proton, a neutron, and an electron in the drawing? *{proton, +; neutron, N; electron, –}*

➤ What does the negative sign tell us about electrons? *{They have a negative charge.}*

➤ Are electrons stationary or moving? *{moving}*

➤ Why do normal atoms usually not have an electrical charge? *{A normal atom has an equal number of protons and electrons.}*

➤ Describe an electron shell. *{a cloudlike layer around the nucleus; Each layer has electrons that are the same average distance from the nucleus.}*

➤ What is the maximum number of electrons any atom can have in its first shell? *{two}*

➤ What is the maximum number of electrons any atom can have in its second shell? *{eight}*

➤ Why is it important for electrons to move freely around the nucleus? *{This movement allows atoms to combine to form other substances.}*

Discussion:
pages 160–61

▶ Which term refers to the number of protons in an atom's nucleus? {atomic number}

💡 How is the atomic number different from the atomic mass? {The atomic mass is calculated by adding together the number of protons and the number of neutrons. The atomic number is the number of protons only.}

▶ Do scientists use the atomic mass or the atomic number to classify each element? {the atomic number}

▶ How many different elements have been identified? {more than 100}

Refer the student to *Meet the Scientist*.

▶ For what was Niels Bohr best known? {his research on atomic structure}

▶ What news did Niels Bohr bring to the United States in 1939? {Germany had successfully split the nucleus of an atom.}

▶ How did this news affect the United States? {It motivated the United States to speed up its research on atomic energy.}

💡 Which war was happening during the time that Bohr had to escape to Sweden? {World War II}

▶ What did Niels Bohr help the United States develop? {the atomic bomb}

Atomic Number

Scientists identify each element by the number of protons in its nucleus. This number is called the atom's **atomic number.** Each element has a different atomic number. Hydrogen has an atomic number of 1. Oxygen's atomic number is 8. Gold has an atomic number of 79. Since the atomic number equals the number of protons in an atom, we know that hydrogen has 1 proton, oxygen has 8 protons, and gold has 79 protons. Presently, scientists have identified and given atomic numbers to more than 100 elements. Some of these elements have been made in laboratories, but most of them occur naturally in the earth.

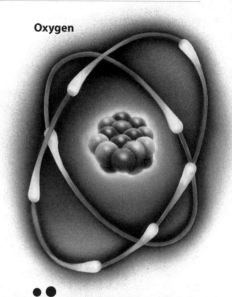
Oxygen

Meet the SCIENTIST — Niels BOHR

Niels Bohr (NEELS BOHR), a Danish scientist born in 1885, is considered to be one of the greatest scientists of the twentieth century. Although he is best known for his research on atomic structure, he also had a key role in developing the atomic bomb.

In 1939 Bohr visited the United States with news that Germany had successfully split the nucleus of an atom. This news motivated the United States to speed up its research on atomic energy. Bohr, however, returned to Denmark to provide a place for scientists escaping from the Nazis. In 1943 Bohr, who was half-Jewish, learned he was to be arrested by the Nazis. He was forced to escape by fishing boat to Sweden.

Shortly afterwards Bohr returned to the United States. His knowledge aided the United States in the development of the atomic bomb. Bohr recognized the potential threat posed by atomic power and worked until his death in 1962 for international control of nuclear weapons.

SCIENCE BACKGROUND

Tiny particles—Neutrinos, muons, gluons, and positrons are some of the other particles that make up an atom. Scientists think there are several hundred types of these tiny particles. Like Mendeleev, these scientists have identified the properties of some of these particles even though they have not actually discovered them yet. Most of these particles exist for such a short period of time (2 millionths of a second for a muon) that it is hard for scientists to track them down even with mammoth computers.

Models of the atom—John Dalton's model showed the atom as a solid sphere. Joseph (J. J.) Thomson's model, called the plum pudding model, has a positive sphere with electrons mixed in like "raisins in a cake." Ernest Rutherford's model was the first to show a nucleus with electrons moving around it like planets. Niels Bohr's model was similar to Rutherford's, but it limited the electrons to specific energy shells. Current models are more complex.

In 1922, when he was 37 years old, Niels Bohr won the Nobel Prize for his work describing the atom. After escaping by boat from Denmark to Sweden, Bohr and one of his sons went to the United States and worked on the nuclear fission bombs. His wife, Margrethe, and their other sons stayed behind in Sweden. After the war, Bohr and his family returned to Copenhagen, Denmark. Bohr died there in 1962. According to Bohr, "An expert is a man who has made all the mistakes which can be made in a very narrow field."

Models of Atoms

Because atoms are so small, no one has actually seen what they look like. We call what scientists think about atoms **atomic theory.** Atomic theory is based on repeated observations of how atoms act in experiments. The models have changed as scientists have discovered more about atoms.

Niels Bohr devised a model of an atom based on the idea that electrons move in shells. The Bohr model of the atom makes understanding the atom easier, but it is a very simplified model of what scientists think happens in an atom.

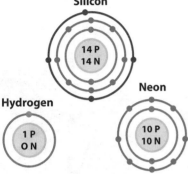

Even Smaller Parts

Once man discovered how to split the nucleus of an atom, it quickly became apparent that there were even smaller particles that were part of the nucleus. Today scientists have special machines called **particle accelerators** that smash atoms. When atoms break apart, these smaller particles can be measured but not seen.

Scientists have given some of these particles strange-sounding names such as quarks, leptons, mesons, pions, and gravitons. Perhaps you have heard these words used in science fiction stories.

As scientists discover smaller and smaller particles, they are less and less able to explain what really makes up all matter. The Bible has an answer for us: "Through faith we understand that the worlds were framed by the word of God, so that things which are seen were not made of things which do appear" (Heb. 11:3). Even though we cannot observe the tiniest level of God's creation, we can rest completely in Him because we know that He is the God who loves and cares for us.

> **QUICK CHECK**
> 1. List and describe the parts of an atom.
> 2. What is an element?
> 3. How do scientists identify elements?

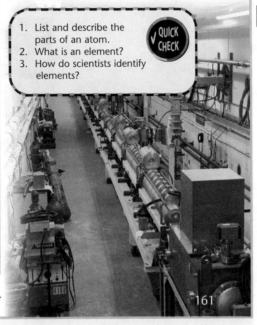

particle accelerator

 A particle accelerator propels parts of the atom to exceedingly high speeds using a series of electromagnets.

Scientists use these accelerated atomic particles in experiments to smash the nucleus of the atoms. Other uses for particle accelerators include industrial x-ray machines that find small flaws in metal and special medical machines used to diagnose and treat cancer.

Some particle accelerators are long and straight, and others are circular. They are very large machines, varying in size from several hundred meters to over twenty kilometers.

Direct an ACTIVITY

Write the word *argon* for display. List under it *18 protons* and *22 neutrons*.

Select several students to draw and label the nucleus and the first two shells of electrons for argon. The illustration should look similar to those shown on Student Text page 161.

> Does this atom need a third shell? How do you know? {Yes; since it has 18 protons, it must have 18 electrons. So far only 10 electrons have been drawn.}

Select another student to complete the third shell with 8 electrons.

Repeat with other elements as time permits.

Discussion:
pages 160–61

> Do scientists know exactly what atoms look like? {no} Why not? {because atoms are so small}

> Which term is used to describe what scientists think about atoms? {atomic theory}

> What is atomic theory based on? {repeated observations of how atoms act in experiments}

> Which scientist devised a model of an atom based on the idea that electrons move in shells? {Niels Bohr}

> Is a Bohr model of an atom what the atom actually looks like? Explain. {No; the Bohr model is a simplified version.}

> 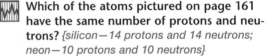 Which of the atoms pictured on page 161 have the same number of protons and neutrons? {silicon—14 protons and 14 neutrons; neon—10 protons and 10 neutrons}

> Hydrogen is the only atom that has no neutrons in its nucleus. How many protons does hydrogen have? {1}

> Which machines do scientists use to smash atoms? {particle accelerators}

> What happens when a particle accelerator smashes an atom? {The atom breaks apart and the smaller particles can be measured.}

> Name some of the subatomic particles. {quarks, leptons, mesons, pions, and gravitons}

> According to Hebrews 11:3, what are all the things we see made of? {things that do not appear, or things that we cannot see}

> Faith allows us to accept things we cannot see. It allows us to believe that God is in control of all things. [Bible Promise: I. God as Master]

Answers

1. The nucleus of the atom contains the protons and neutrons. Protons have a positive charge and neutrons have no charge. Electrons travel around the nucleus in shells. Electrons have a negative electrical charge.

2. An element is a substance containing only one kind of atom.

3. Elements are identified by the atomic number, or the number of protons in the nucleus.

Activity Manual
Reinforcement—page 98

Lesson 75

Student Text pages 162–65
Activity Manual pages 99–100

Objectives

- Identify the terms *period*, *group*, and *family* as they relate to a periodic chart
- Understand that the periodic chart is only a classification system
- Describe the process that Mendeleev used for arranging elements
- Identify the types of information in a square on the periodic table

Vocabulary

chemical symbol
periodic table of the elements
period
group
family

Introduction

Write the following for display: FBI, NHL.

➤ **What do each of these symbols (letters) stand for?** {F-Federal, B-Bureau, I-Investigation; N-National, H-Hockey, L-League}

➤ **Why do we use these symbols instead of words?** {Possible answers: They are easier and faster to write.}

➤ Scientists also use symbols to represent the elements because symbols are easier and faster to write.

Purpose for reading:
Student Text pages 162–65

➤ Why are the chemical symbols for some elements hard to recognize?

➤ How was the periodic table of the elements first organized?

➤ What are three ways that scientists classify elements in the periodic table?

Elements

Classifying Elements

Scientists have devised a system of symbols to use when referring to elements. Symbols are quicker to write and easier to use than full names. And since scientists around the world use the same symbols, communication among themselves is much easier. These **chemical symbols** are abbreviations for the names of the elements.

Many of these abbreviations are based on an element's English name. The symbol for oxygen is O, and the symbol for carbon is C. Since some elements begin with the same letter, a second letter is often added. For example, the symbol for calcium is Ca. Notice that the first letter is capitalized and the second letter is lowercase. The second letter in the symbol is not always the second letter in the element's name. For example, the symbol for chlorine is Cl. Some abbreviations, such as the symbol for iron (Fe), are even more difficult to guess. These strange symbols make sense when you realize that the abbreviation is based on the Latin word for iron, which is *ferrum*. The Latin word for silver is *argentum*, so the symbol is Ag.

UNUSUAL CHEMICAL SYMBOLS

Element	Symbol	Latin word
Tin	Sn	Stannum
Sodium	Na	Natrium
Potassium	K	Kalium
Iron	Fe	Ferrum
Silver	Ag	Argentum
Lead	Pb	Plumbum
Mercury	Hg	Hydrargyrum

162

SCIENCE BACKGROUND

Natural and man-made elements—Most of the first ninety-two elements in the periodic table are found in nature. Most of the elements found after uranium on the chart are man-made elements that do not occur naturally. The names of some of the man-made elements are not universally recognized because scientists from more than one country have claimed their discovery.

Weight, mass, and isotopes—Atomic weight and atomic mass are sometimes used interchangeably, even though they are not quite the same. Atomic weight is the older term, referring to the average weight (protons + neutrons) of all the isotopes of a certain element. Isotopes are atoms of the same element that have different numbers of neutrons. Atomic mass refers to the mass (protons + neutrons) of one isotope. Usually the number given on the periodic table of the elements is the approximate atomic mass and is also the average of the element's isotopes' atomic masses.

Chemical symbols—The symbols used for the elements are also called *atomic symbols*.

172 Lesson 75

Periodic Table of the Elements

As scientists discovered more and more elements, they needed to find a system of classification. In 1869 a Russian chemist named Dmitri Mendeleev (duh-MEE-tree MEN-duh-LAY-uf) came up with a way to organize all of the elements known at that time. Mendeleev put the elements in order based upon their atomic weights. Then he grouped the elements into rows and columns based upon their chemical and physical properties. This classification system is called the **periodic table of the elements.**

Sometimes an element's atomic weight put the element in a location that did not match its chemical and physical properties. When that occurred, Mendeleev put the element where it fit according to properties instead of weight. In doing this, he left some gaps in his chart where he felt sure there existed undiscovered elements. His analysis proved true. Eventually elements filled all of the gaps in the chart.

Scientists later realized that organizing by atomic weight was not the best way to arrange the chart. Instead, arrangement by atomic number (the number of protons) gives a more accurate picture of the common characteristics that elements share. The current table also changes as new elements are discovered or formed in laboratories.

Each square in the periodic table describes one element. Within the square you will find the name of the element, its chemical symbol, its atomic number, and sometimes other information, such as atomic mass. For example, you can look for the atomic number 50 and find out that its name is tin and its chemical symbol is Sn. Or you can look up the name potassium to find its symbol, K, and its atomic number, 19.

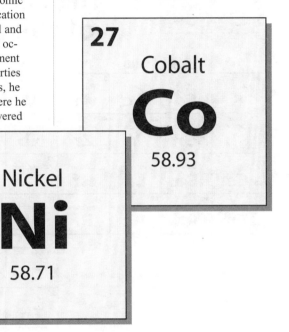

163

Dmitri Mendeleev was born in Tobolsk, Siberia, in 1834. In addition to the periodic table, Mendeleev also wrote a chemistry textbook. He was well-known for wanting to educate people about chemistry. Whenever he traveled by train, he would sit in the third-class section with the Russian peasants and talk to them about his findings.

Mendeleev believed there must be a pattern to the sixty-three known elements of his time. He wrote information on cards about each element's melting point, density, color, atomic mass, and bonding power. Then Mendeleev began the process of arranging the cards in various ways. He discovered that patterns began to appear when he arranged the elements by atomic weight. Next, he placed elements with similar properties underneath one another. After arranging his elements, he found three blank spaces in his table. Mendeleev predicted that these blank spaces would be filled by elements that had not yet been discovered. He even predicted the properties of these missing elements. Within sixteen years, the missing elements, scandium, gallium, and germanium were found.

Discussion:
pages 162–63

▸ **Why do scientists use chemical symbols?** {Symbols are quicker to use, and they make communication with other scientists around the world easier.}

▸ **What is the basis for many of the chemical symbols?** {the element's English name}

▸ **Why is a second letter sometimes included in the chemical symbol for an element?** {Some elements begin with the same letter.}

▸ **What is another basis for some of the chemical symbols?** {the element's Latin name}

▸ **Who came up with a method for classifying the elements?** {Dmitri Mendeleev}

▸ **How did Mendeleev organize his table?** {He put the elements in order based on atomic weight. The rows and columns are organized based on the chemical and physical properties of the elements.}

▸ **What do we call this classification system?** {periodic table of the elements}

▸ **Why did Mendeleev's periodic table have gaps?** {Sometimes an element's atomic weight put the element in a location that did not match its chemical and physical properties. Mendeleev put those elements where they fit according to properties instead of weight. He thought that undiscovered elements would fill in the gaps.}

▸ **What does each square in the periodic table represent?** {one element}

▸ **What kinds of information can be found within each square on the periodic table?** {Possible answers: the name of the element, its chemical symbol, its atomic number, and sometimes its atomic mass}

💡 **If you know the atomic number of an element, what else do you know besides the number of protons?** {the number of electrons that element has}

Look at the picture on page 163. Which element would come first in the periodic table—nickel or cobalt? {cobalt}

▸ **If you arranged these two elements by atomic mass only, which element would come first?** {nickel}

Discussion:
pages 164–65

Refer to *The Periodic Table of the Elements* as each part is discussed.

▶ **What are the horizontal rows of the periodic table called?** *{periods}*

▶ **What do elements in the same period have in common?** *{They all have the same number of shells.}*

▶ **What is the highest number of shells for atoms whose elements are listed in the periodic table?** *{seven}*

▶ **What do we call the columns of elements in the periodic table?** *{groups}*

▶ **What do all the elements in a group have in common?** *{They all have the same number of electrons in their outer shells.}*

▶ **How do you know that sulfur and oxygen have similar physical and chemical properties?** *{They are in the same group.}*

▶ **How do scientists classify families of elements?** *{by their chemical properties and by how they react with other elements}*

▶ **How can you tell which family on the periodic table an element belongs to?** *{The families are color coded.}*

▶ **Based on this periodic table, what are some of the families that scientists use to classify elements?** *{alkali metals, nonmetals, semimetals or metalloids, and transition metals}*

The horizontal rows of the periodic table are called **periods.** Elements in the same row, or period, have something in common. They all have the same number of shells. Both elements in the first row have one shell for their electrons. All the elements in the second row have two shells for their electrons. The table continues all the way to seven shells.

The vertical columns of the table are called **groups.** All the elements in a group have the same number of electrons in their outer shells. Every element in the first column, or group, has one electron in its outer shell. Every element in the second group has two electrons in its outer shell. All the elements in a group also share similar physical and chemical properties. If you want to find the elements that are similar to neon, just look at the elements located in the same column as neon.

Another way scientists classify elements in the periodic table is by

The Periodic Table of the Elements

164

SCIENCE BACKGROUND

Periodic Table—Not all periodic tables look the same. Scientists do not always agree on how the elements should be classified and labeled. The charts are modified as scientific equipment improves and new elements are discovered.

The periodic table shown on Student Text pages 164–65 has been simplified for the student's use. The families shown are general groupings that may be further divided on some charts. Hydrogen is not colored because it has many unique properties and does not clearly belong to one group.

ART

The student can make models of elements 2–36 using foam balls to represent protons and neutrons and smaller foam balls to represent electrons. The balls can be painted to identify the parts they represent. Direct the student to devise a way to show the shells with the appropriate number of electrons using wire. Hang the finished elements from the ceiling.

174 Lesson 75

placing them into **families**. Families of elements possess similar chemical properties and react similarly with other elements. They are often color-coded in the periodic table. The first two columns on the left are the family of elements called the *alkali* (AL kuh LYE) *metals*. Elements may also be placed in categories that include *metals*, *semimetals*, or *metalloids*, and *non-metals*. A stairstep line separates these categories from one another.

Quick Check
1. What are chemical symbols, and why do scientists use them?
2. What is the periodic table of the elements?
3. What do elements in a group in the periodic table of the elements have in common?

Discussion
pages 164–65

 Use *The Periodic Table of the Elements* to answer the questions.

➤ What is the name of element 13? {aluminum}

➤ Which element has K as its chemical symbol? {potassium}

➤ What is the atomic number for neon? {10}

➤ How many protons does an atom of gold have? {79}

➤ What is the chemical symbol for silicon? {Si}

➤ In what family does silicon belong? {semi-metals, or metalloids}

➤ Sodium and calcium have similar properties. In what family are they placed? {alkali metals}

➤ Look at element 57. What is the atomic number of element 57? {57} What is the atomic number of the element to its right? {72}

💡 It appears that some elements are missing. Can you find them on the chart? {Yes; they are the elements in the two rows at the bottom.}

💡 Why do you think these elements are not in the regular chart? {Accept reasonable answers, but conclude that they have properties similar to lanthanum. They are actually called the Lanthanide series.}

📖 Do you think it was an accident that elements can fit so well into an organized system? {Answers may vary, but emphasize that God's creation is a reflection of the fact that God does everything "decently and in order." (I Cor. 14:40)}

Answers
1. Chemical symbols are abbreviations for the names of the elements and are quicker to use than the elements' full names. The abbreviations make communication with scientists around the world easier.

2. a classification system for the elements developed by Dmitri Mendeleev

3. All the elements in a group have the same number of electrons in their outer shells.

Activity Manual
Study Guide—pages 99–100
These pages review Lessons 74 and 75.

Assessment
Test Packet—Quiz 7-A
The quiz may be given any time after completion of this lesson.

Direct an Activity

Many of the elements are named after people, places, or planets.

Ask the student to find the element on the periodic table named after the following.

People: Pierre and Marie Curie (curium), Albert Einstein (einsteinium), Enrico Fermi (fermium), Dmitri Mendeleev (mendelevium), Niels Bohr (bohrium), Alfred Nobel (nobelium), Ernest Rutherford (rutherfordium), Glenn Seaborg (seaborgium).

Places: America (americium), Europe (europium), France (francium), Poland (polonium), Scandinavia (scandium), Germany (germanium), California (californium), and Berkley, CA (berkelium)

Planets: Mercury (mercury), Neptune (neptunium), Pluto (plutonium), Uranus (uranium)

 MATH Atomic number is the number of protons in an element. Atomic mass is the number of protons and neutrons in an atom. Choose an element on the periodic table. Guide the student in finding the average number of neutrons for the element. For example, nickel has an atomic number of 28 and an atomic mass of 58.71. Subtract 28 from 58.71 to get approximately 31 neutrons. Continue with other elements as time allows.

Lesson 76

Student Text page 166
Activity Manual pages 101–2

Objectives

- Research an element
- Construct a visual aid

Materials

- poster board
- other materials to make the poster

Introduction

➤ Have you ever heard a news reporter give a profile on a criminal that the police are looking for?

The criminal may have escaped from police custody, or the police might have a case file on him because he has committed other crimes.

➤ What kind of information is usually included in a case file or on a wanted poster? *{Possible answers: the crime committed, what the criminal looks like, disguises, aliases, last known location of the criminal, how the crime was committed}*

➤ For this exploration you will be preparing a case file for a given element.

Purpose for reading:
Student Text page 166
Activity Manual pages 101–2

The student should read all the pages before beginning the exploration.

Discussion:
page 166

Build enthusiasm for creating the case files. Give each student an element to research. Choose from the elements with atomic numbers 2–36.

Refer the student to the poster pictured on Student Text page 166.

➤ How does the picture show what hydrogen looks like? *{It shows one large proton and a smaller electron.}*

Activity Manual
Explorations—pages 101–2

Assessment
Rubric—TE pages A50–A53

Select one of the prepared rubrics, or design a rubric to include your chosen criteria.

Explorations: Wanted: U or Your Element

There are over ninety naturally occurring elements and over a dozen elements that have been artificially produced in laboratories. Elements have been named for people, places, colors, and a variety of other things. There are even some elements that have no official names because there is disagreement as to who first produced these elements in the laboratory. Some elements have been known since ancient times, and others have been identified only recently (within the last fifty years).

Elements hide in lots of places. Your body, the Sun, and even your food contain many unexpected elements. You would probably be amazed to know all the elements contained in a box of cereal.

You will be given the name of a "wanted" element. Your assignment is to prepare a case file on that "wanted" element. You need to give the element's structure, position on the periodic table, history, and uses. Of course, you also need to provide a poster to help apprehend your element.

What to do

1. After you are given an element to investigate, use the periodic table on pages 164–65 of your text and other resources to complete the case file on pages 101–2 of your Activity Manual.

2. Prepare a "wanted" poster for your element. Attach the case file to your poster.

166

 Teacher Helps

This exploration may be done by individual students or in groups.

There are variations between different periodic tables. Decide if you want other periodic tables used or only the one in the Student Text.

Additional information may include where the element is found in its natural state, its flammability, its color, and its other chemical properties. Chemical properties are the properties and characteristics specific to that element.

Instead of making one large wanted poster, the wanted poster could be prepared on a size of paper that can be placed with the case file in a notebook for future reference.

Compounds

If single atoms were always by themselves, we would never see them because they are so tiny. However, atoms rarely exist alone. Sometimes identical atoms group together to form elements such as gold, silver, copper, and mercury. Atoms may join with other atoms to form units that are called **molecules**. A few atoms chemically join with identical atoms to form units called diatomic molecules. For example, two oxygen atoms may join with one another to form a molecule of oxygen.

However, most atoms join with different types of atoms in a process called a **chemical change**, or **reaction**. Usually the chemical change produces molecules known as **compounds**. They are called compounds because the atoms come from two or more different elements. The basic structures of the atoms in the compound are not altered. The atoms simply rearrange to form a different substance.

Common Compounds

When elements combine through a chemical change, a new substance forms. The properties of this substance are very different from those of the original elements. A combination of hydrogen and oxygen forms the most common compound on earth. Hydrogen is an extremely flammable gas, and oxygen is also a gas that permits things to burn. But when the two combine in a chemical change, they make the compound we call water (H_2O). We use this compound to put out many types of fire. The properties of water are very different from the properties of its original elements.

Another example of a compound that differs greatly from its individual elements is sodium chloride, or table salt. Sodium is an alkali metal that may cause an explosion when it reacts with water. Chlorine is a poisonous, greenish-colored gas. When sodium and chlorine combine, they form a compound that is not only safe for you to digest, but is actually necessary for you to have good health.

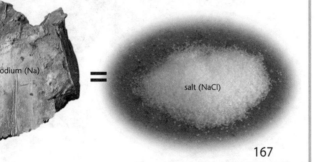

chlorine (Cl) + sodium (Na) = salt (NaCl)

167

Student Text pages 167–69

Lesson 77

Objectives
- Explain that a chemical change occurs when atoms combine
- Give examples of synthesis and decomposition reactions
- Demonstrate how to write a chemical formula

Vocabulary
molecule   compound
chemical reaction   chemical formula

Materials
- trail mix or snack cracker mix
- 500 mL bottle with 125 mL vinegar inside
- balloon with 15 mL baking soda inside

Introduction

Give the student a small amount of trail mix.

➤ **Identify the types of food in the trail mix.** {Possible answers: peanuts, raisins, candy}

➤ **When these foods are mixed together, they keep their original identities. This is called a *mixture* because the parts do not change when they are combined.**

Carefully attach the balloon to the top of the bottle. Lift the balloon quickly, so that the baking soda falls into the vinegar.

➤ **What happened?** {The balloon filled with air.} **Why?** {Elicit that when the baking soda and vinegar combined, a change took place, and a gas was given off, filling the balloon.}

➤ **This type of combination is a chemical reaction.**

Purpose for reading:
Student Text pages 167–69

➤ What is a *compound*?
➤ What happens in a synthesis reaction?

Discussion:
page 167

➤ **What do we call a group of atoms joined together?** {a molecule}

➤ **What occurs when different types of atoms are joined together?** {A chemical change, or chemical reaction, occurs.}

➤ **What is a *compound*?** {a molecule with atoms from two or more different elements}

The properties of the elements that make up a compound are different from the properties of the compound they form.

➤ **What is the most common compound on Earth?** {water}

➤ **Which two elements make up water?** {hydrogen and oxygen}

➤ **What are the properties of hydrogen and oxygen?** {Hydrogen is a flammable gas. Oxygen is a gas that helps things burn.}

➤ **What are some properties of water?** {It is a liquid. We drink it. It puts out many types of fire.}

💡 **If you ate chlorine and sodium individually, what would happen to you?** {You would be poisoned.}

➤ **When chlorine and sodium combine, what compound is formed?** {salt}

Chapter 7   Lesson 77

Discussion:
pages 168–69

➤ **What is a *chemical formula*?** *{the chemical symbols and numbers that show the elements that make up a compound}*

Remind the student of the introduction of the previous lesson where letters represented words.

➤ **Although the individual letters in FBI and NHL stand for words, the symbols together also stand for something. When we hear FBI we do not usually think of the words. Instead we think of our national crime-fighting organization. Scientists also combine the chemical symbols to form chemical formulas that represent molecules.**

➤ **What do the small subscript numbers in a formula tell us?** *{how many atoms of that element are in a molecule}*

➤ **How many atoms of oxygen are in carbon dioxide?** *{2}*

➤ **Compounds can be made of two or more elements. How many elements are in carbonic acid?** *{three}* **What are they?** *{hydrogen, carbon, oxygen}*

➤ **How many atoms of hydrogen are in carbonic acid?** *{2}* **Atoms of carbon?** *{1}* **Atoms of oxygen?** *{3}*

 Refer the student to the chemical formulas for cane sugar, baking soda, and aspirin. The student may use the periodic table on Student Text pages 164–65 to help him identify the elements.

➤ **Which elements make up cane sugar?** *{carbon, hydrogen, and oxygen}*

➤ **Which element in cane sugar has the most atoms?** *{hydrogen}*

➤ **Which four elements combine to form baking soda?** *{sodium, hydrogen, carbon, and oxygen}*

➤ **How many atoms of carbon are in aspirin?** *{9}*

💡 **Look at the formula for carbonic acid in the paragraph and the formulas for cane sugar and aspirin in the chart. What do those formulas have in common?** *{They are all combinations of the same elements—carbon, hydrogen, oxygen.}*

➤ **How are those formulas different?** *{Each compound has different amounts of each element.}*

Chemical Formulas

When scientists want to show the elements that make up a compound, they use a chemical formula. A **chemical formula** uses chemical symbols and numbers to abbreviate the name of a compound. Using a chemical formula is similar to using the letters USA for the United States of America. Often the abbreviation is used for the country's full name.

A chemical formula gives a scientist information about a compound. For instance, the formula for carbon dioxide is CO_2. The subscript 2 tells us that this molecule has two atoms of oxygen. Those two atoms of oxygen are combined chemically with one atom of carbon to form one molecule of carbon dioxide.

Compounds are not always made up of just two elements. Sometimes the atoms of several elements combine to form a compound. For example, the carbonic acid that can cause chemical weathering has more than two elements. The chemical formula for carbonic acid is H_2CO_3. The formula tells us that a molecule of carbonic acid has two atoms of hydrogen, one atom of carbon, and three atoms of oxygen.

Cane sugar	Baking soda	Aspirin
$C_{12}H_{22}O_{11}$	NaHCO	$C_9H_8O_4$

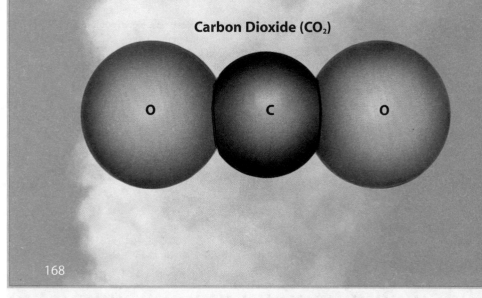

Carbon Dioxide (CO_2)

168

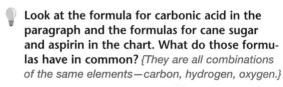

The student should identify the elements in the formula for each compound and the number of atoms of each element. He may use the periodic table for help.

Hydrochloric acid—HCl *{1 hydrogen + 1 chlorine}*

Laughing gas—N_2O *{2 nitrogen + 1 oxygen}*

Liquid rubber—C_5H_8 *{5 carbon + 8 hydrogen}*

Lye—NaOH *{1 sodium + 1 oxygen + 1 hydrogen}*

Sand—SiO_2 *{1 silicon + 2 oxygen}*

The student should write the formula for each compound.

Ammonia—1 nitrogen + 3 hydrogen *{NH_3}*

Chalk—1 calcium + 1 carbon + 3 oxygen *{$CaCO_3$}*

Hydrogen peroxide—2 hydrogen + 2 oxygen *{H_2O_2}*

Moth balls—10 carbon + 8 hydrogen *{$C_{10}H_8$}*

Rust—2 iron + 3 oxygen *{Fe_2O_3}*

TRY IT Yourself

You can demonstrate a chemical reaction. Pour 60 mL of vinegar into a clear plastic cup. Stir 10 mL of salt into the vinegar until the salt is completely dissolved. Place two ¾" copper couplings (pipe joints) into the solution. Add to the solution several iron nails or screws that have not been coated or galvanized. Leave for several hours. What do you observe? *The copper from the pipe replaces the iron on the surface of the nails or screws.*

Chemical Reactions

Some chemical reactions occur when molecules combine to form new substances. This kind of chemical reaction is called a *synthesis* reaction. Let's again use carbonic acid for an example. Carbonic acid forms when carbon dioxide (CO_2) reacts with the water (H_2O) that is in the air. When these molecules combine, they form the new compound H_2CO_3.

Not all chemical reactions result in more complex compounds. Some chemical reactions break down a complex compound into two or more simpler compounds. This kind of chemical reaction is called a *decomposition* reaction. When our bodies digest complex molecules, such as sugars, the sugar is broken down into much simpler molecules, such as water and carbon dioxide.

In other reactions, one element replaces another element in a compound. There are even some chemical reactions that cause compounds to trade elements with each other.

TYPES OF CHEMICAL REACTIONS

Synthesis:	elements combine
Decomposition:	compounds break apart
Single replacement:	one element replaces another
Double replacement:	two elements switch places

QUICK CHECK

1. What is a chemical reaction?
2. What is a compound?
3. What does the symbol CO_2 show about the molecule?
4. What happens in a synthesis chemical reaction?

Discussion:
pages 168–69

- What is a *chemical reaction?* {the process through which different types of atoms join together}
- Which two elements join to make carbonic acid? {carbon dioxide and water}
- Is carbonic acid a different substance than carbon dioxide and water? {yes}
- What is this type of chemical reaction called? {synthesis reaction}
- Look at the chemical symbol for cane sugar found on page 168. Our bodies cannot use this sugar as it is. It must be broken down into simpler molecules that our bodies can use.
- Which type of chemical reaction occurs as the acids in your digestive system break down the food you eat? {decomposition reaction}

Answers

1. the process through which different types of atoms join together
2. molecules formed as a result of a chemical reaction of two or more different elements
3. The compound consists of one atom of carbon and two atoms of oxygen.
4. Elements combine to form new substances.

Direct a DEMONSTRATION
Demonstrate a chemical change

Materials: 60 mL water, 30 g Borax, 60 mL white glue, zip-close plastic sandwich bag

- Describe the appearance and properties of the water, Borax, and glue. {Possible answers: Water is a colorless and tasteless liquid. Borax is a white powder. Glue is a thick white liquid that pours slowly.}
- Borax is made from the element boron. Look at the periodic table of the elements. What is the atomic number for boron? {5}
- What do you think Borax is used for? {It works with laundry detergent to brighten clothing.}
- What do you think will happen when we mix these items together? {Answers will vary.}

In the plastic bag, combine the water and Borax. Add the glue and squeeze (knead) to mix well. Direct the student to observe and record any changes in the appearance of the substance. Remove the substance from the bag. Allow the student to touch the mixture and describe how it feels.

- Describe the properties of this new substance. {Possible answers: smooth, slippery, shiny, stretchy, malleable}
- Hold the substance. Could you pick up the water, Borax, or glue this way? {no}
- Think of a name and uses for your substance.

This substance is similar to Silly Putty.

Chapter 7

Lesson 77

Lesson 78

Student Text pages 170–71
Activity Manual pages 103–4

Objectives

- Compare and contrast ionic and covalent bonding
- Explain what causes an ion

Vocabulary

closed
covalent bond
ion
positive ion
negative ion
ionic bond

Introduction

▶ What do you do when you see that a friend needs to use something that you have? *{Possible answers: share, lend, or give it to the friend}*

Atoms also share electrons.

▶ Today we will talk about how atoms share electrons.

Purpose for reading:
Student Text pages 170–71

▶ How are covalent bonds and ionic bonds different?

Discussion:
pages 170–71

▶ What factor determines how atoms tend to form compounds? *{the number of electrons in the outer shells of the atoms}*

▶ If the last shell of an atom is completely filled with electrons, is the atom considered stable or unstable? *{stable}*

▶ What does *stable* mean? *{the atom is not likely to form compounds with other atoms}*

▶ When is a shell of an atom closed? *{when it is completely filled with electrons}*

Atomic Bonding

Atoms tend to form compounds based on the number of electrons in each atom's outermost shell. Electron shells can hold only a certain number of electrons. If the atom's outermost shell is completely filled, that shell is **closed,** and the atom is considered stable. Stable atoms are not likely to form compounds with other atoms. Noble gases are made up of stable atoms, thus they rarely undergo chemical reactions.

Most atoms have electron shells that are not completely filled. Some atoms' outer shells have electrons that can be either given up or shared to form bonds, or unions. These atoms bond with other atoms so that each atom completes its outer shell.

Atoms bond with each other in several ways. Each kind of bond affects the characteristics of the compound formed by the atoms.

Covalent bonds

Some bonds occur when atoms share pairs of electrons in what is known as a **covalent bond.** By such sharing, the electrons fill the outer shells of both atoms. Gases and liquids are most likely to contain covalent bonds, or bonds formed by sharing electrons. Oxygen in the air is actually a molecule of two oxygen atoms that have joined by sharing electrons. Water (H_2O) is a liquid compound formed by atoms sharing electrons. In water, two hydrogen atoms combine with one oxygen atom to form a water molecule.

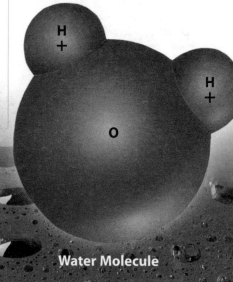

Water Molecule

170

A covalent chemical bond occurs when two or more atoms are joined together. The Bible tells us that Christians have a spiritual bond with other Christians. How are Christians joined together with each other? The Holy Spirit bonds all Christians together in Christ, as Christians are "endeavouring to keep the unity of the Spirit in the bond of peace" (Eph. 4:3).

 God commands Christians to be generous toward others, especially toward other Christians. Read and discuss Hebrews 13:16, Proverbs 22:9, Galatians 6:10, and Psalm 112:5. [BAT: 5b Giving]

SCIENCE BACKGROUND

Metallic bonds—Another type of atomic bonding is called *metallic bonding*. In metallic bonds, free electrons surround a positive ion core. This structure allows molecules to conduct heat and electricity well because the electrons move easily.

Noble gases—Noble gases are the elements listed in Group VIIIA on the periodic table.

Water molecule—Oxygen has an atomic number of 8. That means the outer shell needs two electrons to be complete. That is why it takes two hydrogen atoms, which have only one electron each in their outer shells, to form the covalent bond.

Ionic bonds

An atom that has gained or lost electrons is called an **ion**. Ions occur when the number of electrons in an atom does not equal the number of protons. The atom is positively charged when the number of protons is greater than the number of electrons. A positive charge means that the atom has given away electrons from its outer shell. The atom is then called a **positive ion**. If the number of electrons is greater than the number of protons, the atom is negatively charged. It is then called a **negative ion** because it has gained electrons to fill its outer shell.

Perhaps you remember that the opposite ends—the north and south poles—of magnets attract. The same principle applies to ions. Positively and negatively charged ions attract each other to form **ionic bonds**, resulting in new compounds.

The compound sodium chloride, or table salt, contains ionic bonds. Sodium and chloride transfer electrons to form this compound. Chlorine atoms receive from sodium one electron apiece in order to fill their outer shells. Chlorine is then a negative ion. Each sodium atom gives up one electron to the chlorine and becomes a positive ion. However, you would not find just one molecule made up of these oppositely charged ions. Many of these ions attract each other and bond together. They fit in such a way that they form beautiful geometric shapes called crystals. All ionic compounds form crystals.

1. What are two ways that atoms bond?
2. What is an ion?

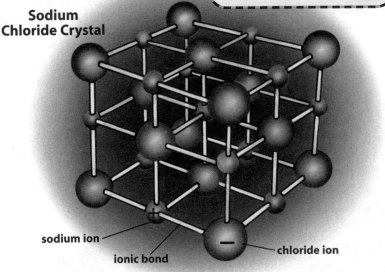

Sodium Chloride Crystal

- sodium ion
- ionic bond
- chloride ion

171

Direct an Activity

Materials: large checkers, marbles, or paper circles to represent electrons

Randomly distribute up to 7 electrons per student. Each student represents an atom, and the checkers, marbles, or paper circles represent the electrons found in the atom's outer shell. For this activity, the outer shell is considered full when it has 8 electrons.

Students should find other atoms with which they can bond. A student may share electrons with another atom (covalent bonding) so that each atom has 8 electrons. Students may also give and receive electrons (ionic bonding).

Direct students to identify each type of bond formed.

Chapter 7

Discussion:
pages 170–71

➤ Which type of bond is formed when atoms share electrons? {a covalent bond}

➤ What are two examples of molecules that form covalent bonds? {oxygen and water}

➤ Do most gas and liquid molecules have a covalent or ionic bond? {covalent}

➤ What do we call an atom that has gained or lost electrons? {an ion}

💡 Why does an atom that has lost an electron have a positive charge? {because it has one more proton than electrons}

➤ What do we call an atom with more electrons than normal? {a negative ion}

➤ What is needed for a molecule to form with an ionic bond? {positive and negative ions}

💡 Why do positive and negative ions join to form bonds? {They are attracted to each other and exchange electrons.}

➤ What shape do all ionic compounds form? {crystals}

➤ Name an example of a molecule that forms from an ionic bond. {Possible answer: sodium chloride (table salt)}

Discuss the diagrams on Student Text pages 170–71.

💡 Why do you think atoms such as oxygen and chloride are pictured larger than the hydrogen and sodium atoms? {Possible answers: Oxygen and chloride are actually larger than the other atoms pictured. They have larger atomic numbers.}

Answers

1. Atoms share electrons in covalent bonds. Atoms give or receive electrons in ionic bonds.
2. An ion is an atom that has a charge due to gaining or losing electrons.

Activity Manual

Reinforcement—page 103

Study Guide—page 104
This page reviews Lesson 78.

Lesson 78

Lesson 79

Student Text pages 172–73
Activity Manual pages 105–6

Objectives

- Recognize whether chemical reactions occur
- Collect data to identify a reaction as endothermic or exothermic

Vocabulary

endothermic reaction
exothermic reaction

Materials

- See Student Text pages

Introduction

➤ Have you ever seen or used sparklers for holidays, such as the Fourth of July? *{Answers will vary.}*

A sparkler burns longer than a firecracker and produces a shower of sparks. It consists of a rigid stick or wire that has been coated with chemicals. When the sparkler is lighted, some of the chemicals decompose to produce oxygen, and other chemicals react with the oxygen.

➤ What do these chemical reactions produce? *{Possible answers: heat, smoke, sparks}*

➤ In today's activity, you will use physical characteristics to determine whether or not a chemical reaction has occurred as different substances are combined.

Purpose for reading:
Student Text pages 172–73
Activity Manual pages 105–6

The student should read all the pages before beginning the activity.

Discussion:
page 172

➤ What are some indications that a chemical change has occurred? *{Possible answers: solid substance appears; change of color}*

📖 A chemical reaction is evidenced by a visible change. A person who accepts Christ as Savior should also show a visible change. What are some visible signs that a person is a Christian? *{Possible answers: shows love to others; studies the Bible; prays}* [BATs: 5a Love; 6a Bible study; 6b Prayer]

ACTIVITY: Hot or Cold

Chemical reactions often show visible signs that something has taken place. Sometimes a solid substance appears in a solution. A color change might also indicate a chemical reaction.

Another sign of a chemical reaction is whether heat is produced or absorbed. An *endothermic* (EN doh THUR mihk) *reaction* uses thermal (heat) energy, so the temperature of a solution experiencing an endothermic reaction decreases. An *exothermic* (EK soh THUR mihk) *reaction* produces heat. The reaction raises the temperature of the solution in which a reaction is occurring.

In this activity, you will combine substances to determine whether the reaction occurring is endothermic, exothermic, or even a chemical reaction at all.

Process Skills
- Predicting
- Observing
- Measuring
- Experimenting
- Recording data

Problem
How can I use temperature to determine whether a chemical reaction has occurred?

Procedure

1. You will need at least three people for this activity—one person to time, one person to read the temperature, and one person to record the information. Be sure everyone is ready before combining the substances.
2. Look at the chart in the Activity Manual. Predict which pairs of substances will cause a chemical reaction. Write a hypothesis based on your prediction.
3. Pour 30 mL of hydrogen peroxide into a cup.
4. Measure and record the temperature.
5. Measure 5 mL of yeast and set it aside until each person is ready.

Materials:
3 plastic cups
measuring spoons
3% hydrogen peroxide
yeast
thermometer
stopwatch or watch with a second hand
water
salt
vinegar
baking soda
goggles (optional)
green, red, and blue colored pencils
Activity Manual

Exothermic means "giving off or producing heat."

Endothermic means "using or absorbing heat."

6. Add the yeast to the hydrogen peroxide and stir.
7. Immediately begin timing 10-second intervals. As one person calls the time every 10 seconds, the second person reads the thermometer, and the third person records the temperature.
8. Using a green colored pencil, graph the recorded information.
9. Repeat the procedure using 30 mL of water and 5 mL of salt. Using a red colored pencil, graph the information.
10. Repeat another time using 30 mL of vinegar and 5 mL of baking soda. Using a blue colored pencil, graph the information.

Conclusions
- Which of the experiments demonstrated a chemical reaction?
- Which reaction demonstrated an exothermic reaction?
- Why was it important to have several people working together on this activity?

Follow-up
- Try changing the amounts of substances used to see if there is a difference in the amount of heat consumed or absorbed.

SCIENCE PROCESS SKILLS
Collecting and recording data

▸ What kinds of data were collected in this experiment? *{temperature and time}*

▸ Could you collect only the data about the temperature and still have a valid experiment? *{Answers will vary, but point out that longer time intervals might have caused the solution to return to room temperature before you could determine whether a chemical reaction had occurred or not.}*

▸ What ways did you record data? *{on the chart and then by graphing}*

▸ How helpful is the graph in showing what took place? *{Answers will vary, but elicit that being able to compare the three on one graph shows the contrast better.}*

Procedure
Help students formulate a hypothesis based on their predictions for each solution.

▸ What kind of reaction uses heat energy? *{endothermic reaction}*

▸ What will happen to the temperature of a solution that experiences an endothermic reaction? *{Possible answers: It will decrease, or get colder.}*

▸ What kind of reaction produces heat? *{exothermic reaction}*

▸ What will happen to the temperature of a solution that experiences an exothermic reaction? *{Possible answers: It will increase, or get warmer.}*

Conclusions
Provide time for the student to evaluate his hypothesis and answer the questions.

Activity Manual
Activity—pages 105–6

Assessment
Rubric—TE pages A50–A53
Select one of the prepared rubrics, or design a rubric to include your chosen criteria.

Lesson 80

Student Text pages 174–77
Activity Manual pages 107–8

Objectives

- Compare and contrast characteristics of acids and bases
- Describe the purpose of an indicator
- Identify products that are acids, bases, or salts
- Explain how a salt is formed

Vocabulary

acid
base
alkali
pH scale
neutral
indicator
neutralize
salt

Materials

- box of baking soda
- container
- water

Introduction

Display the box of baking soda.

➤ **Have you ever used baking soda for a purpose other than baking?** {Answers will vary.}

Read the variety of uses listed on the box of baking soda or those listed in Teacher Helps. Discuss the variety of uses and why the student thinks baking soda works these ways.

Allow a student to examine some dry baking soda.

➤ **What are some characteristics, or properties, of baking soda?** {Possible answers: white, powdery, smooth, odorless}

Allow another student to add a little water to the baking soda and feel it.

➤ **How would you describe the baking soda paste?** {Possible answers: slippery, slimy}

➤ **The compound we call baking soda is classified as a base.** Today we will look at acids and bases to see how these chemicals are useful in our daily lives.

Purpose for reading:
Student Text pages 174–77

➤ **How are acids and bases different?**
➤ **How are salts formed?**

Acids and Bases

Properties

Acids and bases are compounds with properties that make them useful to people in a variety of ways. **Acids** form hydrogen ions (H^+) when they are dissolved in water. The strength of an acid depends on the number of hydrogen ions it forms. Acids have a sour taste. Some weak acids, such as lemon juice and vinegar, can make us pucker when we taste them.

Many stronger acids are dangerous to taste and can burn your skin. Strong acids, such as sulfuric acid, are corrosive. *Corrosive* acids can even dissolve metals. Strangely enough, your stomach produces one of the strongest acids, *hydrochloric* (HY druh KLOR ic) *acid*, to help digest food. However, your gastric acid is *diluted*, or made less concentrated, by the amount of water in the food you eat and drink. If your stomach produces too much acid, you may get a stomachache.

Bases form hydroxide (hy DRAHK SIDE) ions (OH^-) when they are dissolved in water. Similar to acids, the strength of a base depends on the number of hydroxide ions it forms. Bases taste bitter and feel slippery. You have probably experienced this property when you have tried to pick up a piece of soap.

Like acids, some bases are weak and some are strong. Baking soda and antacid tablets are two common weak bases. Just as strong acids can be dangerous, so can strong bases. The base sodium hydroxide is used to make lye and drain cleaners. Bases that dissolve in water are referred to as **alkalis**.

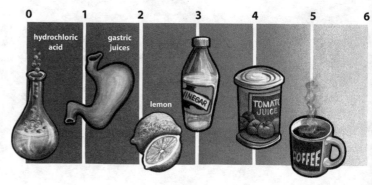

174

Uses for baking soda include cleanser, deodorizer, toothpaste, laundry booster, antacid, and for baking.

Science Background

Definition of an acid—Most definitions of acids are actually lists of properties. The properties of acids include: sour taste, cause blue litmus paper to turn red, neutralize bases, react with metals, produce hydrogen ions when dissolved in water, accept electrons, conduct electricity, and have a pH range of 0–6.9.

Definition of a base—Most definitions of bases are actually lists of properties. The properties of bases include: bitter taste, feel slippery, turn red litmus paper blue, neutralize acids, produce hydroxide ions when dissolved in water, donate electrons, conduct electricity, and have a pH range of 7.1–14.

Bases feel soapy—Bases react with and dissolve natural fatty acids and oils found on skin. This reaction results in the slippery soapiness felt as the friction is reduced. Bases are used to make soap. The high base content of early homemade soaps often irritated the skin. Most modern soaps have a pH value between 9.5 and 10.0.

pH Scale

Scientists use a special scale to determine the concentration, or amount, of an acid or base in a solution. The term *pH* comes from French words meaning "the power of hydrogen." The **pH scale** is numbered from 0 to 14. Acids measure from 0 to 6.9 on the scale, with 0 being highly acidic and 6 being slightly acidic. Bases measure from 7.1 to 14 on the scale, with 14 being highly basic and 8 being slightly basic. Solutions with a pH of 7.0 are **neutral,** meaning they are neither basic nor acidic. Pure water has a pH of 7.0.

Soil pH affects how well plants absorb nutrients from the soil. Most plants grow best if the soil pH is around 6.0–7.0. People can buy pH test kits at most lawn and garden stores. Once a gardener knows the pH of his soil, he can raise or lower the pH by adding substances to the soil. To increase the pH, most gardeners add a type of a base called lime to the soil. To lower the soil pH, many gardeners add peat moss or compost. These items usually have some type of acid in them.

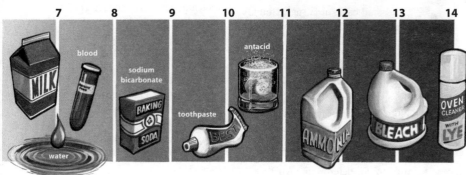

Discussion:
pages 174–75

- What is an *acid*? *{a compound that forms hydrogen ions when dissolved in water}*
- What determines the strength of an acid? *{the number of hydrogen ions it forms}*
- How do acids taste? *{sour}*
- What are some common weak acids? *{Possible answers: lemon juice and vinegar}*
- What are examples of corrosive acids? *{Possible answers: hydrochloric acid and gastric acid}*
- What is a *base*? *{a compound that forms hydroxide ions when dissolved in water}*
- What determines the strength of a base? *{the number of hydroxide ions it forms}*
- What are some characteristics of bases? *{They taste bitter and feel slippery.}*
- What are some common weak bases? *{Possible answers: baking soda and antacids}*
- 💡 How do you dilute an acid or a base? *{add more water to the solution}*
- 💡 What is the name of the scale that is used to show the concentration of an acid or a base in a solution? *{pH scale}*
- How is the pH scale numbered? *{0–14}*
- Why is pure water considered a neutral compound? *{It has a pH of 7, which is neither an acid nor a base.}*
- What numbers on the scale indicate that a compound is an acid? *{0–6.9}*
- Which numbers indicate a strong acid? *{lower numbers}*
- What numbers on the scale indicate that a compound is a base? *{7.1–14}*
- Which numbers indicate a strong base? *{higher numbers}*
- How do gardeners increase the soil pH? *{They add lime or another base.}*
- How do gardeners decrease the soil pH? *{They add compost, peat moss, or an acid.}*
- 💡 Why do gardeners sometimes change the pH of their soil? *{Possible answer: Not all plants grow well in the same type of soil.}*

📖 Discuss the *pH scale* on Student Text pages 174–75. Compare and contrast acid and base characteristics that the student may know about the items pictured.

Discussion:
pages 176–77

▸ **What important property do indicator substances have?** {They change color when in contact with an acid or base solution.}

▸ **Explain how red and blue litmus paper work.** {Red litmus paper turns blue in a base solution. Blue litmus paper turns red in an acid solution.}

▸ **Where does the substance used in litmus paper come from?** {lichens}

▸ **Indicator paper is also used to determine whether a solution is an acid or a base. What else can it show?** {the concentration of an acid or a base in the solution}

▸ **What are some other natural substances that are indicators?** {Possible answers: beets, pears, and red cabbage}

📖 Refer the student to *Creation Corner.*

▸ **What is a unique feature of some hydrangeas?** {They produce blossoms based on the pH of the soil.}

💡 **How is the hydrangea like litmus paper?** {An acidic soil will produce pinkish flowers just as litmus paper turns red in an acid. A basic soil will turn the blossoms blue just as litmus paper turns blue in a base.}

💡 **Why do you think some bushes might have both pinkish and bluish blossoms?** {Answers will vary, but suggest that the soil may be different where different parts of the root system are. As different parts of the root system supply different parts of the bush, the colors may vary.}

▸ God's design includes many things that are for man's enjoyment.

Creation CORNER

Flowering bushes usually produce flowers of the same color year after year. But the blossoms on some hydrangeas will be different depending on the soil pH. Acidic soil produces pink blossoms, while basic soil produces blue blossoms. You can even change the color of the flowers in the middle of a blooming season by adding acid or base substances to change the pH of the soil. What an amazing plant God has created for our enjoyment!

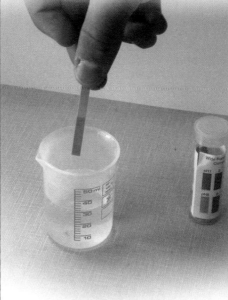

Indicators

Acids and bases are usually found in water solutions. Certain substances called **indicators** change color when exposed to acid or base solutions. Litmus (LIHT muhs) paper is a paper that has been treated with an indicator substance made from lichens. If you dip blue litmus paper into an acid solution, the paper turns red. If you dip red litmus paper into a base solution, the paper turns blue. Litmus paper is a helpful tool for determining whether or not a solution is basic or acidic.

Another helpful tool is indicator paper. Indicator paper will change colors depending on whether the solution is acidic or basic. It also shows by its color the concentration of the acid or base in the solution.

Many other natural substances, such as beets, pears, and red cabbage, can also indicate by changing their color whether a solution is an acid or a base. Their color changes can also indicate how concentrated the acid or base is.

176

SCIENCE BACKGROUND

Antacids—Whether antacids are liquids to drink, tablets to chew, or tablets to be dissolved in water, all antacids are bases. Often they contain a type of flavoring such as mint or fruit to cover the bitter taste of the base. Two common bases found in antacids are sodium hydrogen carbonate and magnesium hydroxide. When you mix baking soda with water, you get sodium hydrogen carbonate. Magnesium hydroxide is also known as milk of magnesia. When combined with the gastric acids in your stomach these bases react chemically and produce water and a harmless salt, thus quickly removing the burning sensation caused by the gastric acid.

Direct an ACTIVITY

Materials: 10 mL of ammonia in a glass container, 10 mL of vinegar in a glass container, red and blue litmus paper, medicine dropper

Label the two containers.

▸ **Which of these is an acid and which is a base? (Students may look at the *pH Scale*.)** {Ammonia is a base and vinegar is an acid.}

▸ **What is one way we can be sure which is an acid and which is a base?** {Test with litmus paper.}

Direct the student to use red and blue litmus paper to test the ammonia. Then use fresh red and blue litmus paper to test the vinegar.

Neutralizing Acids and Bases

Perhaps you have heard someone complain of heartburn, the release of too much acid in the stomach. Sometimes people take antacid tablets to make them feel better. Antacid tablets contain a base material that stops or lessens the effect of the stomach acid.

As we learned earlier, the properties of new substances are often very different from the original substances. When an acid and a base come in contact with each other, they **neutralize** each other. The properties of each substance change. The chemical reaction between an acid and base produces water and a salt. A **salt** is an ionic compound that contains positive ions from a base and negative ions from an acid. Most salts are composed of a metal and a nonmetal. Different combinations of bases and acids form different types of salt. You may be familiar with salts such as table salt (NaCl—sodium chloride), salt substitute (KCl—potassium chloride), and chalk (CaCO₃—calcium carbonate).

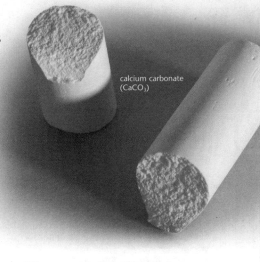
calcium carbonate (CaCO₃)

salt crystals

QUICK CHECK
1. What are the properties of bases?
2. What are the properties of acids?
3. What are some everyday products that contain acids or bases?
4. How is a salt formed?

Continued from page 186

> **What are the results of the tests?** {The ammonia turns the red litmus paper blue, and the blue litmus paper stays blue. The vinegar turns the blue litmus paper red, and the red litmus paper stays red.}

Fill a medicine dropper with ammonia and add five drops to the vinegar. Have another student stir the mixture and check it with blue litmus paper.

Continue adding ammonia one drop at a time and checking with litmus paper until the litmus paper no longer reacts.

> **What change did you observe in the litmus paper?** {At first the blue litmus paper turned red, but after adding ammonia the paper stayed blue.}

> **Is this liquid still an acid? What is it?** {No; it is neutral.}

> **How many drops of ammonia were needed before the solution became neutral?** {Answers will vary.}

> **What formed when you neutralized the acid and base?** {water and a salt}

> **How do we know the solution is neutral?** {Litmus paper does not change color when a solution is neutral.}

Discussion:
pages 176–77

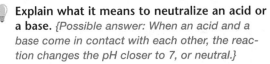 **Explain what it means to neutralize an acid or a base.** {Possible answer: When an acid and a base come in contact with each other, the reaction changes the pH closer to 7, or neutral.}

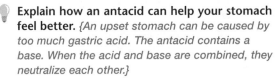 **Explain how an antacid can help your stomach feel better.** {An upset stomach can be caused by too much gastric acid. The antacid contains a base. When the acid and base are combined, they neutralize each other.}

> **What substances form when an acid and a base react chemically?** {water and a salt}

> **What is a *salt*?** {a compound that contains positive ions from a base and negative ions from an acid}

> **Are all salts the same?** {no}

> **What makes salts different from one another?** {the different combinations of bases and acids which form different types of salt}

> **What are some common salts?** {sodium chloride—table salt; potassium chloride—salt substitute; calcium carbonate—chalk}

Answers

1. Bases form hydroxide ions in water, taste bitter, and feel slippery.
2. Acids form hydrogen ions in water and taste sour.
3. Possible answers: Acids include lemon, vinegar, tomato juice, and coffee. Bases include baking soda, toothpaste, antacid, and ammonia.
4. Salts form as a result of the neutralizing of an acid and a base by a chemical reaction.

Activity Manual
Study Guide—page 107
This page reviews Lesson 80.

Bible Integration—page 108
This page looks at the life of Robert Boyle, who was both a scientist and a Christian.

Assessment
Test Packet—Quiz 7-B
The quiz may be given any time after completion of this lesson.

Lesson 81

Student Text pages 178–79
Activity Manual pages 109–10

Objectives

- Identify a solution as an acid or a base by using a pH indicator solution
- Estimate the strength of an acid or base solution by using a pH indicator solution
- Observe the effects of acids and bases on an indicator

Materials

- See Student Text page

Introduction

▶ What is an *indicator*? {a substance that changes color when exposed to acid or base solutions}

▶ In today's activity you will be testing several different substances with an indicator solution to determine whether each is an acid, a base, or neutral.

Purpose for reading:
Student Text pages 178–79
Activity Manual pages 109–10

The student should read all the pages before beginning the activity.

Procedure—Part I

Refer the student to the *Table of Colors* on Activity Manual page 109 as he writes his predictions on the Observation Chart.

▶ What color is the cabbage juice when it is neutral? {purple}

Add lemon juice solution to the cabbage juice in cup one.

▶ What is the color of the cabbage juice now? {bright red/pink}
▶ Is this solution an acid or a base? {an acid}
▶ What is its concentration? {strong}

Continue testing the other four solutions and recording the results on the Observation Chart.

▶ Which solution had the greatest concentration of acid? {lemon juice}
▶ Which solution had the greatest concentration of base? {ammonia}
▶ Is there a way that we could get one of the solutions that have turned green to turn purple again? {If it turned green, it is a base. Add some of one of the acid solutions to change the color back to a red/purple color.}

pH Indicator

ACTIVITY

Just as litmus paper is used as an indicator of the pH of substances, the juice from cooked red cabbage can also act as an indicator for acids and bases. In this activity, you will use red cabbage juice as an indicator to test whether solutions are acids or bases.

Process Skills
- Predicting
- Measuring
- Observing
- Recording data

Problem
How will acid and base solutions change the color of red cabbage juice?

Procedure—Part 1
Caution: Acids and bases can be dangerous. Wear goggles or glasses when handling the solutions. Never try to taste a solution. Avoid spilling the solutions on your skin. If a solution spills on your skin, wash immediately with soap and water.

1. Write the name of the solutions in the *Observation Chart* in your Activity Manual.

2. Use the *Table of Colors* in the Activity Manual to predict whether each solution is an acid or a base and what its approximate concentration is. Record your predictions in the *Observation Chart*.

3. Pour 50 mL of red cabbage juice into each of cups 1–5.

4. Add 15 mL of lemon juice (Solution 1) to the cabbage juice in cup 1.

5. Observe and record the color of the cabbage juice in cup 1.

6. Use the *Table of Colors* to determine whether Solution 1 is an acid or a base and what its approximate concentration is. Record your findings.

178

Materials:
prepared red cabbage juice
goggles
5 plastic or foam cups labeled with numbers 1–5
2 plastic or foam cups labeled with letters A and B
metric spoons and measuring cups
Activity Manual
The following chemicals labeled as indicated:
 lemon juice (Solution 1)
 household ammonia (Solution 2)
 baking soda solution (Solution 3)
 distilled water (Solution 4)
 white vinegar (Solution 5)
 milk of magnesia solution (Solution A)
 colorless carbonated soft drink (Solution B)

Teacher Helps

Prepare the red cabbage juice.
- Chop one head of red or purple cabbage.
- Place cabbage pieces in a stainless steel pan and cover with one liter of water.
- Boil until the water turns a dark color.
- Strain the cabbage, reserving the water.
- Refrigerate the cabbage juice until ready for use.

Note: Beets, pears, blackberries or red onion can be used as a substitute for red cabbage.

Prepare the solutions.
- The solutions made of liquids (ammonia, lemon juice, colorless carbonated soda, distilled water, and white vinegar) do not need to be diluted with water.
- The baking soda and milk of magnesia solutions are made by mixing 15 mL of distilled water with enough of each ingredient to saturate the water.
- The soft drink must be colorless so students can observe a color change in the solution.

188 Lesson 81

7. Repeat steps 4–7 with solutions 2–5. Rinse the measuring spoon or measuring container with water before measuring each different solution.

Procedure—Part 2
1. Based on your findings, write a hypothesis in your Activity Manual that states the results you expect from testing solution A.
2. Pour 50 mL of red cabbage juice into cup A.
3. Add 15 mL of Solution A to the cabbage juice in cup A.
4. Use the *Table of Colors* to determine whether Solution A is an acid or a base and what its approximate concentration is. Record your findings.
5. Repeat steps 1–5 with Solution B.

Conclusions
- Was your hypothesis correct?

Follow-up
- Use another natural indicator, such as red beets, to test the solutions.

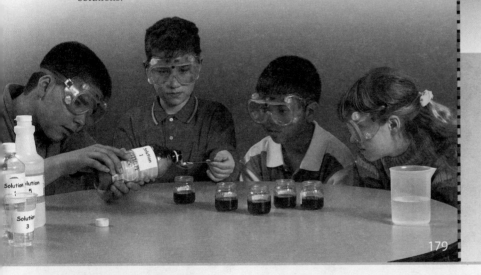

SCIENCE PROCESS SKILLS

Observing

➤ Observing is not always an exact skill. Do you think that your idea of reddish purple might be different from someone else's? *{possibly}*

➤ Would your difference of opinion cause your observation to be flawed? *{possibly}*

➤ How could you make your observations more reliable? *{Possible answers: redo the activity and see if the colors were consistent; come to a group decision about what to call the color; observe other groups}*

➤ How might a scientist in a laboratory make sure his observations were accurate? *{Possible answers: redo his experiment; come to a group decision about the observations; use more advanced instruments to determine the pH of a solution}*

Procedure—Part II
Guide the student in formulating his hypothesis for solution A.

Add the cabbage juice to the solution.

➤ **Is solution A an acid or a base?** *{base}*

➤ **How concentrated is it?** *{Answers will vary.}*

Repeat for solution B.

Conclusions
Provide time for the student to evaluate his hypotheses and answer the questions.

Activity Manual
Activity—pages 109–10

Assessment
Rubric—TE pages A50–A53
Select one of the prepared rubrics, or design a rubric to include your chosen criteria.

Lesson 81

Lesson 82

Student Text pages 180–81
Activity Manual pages 111–12

Objectives

- Hypothesize about the effectiveness of several antacids
- Make and use a model of "upset stomach" acid
- Infer information from the model

Materials

- See Student Text page

Introduction

Purpose for reading:
Student Text pages 180–81
Activity Manual pages 111–12

The student should read all the pages before beginning the activity.

➤ When you have an upset stomach, what kind of medicine do you take? *{Answers will vary.}*

➤ Do you prefer one type of medicine over another? Why? *{Answers will vary.}*

➤ Do you know whether your preferred medicine works better than other medicines? How do you know? *{Answers will vary.}*

➤ Today you will have the opportunity to test which antacids work best for neutralizing an acid.

ACTIVITY: Which Antacid Is Best?

The stomach uses acid to digest food, but the stomach should not be too acidic or too basic. Many antacids can help an upset stomach caused by the presence of too much acid. Some antacids are sold as medications. Other antacids may be food or cooking ingredients found around your house. In this activity you will test the effectiveness of several different antacids on experimental "upset stomachs."

Process Skills
- Hypothesizing
- Experimenting
- Observing
- Inferring
- Recording data

Problem
Which antacid works best to neutralize an acid?

Procedure

1. Write a hypothesis in your Activity Manual explaining which antacid you think will work best to neutralize an acid.

2. Combine 80 mL of water and 40 mL of vinegar in the container to make an *upset stomach mixture*.

3. Place a pH indicator strip into the *upset stomach mixture* for 30 seconds. Immediately compare the color of the strip to the pH chart. Record the pH level in your Activity Manual.

4. Pour 20 mL of the *upset stomach mixture* into each plastic cup. These are your "upset stomachs."

5. Place one dose of each antacid into a different "upset stomach" cup and mix well. (Tablets should first be crushed.)

Materials:
metric measuring cups and spoons
water
vinegar
200 mL or larger container
pH indicator paper (range of 1–14)
6 clear plastic cups
6 spoons or stirring sticks
baking soda
milk
One dose each of four different commercial antacids (or their generic equivalents) such as:
  Alka-Seltzer
  Maalox
  Pepto Bismol
  Rolaids
  Tums
  Milk of magnesia

180

TEACHER HELPS

The pH indicator paper is different from litmus paper.

The activity will be most effective if you ensure that each of your antacids is made of different ingredients. Two antacids with the same ingredients will likely have the same pH reading.

Crush tablets by placing them in plastic bags and pounding them with a hammer.

Red cabbage juice can be used in place of the pH paper, but this will give a less accurate reading. The colors of the cabbage juice and their coordinating pH levels are as follows:

pH 2–red	pH 4–purple
pH 6–violet	pH 8–blue
pH 10–blue-green	pH 12–greenish yellow

SCIENCE BACKGROUND

A normal stomach is actually slightly acidic and has a pH level of 1.6–2.4. The general principles used in this activity are valid; however, the pH numbers used do not approach the true physiologic pH values in the stomach.

An upset stomach may be caused by too much acid in the stomach rather than a lower pH level of acid in the stomach.

Not all antacids work the same way. Since this activity does not involve a real stomach, only the antacids that act as a buffer will give accurate results for this activity. Antacids that reduce the amount of acid in the stomach will appear to be less effective, although they may be more effective in a more acidic stomach.

6. Check the effectiveness of each antacid by placing a fresh piece of pH indicator paper into each solution for 30 seconds. Immediately compare the color of the strip to the pH chart. Record the pH level in your Activity Manual.

7. Compare the pH level with the original pH of the "upset stomach."

 8. Test the effectiveness of baking soda and milk as antacids. Mix each into an "upset stomach" cup. Check with pH paper and record your results.

Conclusions

- Which antacid would you want to use if you had an upset stomach?
- Do you think this is a valid test of an antacid's effectiveness? Why or why not?

Follow-up

- Try some other medicines that are advertised to relieve an upset stomach.
- Prepare an advertising campaign based on your findings.

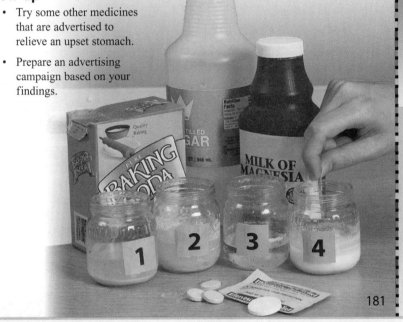

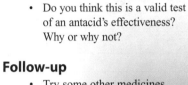

Hypothesizing

Remind the student that a hypothesis is a testable statement that tells what he thinks will happen. Although at this level it is not necessary for all conditions to be defined in a hypothesis, a student's hypothesis should be as specific as possible.

Allow several students to read their hypotheses. Discuss whether the hypotheses are testable and suggest improvements as needed.

Note: The hypothesis for this activity should be straightforward. For example, "Maalox will neutralize the acid the best."

 Instruct the student to make an advertisement for the product he found to work best for neutralizing an acid. The student can choose to write a radio advertisement, a TV commercial, an advertisement for a newspaper, or a billboard advertisement. The advertisements can be presented to the rest of the class.

Procedure

Guide the student in writing a hypothesis. Be sure the student writes a statement that positively states which antacid he thinks will work best.

Explain to the student that the activity requires one dosage of various antacids. Show the student where to look on the package to determine the appropriate dosage.

Conclusions

Provide time for the student to evaluate his hypothesis and answer the questions.

Activity Manual
Activity—pages 111–12

Assessment
Rubric—TE pages A50–A53
Select one of the prepared rubrics, or design a rubric to include your chosen criteria.

Lesson 83

Student Text page 182
Activity Manual pages 113–14

Objectives

- Recall concepts and terms from Chapter 7
- Apply knowledge to everyday situations.

Introduction

Material for the Chapter 7 Test will be taken from Activity Manual pages 99–100, 104, 107, and 113–14. You may review any or all of the material during the lesson.

You may choose to review Chapter 7 by playing "Atomic Construction" or a game from the *Game Bank* on TE pages A56–A57.

Diving Deep into Science

Questions similar to these may appear on the test.

Answer the Questions

Information relating to the questions can be found on the following Student Text pages:

1. page 161
2. page 171
3. pages 164–65

Solve the Problem

In order to solve the problem, the student must apply material he has learned. The answer for this Solve the Problem is based on the material on Student Text page 177. The student should attempt the problem independently. Answers will vary and may be discussed.

Activity Manual

Study Guide—pages 113–14

These pages review some of the information in Chapter 7.

Assessment

Test Packet—Chapter 7 Test

The introductory lesson for Chapter 8 has been shortened so that it may be taught following the Chapter 7 test.

Diving Deep into Science

Answer the Questions

1. Why do scientists call their knowledge of the atom a theory?
   Although repeated observations show the same results, no one can actually see all the properties that scientists think are true of atoms.

2. Which kind of atomic bonding is somewhat like magnetism? Why?
   Ionic bonding is somewhat like magnetism because positive and negative ions attract each other in a way similar to the way the north and south poles of a magnet attract each other.

3. The element cesium has one electron in its outer shell. In which group on the periodic table of the elements would you find cesium?
   Cesium would be in the first group at the far left.

Solve the Problem

A fire ant has just bitten your friend. You remember that a fire ant's bite is painful because of the acid that it contains. Can you think of something from your kitchen or bathroom that would help relieve the pain? Why would it work?

Any answer that recognizes using a base to neutralize the acid would be acceptable. Baking soda, a dab of toothpaste, or even a bit of milk of magnesia might help.

182

Review Game
Atomic Construction

Prepare two sets of the following: 20 small circles each of blue (protons), green (neutrons), and red (electrons) paper.

Divide the class into two teams. Each time a team answers a review question correctly, it may choose an atomic particle to complete a sodium (NA) atom (11 protons, 12 neutrons, 11 electrons). The first team to complete its atom wins.

Additional elements you may use include: oxygen (8 protons, 8 neutrons, 8 electrons), chlorine (17 protons, 18 neutrons, 17 electrons), and fluorine (9 protons, 10 neutrons, 9 electrons).

Student Text page 183
Activity Manual page 115

Lesson 84

Chapter 8: Electricity and Magnetism

GREAT & MIGHTY Things

In the late 1700s a controversy arose about a dead frog's legs. Luigi Galvani found that touching the nerves of a dead frog's legs with two different metals could make the legs twitch. He concluded that animal tissue contains a force called "animal electricity," which Galvani thought flowed throughout the body. But another scientist, Alessandro Volta, thought that the contact between the two different metals produced the electricity. He called his force "metallic electricity."

Further research proved that both men were partly right and partly wrong. As with many other scientific discoveries, God allowed one man's ideas, though not completely right, to inspire another man to further research. The result of Volta's disagreement with Galvani's conclusions was the first battery using two different metals. Electricity became a force that could be controlled enough to do work.

183

Objectives
- Demonstrate knowledge of concepts taught in Chapter 7
- Recognize God's use of man's curiosity

Introduction

Note: This introductory lesson has been shortened so that it may be taught following the Chapter 7 test.

God told the Israelites that He would drive the Canaanites out of the Promised Land. In the same way that the Israelites conquered the land little by little, God also allows man to learn more about His creation little by little. God has given man natural curiosity, and sometimes even man's mistakes help to advance science. One scientist's mistake may inspire another scientist to try another method or hypothesis.

Chapter Outline

I. Electricity
   A. Static Electricity
   B. Current Electricity
   C. Kinds of Circuits
   D. Measuring Electricity
   E. Batteries

II. Magnetism
   A. Magnetic Attraction
   B. Electricity and Magnetism

III. Electronics
   A. Integrated Circuits
   B. Computer Parts
   C. Computer "Intelligence"

SCIENCE BACKGROUND

Luigi Galvani—Galvani's experiments furthered not only the study of electricity, but also the study of anatomy. Scientists concluded from his experiments that the messages of the nervous system were carried by an electrical current rather than through a conduit, like the blood.

Alessandro Volta—Because Volta disagreed with Galvani's conclusion, he built the voltaic pile, a simple wet-cell battery. Volta's original pile was made from zinc and silver discs alternated with pieces of cloth soaked in salt water.

Galvani and Volta controversy—Although the two scientists disagreed about electricity, they did not actually dislike each other. In fact, each highly respected the other. Volta was the man who developed the word *galvanization* to refer to administering electrical shocks.

Activity Manual

Preview—page 115

The purpose of this Preview page is to generate student interest and determine prior student knowledge. The concepts are not intended to be taught in this lesson.

Lesson 85

Student Text pages 184–87
Activity Manual page 116

Objectives
- Explain what causes static electricity
- Identify the two things needed for an electric current to flow
- Describe the characteristics of conductors, resistors, and insulators

Vocabulary
static electricity
current electricity
circuit
conductor
switch
short circuit
insulator
resistor

Introduction

Play the game "Twenty Questions" after giving the clue.

➤ I am not a mineral or an animal, and I am not alive or dead. I am both inside and outside your body and in outer space. I am invisible, tasteless, colorless, odorless, and likely to travel in circles. What am I? {electricity}

Purpose for reading:
Student Text pages 184–87

➤ How is lightning like electricity?
➤ Identify the two kinds of electricity.

Discussion:
pages 184–85

➤ What did the ancient Greeks observe about amber? {It produced sparks when rubbed with fur.}

➤ Why was Thomas Edison's invention significant? {because up to his time, electricity could not be controlled to be useful}

📖 The ability to harness electricity is a great gift from God. Using it wisely is being a good steward of God's gift.

Electricity and magnetism—the two forces are inseparable. In many countries, it would be almost impossible to go through a day without using electricity and magnetism. Yet even complicated electrical equipment and electronics work because of the positive and negative charges that are part of tiny atoms.

Electricity

Man observed electricity long before he understood the parts of an atom. The ancient Greeks noticed that amber, fossilized tree sap, produced a spark when rubbed with fur. But only within the last 150 years has man been able to harness electricity for practical use. When Thomas Edison invented the electric light bulb in 1879, he revolutionized the use of electricity. Finally, electricity could serve practical purposes. But even Edison could not have imagined all of the ways that electricity would change the way we live.

Thomas Edison

Static Electricity

Atoms, the building blocks of matter, generally have a neutral electrical charge, because the protons (+) and electrons (–) balance each other. However, atoms often gain or lose electrons, which causes them to have a temporary negative or positive charge.

Like atoms, objects normally have a neutral charge. However, rubbing two objects together may cause electrons to move from one object to the other. The rubbing action produces static

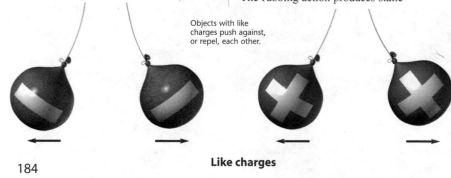

Objects with like charges push against, or repel, each other.

Like charges

184

SCIENCE BACKGROUND

Common lightning safety precautions—During a storm, the main principle to remember is to avoid contact with things that will attract or conduct electricity.

Stay inside a home or other large building when possible.

Avoid using the telephone.

Keep away from windows, and avoid contact with things inside a building that can conduct electricity, such as metal railings and water from metal pipes.

Avoid travel in open vehicles such as motorcycles, golf carts, and bicycles.

If outside, avoid wire fences, metal pipes, power lines, trees, water, open spaces, and high ground.

SCIENCE MISCONCEPTIONS

Electricity cannot be created or lost. According to the Second Law of Thermodynamics, matter and energy can neither be created nor destroyed. They can only be converted into other forms of matter or energy. It only seems like electricity is lost as the electrical energy is converted into other forms of energy, such as heat and light.

electricity. **Static electricity** occurs when electrical charges build up on the surface of an object. One object gains extra electrons and has a negative charge. The other object loses electrons and has a positive charge. An object with a positive or negative charge has an electrical force that will repel or attract other charged objects. Charged objects with like charges repel each other. However, charged objects with unlike charges attract each other.

If you have ever walked across a carpet on a cold, dry day, you have probably experienced static electricity. As your shoes rub on the carpet, they collect charges. These charges build up on your body. When you reach for a doorknob, the electricity discharges, or jumps, from your hand to the doorknob. This discharge of electrons causes a spark and a shock. As the electrons flow from one object to the other, each object returns to a neutral state, having no charge at all. During a thunderstorm, charges build up in the clouds. When these charges discharge, they create the most dramatic kind of static electricity—lightning.

Objects with unlike charges pull toward, or attract, each other.
Unlike charges

Science and HISTORY

In the 1700s Benjamin Franklin began experimenting with electricity. He flew a kite during a thunderstorm because he thought lightning was electricity. Fortunately, lightning did not strike the kite, and Franklin survived his experiment. However, the metal key attached to the silk kite string collected electrical charges from the air. When Franklin touched the key, he felt a small shock because the key was charged. Franklin had demonstrated that lightning is static electricity.

185

LANGUAGE

The word *electron* comes from the Greek word for amber, *elektron*. Scientists used *elektron* to describe the part of the atom that carries a negative charge.

▸ **Which other words are similar to the word *electron*?** {Possible answers: electricity, electronics, electric, electrician}

Circuit comes from a Latin word meaning "to go around."

▸ **Which other words have the prefix *circ-*?** {Possible answers: circle, circular, circulate, circulatory, circumference, circumnavigate, circumstance}

Discussion:
pages 184–85

▸ **Identify the three types of electrical charges.** {neutral, positive, and negative}

▸ **What kind of charge does an atom usually have?** {neutral}

💡 **How do you determine the charge of an atom?** {Count the number of electrons and protons in the atom. If the atom has more electrons than protons, then it has a negative charge. If it has more protons than electrons, then it has a positive charge.}

▸ **How do you represent a positive charge and a negative charge?** {positive charge (+) and negative charge (−)}

💡 **Why would the loss of electrons possibly cause an object to have a positive charge?** {The loss of electrons may cause an object to have more protons than electrons, which will give it a positive charge.}

▸ **Which type of electricity is caused when you rub two objects together?** {static electricity}

▸ **What is *static electricity*?** {the buildup of electrical charges on the surface of an object}

▸ **Describe what an electrical force does.** {attracts or repels other charged objects}

▸ **Why would an object with a negative charge attract an object with a positive charge?** {Objects with unlike charges attract each other.}

▸ **Give some examples of static electricity.** {Possible answers: lightning, static cling, static shock}

Refer the student to *Science and History*.

▸ **Who demonstrated that lightning is static electricity?** {Benjamin Franklin}

▸ **How did Benjamin Franklin show that lightning is static electricity?** {He flew a kite with a metal key attached to it in a thunderstorm.}

💡 **Was this a wise experiment?** {no} **Why?** {He easily could have been struck by lightning.}

Chapter 8

Lesson 85 195

Discussion:
pages 186–87

- What is *current electricity?* {*the continuous flow of electrons around a circuit*}
- Identify the two things needed to make current electricity. {*a circuit and a power source*}
- What is a *circuit?* {*a continuous unbroken path through which electricity can flow*}
- 💡 Explain in your own words the similarities between a power source and a water pump. {*Answers will vary but should include the idea that both push.*}
- What do we call a material through which electricity flows easily? {*a conductor*}
- 💡 Why do you think most electric wires are made of metals rather than of other materials? {*Possible answer: Most metals are good conductors.*}
- 💡 Gold is one of the best conductors of electricity. Why do you think gold is not used to make electric wires? {*Possible answers: too expensive, too soft, too heavy*}
- How can an electric current be turned off without removing the power source? {*with a switch*}
- How does an open switch affect a circuit? {*It causes an open circuit, which does not provide a continuous path for electricity.*}
- What happens if a switch is closed? {*It keeps the circuit closed so that electricity has a continuous path.*}
- When does a short circuit occur? {*when electricity takes an unexpected path*}
- How are short circuits prevented in houses and appliances? {*Houses and appliances have built-in protection, such as circuit breakers or fuses.*}
- How are circuit breakers and fuses switches? {*They can open and close circuits.*}
- How does a fuse work? {*A fuse has a thinner wire that breaks if too much electricity is flowing through the circuit. When it breaks, it opens the circuit.*}
- What is the difference between a conductor and an insulator? {*A conductor allows electricity to flow through it easily. An insulator does not allow electricity to flow through it.*}
- What are some good insulators? {*plastic, wood, glass*}

196 Lesson 85

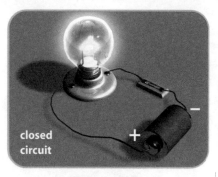

closed circuit

open circuit

Current Electricity

Static electricity may jump from place to place, but its electrical force lasts for only a moment. In order for electrical energy to be useful, it must have a continuous flow of electrons. **Current electricity** is the flow of electrons around a circuit. The moving electrons produce electrical energy that can do work.

Two things are necessary to make current electricity. First, there must be a **circuit**, a continuous unbroken path, through which electricity can flow. Second, the circuit must have a power source that causes the electrons to start moving. The power source pushes electrons through the circuit somewhat like a water pump pushes water through water pipes.

Conductors and switches

A **conductor** is a material that allows electricity to flow through it easily. Electrons in the atoms of these materials move freely from one atom to another. Most metals are good conductors because metals loosely hold their electrons.

186

Using a switch, we can turn most electrical appliances on and off without unplugging them. A **switch** is a conductor that can be moved to either bridge or not bridge the gap in a circuit. When the switch is closed, the circuit is complete, so electricity flows easily. This is a *closed circuit*. An open switch breaks the circuit, so the electrons cannot travel through a complete path. This is an *open circuit*.

Because electricity always flows through the easiest path, sometimes it may take an unexpected path. When this happens, a **short circuit** occurs. Short circuits can cause sparks and fires. To prevent short circuits, many electrical devices, and even house wiring, have fuses and circuit breakers that provide built-in protection. When too much electricity flows through a circuit, a fuse's thinner wire breaks and opens the circuit so electricity can no longer flow. A circuit breaker works in a similar way by popping a switch that opens the circuit. To close the circuit again, someone needs to either replace the fuse or switch the circuit breaker back to the *on* position.

Direct a DEMONSTRATION
Identifying materials as conductors or insulators

Materials: 3 pieces of insulated wire about 8 cm long, small light bulb in a socket, 6-volt battery, electrical tape, conductors to test (copper, iron, steel, aluminum, lead, or other metals) and insulators to test (paper, wood, plastic, rubber, glass, rock, or other organic materials)

Construct a circuit according the diagram.

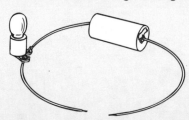

Show the objects you will be testing. Direct the student to list the items on his own paper and record his prediction of whether each item will act as a conductor or as an insulator.

Place the first item between the wires to complete the circuit.

- Did the light bulb glow? Is the item a conductor or an insulator? Was the prediction correct? {*Answers will vary depending on the item.*}

Repeat with the remaining objects.

- What do all the items that were conductors have in common? {*They are made of metal.*}
- What do all the items that were insulators have in common? {*They are made from materials other than metal.*}

wire and insulation

Insulators and resistors

Unlike a conductor, an **insulator** does not allow electricity to flow through it. Plastic, wood, and glass are all good insulators. A wire that conducts electricity would be extremely dangerous without a coating of plastic insulation.

Sometimes we want electricity to flow but not flow easily. A **resistor** reduces the flow of electrons. As the electrons push harder to get through a resistor, friction causes the resistor to become hot. The resistor heats up, and it may begin to glow. The heating element in your toaster is a resistor. The metal wire, or filament, in a light bulb is also a resistor. The filament slows the flow of electrons, creating friction. This friction heats up the filament and causes it to glow and produce light.

QUICK CHECK
1. Why do some objects have a negative charge?
2. Identify two things that are needed to make current electricity.
3. Explain the difference between a conductor, an insulator, and a resistor.

Meet the SCIENTIST — LEWIS LATIMER

Lewis Latimer (1848–1928), the son of former slaves, helped develop electrical lighting. After serving in the Union army during the Civil War, Latimer worked as an office boy. He taught himself to be a draftsman, a person who draws detailed plans and sketches. While working for Thomas Edison's rival, Hiram Maxim, Latimer developed a light bulb filament that lasted longer and was cheaper than Edison's filament. (Edison's filament was the standard for light bulbs for many years.) Latimer supervised the installation of lighting systems in several large cities. Eventually Latimer went to work for Edison's company.

Although Latimer is credited with several inventions, he was also accomplished in many other areas. He wrote a book about electrical lighting, and he also wrote poetry. He played several instruments and taught himself to speak German and French. Latimer contributed greatly to the furtherance of the Industrial Revolution.

187

 Electrical circuits provide many analogies to a Christian's life. Good conductors allow electricity to flow freely. Christians should be good conductors of God's love so that their lives are useful to God and others. [BAT: 5a Love]

Open switches prevent electricity from flowing. Similarly, unconfessed sin prevents Christians' lives from having the power of God. [BAT: 6e Forgiveness]

Resistors allow electricity to flow but not to flow easily. Although this may be good for lightbulbs and toasters, a Christian who resists God's working in his life will also get "hot." His life may be characterized by being under stress.

Discussion:
pages 186–87

- What is meant when we say that a wire is live? *{Possible answer: The wire has electricity flowing through it.}*
- Why would it be harmful for you to touch a bare live wire? *{Possible answers: Body fluids conduct electricity. Electric current is dangerous.}*
- If a friend were receiving an electric shock, would a wooden stick or a metal rod work better to pull him away from the electric current? *{a wooden stick}* Why? *{Metal conducts electricity and would pass the shock through your friend and on to you. Wood is an insulator.}*
- Why are electric wires often covered in plastic? *{The plastic is an insulator and protects from electric shock.}*
- What is the purpose of a resistor? *{to reduce the flow of electrons}*
- What happens as a resistor reduces the flow of electrons? *{As the electrons push harder to get through a material, they create friction. This causes the material to get hot and glow. The electrical energy is changed into heat and light.}*
- What common household items use resistors? *{Possible answers: toaster, incandescent light bulb, hair dryer, electric heater, coffeepot}*

Refer the student to *Meet the Scientist*.
- When did Lewis Latimer live? *{1848–1928}*
- What did Lewis Latimer develop? *{a light bulb filament that lasted longer and was cheaper than Edison's light bulb filament}*
- What were some of Latimer's other accomplishments? *{Possible answers: taught himself to be a draftsman; wrote a book about electrical lighting; wrote poetry; played several instruments; taught himself German and French}*

Answers
1. They have gained electrons.
2. circuit and power source
3. A conductor allows electricity to flow easily. An insulator prevents electricity from flowing. A resistor reduces the flow of electricity.

Activity Manual
Reinforcement—page 116

Lesson 85   197

Lesson 86

Student Text pages 188–89
Activity Manual pages 117–18

Objectives
- Design and build an "unbreakable" circuit
- Experiment to test hypotheses

Materials
- one string of Christmas lights (preferably lights on a series circuit)
- See Student Text page

Introduction

Plug the Christmas lights into a wall socket.

➤ What will happen if I remove one light bulb from the strand?

Remove one light bulb from the strand and plug the lights back into the wall socket.

➤ When the bulb is lit, is it like an open or closed switch? Why? *{It is like a closed switch because it completes the circuit.}*

➤ Is the light bulb a conductor or an insulator? *{It is a conductor because it acts like a closed switch.}*

➤ What happened to the other lights on the strand? Did they all keep working?

➤ Today you are going to try to build an unbreakable circuit, a circuit that keeps working even when one light bulb is missing.

Purpose for reading:
Student Text pages 188–89
Activity Manual pages 117–18

The student should read all the pages before beginning the activity.

ACTIVITY: An "Unbreakable" Circuit

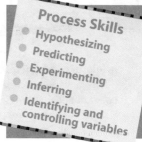

Process Skills
- Hypothesizing
- Predicting
- Experimenting
- Inferring
- Identifying and controlling variables

A circuit is an unbroken path of electricity. The picture shows the continuous flow of electricity from the battery, through two light bulbs, and then back to the battery. In this circuit, if one light bulb goes out, the second light bulb also goes out.

As an amateur electrician, you have the job of designing and building an "unbreakable" circuit. The key to this invention is the arrangement of the bulbs and battery in the circuit.

Materials:
two C- or D-cell batteries
10 wires approximately 6–10 cm long
two light bulbs in their own sockets
electrical tape
one colored pencil or pen
Activity Manual

Problem
How can you set up your circuit with two light bulbs so that you can unscrew one while keeping the other lit?

Procedures
1. List the materials that you will use.
2. Predict a solution to the problem.
3. Draw a sketch of your circuit to use as your hypothesis.

For this discovery activity, do not let the student read ahead in the Student Text about series and parallel circuits.

To determine if your Christmas lights have series or parallel circuits, remove one light bulb and plug the strand of lights into the wall. If the other light bulbs remain lit, your strand has parallel circuits. If part or the entire strand of lights does not light when you plug it in, your strand has series circuits. Some strands have two or three series circuits connected together, so only part of the strand may not work when a bulb is missing.

Insulated wire, 18–25 gauge, will work best.

Remind the students that they are not limited to one circuit. They can attach more than one or two wires to a light bulb or battery.

If some groups are unsuccessful and time permits, allow groups to share information.

4. Tape the two batteries together with electrical tape. The negative end (–) of one battery should be taped to the positive end (+) of the other.
5. Build the circuit.
6. To test your circuit, unscrew one bulb. If you discovered the "unbreakable" circuit, the other bulb will remain lit.
7. If you were unsuccessful, think of a couple of reasons your circuit did not work. Adjust your circuit and retest. Keep adjusting and testing your circuit as time permits.

Conclusions
- What arrangement produced an "unbreakable" circuit?
- What things could you do to improve your circuit?

Follow-up
- Make a circuit where one bulb burns brighter than the others.
- Add more bulbs to your circuit. Observe what happens.

 An electrical current requires an unbroken circuit for electrical power to flow. Much like a broken electrical current, a Christian's fellowship with God can be broken by unconfessed sin, preventing him from having God's power. Only by confessing sins can a Christian close the circuit and regain God's fellowship and power. [BATs: 6d Clear clonscience; 6e Forgiveness]

SCIENCE PROCESS SKILLS
Experimenting

➤ How do you test a hypothesis? *{by experimenting}*

➤ What do you need in order to experiment? *{materials and necessary conditions for testing your hypothesis}*

➤ As a scientist, would you stop if one experiment did not prove your hypothesis? *{probably not}*

➤ What would you do? *{Possible answer: Rethink your hypothesis and try again.}*

➤ Why is experimenting important to scientific discovery? *{Answers will vary, but elicit that without experimentation, ideas are simply people's opinions rather than proven facts.}*

Procedure
Give each group 15–20 minutes to create an unbreakable circuit.

Provide more direction to those who are not having success, or show them a working circuit to build.

The following are questions to ask a struggling group.

➤ What does each bulb need to have to remain lit? *{a complete circuit and a power source}*

➤ Trace the path of electricity around the circuit of the student's diagrams. Does each have a complete path?

To solve the problem, the student will design a type of parallel circuit, such as the one drawn below.

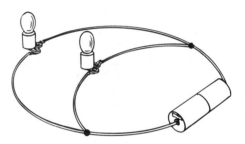

Conclusions
Direct the student to examine successful circuits.

➤ Why did one bulb remain lit when the other one was disconnected? *{Conclude that the lit bulb has a complete circuit.}*

Provide time for the student to answer the questions.

Activity Manual
Activity—pages 117–18

Assessment
Rubric—TE pages A50–A53
Select one of the prepared rubrics, or design a rubric to include your chosen criteria.

Lesson 87

Student Text pages 190–93
Activity Manual pages 119–22

Objectives

- Differentiate between parallel circuits and series circuits
- Distinguish among the three basic units of electrical measurement: volt, ampere, and watt
- Explain how a battery works

Vocabulary

series circuit
parallel circuit
volt
watt
ampere
electric cell
electrolyte
battery

Introduction

➤ Have you ever experienced the power in your house going out because a car hit a power pole down the street?

When a light post is hit and the power goes out, it usually affects the power for more than that pole only. Often several houses and buildings near the pole also lose power, because that pole is connected to one power line that services a group of houses, buildings, and streetlights.

➤ What do you think would happen to the power going to the houses of your neighbors if a branch fell on the line that goes from the street to your house? *{Elicit that nothing would happen to their power.}*

The power in our neighborhoods uses different types of circuits.

➤ Today we will learn about two types of circuits and how they work.

Purpose for reading:
Student Text pages 190–93

➤ What is the difference between a parallel and series circuit?

➤ What are the parts of a battery?

Kinds of Circuits
Series circuits

Have you ever seen a half-lit Christmas tree? Before the invention of the "unbreakable" circuit, all of the Christmas lights on a strand were arranged on one circuit. If one light bulb broke or burned out, the circuit broke, and the entire strand of lights stopped working. To relight the strand, a person would have to test each light bulb until he found and replaced the broken or burned-out bulb, or he would have to replace the whole strand of lights.

A **series circuit** has only one path, or one circuit, for the electricity to travel. If the path has a break at any point, the current cannot complete the circuit. A light bulb attached to a circuit acts like a switch. It closes the circuit, completing the path. However, when the filament in a bulb burns out, the circuit opens and breaks the path of the electricity.

Parallel circuits

Unlike a series circuit, a **parallel circuit** has multiple paths for the electricity to flow. On a parallel circuit, each bulb has its own circuit. If a bulb burns out, it breaks only the one circuit and does not affect the other circuits. All of the bulbs except the broken bulb will light because the electricity can complete its path.

series circuit

parallel circuit

190

SCIENCE BACKGROUND

House circuits—A house has many main circuits. Usually the circuits of a house are arranged so the appliances that use a large amount of power, such as a refrigerator, washing machine, and water heater, are on separate circuits.

200 Lesson 87

electric meter

Measuring Electricity

If you look at the label on a small electrical appliance you will find information about how the appliance uses electricity. One piece of information that the label shows is how many volts the appliance requires. A **volt** is the measurement of the amount of electrical push or force in a circuit. The wiring in a typical house in the United States carries 115–120 volts of electrical force.

An appliance label also indicates how many watts the appliance uses. A **watt** is the measurement of power, or how fast work is done. For example, a 100-watt light bulb uses more power than a 60-watt light bulb. A hair dryer might use 1000 watts of power, but a mixer may use only 200 watts. The electric meter on a house measures the amount of power being used in the house. Power is measured in kilowatts (1kW = 1000 watts).

Another unit of measure for electricity is an ampere. An **ampere** is the unit used to measure how much current flows through a given part of a circuit in one second. In most houses, circuit breakers are designed to allow 15 or 20 amperes of current to pass through a circuit safely. If more electrical current than the circuit can safely handle passes through the circuit, the circuit breaker will open.

The three basic units of measurement—volts, watts, and amperes—are related to each other. If you know two of the measurements, you can use this relationship to find the other measurement. Suppose a small appliance uses 240 watts at the normal household voltage of 120. To find the amperes needed, you divide the number of watts by the number of volts.

> watts ÷ volts = amps

Direct an Activity

Materials: several small appliances

Allow the student to examine the unplugged appliances to find out how many amperes and how many watts each of the appliances uses. The numbers should be written on labels or engraved directly on the appliances.

MATH

Display the formula that shows the relationship between the units of electrical measurement (current x voltage = power, or amps x volts = watts.)

▶ If there are 10 amps and 10 volts, how many watts of power are there? {100}

▶ If there are 5 volts and 20 watts, how many amps are there? {4}

Continue with other examples.

Discussion:
pages 190–91

▶ How many paths does a series circuit have? {one}

▶ How is a light bulb attached to a series circuit similar to a switch? {The light bulb closes the circuit and completes the path. When the bulb burns out, the circuit breaks.}

▶ What happens to the light bulbs on a parallel circuit when one bulb burns out? {The other bulbs stay lit, because the burned out bulb breaks only its own circuit. The electricity has a complete circuit through the other bulbs.}

▶ What basic units measure electricity? {volt, ampere, and watt}

▶ What is a *volt*? {a measurement of the amount of electrical push, or force, in a circuit}

▶ How many volts is most wiring in a house in the United States? {115–120 volts}

▶ Which unit measures power, or how fast work is done? {watt}

▶ Does a 100-watt light bulb use more or less power than a 60-watt light bulb? {more}

▶ What unit of power does the electric meter on a house show? {kilowatts}

▶ How many watts is a kilowatt? {1000 watts}

▶ What is an *ampere*? {the unit used to measure the amount of electric current flowing through any given part of the circuit in one second}

▶ How many amperes do most house circuit breakers allow to pass safely? {15 or 20}

If a circuit has too many watts being required of it, it will cause too much electricity to flow through the circuit. The amperes will exceed the circuit breaker's capacity and cause it to open, thus breaking the circuit.

▶ What is the relationship between amperes, watts, and volts? {the number of watts divided by the number of volts equals the number of amperes}

▶ What does *amps* mean? {It is an abbreviation for amperes.}

💡 Suppose you knew the number of volts and amps but not the number of watts. How would you write the equation to find the number of watts? {volts X amps = watts}

Discussion:
pages 192–93

- **What is needed to make an electric cell?** *{two different metals and an electrolyte}*
- **What is an *electrolyte*?** *{a liquid or paste substance that conducts electricity}*
- **How does an electric cell produce an electrical current?** *{The chemical reaction between the metals and the electrolyte frees electrons to travel around the circuit.}*
- **How does the electricity flow from an electric cell?** *{It flows from the negative terminal through the wire to the positive terminal, through the electrolyte, and back to the negative terminal.}*
- **How many electric cells are contained in a battery?** *{at least one, but there are usually more}*
- **What are two advantages of a wet-cell battery?** *{It can produce a large amount of current for a relatively short time period, and it is rechargeable.}*
- **What is the primary purpose of wet-cell batteries?** *{to supply the quick, powerful charge needed to start an engine}*
- **Name an example of an engine that uses a wet-cell battery.** *{a car engine}*
- **What are some disadvantages of wet-cell batteries?** *{They must remain upright and are not very portable. The electrolyte used in wet-cell batteries is usually corrosive or poisonous and is not sealed.}*
- **Look at the diagram called *Wet-cell Battery*. What are two metals that can be used in a battery?** *{zinc and copper}*

Batteries

Around 1800, scientists discovered that they could use two different metals and a solution to produce an electric current. The device they invented was called an **electric cell.** A wire connected the two metals, and the metals were in contact with an **electrolyte** (ih LEK truh LYTE), a liquid or paste substance that conducts electricity. Scientists found that the chemical reaction between the metals and an electrolyte, such as an acid, allowed electrons to move to complete a circuit.

An electric cell uses chemical energy to produce an electrical current. During the chemical reaction, the freed electrons leave the negative terminal and travel along the wire path of the circuit. They enter the electric cell again through the positive terminal. As long as the chemical reaction continues, the battery will provide a source of electrical current.

A **battery** contains one or more electric cells. The first batteries were wet-cell batteries. A wet-cell battery has the advantage of producing a large amount of current for a relatively short period of time. Wet cells are also rechargeable. Perhaps you have seen someone check the fluid in a car battery. A car battery is a wet-cell battery that provides the quick, powerful charge needed to start an engine.

Wet-cell batteries, however, have some disadvantages. They must remain upright and are not very portable. The electrolyte used in the battery is usually corrosive or poisonous. Because of these problems most of the batteries that we use are dry-cell batteries. Instead of having an electrolyte fluid between the metals, there is an electrolyte

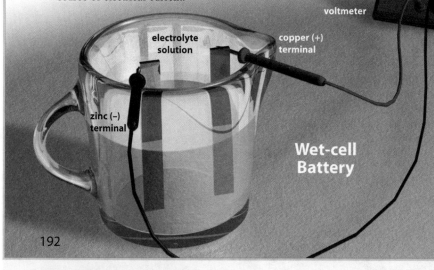

Wet-cell Battery

SCIENCE BACKGROUND

Wet cells and dry cells—The terms *wet cell* and *dry cell* refer to the type of electrolyte that is used in the battery. While all wet-cell batteries are rechargeable to some extent, only certain dry cells are rechargeable. Rechargeable batteries contain electrodes and an electrolyte that can be reconstituted.

Creation CORNER

Did you know that some batteries swim in the ocean? God has designed the electric ray with organs that work similarly to batteries. These organs can store electricity and are connected like parallel circuits. An electric ray can electrocute a large fish with a shock as powerful as 200 volts. Most outlets in your house carry only 110 volts. God has given the electric ray a powerful charge to use to capture prey.

paste. This improvement made batteries safer and easier to use. Dry cell batteries can be used in any position and in most locations. They are sealed and less likely to expose a person to the corrosive or poisonous electrolyte. Scientists continue to experiment with metal and electrolyte combinations to try to make batteries that last longer.

In a "dead" battery, the metals no longer react with the electrolyte. This happens when one of the metals is used up or the electrolyte is depleted. As a result, no electrons are free to move around the circuit. A rechargeable battery uses electricity to make the metals react with the electrolyte again.

QUICK CHECK
1. What is the difference between a series circuit and a parallel circuit?
2. What are three units used to measure electricity? What does each of them measure?
3. What is needed to produce an electric cell?

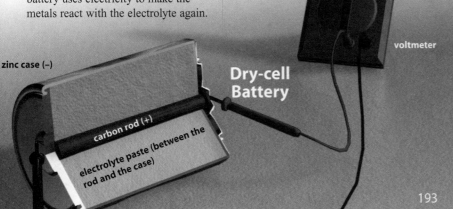

zinc case (−) · carbon rod (+) · electrolyte paste (between the rod and the case) · Dry-cell Battery · voltmeter

193

Direct a DEMONSTRATION

Demonstrate making a battery

Materials: small piece of copper, 2 plastic cups, 20 mL vinegar, 10 mL salt, 20–30 mL water, 5–10 cm strip of aluminum foil, 5–10 cm strip of paper towel, voltmeter

Copper and aluminum should react when an electrolyte is between them.

Soak the copper in vinegar while setting up the demonstration.

Mix the salt and water together.

Fold the paper towel into a square and immerse it in the salt water.

Fold the piece of foil into a small square.

Make a battery cell by sandwiching the wet paper towel between the copper and foil.

Set the voltmeter for DC (direct current). Using the voltmeter, place the red lead (+) on the copper and the black lead (−) on the aluminum foil.

Direct the student to observe how much electricity the battery generates.

➤ What was the liquid electrolyte? *{salt water}*

➤ Which two metals reacted with each other? *{copper and aluminum}*

➤ What change occurred to the piece of copper? *{It may have dark blotches of corrosion on it.}*

If the voltmeter does not register that any electricity has been produced, check to see if the leads are touching the desired objects. If needed, touch the black lead to the copper and the red lead to the aluminum foil.

Discussion:
pages 192–93

➤ How is the electrolyte of a dry-cell battery different from that of a wet-cell battery? *{A dry-cell battery uses a paste instead of a fluid.}*

➤ What are three advantages of a dry-cell battery? *{It can be used in any position. It can be used in most locations. It is sealed and is less likely to expose a person to the corrosive or poisonous electrolyte.}*

➤ What happens when one of the metals in a battery is used up or when the electrolyte is depleted? *{No electrons are freed to move around the circuit, and the battery is "dead."}*

➤ What kind of battery uses electricity to make the metals react with the electrolyte again? *{a rechargeable battery}*

📖 Refer the student to *Creation Corner*.

➤ How is an electric ray similar to a battery? *{It has organs that can store electricity and are connected like a parallel circuit.}*

➤ How did God design for the electric ray to use its stored electricity? *{to capture prey}*

➤ Is the electric ray's shock greater or less than a household circuit? *{greater}*

Answers

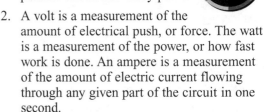

1. A series circuit has one path. A parallel circuit has many paths.
2. A volt is a measurement of the amount of electrical push, or force. The watt is a measurement of the power, or how fast work is done. An ampere is a measurement of the amount of electric current flowing through any given part of the circuit in one second.
3. An electric cell needs two different metals and an electrolyte.

Activity Manual

Reinforcement—page 119

Study Guide—page 120

This page reviews Lessons 85 and 87.

Expansion—pages 121–22

These pages identify and use drawings of electric circuits.

Assessment

Test Packet—Quiz 8-A

The quiz may be given any time after completion of this lesson.

Lesson 88

Student Text pages 194–96
Activity Manual pages 123–24

Objectives
- Describe what happens to magnets at their poles
- Explain the relationship between magnetism and electricity
- Identify and describe the parts of a generator
- Explain how a generator works

Vocabulary
magnet
magnetism
magnetic field
electromagnet
generator

Materials
- voltmeter (or current detector if voltmeter is not available)
- batteries with different voltages (1.5, 6, 9)
- black paper

Introduction
Prior to class wrap black paper around each of the batteries to hide the labels. Display the voltmeter and attach it to one of the batteries.

➤ How many volts did this battery have?

Remove the black paper and compare the number identified by the voltmeter with the number written on the battery. Repeat this test for each battery.

➤ How do you think the voltmeter got its name? *{Meter means "to measure," so a voltmeter measures the number of volts.}*

➤ How does a voltmeter detect current? *{The magnetic field produced by electricity attracts the magnet in the compass.}*

Purpose for reading:
Student Text pages 194–96

➤ How are electricity and magnetism similar?
➤ What are the parts of a generator?

Magnetism

A **magnet** is any material that has the ability to attract iron. In addition to attracting iron, magnets also attract objects made from nickel and cobalt. Magnets can be natural or manmade. Even though magnets come in many shapes and sizes, every magnet has two poles, a north pole and a south pole.

Magnetic Attraction
Another name for magnetic force is **magnetism**. By sprinkling iron filings in the **magnetic field**, the area of magnetic force, we can see that the magnetic force is strongest at the poles of a magnet. The middle of a magnet has the weakest magnetic force.

Magnets and static electricity act in somewhat similar ways. Both can attract and repel other objects. With static electricity, objects that have different charges attract each other. In the same way, the opposite ends, or poles, of magnets attract each other. The south pole of one magnet will be attracted to the north pole of another magnet. However, two south poles will repel each other.

Electricity and Magnetism
Electricity and magnetism are related. A flow of electricity can produce a magnet, and a magnet can produce electricity. In 1820 the Danish scientist Hans Christian Oersted (UHR sted) discovered that current traveling though a wire produces a weak magnetic field in the wire. Then in 1831 American scientist Joseph Henry and British scientist Michael Faraday made similar discoveries. Although they did not work together, both discovered that moving a magnet around or through a loop of wire produces electricity in the wire.

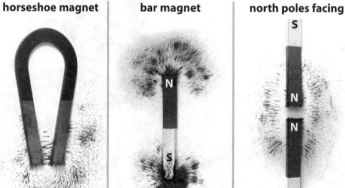

horseshoe magnet | bar magnet | north poles facing | north and south poles facing

Make a current detector—A current detector does not measure the number of volts, but it can detect electricity.

Use 2–3 m of 25 gauge insulated wire and a compass.

Leave a length of 10 cm on both ends of the wire. Coil the remaining wire around a compass at least 40 times so that the coil forms an X with the compass needle.

When the ends of the wire are attached to the poles of a battery, the compass needle should move.

SCIENCE BACKGROUND

Detecting current—A voltmeter, galvanometer, and homemade current detector each detect current using a coil of wire and a pivoting needle. The coil has a magnetic field when electricity flows through it that causes the needle to move.

Poles of a magnet—Every magnet has two poles, a north pole and a south pole. The location of the poles depends on the shape of the magnet. A bar magnet has its poles on either end of the magnet. A horseshoe magnet has poles at the ends of the magnet. Donut magnets and disc-shaped magnets have one pole on the top of the magnet and the other pole on the bottom.

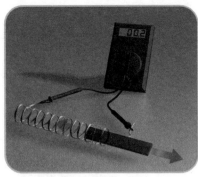

They also found that moving a wire between the north and south poles of two magnets produces electricity in the wire. However, a wire resting between two magnets does not produce electricity. Either the wire or the magnet must be moving. Modern machines, such as generators, use this principle to make electricity.

Electromagnets

The magnetic field of an electrical wire in your home is too weak to do work. Coiling a current-carrying wire increases the strength of the magnetic field. In 1825 William Sturgeon discovered that adding a metal core to a coil of wire increases the magnetism even more. An **electromagnet** is a coil of wire with a core attached to an electrical source. When electricity travels through the coil of wire, the core acts like a magnet. The core has both north and south poles. However, if no electricity travels through the coil of wire, the core is not magnetized. An electromagnet with an iron core can lift a piece of iron 20 times heavier than the magnet's iron core. When it has a larger core, the magnet is able to do even more work. The shape of the core and the distance between the poles also affect the strength of the electromagnet.

An electromagnet has several advantages over other magnets. It is made from inexpensive, common materials. Since electromagnets can turn on and off, they can power machines. The strength of an electromagnet can be increased or decreased by increasing or decreasing the number of coils. An electromagnet is much stronger than a natural magnet. Doorbells, microphones, radio speakers, and even the particle accelerators used to study atoms all use electromagnets.

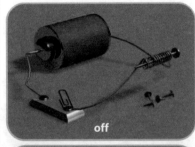
off

on

Direct a Demonstration
Demonstrate magnetic fields

Materials: one blank overhead transparency, two bar magnets, one horseshoe magnet, iron filings

To see the magnetic field of a bar magnet, place the magnet under a sheet of overhead transparency. Sprinkle iron filings on the transparency. Observe the shape of the magnetic field. Repeat the demonstration with the horseshoe magnet.

This demonstration can also be done with two magnets. Place two bar magnets under an overhead transparency with opposite poles facing each other but not touching. Sprinkle the iron filings onto the transparency. Repeat with like poles facing each other.

Note: If iron filings are not available for purchase, a substitute can be made by cutting a piece of steel wool.

Discussion:
pages 194–95

- What is a *magnet*? {any material that has the ability to attract iron}
- What are the two types of magnets? {natural magnets and man-made magnets}
- What other materials will a magnet attract? {Possible answers: nickel, cobalt}
- What is *magnetism*? {magnetic force}
- How many poles does every magnet have? {two} What are these poles called? {north pole and south pole}
- How can we demonstrate that a magnetic field is strongest at a magnet's poles? {When iron filings are sprinkled around the magnet, we can see that more iron filings go to the poles than to the center of the magnet.}
- What is the relationship between magnetism and electricity? {Electricity can produce magnetism, and magnetism can produce electricity.}
- What do magnets and static electricity have in common? {Both can attract and repel other objects.}
- Who discovered that a wire has a magnetic field? {Hans Christian Oersted}
- Which men discovered how to make electricity using a magnet? {Joseph Henry and Michael Faraday}
- Describe how a magnet produces electricity. {Possible answers: moving a magnet around or through a loop of wire; moving a wire between magnets}
- How can you increase the magnetic field of a current-carrying wire? {coil the wire}
- What is an *electromagnet*? {An electromagnet is a coil of wire attached to an electrical source. The coil of wire has a metal core.}
- Who discovered the electromagnet? {William Sturgeon}
- Which part of an electromagnet acts like a magnet? {the core}
- What are two ways to change the strength of an electromagnet? {Possible answers: change the number of coils; change the core}
- What are three advantages of an electromagnet? {inexpensive; can turn on and off; strength can be increased or decreased}
- Identify two common uses of an electromagnet. {Possible answers: doorbell, microphone, radio speaker, particle accelerator, videotape}

Discussion:
page 196

➤ **What is a *generator*?** *{a machine that converts motion into electrical energy}*

➤ **Describe how a water-powered generator works.** *{Water turns the turbine, and the turbine turns the shaft and the magnet. A coil of wire or a metal ring surrounds the spinning magnet. The spinning magnet generates electricity in the coil.}*

➤ **Is it possible to generate electricity by moving the coil of wire instead of the magnet?** *{yes}* **Explain.** *{Electricity will be generated if either the wire or the magnet is moving.}*

➤ **Why do most power plants have a moving magnet and a stationary coil?** *{It takes less energy to spin the magnet than it does to rotate the heavy coil around the magnet.}*

According to the Second Law of Thermodynamics, matter (energy) is neither created nor destroyed. Energy simply changes its form. Generators change motion into electrical energy. A motor does just the opposite. It converts electrical energy into motion, such as that of a mixer, electrical pencil sharpener, or fan.

 Refer the student to the diagram on Student Text page 196. Discuss the path of electricity from the power plant to the house.

Answers

1. Possible answers: Moving a magnet around or through a loop of wire produces electricity in the wire. Moving a wire between the north and south poles of two magnets produces electricity in the wire.
2. An electromagnet is a coil of wire with a core attached to an electrical source.
3. A generator is a machine that converts motion into electrical energy.

Activity Manual

Study Guide—page 123
This page reviews Lesson 88.

Bible Integration—page 124
Like magnets, people can attract or repel friends. This page examines what the Bible says about gaining and losing friends.

Generators

The movement of a magnet within a coil of wire or of a coil of wire within a magnet causes an electrical current. This principle is used by a **generator**, a machine that converts motion into electrical energy.

Some generators are very small. You may even have seen one attached to a bicycle. As the bicycle wheel turns, it turns a small wheel on the generator. The small wheel turns a shaft attached to a magnet. As the magnet rotates in a coil of wire, it causes an electrical current that may be used to power the bicycle's light.

Most electric power plants use turbines to power generators. The turbine has blades that are moved by water, wind, or steam. As the blades turn, a shaft turns a magnet within a coil of wire. The spinning magnet creates an electric current in the coil of wire.

Another way to generate electricity is to attach the shaft of the turbine to the coil of wire. The moving turbine rotates the coil around the magnet or spins the coil in a magnetic field. However, spinning the heavy coil around the magnet requires more energy than rotating the magnet requires. For this reason, most electrical power plants have stationary coils and moving magnets.

1. How can you produce an electric current with a magnet and a coil of wire?
2. What is an electromagnet?
3. What is a generator?

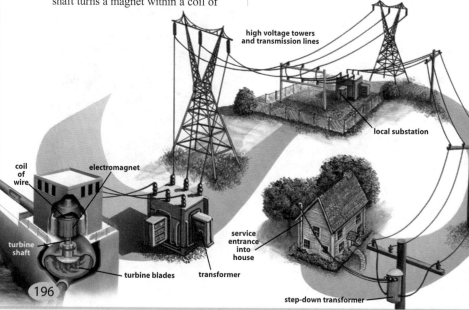

 Although the Student Text does not elaborate on all the parts of the path of electricity from power plant to house, help the student identify parts of the system that he is familiar with.

Explorations: Famous Inventors

A whole new door of opportunity for inventors opened when man learned how to use magnetism and electricity to power devices. Men such as Thomas Edison became famous for the hundreds of devices that they invented and patented. A person can obtain a patent by registering his or her invention with the government so that no one else can claim credit for the invention.

Imagine that you are at a dinner honoring one of the men who invented a device that furthered the use of electricity and magnetism. You have been asked to introduce the guest of honor.

What to do

1. Choose an inventor from the list your teacher provides.
2. Research the person that you choose. Your introduction should include the following: the inventor's name, his country, when he lived, his family background, and what invention he is famous for. Also include how the invention improved people's lives, how it furthered research and use of electricity and magnetism, and any other inventions that the honoree is credited with inventing.
3. Prepare a one-to-two minute speech honoring your chosen inventor.

197

Encyclopedias, trade books, and the Internet are all sources of information.

Encourage the student to make a checklist of the criteria required for the speech. He can mark each item as he finds the information.

Encourage the student to practice eye contact and clarity as he speaks.

For some students, public speaking is a frightening experience. Although the student should do adequate and accurate research, a lighthearted approach to the speech may ease some students' fears.

Suggest that the student prepare a note card with the basic information for his speech. Encourage the student to refer to the card, but discourage his reading it word for word.

Student Text page 197

Lesson 89

Objectives
- Research an inventor
- Present a speech honoring an inventor

Materials
- copy of TE page A26, *Magnetic Personalities and Shocking Discoveries*, for display

Introduction

▸ Have you ever heard your pastor introduce a missionary, evangelist, or other speaker? *{Answers will vary.}*

▸ Why would a pastor introduce a special speaker? *{Possible answers: to help the audience get to know the speaker; to get the audience interested in listening}*

▸ What types of details might a pastor include in his introduction? *{Possible answers: where the person had been; important things that the person did; the length of time that the person has been involved in the ministry; a brief family background; awards earned; special positions held}*

▸ The Bible tells us that it is appropriate to honor those who deserve honor. Giving an introduction is a way of acknowledging a person who has earned honor (Rom. 13:7).

Purpose for reading:
Student Text page 197

The student should read the page before beginning.

Discussion

▸ What is a *patent*? *{the registration of an invention so that no one else can claim credit for it}*

Thomas Edison had over 1,000 patents on his inventions.

Procedure

Display *Magnetic Personalities and Shocking Discoveries*. Guide the student in choosing a person to research. Encourage the student to consult a variety of resources.

Assessment

Rubric—TE pages A50–A53

Select one of the prepared rubrics, or design a rubric to include your chosen criteria.

Lesson 90

Student Text pages 198–99
Activity Manual pages 125–26

Objectives

- Identify ways to increase a wire's magnetism
- Predict ways to strengthen an electromagnet
- Experiment to test predictions

Materials

- See Student Text page

Introduction

➤ Today you will build an electromagnet and experiment with ways to increase its strength.

Purpose for reading:
Student Text pages 198–99
Activity Manual pages 125–26

The student should read all the pages before beginning the activity.

ACTIVITY: Build an Electromagnet

Electromagnets are a vital part of our everyday lives. From the doorbell on your house to the mega magnet at the junkyard, magnetism and electricity provide many useful tools for our daily lives. In this activity you will build an electromagnet. It is your job to determine how to make the magnet stronger.

Process Skills
- Hypothesizing
- Predicting
- Experimenting
- Observing
- Inferring
- Identifying and controlling variables
- Recording data

Problem

How can you make an electromagnet stronger?

Procedure

Note: Disconnect your electromagnet from the battery after each test.

1. Remove five centimeters of insulation from the tips of each piece of wire. Wrap the longest piece of wire five times around a straightened paper clip. Attach the ends of the wire to the 1.5-volt battery. See how many staples your electromagnet (bare tip of the paper clip) attracts. Record the results in your Activity Manual.

2. Write a hypothesis about an electromagnet and the amount of staples it can pick up. You may change the object used as the core of the electromagnet, or you may change the amount of times the wire is wrapped around the core.

3. Build your electromagnet and test it. Record the result in your Activity Manual.

4. If your electromagnet was not stronger, choose one variable to change. Explain the adjustment you make. Retest your electromagnet and record the test results.

Materials:
2–3 meters of insulated wire
1 large metal paper clip (straightened)
2 D-cell (1.5-volt) batteries or 1 six-volt battery
1 bar of staples (separate into the individual staples)
various materials such as different sized nails, wooden pencils, plastic rods, additional batteries and wire, and electrical tape
Activity Manual

198

 Use fresh batteries. Avoid using more than a 6-volt battery. A stronger battery may cause the wires to get hot.

Staples are easily separated by discharging them through a stapler.

The first electromagnet that each group makes may not be strong enough to attract one staple.

A strong electromagnet easily magnetizes the staples. Once magnetized, the staples will clump together. This clumping is not a problem unless you are trying to test a weak electromagnet using the magnetized staples. To test a weak electromagnet, use staples that are not magnetized.

208 Lesson 90

 5. If your electromagnet was stronger, keep trying to increase the strength until it can pick up 50 or more staples. Remember to change only one variable at a time. Use the chart to record your information.

Conclusions
- Was your hypothesis correct?
- Can you see any direct relationship between the changes you made and the strength of the electromagnet? For example, how many more staples could an additional battery pick up?

Follow-up
- Change other variables to try to strengthen your electromagnet.

199

SCIENCE PROCESS SKILLS

Predicting

▶ If your electromagnet did not pick up the number of staples in your hypothesis, how did you determine what to do to increase its strength? *{Answers will vary, but elicit that the student knows that some actions, such as increasing the number of coils, will increase the strength of the electromagnet.}*

▶ Could you predict the result of changing the core? *{Answers will vary, but elicit that unless the student had tried a different core or had additional knowledge outside of the Student Text, he would not have any basis for predicting.}*

Predicting is based on knowledge and past experiences.

▶ After trying different cores, could you predict how one of those cores and how additional coils would affect the electromagnet? *{yes}* Why? *{because the prediction would be based on knowledge and experience}*

Chapter 8

Procedure
To test the electromagnet, the student should touch an end of the core to the staples. Encourage the student to wrap the wire tightly against the core. Batteries may be taped together.

Use the following questions to help the student think of ideas to try.

▶ Did the information about electromagnets in the Student Text give any ideas for increasing the strength of an electromagnet?
▶ How many times did you coil the wire?
▶ Would a wooden or plastic core work? Why or why not?
▶ Do you think it makes a difference if the wires are wrapped smoothly and evenly or wrinkled and criss-crossed?
▶ Could you increase the size of the iron core?
▶ Is there a way that you could increase the voltage or amount of electricity?

Conclusions
Provide time for the student to evaluate his hypothesis and answer the questions.

▶ **What are some ways to strengthen an electromagnet.** *{Possible answers: use a bigger core; increase the number of coils; increase the amount of electricity}*

Activity Manual
Activity—pages 125–126

Assessment
Rubric—TE pages A50–A53
Select one of the prepared rubrics, or design a rubric to include your chosen criteria.

Lesson 90   209

Lesson 91

Student Text pages 200–203
Activity Manual pages 127–28

Objectives

- Explain the difference between electricity and electronics
- Identify the benefits of an integrated circuit
- Identify some of the parts of a computer

Vocabulary

electronic device
electrical signal
binary number system
integrated circuit
semi-conductor
CPU
ROM
RAM

Introduction

Electronic devices use electricity to communicate information.

➤ What are your favorite electronic devices?
➤ What are some electronic devices that you would find at home?
➤ How would you describe life without electronics?

Purpose for reading:
Student Text pages 200–203

➤ Identify why the invention of the integrated circuit was important.
➤ Does a computer have intelligence? Why or why not?

Electronics

For one moment, imagine life without electronics. There would be no computers, radios, televisions, digital watches, or even cell phones. Each of these objects is an **electronic device** that uses electricity to communicate information.

Although not identical, electronic devices are similar to electrical devices. Both use current electricity, and both can do work. However, electronic devices can communicate information, but electrical devices cannot. For example, your toaster is an electrical device. It uses electricity to produce heat. But your television can communicate the evening news. A toaster cannot communicate information.

To carry information, the current must vary, or change, in some way. An **electrical signal** is an electric current that carries information. Televisions, like other electronic devices, use electrical signals to broadcast programs. Sound, motion, pictures, letters, and words are communicated to your television set by electrical signals.

Electronic devices use a code in much the same way as a telegraph operator uses the Morse code. Instead of dots and dashes, electronic devices use a code called the binary system. The **binary number system** uses open and closed circuits to communicate the number 0 and the number 1. An open circuit means 0, and a closed circuit means 1. Using this code, electronic devices can communicate almost any piece of information.

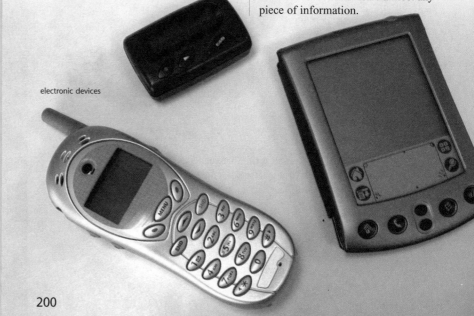

electronic devices

SCIENCE BACKGROUND

Technical terms—An integrated circuit has four main parts. The terms *switch, resistor, insulator,* and *conductor* are used to describe the functions of the parts. The technical names for the parts of an integrated circuit are diode, transistor, resistor, and capacitor. For this lesson the function of the parts is more important than the actual terms.

Silicon—The parts of most integrated circuits are made from silicon. The location of the impurities in the silicon determines when the silicon acts like a diode, transistor, resistor, or capacitor. By varying where the impurities are located and putting them in a pattern, scientists can determine where each resistor, transistor, and capacitor is placed. Adding impurities to silicon is called *doping.*

 The integrated circuit was independently developed by two men, Jack Kilby and Robert Noyce. In 1959 both men applied for and received patents for miniaturized electronic circuits. One of the first organizations to use the new computer chips was the U.S. Air Force.

Jack Kilby also invented the portable calculator in 1967. In 1968 Robert Noyce started Intel, a company that led the way in microprocessors.

Integrated Circuits

Today most electronic devices use computer chips to signal information quickly. A computer chip or microchip is really an **integrated circuit** (IC), a very small circuit with all of its parts built into it. Integrated circuits are called *microchips* because they are made from chips of silicon. Each microchip can have over ten million parts inside, including switches, resistors, insulators, and conductors.

Silicon, a major ingredient of sand, is an excellent semiconductor. A **semiconductor** is a material that conducts electricity better than insulators but not as well as conductors. It acts somewhat like a switch. Under some conditions, a semiconductor conducts electricity and keeps the circuit closed. Under other conditions, it does not conduct electricity and causes an open circuit. At other times, silicon acts like a resistor.

Do you remember Paul Revere's light signal? The lights Paul Revere saw in the Old North Church communicated information. If one lantern was lit, the British were coming by land. If two lanterns were lit, the British were attacking by sea. More information could have been communicated through a sequence of turning the lanterns on and off or by making the flame brighter or dimmer. In a similar way, integrated circuits signal information by varying the amount of electricity by opening and closing circuits in sequence.

Integrated circuits have many benefits that other circuits do not have. They are small in size, inexpensive to make and put together, and durable. To make an IC durable, a piece of plastic or ceramic insulates the whole chip.

Microchips are getting smaller and smaller in size. Because the chips are so small, electricity does not have to travel far. One electronic signal can go as fast as a lightning bolt.

To compete in the space race, the United States of America needed faster, smaller computers. Some early computers were as big as a warehouse! The invention of the microchip solved these problems. The computer that sits on your desk, resides in your cell phone, or rests in the palm of your hand is probably more powerful than the computer that sent the first men into space.

201

Direct an ACTIVITY

Materials: six coins

Direct the student to use coins to show the numerous combinations available using a binary code.

Pretend that each coin is a circuit. Heads represents a closed circuit. Tails represents an open circuit.

The student should see how many combinations of open and closed circuits (heads and tails of each coin) he can make in five minutes and write each combination on paper.

For example, one combination would be having the first circuit closed (heads) and the other circuits open (tails).

Write this combination as follows:
h t t t t t.

Without repeating any combination, there are 64 possible combinations.

Discussion:
pages 200–201

▶ Identify the similarities and differences between an electrical device and an electronic device. {Both use electricity and do work. Electronic devices communicate information.}

▶ Define an *electrical signal*. {an electric current that carries information}

▶ What is the name of the code that integrated circuits use to communicate information? {binary system}

💡 How does the binary system work? {The binary system uses open and closed circuits to communicate the numbers 0 and 1.}

▶ What is an *integrated circuit*? {a very small circuit with all of its parts built into it}

▶ What is a *computer chip* or *microchip*? {an integrated circuit}

▶ Why is an integrated circuit called a microchip? {because it is made from a chip of silicon}

▶ Other than integrated circuits, what substance is silicon a major ingredient of? {sand}

▶ How is a semiconductor different from a conductor and an insulator? {A semiconductor conducts electricity better than insulators but not as well as conductors.}

▶ How does the integrated circuit communicate information? {by opening and closing circuits in sequence and by varying the amount of electricity going through the circuit}

▶ What are some benefits of using integrated circuits? {Possible answers: They are durable, small, inexpensive to make and put together, and fast.}

▶ How can integrated circuits be so fast? {The electricity has very little distance to travel.}

▶ How big were some of the computers used for the space race? {as big as warehouses}

▶ What is amazing about the size of computers today? {Even small modern computers are probably more powerful than the computers needed to compete in the space race.}

Discussion:
pages 202-3

- What is a *computer*? {*an electronic device that uses many integrated circuits connected together*}

- What are some things a computer can do? {*Possible answers: It can communicate information and save, process, and recover that information.*}

- Name three examples of integrated circuits found in a computer. {*ROM, RAM, and CPU*}

- What does CPU stand for? {*central processing unit*}

- What is the CPU compared to? {*the brains of the computer*}

- What is *ROM*? {*the read only memory; It contains the built-in memory and programs.*}

- What are programs? {*sets of instructions that tell the computer what to do*}

- How is ROM like your long-term memory? {*ROM remembers facts for a long time.*}

- What does RAM stand for? {*random access memory*}

- What is RAM? {*temporary memory storage*}

- Which can be changed, ROM or RAM? {*RAM*}

- Which part of the computer's memory is lost forever when the computer is turned off? {*RAM*}

- Can a computer save information to RAM or ROM? Why or why not? {*No; ROM cannot be changed, and RAM is lost when the computer turns off. A computer can save information to disks.*}

- What are some devices that store retrievable data on computers? {*Possible answers: CD-ROM, DVD, hard drive, floppy disk, memory stick*}

- Which kind of storage device is a permanent part of the computer? {*hard disk or hard drive*}

- What are some input devices of a computer? {*Possible answers: keyboard, mouse, modem, scanner*}

- What are some output devices? {*Possible answers: monitors, printers*}

Computer Parts

The electronic device called a computer uses many, many integrated circuits connected together. A computer can communicate information and save, process, and recover that information. The CPU, ROM, and RAM are examples of integrated circuits in a computer.

The **CPU (central processing unit)** acts like the brain of the computer. It is the part of the computer that tells the other parts of the computer what to do. The CPU is part of the motherboard.

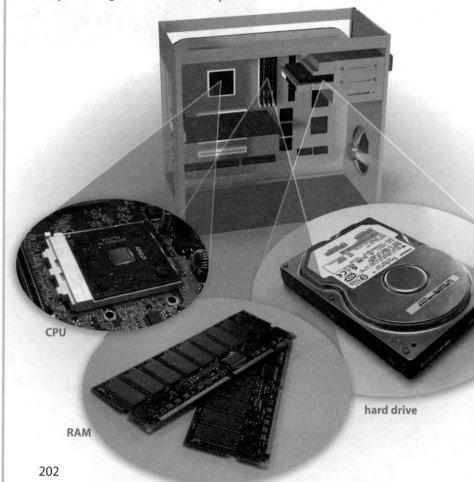

CPU

RAM

hard drive

202

SCIENCE BACKGROUND

Computer parts—The hard drive is only one part of a computer. A computer actually has three main categories of parts: input devices, output devices, and a system unit.

Input devices, such as the keyboard and mouse, take your message and translate it into computer language. Then the computer can save, process, and retrieve that information.

An output device, such as a computer monitor, translates the information from computer language to a language that you can read.

The CPU, RAM, and ROM are part of the system unit. This part of the computer does the actual saving, processing, and retrieving of data. The hard disk is a type of secondary, or permanent, storage. It can be stored inside the hard drive or in the system unit.

 Credit for inventing the first computer is given to Charles Babbage. Although his computer in 1884 was a mechanical device and not an electronic device, many of the principles he used are still used for modern computers.

212 Lesson 91

ROM (read only memory) is built-in memory and programs, sets of instructions that tell the computer what to do. ROM cannot be deleted or changed by the CPU. When someone asks your name, you can easily recall it. It is part of your long-term memory. Similarly, ROM remembers facts for a long time.

RAM (random access memory) stores facts temporarily. The CPU can change information stored in RAM. Facts stored in RAM are not remembered forever. They are lost when you turn the computer off.

The computer cannot save information in RAM or ROM. The CPU can save information only to data storage devices. A hard drive, floppy disk, CD-ROM, DVD, and memory stick are examples of data storage devices. A hard drive, sometimes called the hard disk, is a permanent part of the computer. The other devices can be put in and taken out of the computer.

Computers gain information through input devices such as a keyboard, mouse, modem, or scanner. It gives information through output devices such as monitors and printers.

Computer "Intelligence"

A computer knows only two things, open and closed circuits. An open circuit can mean no or 0, while a closed circuit can mean yes or 1. By sequencing the open and closed circuits, a computer can signal information using the binary number system.

A computer can solve problems so fast that it seems like it is actually thinking. But it isn't. A computer only seems to have intelligence. It can only do what the programmer and designer tell it to do. It can recall facts quickly and accurately, but it does not understand the facts. A programmer could tell the computer that $2 \times 2 = 4$, and the computer would "know" this fact. But it would not know why two times two equals four. A programmer could also tell the computer that $2 \times 2 = 5$. The computer would not know that this is incorrect unless it was programmed with that information.

The Bible says that true intelligence is more than just knowing facts. Wisdom is knowing, understanding, and remembering the facts. Understanding is knowing when and how to use the facts. Discerning what is true and what is a lie is also a part of understanding. The greatest of all wisdom is to know and understand God (Jeremiah 9:23–24).

"Get wisdom; get understanding: forget it not." (Proverbs 4:5)

QUICK CHECK
1. How does an integrated circuit signal information?
2. What are four benefits of an integrated circuit?
3. How much information does a computer really know?

Discussion:
pages 202–3

▶ **How much does a computer really know?** *{It knows two things—open and closed circuits.}*

▶ **What information does a closed switch communicate?** *{1 or yes}*

▶ **Does a computer really think and understand?** *{No; it knows only what the programmer and designer tell it. It does not understand facts and cannot tell what is the truth and what is a lie.}*

📖 **Is it enough to only know the facts?** *{no}* **Why?** *{Possible answer: The Bible says to get wisdom and understanding and to not forget what you have learned.}*

▶ **What does the Bible mean when it says to get wisdom and understanding?** *{Having wisdom and understanding is to use and discern what is known.}*

Discerning is judging whether something is right or wrong.

▶ **What is the greatest wisdom?** *{to know and understand God}*

Answers

1. It varies the amount of electricity and opens and closes circuits in sequence.
2. small, inexpensive, durable, fast
3. It knows only two things—open and closed circuits. (What it is programmed to know is also an acceptable answer.)

Activity Manual
Expansion—page 127
This page categorizes the types and uses of electronic and electrical devices around the house.

Study Guide—page 128
This page reviews Lessons 88 and 91.

Assessment
Test Packet—Quiz 8-B
The quiz may be given any time after completion of this lesson.

Direct an ACTIVITY

Encourage the student to try going without using any electronic devices for a day. He can record his experiences in a journal.

The student can record the electronic devices he would have used and their purposes in his life. He can also write about frustrations, humorous experiences, and how often he was successful at avoiding electronics.

Lesson 92

Student Text page 204
Activity Manual pages 129–30

Objectives

- Recall concepts and terms from Chapter 8
- Apply knowledge to everyday situations

Introduction

Material for the Chapter 8 Test will be taken from Activity Manual pages 120, 123, and 128–30. You may review any or all of the material during the lesson.

You may choose to review Chapter 8 by playing "Complete the Circuit" or a game from the *Game Bank* on TE pages A56–A57.

Diving Deep into Science

Questions similar to these may appear on the test.

Answer the Questions

Information relating to the questions can be found on the following Student Text pages:

1. page 190
2. page 191
3. page 200

Solve the Problem

In order to solve the problem, the student must apply material he has learned. The answer for this Solve the Problem is based on the material on Student Text page 191. The student should attempt the problem independently. Answers will vary and may be discussed.

Activity Manual

Study Guide—pages 129–30

These pages review some of the information from Chapter 8.

Assessment

Test Packet—Chapter 8 Test

The introductory lesson for Chapter 9 has been shortened so that it may be taught following the Chapter 8 test.

Diving Deep into Science

Answer the Questions

1. Why would wiring in a house be arranged in parallel rather than series circuits?
   Answers will vary, but elicit that if the whole house was on a series circuit, every time a break in the circuit occurred, every electrical device would stop working. Without parallel circuits, any problem with the wiring would be harder to find as well.

2. Some bread machines use 480 watts of power. If the voltage supplied to the house is 120 volts, how many amperes of current do these bread machines require?
   4 amperes (480 watts ÷ 120 volts = 4 amperes)

3. How are electrical devices different from electronic devices?
   Electrical devices usually change electrical energy into some other form of energy, such as heat or motion. Electronic devices use electricity to communicate information.

Solve the Problem

Every morning in her bedroom your sister uses a blow dryer and hot curlers to get ready for school. When her 750-watt blow dryer stopped working, she replaced it with one that uses 1500 watts. The next morning she plugged in the blow dryer. As soon as she turned it on, the electricity in her room went out. What might have happened, and why did it happen? Can you think of a way to prevent it from occurring again?

Accept any answer that demonstrates that additional wattage results in more amperes being used, which could cause a circuit to overload and the circuit breaker to trip. The circuit in her room could not handle the additional 750 watts. She may need to unplug some electrical devices on that circuit or find another circuit that has fewer amperes being used.

Review Game
Complete the Circuit

Divide the class into teams of four members each. Tell the teams that they can choose to be a parallel or a series circuit. If they choose a series circuit, all four members must answer a question correctly. If the team completes the series circuit, they receive three points. If the team chooses a parallel circuit, the team may miss one question and still complete the circuit. However, the team will receive only one point. If the team cannot complete the circuit it chooses, then the team has "short circuited" and receives no points. A team may choose a different circuit each time it completes a circuit or "short circuits." Alternate questions between teams.

Student Text page 205
Activity Manual page 131

Lesson 93

Motion and Machines

GREAT & MIGHTY Things

Over the years some scientists, even famous ones such as Leonardo da Vinci, have tried to make perpetual motion machines. These machines, once started, would run without any additional forces or energy being applied to them. Often they were based on the laws of motion. But not until man more clearly understood one of God's basic energy laws did it become obvious that no machine can be totally efficient. Machines always lose some energy, even if only a small amount. So in order for a machine to keep moving, that machine would have to create energy from nothing. But only God can create something out of nothing. Man is limited to using and converting only the energy and matter that God created in the beginning.

205

Objectives
- Demonstrate knowledge of concepts taught in Chapter 8
- Recognize that only God can create energy

Materials
- string with weight attached

Introduction

Note: This introductory lesson has been shortened so that it may be taught following the Chapter 8 test.

Direct a student to hold the string with the weight hanging from it. Start the string swinging. Tell the student to keep his arm still while the string swings.

➤ What did he have to do to start the string moving? *{Use his hand to move it.}*
➤ What eventually happens to the swinging? *{It stops.}*
➤ What would he have to do in order to get it swinging again? *{Move it again.}*
➤ Men have tried to make machines that would run without additional force being added.

Chapter Outline
I. Motion
  A. Distance and Speed
  B. Velocity and Acceleration
  C. Momentum
II. Laws of Motion
  A. First Law of Motion
  B. Second Law of Motion
  C. Third Law of Motion
III. Machines
  A. Work
  B. Simple Machines
  C. Compound Machines

Activity Manual
Preview—page 131

The purpose of this Preview page is to generate student interest and determine prior student knowledge. The concepts are not intended to be taught in this lesson.

SCIENCE BACKGROUND

Leonardo da Vinci—(1452–1519) Known primarily for his famous paintings, da Vinci was also a scientist and an engineer. Da Vinci recorded his ideas and observations in notebooks. Today this information, along with his detailed drawings, is a resource for scientists and inventors.

Perpetual motion—Perpetual motion is a theorized condition in which a mechanical device or system operates indefinitely without a continuing energy source.

Chapter 9

Lesson 93   215

Lesson 94

Student Text pages 206–9
Activity Manual page 132

Objectives

- Differentiate between speed and velocity
- Explain why a reference point is needed to observe motion
- Explain the relationship of mass and velocity to momentum

Vocabulary

energy
potential energy
kinetic energy
mechanical energy
motion
reference point
distance
speed
instantaneous speed
velocity
acceleration
force
friction
momentum

Introduction

▸ **Do your parents have a growth chart hanging on the wall for them to measure and mark your height?**

▸ **The chart is a type of reference point to keep track of your growth progress.**

▸ **What are some characteristics of a reference point?** *{It does not move. It does not change.}*

▸ **Name some other reference points that you have seen or used.** *{Possible answers: the lights on an airport runway; the finish line of a race; a target for arrows; a lighthouse; a basketball or soccer goal}*

A reference point is something that remains still.

Purpose for reading:
Student Text pages 206–9

▸ **What is the difference between a moving object's speed and its velocity?**

▸ **How do mass and velocity affect momentum?**

Discussion:
pages 206–7

▸ **What is *energy*?** *{the ability to do work}*

▸ **What is the main difference between potential energy and kinetic energy?** *{Potential energy is stored. Kinetic energy is due to motion.}*

▸ **Why are both potential and kinetic energy forms of mechanical energy?** *{Both have the ability to make something move.}*

Whether they are roller coaster cars speeding down a track or leaves fluttering in the wind, objects are constantly in motion. As they move from one location to another, objects convert energy from one form to another. Perhaps you remember that scientists describe **energy** as the ability to do work. If a wagon is sitting on the top of a hill and you give it a tiny push, what happens? A force called gravity causes the wagon to move down the hill. The wagon at the top of the hill has **potential energy**, or stored energy, because of its position. But as the wagon starts to roll down the hill, its potential energy changes into **kinetic** (kuh NET ik) **energy**, or energy due to motion.

Potential and kinetic energy are forms of **mechanical energy**, or the ability to make something move. Other forms of energy can also be changed into mechanical energy. For example, our bodies convert the chemical energy from food into the ability to move. We use electrical energy to move motors to help us perform tasks. We also use machines to increase the force of mechanical energy. From a simple lever to a complex airplane, mechanical energy and machines are essential parts of our everyday lives.

Motion

What exactly is motion? **Motion** is the change of an object's position. To determine whether an object has changed position, we need a fixed, unmoving object or location to use as a **reference point**. Think about riding in a car. You can tell the car is moving because you can see its position changing in relation to stationary objects outside of the car. However, when you look at the other people inside your car, they look like they are not moving even though they are. Your reference point inside the car is not fixed. It moves as the car moves. At night when it is hard to see reference points outside the car, another car's headlights can be very confusing. Is the other car still and your car moving? Or are both cars moving? Is one car moving faster than the other car? Without a reference point, you cannot tell visually.

Potential energy

Kinetic energy

206

SCIENCE BACKGROUND

Writing about speed—The slash mark replaces the word *per* when abbreviating rates of speed. Kilometers per hour is written as km/h.

216 Lesson 94

Distance and Speed

By measuring an object's position at different times, you can determine the **distance** it travels. Suppose you are in a sprint race. When the starter gun sounds, you start running. At 6 seconds, you are 48 meters from the starting line. The 48 meters is the distance that you have traveled.

To know how fast you ran those 48 meters, we need to add another factor—time. By dividing the distance traveled by the amount of time it takes to travel that distance, we can determine the **speed,** or rate, that you ran. Dividing 48 m by 6 seconds tells us that your average speed was 8 m per second (8 m/s).

Distance ÷ Time = Speed
48 m ÷ 6 s = 8 m/s

Speed is measured in length units per time units, but the units are not always the same. For example, you would probably measure a runner's speed in meters per second (m/s). Your car's speed, though, is measured in kilometers per hour (km/h) or miles per hour (mi/h).

speedometer

Have you ever heard someone say, "Well, I'm just as good as that person"? In the Christian walk, it is essential to have a reference point other than man. In II Corinthians 10:12, Paul says we are foolish when we compare ourselves with other people. Christians cannot tell whether they are moving toward Christlikeness if their reference point is sinful, changeable man. That reference point is always moving. Instead they should follow the instructions of Hebrews 12:2, "Looking unto Jesus the author and finisher of our faith." When Christians use Jesus as the unmoving reference point for their lives, they can measure their movement toward the goal of Christlikeness.

You are probably familiar with the speedometer in your family car. The speedometer registers how fast a car can go in kilometers or miles an hour. The speedometer measures the **instantaneous speed,** the car's speed at one particular moment. However, we usually use the average speed to calculate distance and time. To determine the distance a car travels, we multiply its average speed by the amount of time it travels. For example, a car going 100 km/h (62 mi/h) for 2 hours would travel 200 km (124 mi).

Speed \ Time = Distance
100 km/h × 2 h = 200 km

Direct a student to determine the average speed of an object by dividing the distance traveled by the amount of time the object takes to travel that distance.

➤ **If a baseball travels 100 meters in 5 seconds, what is its average speed?** {20 m/sec}

➤ **If a football player runs 50 yards in 6 seconds, what is his average speed?** {8.3 yd/sec}

Continue with other examples as time allows.

Discussion:
pages 206–7

The Second Law of Thermodynamics states that matter cannot be created or destroyed.

💡 **If energy is not created or destroyed, where does energy come from?** {It changes from one form to another.}

➤ **Potential energy changes into kinetic energy. What are some other types of energy changes?** {Possible answer: Chemical energy and electrical energy can change to mechanical energy.}

💡 **Why must a reference point be an object that is not moving?** {A moving object is unreliable as a reference point.}

💡 **What happens when your reference point is a moving object?** {Your perception is wrong.}

➤ **How do you determine the distance an object travels?** {by measuring an object's position at different times}

➤ **How is speed calculated?** {by dividing the distance the object travels by the time it takes the object to travel that distance}

➤ **What is another word for speed?** {rate}

➤ **How is speed measured?** {in length units per time units}

💡 **Give an example of speed.** {Accept any answer that demonstrates distance divided by time.}

➤ **What is the difference between instantaneous speed and average speed?** {Instantaneous speed is the speed of an object at any one moment, and average speed is the average speed an object travels during an entire trip.}

➤ **What does the speedometer in a car measure?** {instantaneous speed}

📖 Refer the student to *Science and the Bible.*

➤ **Why would it be foolish to judge your actions by what your friend does?** {Even a good friend makes mistakes.}

➤ **What reference point should you use?** {Christ}

➤ **Where do we learn about Christ?** {in the Bible} [BAT: 6a Bible study]

Discussion:
pages 208–9

▸ **What is *velocity*?** *{the speed of a moving object in a certain direction}*

💡 **How is the velocity of an object different from its speed?** *{Velocity is an object's speed in a certain direction.}*

Instruct the student to determine whether the following are examples of velocity and to explain why.

▸ **Dad is driving west at 105 km/h (65 mi/h).** *{Yes, because it gives the speed in a certain direction.}*

▸ **Jim runs 10 km/h (6 mi/h).** *{No, because it does not tell direction.}*

▸ **What is *acceleration*?** *{a change in velocity during a period of time}*

▸ **Name three different ways you could accelerate while riding your bike.** *{speed up, slow down, change direction}*

Discuss relationships among speed, velocity, and acceleration.

▸ **Lindy the racecar driver drives at 150 mi/h. Is 150 mi/h her speed or velocity?** *{speed}*

▸ **Lindy maintains the 150 mi/h speed as she circles the track. Does her velocity remain the same during each lap? Why?** *{No, because she changes direction at each turn.}*

▸ **Rodney has come alongside and matched Lindy's speed. Lindy increases her speed to get to the finish line before Rodney. What term best describes her action?** *{acceleration}* **Why?** *{The speed changed.}*

Refer the student to *Fantastic Facts*.

▸ **Why can a plane accelerate in ways that a car usually does not?** *{It can move up and down.}*

▸ **What instrument helps pilots know whether they have adequate acceleration to turn?** *{an accelerometer}*

▸ **How has this technology been developed for use in cars?** *{It is used to activate air bags.}*

Velocity and Acceleration

The speed of an object refers to the distance the object moves over a given amount of time. However, **velocity** refers to the distance an object moves over a given amount of time *in a certain direction*. Velocity refers to both the speed *and* direction of a moving object. Two objects may move at the same speed, but if they are going different directions, they will have different velocities. For example, a car traveling north at 80 km/h (50 m/h) has a different velocity than a car traveling south at 80 km/h (50 m/h). The velocity of an object changes if the speed or the direction changes or if both the speed and direction change. A car traveling around a curve changes velocity even if its speedometer shows a constant speed.

Acceleration is a change in velocity during a period of time. The faster the velocity changes, the greater the acceleration. Usually we say that an object accelerates when it speeds up and decelerates when it slows down.

Fantastic FACTS

Think about all the ways that an aircraft can accelerate. In addition to the ways a car accelerates, an aircraft also moves up and down. Pilots need to maintain a certain amount of acceleration to keep a plane flying. Aircraft have instruments called accelerometers in them. Just like cars, airplanes need greater acceleration to turn than to fly straight. Accelerometers show a pilot whether he has adequate acceleration to perform a certain maneuver. Some cars also use the technology of accelerometers. When a sudden decrease in acceleration occurs, such as in a collision, an accelerometer activates the airbags that protect the occupants of the car.

However, in scientific language, acceleration occurs whenever an object speeds up, slows down, or changes direction.

Most people like having cars that can accelerate quickly. They want the car to go faster in a shorter amount of

Velocity refers to both speed and direction.

208

SCIENCE MISCONCEPTIONS

Velocity, acceleration, and *deceleration* each have a common meaning and a scientific meaning. Thinking of the common meanings may cause confusion when trying to learn the scientific meanings.

Velocity is commonly used interchangeably with *speed,* leaving out direction as a factor.

Acceleration is usually thought to mean only "to go faster," rather than meaning "a change of velocity," which includes a change of speed and/or direction.

Deceleration commonly means "to slow down," but it is really a negative change in velocity, or a change opposite of acceleration.

time. But actually, the most important acceleration of a car is how fast it stops. Good tires and brakes help decelerate a car in a safe amount of time and distance.

In order for acceleration to occur, a **force**, a push or pull, must be applied to an object. If you want to make a wagon move, you must push or pull it. If you want to make it turn, you must apply a force to the handle. If you want to make it stop, you must also push or pull it. If you do nothing to a moving wagon, it will eventually stop, because a force, friction, is working on the wagon. **Friction** is a force that keeps objects from moving against other objects.

Momentum

The mass and the velocity (speed and direction) of an object determine its **momentum**. This concept sounds complicated, but you undoubtedly have experienced the results of momentum. If someone throws a baseball and a tennis ball toward you at the same speed, which one would you try hardest to avoid? Probably the tennis ball would be less painful than the baseball. It has less momentum because it has less mass. For the same reason, you would try to avoid being hit by a baseball thrown by a professional baseball player. But you probably would not run too hard to avoid a baseball thrown by a little child. Although the ball's mass is the same, the velocity of the professional player's ball is much greater than that of the little child.

A train is an example of a moving object that has a lot of mass. Its great mass gives it great momentum. A train takes much longer to slow down than a car. Depending on its velocity, the train could take several kilometers to stop after it has applied its brakes.

1. What is energy?
2. How is velocity different from speed?
3. Why would a fast-moving train not be able to stop quickly?

209

The word *accelerometer* combines the terms *accelerate* and *meter*.

> **What does the word *meter* mean in this context?** {Elicit that it is a device for measuring.}

> **What other words use *meter* in this way?** {Possible answers: thermometer, barometer, speedometer}

> **Since the word *meter* has to do with measuring, what would be the definition of *accelerometer*?** {a device that measures acceleration}

Discussion:
pages 208–9

▸ What is *force*? {a push or a pull}

▸ Force is the main element in a game of tug of war. The game begins with the forces balanced and the rope not moving.

💡 How does each side exert force on the rope? {by pulling it}

💡 Is the force on each side usually equal? {no} How do we know this? {One side wins and the other side loses.}

💡 If I pull on a wagon of bricks and it moves, are the forces equal? {no} What has the greater force? {my pull}

💡 If the wagon remains stationary because I cannot move it, are the forces equal? {probably}

💡 If the wagon is on a hill and it starts to roll, pulling me along with it, are the forces equal? {no} What has the greater force? {the rolling wagon}

▸ What is *friction*? {a force that keeps objects from moving against other objects}

▸ What affects the momentum of a moving object? {the mass and velocity of the object}

💡 What is *mass*? {the quantity of matter in an object}

💡 What is the relationship between the mass of an object and its momentum? {The greater the mass is, the more momentum the object will have.}

▸ What are some examples that show the relationship between mass and momentum? {Possible answers: a moving train, a baseball, a tennis ball}

Answers

1. Energy is the ability to do work.
2. Speed is how fast an object is moving. Velocity is the speed of the object in a certain direction.
3. A fast-moving train would not be able to stop quickly because not only does its great mass give it great momentum, but the greater velocity also increases the momentum.

Activity Manual
Reinforcement—page 132

Lesson 95

Student Text pages 210–13
Activity Manual pages 133–34

Objectives

- Identify Newton's three laws of motion
- Explain that both gravity and friction work against inertia

Vocabulary

first law of motion
inertia
gravity
second law of motion
third law of motion

Introduction

Imagine that you are the passenger in a car with your dad driving. Suddenly, your dad notices that the car in front of him has stopped. Your dad slams on the brakes.

➤ How does your body respond? Does your body jolt forward, stay still, or pull backward? Show me. *{jolts forward}*

➤ How does your body respond when you are in a stopped vehicle and all of a sudden the driver stomps on the gas pedal? Show me. *{pushes backward}*

➤ You may not have ever stopped to wonder why your body responds this way. Today we will learn why.

Purpose for reading:
Student Text pages 210–13

➤ Is friction helpful or harmful?
➤ What is the relationship between the mass of an object and the force needed to move the object?

Discussion:
pages 210–11

➤ Were the observations of medieval scientists correct or incorrect? *{correct}*

📖 Why did they draw incorrect conclusions? *{Answers will vary. Elicit that man's ovservations are limited. Only God knows all things and is not limited by lack of wisdom and knowledge (Rom. 11:33–36).}*

➤ What is the first law of motion? *{An object tends to stay at the same velocity unless another force causes it to change.}*

Remind the student that acceleration is a change in velocity during a period of time.

Laws of Motion

Medieval scientists believed that all things tend to slow down and come to a rest. Though these scientists made correct observations, their conclusions were wrong. Galileo Galilei (GAL-uh LEE-oh GAL-uh-LAY) (1564–1642) discovered that, contrary to what the scientists of his day believed, a moving object does not come to rest unless an outside force acts on the object. As Sir Isaac Newton (1642–1727) continued Galileo's study of how and why things move, he formulated the three laws of motion.

First Law of Motion

The **first law of motion** says that an object tends to stay at the same velocity unless another force causes it to change. An object at rest, zero velocity, tends to stay at rest. An object traveling in a straight line at a constant speed tends to keep moving that way.

The resistance to a change in motion is called **inertia** (ih NUHR shuh). If you

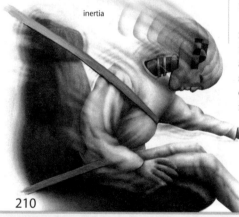

inertia

210

have ever been in a car that stopped suddenly, you experienced inertia. You were thrown forward in the car because your body was in motion, and it stayed in motion even after the car stopped. When you push someone on a swing, it is harder to get him started than it is to keep him going. You must first overcome his body's tendency to stay at rest.

Velocity involves not only speed but also direction. If you have gone around a sharp turn in a car, you know that your body wants to go straight even though the car turns. You feel like you are being pushed to one side. Perhaps you have swung an object on the end of a rope around and around. When you release the rope, the object goes in a straight line from the point that you let go. It does not continue circling.

Obviously, objects in the universe do not remain still or keep moving in straight lines. If you jump up to shoot a basketball, you will not continue floating up into the air. If you roll a toy car across the carpet, it will roll for a while and then come to a stop. Certain forces are operating against inertia.

One force that keeps an object from moving indefinitely in a straight line is **gravity**, the pull of one object on another. Objects with greater mass have stronger gravitational pulls. Since the earth has more mass than anything on it, it pulls objects toward itself. If you throw a softball through the air, gravity will pull the ball to the earth. At the same time, the ball pulls the earth toward

Direct a DEMONSTRATION
Demonstrate inertia

Materials: a clear plastic or glass cup, a coin, and a note card large enough to cover the cup opening

Place the note card on top of the cup, and place the coin on top of the note card. With your finger, flick the card off the cup. Observe that the coin falls into the cup instead of moving with the note card.

➤ Why did the coin fall into the cup rather than stay on the note card? *{The coin tended to stay at rest over the center of the cup, so it fell when the note card was moved.}*

Variation: Try a similar demonstration using plastic dishes on a tablecloth.

itself. But because the mass of the earth is so much greater than that of the ball, the pull of the ball on the earth is very, very tiny.

When we talk about the *mass* of an object, we are referring to the amount of matter in that object. However, when we talk about the *weight* of an object, we are referring to the gravitational force on that object. So weight is actually a measure of force. An object that weighs less has less gravitational force being applied to it. For example, you would weigh less on the Moon because the gravitational force pulling on you would be less than that on Earth.

Friction also works against inertia and is helpful in many ways. Without the force of friction you would not be able to walk. Your foot would not be able to grip the floor and push you forward. Without friction, turning over in bed or stopping your car would be impossible.

Sometimes people increase the friction of surfaces. Rubber cutouts stuck to bathtub floors increase friction and keep people from falling. Baseball players wear batting gloves to keep their hands from slipping on the bat. Rubber tires give more friction against the road and help keep a car from slipping on the road.

In other situations, friction is not helpful, and people try to reduce it. Winter sports such as sledding, skiing, and ice-skating are fun only when there is less friction. Snow skiers wax their skis to reduce friction and to allow the skis to glide quickly along the snow. People put lubricants, such as grease and oil, on machine parts to reduce friction on the parts that rub together in movement.

Fantastic FACTS

Air resistance is an example of helpful friction. If raindrops met no resistance, they would be falling faster than the speed of sound by the time they reached the earth. Rain showers would be painful indeed! Gravity pulls the drops earthward, and air resistance limits their maximum speed. The two forces soon balance each other, and the raindrops fall at a constant speed.

WRITING The student could use the information learned about momentum and inertia to write a persuasive letter encouraging others to wear seat belts.

Discussion:
pages 210–11

💡 **According to Newton's first law of motion, what would have to happen to an object in order for it to accelerate?** *{Answers will vary, but elicit that a force would have to be applied. Otherwise the object would continue at the same velocity (speed and direction).}*

▸ **What is *inertia*?** *{the resistance to a change in motion}*

▸ **Why is it important that you wear your seat belt in a moving car?** *{A seat belt will keep your body from moving forward if the car stops suddenly.}*

▸ **What are two forces that work against a moving object?** *{gravity and friction}*

💡 **Do you think gravity and friction work for or against an object that is starting to move? Why?** *{against; because the object tends to resist the movement}*

▸ **How is mass different from weight?** *{Mass is the amount of matter in an object. Weight is the gravitational force on an object.}*

▸ **Why would your mass on the Moon be the same as your mass on Earth but your weight different?** *{Your body would still have the same amount of matter on the Moon, but the gravitational pull on your body would be different.}*

▸ **Why do moving objects not keep moving?** *{Friction slows down a moving object and brings it to a stop.}*

💡 **Other than the ways listed in your book, what are some ways that friction is helpful?** *{Possible answers: opening a jar; climbing stairs; turning a screwdriver}*

💡 **Other than the ways listed in your book, where would we want to reduce friction?** *{Possible answers: on a bike chain, door hinge, pulley}*

💡 **Do you think a bowler would rather increase or reduce friction on a bowling lane?** *{reduce}* **Why?** *{A lack of friction allows the bowling ball to move quickly down the lane in the direction the bowler rolls it.}*

💡 **Why do work crews sometimes put sand on icy roads?** *{to increase friction}*

Refer the student to *Fantastic Facts*.

▸ **What force is air resistance?** *{friction}*

▸ **What other force affects how raindrops fall?** *{gravity}*

Discussion:
pages 212–13

➤ A formula is the mathematical sentence that shows how something works. What is the formula for the second law of motion? *{mass x acceleration = force}*

➤ Why does it take more force to pick up a full carton of milk than it takes to pick up an empty carton? *{The full carton has a greater mass.}* **Identify the principle that supports your answer.** *{the greater the mass of the object, the greater the force needed to move it}*

➤ Which usually travels faster—a baseball that is thrown or a baseball that is hit with a bat? *{the baseball that is hit with the bat}* **Why?** *{the greater the force, the greater the acceleration}*

➤ Jamal and his dad were target practicing with a bow and arrows. Jamal's arrows usually flew at 45 m/sec, and his dad's flew at 60 m/sec. Whose arrows had the greatest force exerted on them? *{his dad's}* **How do you know?** *{The arrows with the greater acceleration had the greater force.}*

➤ At Saturday's game the Tigers kicked a 45-yard field goal. The Bears kicked a 20-yard field goal. Whose ball probably had the greater force exerted on it? *{the Tigers' ball}* **How do you know?** *{It traveled farther.}*

➤ Your Mom wants to rearrange some furniture. Your job is to help move the bookcase, but you want to use as little force as possible. Will you move the bookcase with or without the books on the shelves? *{without the books}* **Why?** *{Less mass will require less force.}*

Discuss other possible examples of the second law of motion.

➤ What does the third law of motion say? *{When one object exerts a force on a second object, the second object reacts by exerting an equal force back on the first object.}*

➤ What is another name for Newton's third law of motion? *{the law of action and reaction}*

Second Law of Motion

Most of us recognize that moving a 50 kg (110 lb) box is easier than moving a 500 kg (1,102 lb) box. Newton's **second law of motion** states that force is equal to an object's mass and the acceleration of the object. The following formula illustrates the second law of motion.

$$mass \times acceleration = force \quad (ma = F)$$

Several principles result from this formula. First, *the greater the mass of the object, the greater the force needed to move it.* For example, a little bit of force can move an empty cardboard box, but moving a cardboard box filled with encyclopedias requires much more force. Moving a golf ball with a golf club is quite simple, but moving a bowling ball with a golf club would be quite difficult!

Another principle we can derive is that *the greater the force exerted on the object, the greater its acceleration will be.* Athletes use this principle often. The harder you swing a tennis racket or baseball bat, the faster the ball accelerates after you hit it.

We can also conclude from the formula that *the greater the acceleration, the greater the force exerted on an object.* This is why it takes more force to throw a fast baseball pitch than to throw a slow baseball pitch. The fast ball has more acceleration, so it requires more force.

◄— Force exerted

◄— Greater force exerted

The Second Law of Motion

212

Allow the student to practice using the formula for Newton's second law of motion: mass x acceleration = force. The following problems illustrate the formula and provide practice identifying the unknown part.

➤ If the mass of an object is 10 kg and the acceleration is 10 km/h, how many units of force were applied? *{10 kg x 10 km/h = f; f = 100}*

➤ If there are 20 units of force and the object weighs 10 kg, what is the acceleration? *{10 kg x a = 20; a = 2 km/h}*

➤ If the acceleration of an object is 10 km/h and 5 units of force were applied, what is the mass of the object? *{m x 10 km/h = 5 kg; m = 0.5 kg}*

Direct an ACTIVITY

Conduct an arm wrestling competition.

Observe various matches and discuss how force, mass, and acceleration affect the results.

Enrich the activity by charting and graphing the results.

Third Law of Motion

The **third law of motion** is sometimes called "the law of action and reaction." All forces come in pairs. When one object exerts a force on another object, the second object reacts by exerting an equal force back on the first object. For example, a person paddling in a kayak pushes the paddle backward against the water, and the water reacts by pushing the boat forward. The boat moves in the opposite direction of the force of the paddle. If you sit in a chair that has wheels and push against your desk, the desk pushes you back and sends you rolling backwards.

If you have ever ridden on bumper cars at an amusement park, you have felt Newton's third law of motion. As you bump into another car, the other car bumps you back with an equal amount of force.

Rockets ascend because of Newton's third law. As the rocket expels gases downward, an equal and opposite reaction occurs. The gases exert a force that causes the rocket to move upward.

QUICK CHECK
1. What is the difference between mass and weight?
2. What is Newton's second law of motion?
3. Why is the third law of motion sometimes called the "law of action and reaction"?

SCIENCE MISCONCEPTIONS

Cause and effect—Newton's third law of motion deals with the action and reaction of force. Sometimes examples of cause and effect not relating to force are incorrectly identified as action and reaction.

An example of action and reaction—Two marbles bounce away from each other when one is rolled into the other.

An example of cause and effect—You shout when your finger is hit with a hammer.

 In the Bible, Jesus gives examples of how God desires Christians to respond to the words and actions of others. Christ's reactions to harsh words, beatings, and ultimately death were filled with love and compassion for those mistreating Him. [BATs: 3c Emotional control; 5a Love; Bible Promise D. Identified in Christ]

➤ How do you react when someone wrongs you? {Answers will vary.}

Discussion:
pages 212–13

➤ Each competitor at the swim meet waits in the water at the end of his lane. When the starting gun sounds, each uses his feet to push off from the side of the pool. How does this demonstrate Newton's third law of motion? {The action of pushing against the wall produces the reaction of moving the swimmer forward.}

Remind the student of the discussion on balanced and unbalanced forces.

💡 What would happen if you exerted pressure on a chair by sitting in it and the chair did not exert an equal force back; instead, the chair exerted less force back? {The chair would collapse or break, and you would fall.}

💡 Can you think of other examples of Newton's third law in everyday life? {Possible answers: bouncing a ball; swinging on a swing set; rowing a boat; recoil from firing a gun}

📖 People also react to the world around them. Unlike the laws of motion, Christians can control their reactions through the help of the Holy Spirit. [BATs: 3c Emotional control; 6c Spirit-filled]

Answers

1. Mass is the amount of matter in an object. Weight refers to the amount of gravitational force on that object.
2. Newton's second law states that force is equal to an object's mass and the acceleration of the object.
3. Newton's third law is called the "law of action and reaction" because forces work in pairs. If an object exerts a force on another object, the second object reacts by exerting an equal force back on the first object.

Activity Manual

Study Guide—pages 133–34

These pages review Lessons 94 and 95.

Assessment

Test Packet—Quiz 9-A

The quiz may be given any time after completion of this lesson.

Lesson 96

Student Text pages 214–15
Activity Manual pages 135–36

Objectives

- Plan a demonstration to illustrate the laws of motion
- Experiment to show each of the laws of motion with toy cars
- Identify the laws of motion in real-life situations

Materials

- See Student Text page

Introduction

➤ Have you ever wondered how a policeman determines what happened at an accident scene? Today you will study and practice one of the ways a policeman can discern what happened.

Purpose for reading:
Student Text pages 214–15
Activity Manual pages 135–36

The student should read all of the pages before beginning the activity.

Discussion:
pages 214–15

➤ What do we call people who try to figure out what happened at an accident scene? *{accident reconstructionists}*

➤ What skills do they use? *{math and physics (mainly Newton's laws of motion)}*

➤ Physics is the branch of science that deals with matter and energy and their relationships to each other.

📖 God's laws of motion are consistent. Because the laws do not change, man can build cars and evaluate accident sites based on the laws that he knows will always be the same.

ACTIVITY: Mini Cars in Motion

Process Skills
- Experimenting
- Making and using models
- Observing
- Communicating

Have you ever passed a car accident and wondered how the cars ended up where they were? Eyewitnesses often help sort out the events of accidents, but what if there are no eyewitnesses? People called accident reconstructionists use skills such as mathematics and physics (mainly Newton's laws of motion) to try to figure out what happened at an accident scene. A few engineering firms even specialize in accident reconstruction. Most of these people use very advanced computers to recreate accident scenes.

In this activity you will use small objects, including toy cars, to demonstrate each of Newton's laws of motion. Then you will apply each law to a simulated accident to try to analyze what happened at the accident scene.

Problem

How can I use models to demonstrate Newton's laws of motion?

Materials:
two small toy cars
other items, such as books, pennies, rubber bands, blocks, ruler, spools of thread, and tape
Activity Manual

 If a student chooses to demonstrate Newton's first law by placing the toy car on the table and saying that inertia keeps it from moving, acknowledge that he is correct, but encourage a more active demonstration.

You may choose to limit the amount of time available for the student to produce demonstrations.

Although experimenting usually involves proving a hypothesis, in this activity the students will be experimenting to show the laws of motion.

224 Lesson 96

Procedure

1. Draw a plan for your demonstration of Newton's first law of motion. (Remember: You are not trying to demonstrate the accident scenario. You are trying to show how the law of motion works.)
2. Choose and list the materials you need for your demonstration.
3. Test your demonstration. Continue experimenting until the demonstration shows Newton's first law of motion adequately.
4. Read the first simulated accident scenario on page 136 of your Activity Manual. Complete the accident report by using your knowledge of Newton's first law of motion.
5. Repeat the procedure for the other laws of motion.
6. Be prepared to share your demonstration and conclusions.

Conclusions
- Which of the laws of motion was the easiest to demonstrate? Which was the hardest to demonstrate?

Follow-up
- Do additional research about accident reconstructionists.

SCIENCE PROCESS SKILLS

Experimenting—

- What was the purpose of this experiment? *{to show the laws of motion}*
- Do all the demonstrations for the first law of motion have to be the same in order to show the first law of motion? *{no}*
- Is it harder to conduct an experiment or to plan and set up an experiment? *{Answers will vary, but probably to plan and set up an experiment is harder.}*
- What part of an experiment is the hardest to set up? *{Answers will vary, but probably controlling the variables is hardest.}*

Procedure
Guide the student as he sets up each demonstration. Ask leading questions that will encourage the student to identify what he is trying to do. A real-life example may help him visualize what he needs to try.

If needed, use these possible solutions to demonstrate each law.

First law—Place a thin book on a table several inches away from the toy car. The spine of the book should face the car. Place a penny on top of the toy car. Roll the car into the book. The car will hit the book and stop moving, but the penny will continue moving. Although the car had an outside force acting on it (the book), the penny did not, so it continued to move even when the car stopped. Gravity eventually stopped the penny.

Second law—Increase the mass of one car by taping four pennies to the top. Line the two cars up side by side on a table. Use a ruler to quickly push both cars at the same time. The car without the penny weights should accelerate more. The same force acted on both cars, but the object with less mass was able to accelerate more.

Third law—While holding one car still with one hand, roll the second car into the first car. After hitting the first car, the second car will roll backward. The action of hitting the first car caused the reaction of the first car hitting the second one back.

Conclusions
Discuss the accident scenarios. The student's answers may vary. Allow time for the student to explain the reasons for the answers he chose.

Activity Manual
Activity—pages 135–36

Assessment
Rubric—TE pages A50–A53
Select one of the prepared rubrics, or design a rubric to include your chosen criteria.

Lesson 97

Student Text page 216
Activity Manual pages 137–38

Explorations: Roller Coaster

Objectives
- Design and make a model roller coaster
- Discover relationships between slope, speed, and momentum

Materials
- photos, pamphlets, or books about roller coasters
- See Student Text page

Introduction
Discuss roller coasters the students may have seen or ridden. Discuss the photos or other materials to see the variety of roller coasters and their use of hills and loops.

Purpose for reading:
Student Text page 216
Activity Manual pages 137–38

➤ What are some forces that cause a car to roll along a roller coaster? *{gravity, inertia, and momentum}*

The student should read all the pages before beginning the exploration.

Procedure
If the student has difficulty making a successful model, ask the following questions.

Problem with the "car" hitting a dead space and slowing or stopping:
➤ Is the slope steep enough to keep the "car" moving?
➤ Is the slope too steep, causing the "car" to drop and stop rather than roll?

Problem with the "car" making it through a loop:
➤ Is enough distance given along the slope for the "car" to gain enough momentum and speed?
➤ Is the slope steep enough or too steep for the "car" to gain enough momentum and speed?

Problem with the "car" making it through several loops in a row:
➤ Should the size of each loop increase or decrease along the ride?
➤ If the loops stay the same size, is more distance needed between them for the "car" to gain momentum and speed before entering the next loop?

Activity Manual

Many roller coasters are called gravity coasters because once the car is at the top of the first hill, the main force causing it to roll is gravity. The height and placement of hills, loops, and turns make use of inertia and momentum that cause the car to continue to the end of the ride. In this activity you will use gravity, inertia, and momentum to cause your BB "car" to complete your "roller coaster."

What to do
Plan your roller coaster with the following requirements.

- The start must be 1 meter above the finish.
- The roller coaster must contain at least 1 loop.
- A plastic bag must be attached to the end of the roller coaster to catch the BBs.

1. In your Activity Manual, draw and label a diagram of your idea for a "roller coaster."
2. List the materials you will use.
3. Make your roller coaster by taping the tubing to the wall according to your diagram.
4. Test your roller coaster with a BB "car."
5. Record the results. Adjust your roller coaster as needed to get the BB to the end of the ride and into the plastic bag.
6. When you have made the roller coaster work as you want it to, draw a diagram of the completed roller coaster.

Materials:
10 ft of plastic tubing
tape
plastic bag
BBs
Activity Manual

Cedar Point, Ohio

216

Explorations—pages 137–38

Assessment
Rubric—TE pages A50–A53
Select one of the prepared rubrics, or design a rubric to include your chosen criteria.

Teacher Helps

This activity is designed to have the tubing attached to a smooth surface, such as a wall or chalkboard. You may choose to have the student hold the tubing or use other means of support.

Once successful models are made, you may choose to have the student time the "cars." Compare results and draw conclusions as to the reasons for slower and faster times.

Discuss the places along the roller coaster where gravity and momentum give the "car" the greatest speed. Decide where gravity and friction slow the "car."

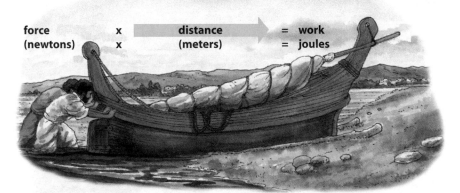

force (newtons) × distance (meters) = work (joules)

"Let him labour, working with his hands the thing which is good." Ephesians 4:28

Machines

Work

Suppose you push as hard as you can on a brick wall. Have you done any work? You might be sweating and exhausted, but you have not really done any work unless you have moved the wall. **Work** is defined as a force acting on something as it moves a certain distance. Applying a force to an object is not enough. You must make the object move.

> force (newtons) × distance (meters) = work (joules)

Force is measured in **newtons** (N). Suppose we say that an object weighs 3 newtons. We are saying that 3 newtons of gravitational force are being exerted on the object. So newtons are a measure of both force and weight. A spring scale can measure an object's force in newtons. If you multiply the object's force with the distance the object travels,

you will determine the amount of work done.

The unit used to measure work is called the **joule** (JOOL). If you lift an object that weighs 100 newtons upward 5 meters, you have done 500 joules, or newton-meters, of work. But pushing a box against 20 newtons of friction over a distance of 25 meters also equals 500 joules of work. Because work equals force times distance, applying a greater force over a shorter distance involves the same amount of work as applying a lesser force for a greater distance.

spring scale

> 100 newtons × 5 meters = 500 joules

> 20 newtons × 25 meters = 500 joules

217

SCIENCE BACKGROUND

Weight and force—The words *weight* and *force* can be used interchangeably when measuring in newtons.

Joules—When multiplying *newtons* times *meters* to find the amount of *work*, the answer is in *newton-meters*, or *joules*.

 Read II Thessalonians 3:10. What does the Bible say about work? *{If you do not work, you should not eat.}*

Christians are to be busy doing God's work with enthusiastic attitudes. [BATs: 2e Work; 2f Enthusiasm]

Student Text pages 217–19
Activity Manual pages 139–40

Lesson 98

Objectives

- Explain that *work* equals *force* times *distance*
- Differentiate among the three classes of levers

Materials

- ruler
- chalkboard eraser
- cardboard tube or other object to use as a fulcrum

Vocabulary

work
newton
joule
machine
effort force
resistance force

lever
fulcrum
first-class lever
second-class lever
third-class lever

Introduction

▶ How hard do you think you are working while reading your textbook? *{Accept any answer.}*

▶ Scientifically speaking, you are not doing any work when you read your textbook. Today we will learn what work is and what it is not.

Purpose for reading:
Student Text pages 217–19

▶ Why is playing basketball considered work but doing homework is not?

▶ What is a *fulcrum*?

Discussion:
page 217

▶ What is the definition of *work?* *{a force acting on something and moving it a distance}*

▶ What is the formula for determining how much work is done? *{force × distance = work, or newtons × meters = joules}*

▶ Which unit measures force? *{newton}*

▶ Which unit measures work? *{joule}*

▶ How much work is done if you lift a box that weighs 4 newtons from the floor onto a shelf that is 3 feet off the ground? *{12 joules}*

Chapter 9

Lesson 98 227

Discussion:
pages 218–19

▸ What is a *machine*? {*any object that makes work easier*}

▸ How does a simple machine make work easier? {*It strengthens the force used to do the work or changes the direction of the applied force.*}

💡 If it takes 10 newtons of force to lift 1000 grams without a simple machine, would it take more or less force to lift 1000 grams with a simple machine? Why? {*Elicit that it should take less force, because a simple machine strengthens the force. A simple machine makes it possible to do more work using less force.*}

▸ Name and explain the three parts of a lever. {*The effort is the force applied to the lever. The resistance is the force on the lever that acts against the effort. The fulcrum is the place on the lever where the lever pivots.*}

▸ How are levers classified? {*Scientists classify levers into three categories, depending on where the effort and resistance are located in relation to the fulcrum.*}

▸ What are the classifications of levers? {*first-class, second-class, and third-class*}

▸ Where is the fulcrum on a first-class lever? {*It is between the effort and the resistance.*}

Use the ruler, eraser, and cardboard tube to demonstrate how a first-class lever works. Direct the student to identify the fulcrum, effort, and resistance.

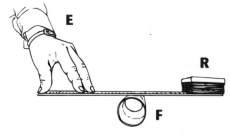

▸ Name examples of first-class levers. {*Possible answers: seesaw, balance scale, crowbar, boat oar, scissors, head movement, some automobile jacks, nail clippers, pry bar*}

📖 Refer the student to *Creation Corner*.

▸ Who originally designed levers? {*God*}

▸ The joints and muscles in our bodies are so complex that often the same joint can function as different classes of levers, depending on which muscles are providing the effort.

▸ Why are the levers in our bodies important? {*Without the levers, we could not do many of our daily tasks.*}

Creation CORNER

Man has used mechanical levers for thousands of years. But God created the original design. Many of the joints in the human body are levers. Your head pivots on a first-class lever. If you drop your head down, the resistance is your head, the fulcrum is the top of your spine, and the muscles in your back provide the effort. Your toe joints provide the fulcrums for second-class levers. The resistance is your weight, and the effort is your leg muscles. Your elbow is an example of a third-class lever. The joint is the fulcrum, the lower arm muscles provide the effort, and whatever you carry or throw is the resistance. It takes only a small movement at the joint to give a great force. Without these God-given levers, our bodies would not be able to perform many of the tasks that they normally perform.

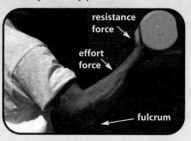

Simple Machines

A **machine** is any object that makes work easier. Simple machines do not reduce the amount of work done. However, a simple machine makes the work easier by strengthening the force used to do the work or by changing the direction of the applied force.

Remember, a force is a push or a pull. The force applied to a simple machine is called the **effort force**, or effort. The **resistance force**, or resistance, is the force that works against the effort. Sometimes the resistance force is an actual push or pull, but often it is the weight of the object being moved.

There are six simple machines: lever, pulley, wheel and axle, inclined plane, wedge, and screw. All machines, even complex ones, are usually made of one or more of these simple machines.

218

Lever

A **lever** is any bar that turns on a point, such as a seesaw, a wheelbarrow, or a broom. The spot where the bar turns, or pivots, is called the **fulcrum.**

We use levers to lift objects that are otherwise too heavy for us. Archimedes, a mathematician who lived more than two hundred years before Christ, was one of the first men to study simple machines. He once said that if he had a long enough lever and a fixed fulcrum, he could move the earth. Of course, Archimedes could never prove his statement, but he understood that levers make work easier.

Scientists classify levers into three categories, depending on where the effort and resistance are located in relation to the fulcrum. On a **first-class lever** the fulcrum is located between the effort and the resistance. Seesaws

 You may want the student to do his own demonstration of each type of lever.

 Words such as *work* and *machine* have scientific definitions as well as common use definitions. Direct the students to come up with different sentences using either meaning of each word, and call on another student to identify whether the word is used in its scientific context or in its common usage context.

228 Lesson 98

First-class lever

and crowbars are both first-class levers. If you use a paint can opener, you are using a first-class lever. The fulcrum is the rim of the can. The lid provides the resistance force, and your pushing down on the other end produces the effort force. A pair of scissors is an example of two first-class levers working together.

A **second-class lever** has the resistance between the effort and the fulcrum. Some examples of second-class levers are a door and a wheelbarrow. Almost everything that has a hinge is a second-class lever. An example of two second-class levers working together is a nutcracker.

In a **third-class lever**, the effort is between the resistance and the fulcrum. In these levers, a little movement by the effort gives greater movement at the resistance. Some examples of third-class levers are a broom and a fishing pole. You can move the end of a broom a little, but at the floor the broom moves much more than the little that you moved it. An example of two third-class levers working together is a pair of tweezers.

Third-class lever

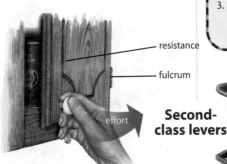

Second-class levers

QUICK CHECK
1. What is work?
2. What is the function of a machine?
3. Describe the location of the effort, resistance, and fulcrum of each class of lever.

Direct a DEMONSTRATION
Test various levers
Have several examples of levers available for the students to see and touch. Levers you could use are a broom, boat oar, scissors, crowbar, nutcracker, fishing pole, or tweezers. Direct the student to demonstrate the movement of each lever, identify its effort, fulcrum, and resistance, and then classify it.

Discussion:
pages 218–19

➤ **How are the parts of a second-class lever arranged?** {The resistance is between the fulcrum and the effort.}

Use the ruler and eraser to demonstrate how a second-class lever works. Direct the student to identify the fulcrum, effort, and resistance. Emphasize that the fulcrum is at one end and the resistance is in the middle.

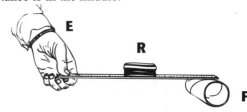

💡 **Name some examples of second-class levers.** {Possible answers: wheelbarrow, door on a hinge, lower leg movement when standing on toes, other types of hinges, paper cutter, garlic press, bottle opener, pocket knife}

➤ **How are the parts of a third-class lever arranged?** {The effort is between the fulcrum and the resistance.}

Use the ruler and eraser to demonstrate how a third-class lever works. Direct the student to identify the fulcrum, effort, and resistance.

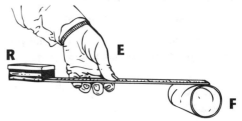

💡 **Name some examples of third-class levers.** {Possible answers: tweezers, broom or most long-handled tools, fishing pole, most joints in the body}

Answers
1. a force acting on something as it moves a certain distance
2. to make work easier
3. A first-class lever has the fulcrum between the effort and the resistance. A second-class lever has the resistance between the effort and the fulcrum. A third-class lever has the effort between the fulcrum and the resistance.

Activity Manual
Reinforcement—pages 139–40

Lesson 99

Student Text pages 220–23
Activity Manual pages 141–44

Objectives

- Describe a pulley, wheel and axle, inclined plane, wedge, and screw
- Discern between a fixed pulley, a moveable pulley, and a block and tackle
- Explain what a compound machine is

Vocabulary

pulley
fixed pulley
moveable pulley
block and tackle
mechanical advantage
wheel and axle
inclined plane
wedge
screw
threads
compound machine

Materials

- simple machines, such as a pulley, can opener, wedge, screw, and screwdriver

Introduction

Display and identify the types of simple machines and their uses.

➤ Today you will learn how each of these objects makes work easier.

Purpose for reading:
Student Text pages 220–23

➤ What is *mechanical advantage*?

➤ How are compound machines similar but different from simple machines?

Discussion:
pages 220-21

➤ What is a *pulley*? {a grooved wheel with a chain or rope wrapped in the groove}

➤ How does a pulley make work easier? {by changing the direction of the force or by reducing the amount of work needed}

➤ What are two kinds of pulleys? {fixed pulley and moveable pulley}

💡 Look up the word *fixed* in the dictionary and read aloud the definition that applies to a fixed pulley. {nonmoving}

➤ Which type of pulley changes the direction of the force without reducing the amount of work? {fixed pulley}

➤ Which type of pulley reduces the amount of force needed to move an object or produce a gain in force? {moveable pulley}

Pulley

A **pulley** is a grooved wheel with a chain or rope wrapped in the groove. Pulleys are used to raise and lower things. To pull something up, you pull down on the rope. If you want to lower something, you release the rope. Pulleys make work easier by changing the direction of the force or by reducing the amount of force needed to move an object.

A **fixed pulley** is attached to something, so it does not move. It makes work easier by changing the direction of the force but does not reduce the amount of force needed to move the object. Flagpoles use fixed pulleys.

A **moveable pulley** moves with the load or resistance. It produces a gain in force but does not change the direction of the force.

A **block and tackle** combines multiple fixed and moveable pulleys. The fixed pulleys change the direction of the force, while the moveable pulleys produce a gain in force. People use block and tackle pulleys to raise sails on sailboats. Big construction cranes often use a block and tackle to lift heavy pieces of equipment.

Multiple fixed and moveable pulleys combined provide greater mechanical advantage. **Mechanical advantage** is the decrease in effort that is needed to move an object. All simple machines give some mechanical advantage. However, remember our definition of work. When you apply less force, you must apply it

Fixed pulley

Direction of the force changed but not the amount of force needed to lift the barrel

Fixed and moveable pulleys

Moveable pulley moves up as force is applied

Produces a gain in force, making it easier to lift the barrel

Block and tackle

Combination of two fixed pulleys

1 moveable pulley

Combination of multiple fixed and moveable pulleys, which reduce the force needed to move an object

SCIENCE BACKGROUND

Further information about turbines is in Chapter 8, *Electricity & Magnetism*.

SCIENCE MISCONCEPTIONS

Be sure that the student understands that simple machines make work easier but do not reduce the amount of work. Work is a result of force and distance. Simple machines can reduce the amount of force required by moving the object through a longer distance, or they can reduce the distance as more force is applied.

Moveable pulleys are seldom used alone, but the fixed pulley that is usually used with a moveable pulley is not necessary to produce an increase in force. The fixed pulley makes the work easier only because pulling down is easier than pulling up.

TEACHER HELPS

A pulley can be made from an empty thread spool and a wire clothes hanger. Cut and bend the hanger.

Insert the ends of the hanger into the spool.

through a greater distance to get the same amount of work. Although a block and tackle requires less effort than a simple fixed pulley does, it takes a much longer piece of rope to lift a load.

Wheel and axle

A **wheel and axle** consists of a wheel and a rod, or axle, running through the wheel. Sometimes we apply force to the axle to make the wheel move a greater distance. This is how the wheels on a car work.

At other times people turn the wheel to make the smaller axle turn with greater force. Have you ever tried unscrewing a screw holding just the shaft rather than the handle of a screwdriver? You can get very little force and probably cannot move the screw very much. But when you turn the handle (the wheel), you can get greater force on the shaft (the axle).

Think about a turbine. A turbine is another example of a wheel and axle. Water, wind, or steam turns the "wheel" that turns the axle. This turning rotates a magnet inside a coil of wire to generate electricity.

A *gear* is a type of wheel and axle. A gear has toothlike projections around the wheel. Some gears have teeth that interlock with the teeth on another gear. When one gear turns, it moves the other gears that it touches in the opposite direction. However, on a bicycle the gears are connected with a chain, and both move the same direction. Gears are used in many mechanical devices. People use them in simple things, such as can openers, and also in things as complex as automobiles.

Types of Gears

Discussion:
pages 220-21

- What is a *block and tackle*? {a combination of moveable and fixed pulleys}
- What is *mechanical advantage*? {a decrease in effort needed to move an object}
- Why does a block and tackle require a longer piece of rope in order to use less force? {Answers will vary, but elicit that work is equal to force times distance, so less force requires a greater distance for the same work.}

Remind the student of the formula for work, force × distance = work.

Discuss the pictures on Student Text page 220.

- What is a *wheel and axle*? {a wheel with a rod, or axle, running through it}
- Why do large items such as refrigerators and pianos have wheels? {to make moving them easier}
- What other simple machine is a type of wheel and axle? {a pulley}
- What are two ways work is done using a wheel and axle? {One way is to apply force to the axle to make the wheel move a greater distance. Another way is to turn the wheel to make the axle turn with greater force.}
- What part of a wheel and axle is a screwdriver handle? {the wheel}
- What is a *gear*? {a wheel and axle that has toothlike projections around the wheel}

Direct a DEMONSTRATION

Compare the amount of work done with pulleys

Materials: book, spring scale, pulley, string or thin rope, place to attach pulley

Direct a student to lift the book 30 cm using the spring scale. Record the measurement.

Set up a fixed pulley as illustrated. Direct a student to lift the book 30 cm. Record the measurement.

Set up moveable and fixed pulleys as illustrated. Direct a student to lift the book 30 cm. Record the measurement. Compare the measurements.

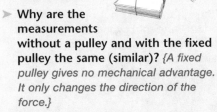

- Why are the measurements without a pulley and with the fixed pulley the same (similar)? {A fixed pulley gives no mechanical advantage. It only changes the direction of the force.}
- Was more or less work used with the moveable and fixed pulleys working together? {less}

Chapter 9

Lesson 99 231

Discussion:
pages 222–23

▸ What is an *inclined plane?* {*a flat, slanted surface, such as a ramp*}

▸ Is more force required to push an object up a steep slope or push the same object up a gradual slope? {*up a steep slope*}

▸ For an inclined plane, what is the relationship between the steepness of the slope, the distance of the plane, and the amount of work done? {*A more gradual slope requires less force to move an object, thus making the work easier; however, it has a longer distance for the object to travel.*}

 What do we call the decrease in effort needed to move an object? {*mechanical advantage*}

Refer the student to the pictures at the top of Student Text page 222.

▸ Which action requires the most force? {*lifting straight up*}

▸ Which action requires the least force? {*pushing up the longest ramp*}

Point out that in each picture the mass of the object being moved is the same, but as the distance increases, the amount of effort decreases.

▸ What is a *wedge?* {*two inclined planes back to back*}

▸ What is the purpose of a wedge? {*It splits or lifts objects.*}

 Sin can act like a wedge that splits or separates. Sin separates a non-believer from God and salvation. Unconfessed sin can separate a Christian from fellowship with God and fellowship with other believers. [BAT: 1b Repentance and faith; Bible Promise A. Liberty from Sin]

▸ Which force helps a wedge do work? {*friction*} Why? {*It keeps the wedge from sliding.*}

▸ Give examples of common types of wedges. {*Possible answers: knives, hatchets, axes, nails*}

▸ What activities would be impossible without a wedge? {*Possible answers: cutting meat with a knife; digging in the dirt with a shovel to plant flowers*}

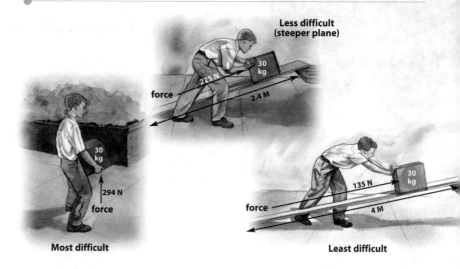

Inclined plane

An **inclined plane** is a flat, slanted surface, such as a ramp. This type of simple machine makes moving an object up a distance easier than lifting the object straight up. The mechanical advantage of an inclined plane is based on its slope and the length of the inclined plane. A more gradual slope requires less force to move the object, thus making the work easier. A more gradual slope, however, has a longer distance for the object to travel.

Wedge

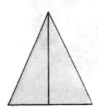

A **wedge** is two inclined planes placed back-to-back. Knife blades, axes, hatchets, and even the end of a nail are all wedges. A wedge splits or lifts objects. Long, thin wedges cut and split better than short, fat ones. This is why a sharpened blade works better than a dull blade.

Friction keeps wedges in place. Without friction a wedge would slide off the object that it is trying to split or lift.

Sometimes people pound wedges under objects to separate the objects from their surroundings. A wedge may be used to lift a heavy object just a little so that a lever can be used to move it farther.

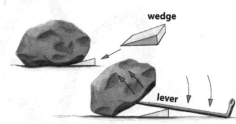

222

 Many historians believe that the Egyptians used inclined planes to help in the building of the pyramids. A ramp wound around the outside perimeter of a pyramid would have allowed huge stones to be pushed up the incline. Historians also think that levers may have been used to raise the huge obelisks in Egypt.

Although there may not be documentation that ancient Egyptians actually used these simple machines in these ways, there seems little doubt that ancient civilizations knew how to use simple machines to build enormous structures.

Screw

A **screw** is an inclined plane wound around a cylinder or a cone. The ridges in the screw are called **threads.** As these threads are turned, they cut into a material and hold the screw firmly in place. Pulling a screw out of a piece of wood without unwinding the screw is very difficult. Screws depend on friction to stay in place. Without friction, you could remove a screw easily, since it would not grip the material it was screwed into.

You have probably seen screws holding together wood or metal pieces many times. Another example of a screw is the top of a jar. The lid also has a screw, and the two parts fit together to close the jar. Some winding mountain roads are also types of screws. Instead of going straight up a mountain, the road has a slight incline that goes around and around. It is easier to travel than a sharp incline, but it takes a lot longer.

Compound Machines

A **compound machine** combines two or more simple machines to make work even easier. Scissors are an example of a compound machine. The blades are two levers working together, and the cutting edges are wedges. A screw and screwdriver working together make another compound machine. The screw itself is a simple machine, and the screwdriver is a wheel and axle. Most machines we use daily are made up of at least two simple machines.

1. In what two ways can pulleys make work easier?
2. What is mechanical advantage?
3. How is a wedge different from an inclined plane?

Direct the student to choose an ancient civilization (Greece, Rome, Aztec, Maya, Inca, Babylon, etc.) and research how simple machines may have been used to construct some of their monuments.

Discussion:
pages 222–23

▸ **What is a *screw*?** *{an inclined plane wound around a cylinder or a cone}*

▸ **What is another name for the ridges in a screw?** *{threads}*

▸ **What is the purpose of these threads?** *{They cut into a material and hold the screw firmly in place.}*

▸ **How are some winding mountain roads types of screws?** *{They are slight inclined planes that go around the mountain.}*

▸ **Why would engineers design roads this way?** *{because it is easier to travel on a slight incline on a winding mountain road than on a steep incline}*

▸ **What is a *compound machine*?** *{A machine that combines two or more simple machines.}*

▸ **What are some examples of compound machines? Identify the simple machines in each compound machine.** *{Possible answers: scissors: levers and wedges; screw and screwdriver: screw and wheel and axle; wheelbarrow: levers and wheel and axle}*

Answers

1. They can change the direction of the force or reduce the amount of work needed.
2. the decrease in effort that is needed to move an object
3. A wedge is two inclined planes placed back to back. An inclined plane is a single, flat, slanted surface.

Activity Manual
Reinforcement—pages 141–42

Study Guide—pages 143–44
These pages review Lessons 98 and 99.

Assessment
Test Packet—Quiz 9-B
The quiz may be given any time after completion of this lesson.

Lesson 100

Student Text pages 224–25
Activity Manual pages 145–46

Objectives

- Experiment to show that an inclined plane reduces the amount of force needed to do work
- Measure metrically in newtons and centimeters
- Define operationally the results of the activity

Materials

- See Student Text page

Introduction

➤ What is *work*? {a force acting on something and moving it a distance}

➤ What is the formula for work? {force x distance = work}

➤ Today we will be measuring and calculating the work required to move an object a distance.

Purpose for reading:
Student Text pages 224–25
Activity Manual pages 145–46

The student should read all the pages before beginning the activity.

ACTIVITY: How Much Force?

Process Skills
- Measuring
- Observing
- Defining operationally
- Recording data

Work is a force acting on something and moving it a distance. We calculate the amount of work by multiplying the amount of force on the object and the distance the object is moved.

Force *(newtons)* × **Distance** *(meters)* = **Work** *(joules)*

In this activity you will calculate the amount of work needed to move an object along an inclined plane. The force will be measured using a spring scale. You will measure distance in centimeters and convert the measurements to meters before calculating the amount of work done.

Problem
How can an inclined plane reduce the amount of force needed to do work?

Procedure

1. Measure from the end of the board. Place pieces of tape at 20 cm, 40 cm, and 60 cm from the end of the board. Place the stack of books (15 cm high) under the board at the place marked 60 cm. Make sure the books are straight. The board will be acting as an inclined plane.

2. Attach the string to the object being used as the weight if needed.

3. With the spring scale attached to the object or string, place the bottom of the object at the bottom of the board. Slowly and steadily pull the object up the board until the bottom of the object passes the 60 cm mark. As you pull the object up the inclined plane, read the spring scale to see how many newtons of force are being exerted. Record this measurement in the *Force in newtons* column on the chart. Also record the 60 cm in the *Distance in centimeters* column on the chart.

224

Materials:
spring scale (newton)
object, such as a book or bag of candy
string 1 meter long
masking tape
meter stick
stack of books 15 cm high
board approximately 30 cm × 80 cm
calculator (optional)
Activity Manual

Teacher Helps

If your spring scale measures in grams rather than newtons, convert the grams to newtons. One newton equals about 100 grams.

If the students are working in groups, you may want to assign each person in the group a particular job. This will keep all students on task during the activity. All students should observe each step and work together cooperatively in the group. Here are suggested job assignments for a group of four students.

Job 1: Measure and mark the board with tape.

Job 2: Stack the books and adjust the board each time.

Job 3: Pull the spring scale and observe the newton measurements.

Job 4: Record the measurements. Do the calculator calculations.

4. Move the stack of books to the 40 cm mark. Repeat step 3.

 5. Repeat step 3, moving the stack of books to 20 cm.

 6. Convert the number in the *Distance in centimeters* column to meters and record the answer in the *Distance in meters* column.

 7. Multiply the *Force in newtons* by the *Distance in meters* to calculate the amount of work done. Your answer will be in newton-meters (N•m), or joules.

Conclusions
- Why was the amount of work done at each distance about the same?
- At which distance was the most force required to do the work?

Follow-up
- Set up an experiment to show how much work is done rather than how much force is needed to do the work.
- Change variables such as the length of the board or height of the books.

Science Process Skills

Defining operationally—In order to define operationally, the student should use terms correctly to explain his observations or conclusions in his own words.

▶ **What was the purpose of the activity?** *{to discover whether an inclined plane reduced the amount of force needed to do work}*

▶ **What did you discover?** *{Answers will vary but should include the idea that less force was needed for the object to move a greater distance.}*

Help the student explain his results in his own terms. The key is that he understands what is happening and is able to use the scientific terms accurately.

Procedure
Explain the importance of pulling the object carefully. If the movement is not slow and steady, the scale may give an incorrect reading.

Emphasize that the bottom of the object should start at the bottom of the board. The reading is taken when the bottom of the object crosses the designated mark. To measure accurately, establish set conditions and expect them to be followed carefully.

Conclusions
Provide time for the student to evaluate his results and answer the questions.

Activity Manual
Activity—page 145

Bible Integration—page 146
This page looks at some machines used in Bible times.

Assessment
Rubric—TE pages A50–A53
Select one of the prepared rubrics, or design a rubric to include your chosen criteria.

Lesson 101

Student Text page 226
Activity Manual pages 147–50

Objectives

- Recall concepts and terms from Chapter 9
- Apply knowledge to everyday situations

Introduction

Material for the Chapter 9 Test will be taken from Activity Manual pages 133–34, 143–44, and 147–48. You may review any or all of the material during the lesson.

You may choose to review Chapter 9 by playing "Moving Along" or a game from the *Game Bank* on TE pages A56–A57.

Diving Deep into Science

Questions similar to these may appear on the test.

Answer the Questions

Information relating to the questions can be found on the following Student Text pages:

1. page 217
2. page 212
3. page 209

Solve the Problem

In order to solve the problem, the student must apply material he has learned. The answer for this Solve the Problem is based on the material on Student Text pages 218–23. The student should attempt the problem independently. Answers will vary and may be discussed.

Activity Manual

Study Guide—pages 147–148

These pages review some of the information in Chapter 9.

Expansion—pages 149–50

Students examine and design imaginative inventions.

Assessment

Test Packet—Chapter 9 Test

The introductory lesson for Chapter 10 has been shortened so that it may be taught following the Chapter 9 test.

Diving Deep into Science

Answer the Questions

1. If you stand holding a 10 kg bag of grass seed, how much work are you doing? Explain your answer.
   You are not doing any work unless you move the bag of grass seed a distance. Work requires a force and a distance.

2. Tornadoes sometimes have winds greater than 500 km/h. Occasionally, after a storm people find straw stuck in trees like arrows. Explain how such a fragile object as straw could have enough force to stick in a tree?
   Force is equal to mass multiplied by acceleration. Although the straw has very little mass, it would have tremendous acceleration due to the wind. Therefore, it would have great force.

3. Train A has three engines and is pulling one hundred full boxcars. Train B has two engines and is pulling fifty empty boxcars. Both trains are travelling at 80 km/h. Which train will take longer to stop and why?
   Train A will take longer to stop because it has greater momentum.

Solve the Problem

Your family just bought a new washing machine. When you bring it home, you suddenly remember that the washing machine will need to be lifted up three steps. What simple machines might you use to help you move the washing machine into your house?
Answers will vary: inclined planes, pulleys, wheels, levers.

226

Review Game
Moving Along

Prepare 25 index cards by labeling 5 cards *Friction*, 5 cards *Gravity*, 5 cards *Force-1*, 5 cards *Force-2*, and 5 cards *Force-3*. Display a number line numbered from 0 to 15. Divide the class into two teams. Shuffle the cards and assign each team a symbol or marker to use with the number line.

Each time a team answers a question correctly, a member of the team draws one of the cards. If a type of force card is drawn, the team moves their marker forward 1, 2, or 3 spaces as indicated by the number on the card. If a gravity or friction card is drawn, the team's inertia is stopped and the team does not move that turn. The team cannot move again until they draw a force card to overcome their zero velocity. The first team to reach 15 wins.

236 Lesson 101

Student Text pages 227–29
Activity Manual page 151

Lesson 102

Objectives

- Demonstrate knowledge of concepts taught in Chapter 9
- Understand the interrelationship of science concepts
- Recognize God's miraculous control over nature

Unit Introduction

Note: This introductory lesson has been shortened so that it may be taught following the Chapter 9 test.

The purpose of this unit is to showcase God's marvelous universe. It is not intended to be an exhaustive study, but to provide basic knowledge that will be further developed in secondary science.

Both chapters emphasize the importance of gravity on celestial bodies and the vast distances in the universe.

Chapter 10 discusses stars and star groups as well as asteroids, meteoroids, and comets. It explains some of the ways that man learns about the universe.

Chapter 11 narrows the focus of the universe to the solar system. The most important star to us, the Sun, is discussed, as are some of the distinctive features of Earth, the Moon, and other planets. The chapter concludes with a lesson on how man learns about the solar system.

➤ **Look through Unit 4. What kinds of topics do you think you will be studying?** *{Possible answers: stars, telescopes, planets, space exploration, Earth, the Moon}*

➤ **How do the chapters relate to each other?** *{Answers will vary, but elicit that both are mainly about astronomy.}*

Beyond Our Earth

Additional information is available on the BJU Press website.
www.bjup.com/resources

Project Idea

The project idea presented at the beginning of each unit is designed to incorporate elements of each chapter as well as information gathered from other resources. You may choose to use the project as a culminating activity at the end of the unit or as an ongoing activity while the chapters are taught.

Unit 4—Moon Base

The student should design a Moon base. After choosing what scientific function the base will have, he needs to plan transportation (both to and from Earth, as well as on the Moon), food, and shelter needs. The student should consider gravity, threat of meteorites, lack of atmosphere, and other features of the Moon as he plans his Moon base.

Technology Lesson

A correlated lesson for Unit 4 is provided on TE pages A10–A11. This lesson may be taught with this unit or with the other Technology Lessons at the end of the course.

Bulletin Board

Note: This suggested bulletin board idea may be modified for use in a learning center.

Divide the bulletin board into five timeline columns. Write the years 1960, 1970, 1980, 1990, and 2000 at the top of the columns. The student should collect pictures and articles about space exploration. Place the pictures and articles underneath the correct decade. If the student is having trouble finding space information, refer him to resources such as the NASA Internet site.

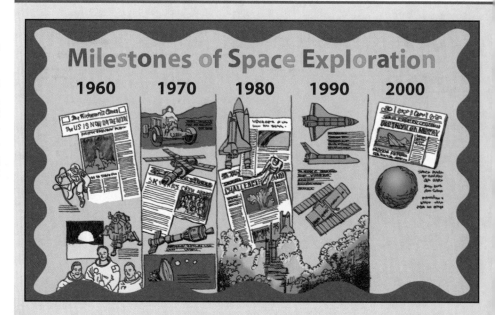

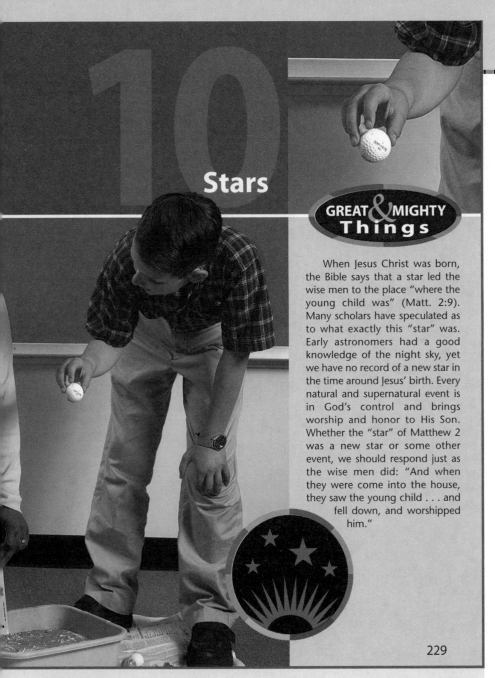

10 Stars

GREAT & MIGHTY Things

When Jesus Christ was born, the Bible says that a star led the wise men to the place "where the young child was" (Matt. 2:9). Many scholars have speculated as to what exactly this "star" was. Early astronomers had a good knowledge of the night sky, yet we have no record of a new star in the time around Jesus' birth. Every natural and supernatural event is in God's control and brings worship and honor to His Son. Whether the "star" of Matthew 2 was a new star or some other event, we should respond just as the wise men did: "And when they were come into the house, they saw the young child . . . and fell down, and worshipped him."

Introduction

God is not limited by anything. He can choose to use natural events or perform miraculous acts that go beyond His natural laws.

➤ What led the wise men to Jesus Christ? {a star}

➤ Would a new star have been seen by astronomers all around the world? {probably}

➤ Is it important to know what the "star" actually was? {no} Why? {It is only important to know that the star was used to glorify God.}

Chapter Outline

I. Our Closest Star

II. Characteristics of Stars
   A. Brightness
   B. Colors of Stars
   C. Sizes and Distances of Stars

III. Kinds of Stars
   A. Variable Stars
   B. Novas
   C. Supernovas
   D. Neutron Stars
   E. Black Holes

IV. Observing the Heavens
   A. Constellations
   B. Astrology
   C. Telescopes
   D. Spectroscopes

V. Star Groups
   A. Small Groups of Stars
   B. Star Clusters
   C. Galaxies
   D. The Local Group

VI. Other Space Objects
   A. Asteroids
   B. Meteoroids
   C. Comets

Activity Manual

Preview—page 151

The purpose of this Preview page is to generate student interest and determine prior student knowledge. The concepts are not intended to be taught in this lesson.

SCIENCE BACKGROUND

Early astronomers—Most ancient civilizations studied astronomy. They used the movement of the Moon and the stars to make calendars. The Egyptians used the movement of the stars to predict when the Nile would flood. The ancient Polynesians were sailors who used stars for navigation. Both the Babylonians and the Chinese kept detailed astronomical records. The Chinese have records of a solar eclipse that occurred in 2136 B.C.

Lesson 103

Student Text pages 230–33
Activity Manual page 152

Objectives

- Explain how stars produce their own light
- Distinguish between apparent magnitude and absolute magnitude of stars
- Recognize that stars are classified according to color
- Explain ways distance is measured in space
- Read a graph

Vocabulary

magnitude
apparent magnitude
absolute magnitude
dwarf
giant star
supergiant
light-year
parallax

Materials

- meter stick

Introduction

Choose students to measure the length and width of the room using centimeters. Record the measurements. Multiply the numbers to find the area of the room.

➤ Are centimeters the best unit of measurement to find the area of this large room? {no} Why? {Possible answer: Using larger units to measure the room would be easier because larger units of measurement give smaller numbers to multiply.}

➤ What would be the best unit to use for measuring the room? {meter}

➤ Would a meter be a good unit to use to measure distances in space? {no} Why? {Possible answer: The measurements would have very large numbers.}

📖 Only God knows the size of the space He created. Even a kilometer is too small to measure space. Scientists have developed other units of measure to help with space exploration and study.

Purpose for reading:
Student Text pages 230–33

➤ What are some ways distances are measured in space?

➤ Does the brightness of a star tell us accurately how far the star is from Earth?

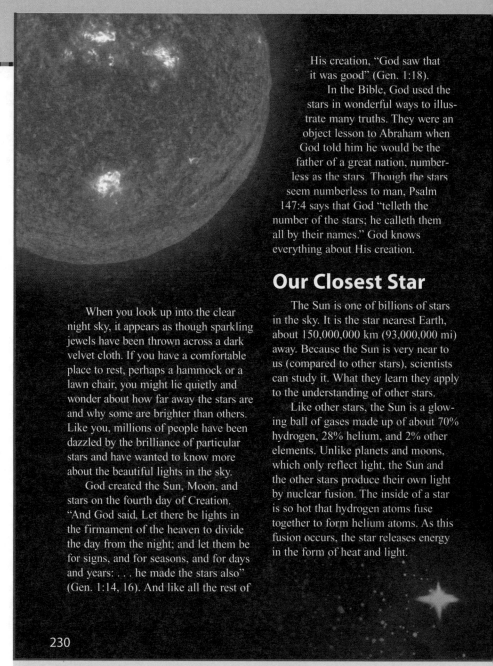

When you look up into the clear night sky, it appears as though sparkling jewels have been thrown across a dark velvet cloth. If you have a comfortable place to rest, perhaps a hammock or a lawn chair, you might lie quietly and wonder about how far away the stars are and why some are brighter than others. Like you, millions of people have been dazzled by the brilliance of particular stars and have wanted to know more about the beautiful lights in the sky.

God created the Sun, Moon, and stars on the fourth day of Creation. "And God said, Let there be lights in the firmament of the heaven to divide the day from the night; and let them be for signs, and for seasons, and for days and years: . . . he made the stars also" (Gen. 1:14, 16). And like all the rest of His creation, "God saw that it was good" (Gen. 1:18).

In the Bible, God used the stars in wonderful ways to illustrate many truths. They were an object lesson to Abraham when God told him he would be the father of a great nation, numberless as the stars. Though the stars seem numberless to man, Psalm 147:4 says that God "telleth the number of the stars; he calleth them all by their names." God knows everything about His creation.

Our Closest Star

The Sun is one of billions of stars in the sky. It is the star nearest Earth, about 150,000,000 km (93,000,000 mi) away. Because the Sun is very near to us (compared to other stars), scientists can study it. What they learn they apply to the understanding of other stars.

Like other stars, the Sun is a glowing ball of gases made up of about 70% hydrogen, 28% helium, and 2% other elements. Unlike planets and moons, which only reflect light, the Sun and the other stars produce their own light by nuclear fusion. The inside of a star is so hot that hydrogen atoms fuse together to form helium atoms. As this fusion occurs, the star releases energy in the form of heat and light.

SCIENCE BACKGROUND

Apparent magnitudes—Many heavenly bodies vary in their brightness; therefore their location on the Hipparchus scale may vary as well.

The approximate apparent magnitudes of the Sun, Moon, and Venus are: Sun –27, Moon –12, and Venus –4.

 LANGUAGE The words *apparent* and *appear* come from the Latin word *appārēre*, which means "to show." *Apparent* is an adjective form of the verb *appear*.

The word *absolute* means "complete" or "pure." Adjusting the viewing distances of all the stars as if they are equal removes the variable of distance and makes the measurements "pure."

 BIBLE Today one of the ways a person can honor another person is to name a star after him.

➤ Do you think astronomers will run out of stars to name? Why? {no; The stars seem numberless to man.}

➤ The Bible says that God knows the number of the stars and the names of each one (Ps. 147:4).

➤ How should knowing that God knows each of the stars be a comfort to us? {Answers will vary, but elicit that if God knows the stars, then He also knows all about us.}

Characteristics of Stars

Brightness

The brightness of a star is called its **magnitude**. The magnitude of a star depends on the star's size, temperature, and distance from Earth.

When we look at the stars and say that one is brighter than another, we are talking about how bright each star appears to us. Astronomers call this a star's apparent brightness, or **apparent magnitude**. Some stars, such as our own Sun, appear brighter because they are larger or closer to Earth than other stars. Stars that are farther away appear as faint lights in the sky.

Astronomers use a set of numbers to represent apparent magnitude. Lower numbers represent brighter stars. Hipparchus (hih PAHR kuhs), a Greek who lived 130 years before the time of Christ, devised the system still used today to classify stars by their brightness. In Hipparchus's day, no telescopes existed. Hipparchus classified the brightest stars he could see as +1 on his scale. He classified the faintest stars as +6. Since telescopes now allow men to see much farther into space, astronomers have had to adjust Hipparchus's scale. With huge telescopes today we can see stars as faint as magnitude +29 or greater. Astronomers have also added negative numbers to represent objects that are even brighter than many stars.

The true brightness of a star, called its **absolute magnitude**, measures how bright a star really is, not just how bright it appears to be. Astronomers determine absolute magnitude by imagining that all stars are the same distance from Earth. How bright a star would appear at that distance is the star's absolute magnitude. Special measurements and mathematics help astronomers calculate the absolute magnitude of a star.

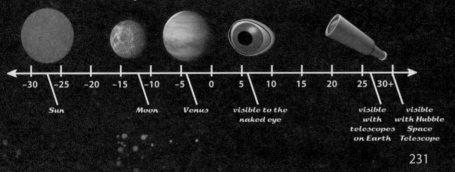

Apparent Magnitudes

Discussion:
pages 230–31

- On what day did God create the Sun, Moon, and stars? {the fourth day of Creation}
- Name some times that God used stars in a specific way in the Bible? {Possible answers: as a lesson to Abraham; to announce Jesus' birth to the wise men}
- Which star do scientists study the most? {the Sun} Why? {Possible answer: It is the closest star to Earth.}
- What are stars? {glowing balls of burning gases}
- What are the percentages of the main elements that make up most stars? {70% hydrogen, 28% helium, and 2% other elements}

Direct the student to locate the elements on *The Periodic Table of the Elements* on Student Text pages 164–65.

- How do stars produce their own light? {by nuclear fusion}
- Name two forms of energy given off by the Sun. {heat and light}
- What does *magnitude* of a star mean? {the star's brightness}
- How is apparent magnitude different from absolute magnitude? {Apparent magnitude is how bright a star appears from Earth. Absolute magnitude is how bright the star actually is.}
- Why might one star appear brighter than another star? {It may actually be brighter, but it may also be only closer or larger than other stars.}
- How do scientists determine absolute magnitude? {They treat all the stars as if they were the same distance from Earth.}
- When Hipparchus classified the stars that he could see, he used the numbers 1–6. Later, scientists wanted to rank other objects, such as the Sun, Moon, and planets that are brighter than stars with a magnitude of 1. Scientists assigned these objects negative numbers.
- Look at the *Apparent Magnitudes* scale. What is the approximate magnitude of the Moon? {–12}
- What is the approximate magnitude of the faintest stars seen by the Hubble Space Telescope? {30+}
- Do you think fainter stars will ever be seen? {Answers will vary, but elicit that scientists continue to improve telescopes, so they will probably continue to find fainter stars.}

Direct a Demonstration

Demonstrate that star size and distance affect apparent magnitude

Materials: 2 flashlights of the same size, a piece of foil or dark paper with a 0.5 cm diameter hole in the middle, a rubber band

Direct students to stand at the side of the room. Choose two students to hold flashlights and stand two meters in front of the other students.

Darken the room and turn on the flashlights. Instruct the two students to hold the flashlights so that the other students can see that the lights are the same.

- How do the apparent magnitudes of the lights appear? {the same}

Cover one flashlight with the piece of foil and secure with the rubber band.

- How do the apparent magnitudes of the lights appear? {The covered light looks smaller or weaker.}

Uncover the flashlight and move one student with a flashlight to the opposite side of the room.

- How do the apparent magnitudes of the lights appear? {The light farther away looks fainter, or weaker.}

Conclude that both the size and distance affect apparent magnitude.

Discussion:
pages 232–33

➤ What is the color of a star related to? *{the star's surface temperature}*

 Refer the student to the diagram called *Stars* on Student Text page 232.

💡 Which term could be used in place of *actual brightness*? *{absolute magnitude}*

➤ Which stars have a low actual brightness but are hot stars? *{white dwarfs}*

➤ Which has a brighter absolute magnitude—a white dwarf or a blue giant? *{blue giant}*

➤ Why would a supergiant have a brighter absolute magnitude than a white dwarf? *{Possible answer: The supergiant is larger.}*

➤ Does a supergiant have a hotter or cooler surface temperature than a white dwarf? *{cooler}*

➤ What is the general relationship between size and magnitude? *{Larger stars have a brighter absolute magnitude.}*

➤ Using the chart, how would you describe the Sun? *{average brightness, average surface temperature, average size}*

➤ Describe Betelgeuse. *{cool surface temperature, very bright, huge supergiant}*

➤ What kind of star is Aldebaran? *{a red-orange giant}*

➤ The line of stars in the middle of the chart is called the main sequence. Most stars are part of the main sequence.

➤ Besides the Sun, what is another named star that is part of the main sequence? *{Sirius A}*

Colors of Stars

When you first look at the stars, they all may appear to be white. However, stars actually are many colors. A star's color is closely related to its surface temperature. The coolest stars are a dull red, and the hottest stars are blue.

On a clear night you may be able to see the colors of some large stars. You need to be away from lights. Give your eyes time to adjust to the dark. Some stars will show a faint color. Study the stars in the sky to find out whether you can see any colors.

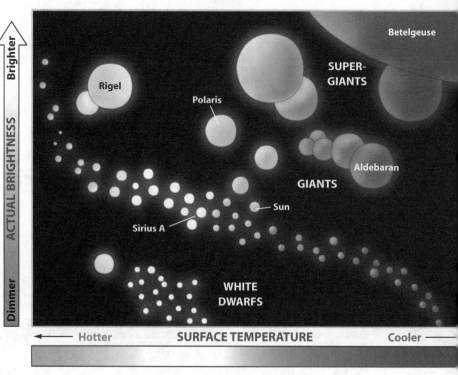

232

Direct a DEMONSTRATION
Relate the color of a flame to its hotness

Materials: candle in a holder, Bunsen burner or other gas flame source, matches

Note: Check on any open-flame restrictions for the classroom.

1. Light the candle and the gas flame.
2. Direct the student to observe the color differences between the two flames.

➤ What is the color of each flame? *{The candle flame is yellow. The gas flame is blue.}*

➤ According to the diagram on page 232, which flame is hotter? *{the blue gas flame}*

➤ Without using a thermometer, how could we test to find out which flame is hotter? *{Possible answer: Place containers of equal amounts of water over the flames and observe which boils first.}*

242 Lesson 103

Sizes and Distances of Stars

Stars come in many sizes. The small and medium-sized stars are called **dwarfs.** Our Sun has a diameter of 1,400,000 km (865,000 mi). Though it seems large to us, the Sun is only a medium-sized yellow star. **Giant stars** are tens to hundreds of times larger and hundreds of times more luminous than the Sun. **Supergiants** are hundreds of times larger than the Sun and thousands of times brighter. If placed where our Sun is, the supergiant star named VV Cephei (SEE fee ee) would span all the way to Jupiter's orbit.

After the Sun, the next closest star to Earth is Proxima Centauri (PRAHX-ih-muh sen-TAW-ree). This star is 270,000 times farther away from Earth than the Sun! Because distances in space are very great, measuring in kilometers (or miles) requires enormous numbers. Astronomers solve this problem by using other units of measurement. One of these is the **light-year,** the distance that light travels in one year. Proxima Centauri is 4.3 light-years away.

Astronomers can determine how far away from Earth some stars are by measuring how far stars appear to move compared to even more distant stars. To determine the distance a star appears to move, astronomers take pictures of the star at six-month intervals. These pictures enable them to view the star from opposite points in Earth's orbit around the Sun. A star that is close to Earth will appear to move more than a star that is far away. Scientists examine the photographs, noting a star's change in position in relationship to more distant stars. The apparent movement or change in position of one star in relationship to other stars is known as **parallax.**

> **QUICK CHECK**
> 1. Describe the difference between apparent and absolute magnitude.
> 2. What color is the star with the hottest surface temperature?
> 3. Name one unit of measurement scientists use for distances.

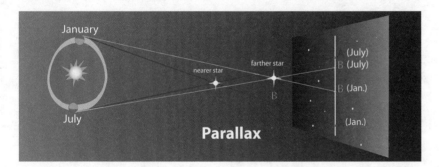

Parallax

233

Use the following exercise to help students understand how huge a light-year is. You will need to use a calculator.

1. Light travels at approximately 300,000 km per second.
2. To find out how far it travels in a minute, multiply by 60 (seconds per minute). *{18,000,000 km per minute}*
3. To find out how far it travels in an hour, multiply by 60 (minutes per hour). *{1,080,000,000 km per hour}*
4. To find how far it travels in a day, multiply by 24 (hours per day). *{25,920,000,000 km per day}*
5. To find how far it travels in a year, multiply by 365.25 (days per year). *{9,467,280,000,000 km per year}*

Even if you cannot complete the exercise, students will quickly see how huge a number will be formed by the calculations.

Discussion:
pages 232–33

➤ What are small and medium stars called? *{dwarfs}*

➤ How large are giant stars compared to the Sun? *{tens to hundreds of times larger}*

➤ How much brighter are giant stars than the Sun? *{hundreds of times brighter}*

➤ How large and bright is a supergiant compared to the Sun? *{hundreds of times larger and thousands of times brighter}*

➤ Other than the Sun, what is the next closest star to Earth? *{Proxima Centauri}*

➤ Why would kilometers not be a good way to measure distance in space? *{the distances are too great}*

➤ What have scientists done to solve this problem? *{They have created another measuring unit—a light-year.}*

➤ What do we call the apparent change in a star's position that allows scientists to determine the star's distance from Earth? *{parallax}*

➤ Look at the *Parallax* diagram. Which star appears to have moved the most? *{star A; the closer star}*

➤ How do scientists measure parallax? *{They compare the distance that a star appears to move over a six-month period.}*

Answers
1. A star's apparent magnitude is how bright the star appears to be from Earth. A star's absolute magnitude is the star's true brightness.
2. blue
3. a light year

Activity Manual
Reinforcement—page 152

Lesson 104

Student Text pages 234–37
Activity Manual pages 153–54

Objectives

- Differentiate between a pulsating variable star and an eclipsing variable star
- Describe the causes of a nova and of a supernova
- Explain how a neutron star or black hole is formed

Vocabulary

variable star
pulsating variable star
eclipsing variable star
nova
nebula
supernova
neutron star
pulsar
black hole

Introduction

➤ Have you ever observed a flower, such as a rose, slowly open from a bud to an open bloom?

➤ You could watch for hours and never see the petals move. This action happens so slowly that you may not detect it.

➤ How can you tell the flower has bloomed? *{Possible answer: by observing the flower before it blooms and after it blooms}*

➤ Another way to observe a flower bloom is to videotape the flower as it blooms. Then, if you watched the film on fast forward, you would actually see the petals unfurl. In a similar way, astronomers are able to detect changes in stars.

Purpose for reading:
Student Text pages 234–37

➤ What are some ways that stars change?

➤ Which kind of violent events can occur during the life of a star?

➤ Which kind of objects may remain after a star "dies"?

Discussion:
pages 234–35

➤ What is a *variable star?* {a star that regularly or repeatedly changes in magnitude}

➤ What is a *pulsating variable star?* {a star that goes through periods of swelling and brightening, then shrinking and dimming}

Kinds of Stars

Variable Stars

Over the centuries astronomers have discovered that some stars do not have consistent magnitudes. Some regularly change in brightness, and others flare up suddenly and then slowly return to their original size and brightness. Stars that regularly or repeatedly change in magnitude are called **variable stars.**

One kind of variable star is a **pulsating variable star.** Pulsating variable stars go through periods of swelling and brightening, then shrinking and dimming. The absolute magnitude of the star changes during this cycle. Some pulsating variable stars change in regular patterns. Others seem to have no pattern at all.

Another kind of variable star does not change its absolute magnitude but does change its apparent magnitude. These stars are called **eclipsing variable stars.** Eclipsing variable stars are actually pairs of stars that orbit each other because of their gravitational pulls on one another. The apparent brightness of the stars is greater when both stars can be seen. However, when one star eclipses, or moves between the earth and the other star, the reduction in light causes the apparent brightness to dim.

pulsating variable

eclipsing variable

SCIENCE BACKGROUND

Going nova—Some binary stars repeatedly "go nova."

SCIENCE MISCONCEPTIONS

Some astronomers believe that new stars form from nebulae; however, astronomers have never witnessed a star birth, and there are many scientific reasons why this theory is unlikely. Some scientists want it to be possible for stars to form from nebulae to justify their beliefs in the origin of the universe without the need for God. The Bible says that God created the stars on the fourth day of Creation (Gen. 1:16–18). [BAT: 8b Faith in the power of the Word of God]

LANGUAGE

The Latin word for star is *stellar*. *Interstellar* means "between or among stars."

The plural of *nebula* is *nebulae*.

WRITING

Encourage the student to write a short story describing the reactions of people in a small village in the past to the sight of a nova.

244   Lesson 104

Science and the BIBLE

In I Corinthians 15:41 the Bible says, "For one star differeth from another star in glory." Before the invention of the telescope, all stars appeared to be relatively similar. As technology has improved, man has learned that the stars are actually very different. God's Word, however, has said all along that stars are different.

Novas

Sometimes scientists will notice a star where one was not visible before. The star then fades over the next few nights, weeks, or months. In the past when such a phenomenon lit the skies, many regarded it as a signal from the heavens that an important or disastrous event would soon occur. Some believed that it signaled the birth of a star. Astronomers called such a star a **nova** (NOH vuh), which means "new" in Latin. Scientists now believe that a nova forms when an existing star suddenly flares up and becomes hundreds or thousands of times brighter than normal.

Novas are part of pairs of stars and flare because one star's gravity pulls gases from the other star.

Novas are not common occurrences. Seeing a nova without the help of a telescope is very rare. Even with telescopes, people spot only about two to three novas in our galaxy each year.

On February 19, 1992, astronomers were excited to observe Nova Cygni (SIG nee) 1992. It was the brightest nova in recent history and could be seen without a telescope.

When a star "goes nova," it spews dust and gases into space. Its outer layers gradually float off into space, usually leaving a smaller, dimmer star behind. A cloud of interstellar gases and debris is called a **nebula.** Nebulas can be seen either because they glow from light within or because they block light from behind them and look like dark clouds.

Nova Cygni 1992
Taken by Hubble Space Telescope faint object camera

Pre-COSTAR Raw Image
May 1993

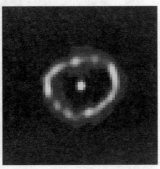
With COSTAR Raw Image
Seven months later

Direct a DEMONSTRATION
Demonstrate pulsating variable stars and eclipsing variable stars

Materials: two unshaded lamps (one with a three-way switch), one 3-way light bulb, two 40-watt light bulbs

Place the 3-way bulb in one lamp. Darken the room. Turn the switch from the lowest setting to the medium setting and then to the highest setting.

▶ Which kind of variable star does this represent? *{pulsating variable star}*

▶ In what way does our model differ from a pulsating variable star? *{The lamp can go only from low to medium to high, then off and back to low. A pulsating star can gradually become brighter and then gradually get dimmer.}*

Place the 40-watt bulbs in both lamps. Darken the room. Rotate one lamp around the other.

▶ When does the brightness of the bulbs appear greatest? *{when you can see both bulbs}*

▶ How often does the brightness dim? *{each time one lamp eclipses (hides) the other}*

Often in a pair of stars, one star is larger than the other. Therefore, one star may completely eclipse the other, but the smaller star may not completely eclipse the larger star.

Discussion:
pages 234–35

▶ Does a pulsating variable star change its absolute or apparent magnitude? *{absolute}* Why? *{The star actually changes in brightness.}*

▶ What are eclipsing variable stars? *{A pair of stars that orbit each other. The apparent brightness is greater when both stars are seen. But when one star eclipses, the light appears to dim.}*

▶ What kind of magnitude changes for an eclipsing variable star? *{apparent}* Why? *{The star stays the same brightness. The brightness appears to change because one star blocks the light of the other.}*

▶ What is a *nova*? *{a star that appears where none was visible before and then fades away}*

▶ How did some people in the past respond when they saw a nova? *{They thought that an important or disastrous event was about to occur.}*

💡 Why would people have thought that? *{Possible answer: It was an unusual occurrence and caused fear because people could not explain it.}*

▶ Why did astronomers give this phenomenon the name *nova*? *{Some thought that it was the birth of a new star, and nova means new.}*

▶ How does a nova form? *{A nova forms when an existing star suddenly flares up and becomes hundreds or thousands of times brighter than normal.}*

▶ What do astronomers believe causes a nova? *{One star's gravity pulls gases from another star in a binary system.}*

▶ What results when a star "goes nova?" *{Dust and gases are spewn into space, leaving a smaller star behind.}*

💡 Why can scientists not be sure exactly what causes a nova? *{Possible answers: It is so far away. It occurs only a few times a year in our galaxy.}*

▶ What is a *nebula*? *{a cloud of interstellar gases and debris}*

💡 Compare the two pictures on page 235. What does the second picture show you about Nova Cygni? *{Possible answer: After seven months the debris spread, leaving a much smaller star in the center.}*

📖 Refer the student to *Science and the Bible*.

▶ What truth of the Bible has man proven as his technology has improved? *{that stars are different from each other}*

Point out that man's proving the scientific accuracy of the Bible does not make the Bible any more true. The Bible has always been true.

Discussion:
pages 236–37

➤ **What is a *supernova*?** *{the death explosion of a star}*

➤ **What causes a supernova?** *{A massive star uses up its hydrogen fuel. The star starts to collapse, but the pressure of the star's gravity causes the star to heat up quickly and explode. Then the star fades and collapses.}*

➤ **What is left of the star after a supernova occurs?** *{a neutron star or a black hole}*

➤ **How often does a supernova occur in our galaxy?** *{less than once a century; However, there has not been one since 1604.}*

➤ **Who recorded a supernova in A.D. 1054?** *{Chinese astronomers}* **What was the result of this supernova?** *{the Crab Nebula}*

Point out the picture of the Crab Nebula on Student Text page 236.

➤ **Do you think it would be a special event to witness the occurrence of a supernova?** *{Answers will vary.}*

➤ **How do astronomers believe a neutron star is formed?** *{When a supergiant collapses, the pressure is so great that the protons and electrons of the star's core are crushed to form neutrons.}*

💡 **Why can astronomers only theorize about what forms a neutron star?** *{Possible answers: They cannot see it happen. They cannot duplicate the process on Earth.}*

➤ **What do scientists believe about the weight of the core of a neutron star?** *{The core is so tightly packed that a piece the size of a teaspoon would weigh one billion tons on Earth.}*

Point out the cartoon on Student Text page 236.

💡 **What do you think the cartoon is showing?** *{Answers will vary, but elicit that the "superhero" is trying to lift the weight of a neutron star compacted into the area of a teaspoon.}*

Crab Nebula

Supernovas

The death explosion of a star is called a **supernova**. Astronomers believe that a supernova occurs when a massive star has used up its hydrogen fuel. The star starts to collapse, but the tremendous pressure created by the star's gravity causes the star to heat up quickly and explode. The star increases in size and brightness and can be brighter than a galaxy. Sometimes a supernova can even be seen during the day. After brightening, the star fades and collapses. This explosion usually results in the complete destruction of the star, and the remnant often becomes a neutron star or a black hole.

Supernovas occur less than once a century in our galaxy. The last supernova in our galaxy occurred in 1604. Some supernovas that occur in other galaxies can be seen from Earth with telescopes.

In A.D. 1054 Chinese astronomers observed a supernova. The remnants from this supernova explosion form the Crab Nebula, the closest nebula to Earth caused by a supernova.

Neutron Stars

Astronomers think that when some supergiants collapse, the extreme pressure in the star's core crushes the protons and electrons together to form neutrons. The core is then made up mostly of neutrons. A star that began as a supergiant hundreds of millions of kilometers in diameter may become as small as a few kilometers in diameter. Because the neutrons are very tightly packed in this **neutron star,** just one teaspoon of the star's core might weigh one billion tons.

236

SCIENCE BACKGROUND

Black holes—At the center of a black hole is a point called a *singularity*. Around that point is an area where the gravitational pull is so great that even light cannot escape. This area, the *event horizon*, is what astronomers are talking about when they speak of the size of a black hole.

Gravitational pull—Both neutron stars and black holes have approximately the same mass as the star that collapses. That is why only the collapse of a massive supergiant can cause a black hole.

 TECHNOLOGY
Some astronomers think there is a phenomenon called *gravitational waves*. They think that large disruptions in space, such as supernovas, neutron stars, and black holes cause gravitational waves.

These gravitational waves are so tiny that no instrument available today can measure them. However, scientists are developing new instruments that should be able to measure these gravitational waves.

A **pulsar** is a neutron star that spins rapidly on its axis. If a neutron star's core continues to collapse, it starts to spin rapidly and fling pulses of energy into space. From this action pulsars got their name. It is also how astronomers find them. They trace the pulses of energy back to the star emitting them.

Black Holes

Some astronomers believe that when a massive supergiant star runs out of fuel, its gravitational force is so great that the core cannot stop collapsing. It essentially disappears from space. Astronomers call this a **black hole.** The gravitational force of a black hole is so great that it pulls everything into it—even light. Astronomers cannot see a black hole, but they have seen the effects of its gravitational pull on other matter. Sometimes when two stars are near each other, gases from one star appear to spiral into a black hole. The light from the gases seems to just disappear.

1. What do we call stars that regularly or repeatedly change in brightness?
2. What usually causes a nova?
3. Why can't we see a black hole?

Discussion:
pages 236–37

▶ What is a *pulsar*? {a neutron star that spins rapidly on its axis, emitting pulses of energy}

▶ How do astronomers find pulsars? {They trace the pulses of energy back to the stars emitting them.}

▶ How do astronomers believe a black hole forms? {They believe that a massive supergiant collapses with such great gravitational force that its core cannot stop collapsing.}

Gravity is related to the mass of an object. Therefore, a massive supergiant star would have a massive gravitational pull.

▶ What happens to light and matter around a black hole? {The gravitational force pulls light and matter into the black hole.}

▶ How do scientists detect a black hole? {Although they cannot see a black hole, they detect its gravitational pull on other matter.}

Answers

1. variable stars
2. Novas are usually part of a pair of stars and flare up because one star's gravity pulls gases from the other star.
3. Scientists think the gravitational force of a black hole is so great that even light is pulled into it. Without light, people cannot see.

Activity Manual
Reinforcement—page 153

Study Guide—page 154
This page reviews Lessons 103 and 104.

Assessment
Test Packet—Quiz 10-A
The quiz may be given any time after completion of this lesson.

Direct a DEMONSTRATION
Demonstrate that the mass of a star is compacted

Materials: foam ball, tape measure (or string and yardstick), scale, plastic sandwich bag

Direct the student to follow the procedure. Measure and record the circumference of the foam ball and place it in the plastic bag. Measure and record the mass of the ball and the bag. Crush the ball. Compress the pieces into a ball shape in one corner of the bag. Measure and record the approximate circumference. Measure and record the mass. Compare the measurements.

▶ Has the size (circumference) changed? {yes}

▶ Has the mass changed? {no}

▶ How does this compare to what happens when a supergiant collapses? {In the same way that the mass of the foam does not change when the size changes, the resulting neutron star has about the same mass as the original supergiant before it collapsed.}

Lesson 105

Student Text pages 238–41
Activity Manual page 155

Objectives

- Identify various constellations
- Explain why a Christian should not be involved in astrology
- Describe the difference between a reflecting telescope and a refracting telescope
- Identify instruments used to study the stars

Vocabulary

constellation
circumpolar constellation
astrology
astronomy
refracting telescope
reflecting telescope
radio telescope
spectroscope
redshift

Materials

- magnifying glass

Introduction

Allow several students to look at a text page with the magnifying glass.

➤ **How do you think this magnifying glass produces a larger image of the text page?** *{Accept any answer, but explain that the light is bent as it passes through the glass, producing a larger image of the page.}*

Principles of magnification are also used in telescopes.

Purpose for reading:
Student Text pages 238–41

➤ **Why is the practice of astrology wrong?**
➤ **What are some instruments that are used to observe the heavens?**

Discussion:
pages 238–39

➤ **What is a *constellation*?** *{a group of stars that seem to form patterns}*
➤ **How many official constellations are there today?** *{eighty-eight}*
➤ **Have you ever found the Big Dipper?** *{Answers will vary.}*
➤ **What constellation is the Big Dipper part of?** *{the Great Bear, or Ursa Major}*

Observing the Heavens

Constellations

For thousands of years stargazers have watched the night sky. They found groups of stars in patterns that reminded them of ancient heroes, animals, mythological characters, and objects in nature. These groups of stars, called **constellations** (KAHN stuh LAY shunz), make stars easier to find. Three hundred years before Christ, Aratus of Soli listed forty-four constellations. Today there are eighty-eight official constellations.

In the northern hemisphere, many people can find the Big Dipper. These seven stars form the back and tail of the constellation called the Great Bear, or Ursa Major. If you follow a straight line from the two stars at the front of the Big Dipper's bowl, you will be able to locate the North Star, or Polaris (puh LEHR ihs), which sits over Earth's gravitational North Pole. The North Star is also part of the group of stars called the Little Dipper. These stars are part of the constellation Little Bear, or Ursa Minor.

Once you find one or two constellations, you can use them as markers to find others. Draco the Dragon winds its way like a huge serpent between the Big and Little Dippers. The stars in the nearby constellation Cassiopeia (KASS ee uh PEE uh) appear as a giant *W*.

During the year, the Great and Little Bears, Draco, and Cassiopeia seem to revolve around the North Star (Polaris). For that reason, astronomers call these constellations **circumpolar** (SUHR kuhm POH luhr) **constellations**.

When looking for Orion (oh RY uhn) the Hunter, you should locate three stars lined up closely together in the sky. These stars mark Orion's belt. Around the three stars is a larger box of four stars. The top left star is the orange-red supergiant Betelgeuse (BEET uhl JOOZ), the shoulder of Orion. The bottom right star is brilliant blue-white Rigel (RY juhl). This star marks Orion's ankle.

238

SCIENCE BACKGROUND

Stars in constellations—Sometimes constellations may be difficult to find because other stars around them are not part of the "picture."

Modern charts—Modern constellation charts show constellations as polygonal regions rather than as pictures, such as classical constellation charts show.

Pleiades—The star cluster Pleiades is called the Seven Sisters. However, only six of the seven stars can be seen with the naked eye.

 Several constellations are mentioned in Scripture. Examples are Orion (Job 9:9), Pleiades (Job 38:31), and the crooked serpent Draco (Job 26:13).

SCIENCE MISCONCEPTIONS

The stars in a constellation do not always outline the picture as dots in a dot-to-dot puzzle, but are merely the key points of the picture.

 Not all countries and cultures use the same names for the constellations. Encourage the student to use the library and Internet to find the names for constellations used in other countries.

➤ **Why was the group of stars given that name?** *{Answers will vary.}*
➤ **Which name do you prefer?** *{Answers will vary.}*

248   Lesson 105

If you look above and to the right of Orion, you may be able to see the star Aldebaran (al DEB uhr uhn). This star is the right eye of the constellation Taurus the bull. A *V* made of stars marks his horns and nose. On the shoulder of the bull is the star cluster called Pleiades (PLEE uh DEEZ).

Astrology

Many people believe that stars control the lives of people. They teach that the positions of the Sun, Moon, planets, and stars at the moment of a person's birth determine his destiny. These beliefs are part of the practice of **astrology** (uh STRAHL uh jee). People who believe in astrology often consult a daily horoscope to give them guidance for the future.

Astrology is different from astronomy. **Astronomy** deals with the scientific study of the stars, but astrology is the practice of trying to find guidance from them. The Bible tells us to look at the stars to appreciate God's creation and to realize how small we are compared to God. But astrology is a distortion of what God intended for us to enjoy as we worship Him. Instead of worshiping the God who created the stars, those who practice astrology worship the creation by looking to it for guidance in place of the Creator.

Why is astrology a dangerous belief? It is dangerous because those who deal in astrology look for guidance apart from God and His Word. The Bible tells us in John 16:13 that Christians' lives are to be led by the Holy Spirit and that He will guide them into all truth. The Bible, rather than the stars, is to be a Christian's guide. Psalm 119:105 says, "Thy word is a lamp unto my feet, and a light unto my path." Looking to the stars for guidance is futile and worthless. Daniel 2:27–28 says, "The secret which the king hath demanded cannot the wise men, the astrologers, the magicians, the soothsayers, shew unto the king; But there is a God in heaven that revealeth secrets." We need to look to God's Word as we make decisions.

239

Direct an Activity

Materials: one index card per student

Constellations are simply man's imaginations. Different cultures have looked at the same stars and found different images.

Give each student an index card. Each student is to draw seven stars in random order on the unlined side of his index card.

Students exchange index cards and design a "constellation," using the seven stars on the cards they receive. On the back of the card, they should write the name of the constellation and the reason for the picture and name.

Language

The prefix *circum-* means "around." *Polar* means "at or near the pole of the earth." Circumpolar constellations appear to be moving around the North Pole.

The Greek root *hora* means "season" or "hour," and the root *skopos* means "observer." The word *horoscope* could be translated to mean "a season observer."

Discussion: pages 238–39

▸ How can you find Polaris in the sky? {Follow a straight line up from the two stars at the front of the Big Dipper's bowl to the handle of the Little Dipper.}

▸ What is another name for Polaris? {the North Star}

💡 If you were lost at night, how could you use Polaris to help you find your direction? {Polaris sits over the gravitational North Pole, so if you went toward Polaris, you would be going north.}

▸ What constellation is the Little Dipper part of? {the Little Bear, or Ursa Minor}

▸ What are some other constellations that you might see in the Northern Hemisphere? {Possible answers: Draco, Cassiopeia, Orion, Taurus}

▸ What does *circumpolar* mean? {revolves around Polaris}

▸ Which stars form Orion's shoulder and ankle? {Betelgeuse—shoulder; Rigel—ankle}

▸ Where is the star cluster Pleiades located? {on the shoulder of Taurus the Bull}

As specific stars are discussed, refer the student back to the *Stars* diagram on Student Text page 232 to help him gain a perspective of the sizes of the stars.

▸ What is *astrology*? {the belief that the positions of the Sun, Moon, planets, and stars at the time of a person's birth determine his destiny}

▸ How is astrology different from astronomy? {Astronomy is the study of the stars and other objects in space. Astrology is the practice of seeking guidance for the future by looking at the arrangement of the stars.}

📖 Why is it unwise for a Christian to participate in astrology? {Those who practice astrology are seeking guidance apart from God and His Word.}

▸ Where should Christians look for guidance about the future? {the Bible} [BATs: 8a Faith in God's promises; 8b Faith in the power of the Word of God]

▸ What does Daniel 2:27–28 say about the source of guidance? {For a Christian, God is the source of guidance.}

Discussion:
pages 240–41

➤ **What does a refracting telescope use to magnify images?** *{convex and concave lenses}*

➤ **What is the difference between a refracting telescope and a reflecting telescope?** *{A refracting telescope bends light to make an object seem larger. A reflecting telescope reflects light to make an object seem larger.}*

➤ **What is the disadvantage of a refracting telescope?** *{As light is refracted (bent), the colors are distorted.}*

➤ **What does a reflecting telescope use to magnify images?** *{mirrors}*

➤ **In which way is a reflecting telescope better than a refracting telescope?** *{The light is reflected rather than bent, so the colors are not distorted.}*

➤ **Why do telescopes in space have clearer images than telescopes on Earth?** *{Earth's atmosphere contains dust and water droplets that prevent telescopes on Earth from showing clear images.}*

➤ **Which telescope solved some of these problems?** *{the Hubble Space Telescope}* **Why?** *{because it orbits above Earth's atmosphere}*

💡 **How do you think viewing images taken from the Hubble Space Telescope has caused astronomers to change their views of the universe?** *{Accept any answer, but suggest that additional information usually causes man to reevaluate his theories.}*

Telescopes

Ever since Galileo made his first telescope in 1609, the telescope has been the most important instrument astronomers use to find new stars. Early telescopes were **refracting telescopes** that bend, or refract, light to make objects seem larger. The light enters a convex (curved outward) lens and then travels through a concave (curved inward) lens to the eyepiece. The convex lens makes the image look bigger but blurry, and the concave lens makes the object look smaller but clear.

refracting telescope

The combination of the two lenses produces a clear, magnified image. However, a refracting telescope causes color distortions, because the light bends at different angles as it is refracted.

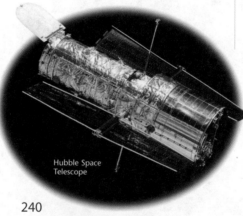
Hubble Space Telescope

About seventy years after Galileo made his telescope, Sir Isaac Newton invented a reflecting telescope that solved some of the problems of color distortion caused by refracting telescopes. A **reflecting telescope** produces a clearer magnified image than the refracting telescope because light is reflected rather than bent. Light enters the telescope and reflects off a large concave mirror to a smaller flat mirror. It then enters the eyepiece.

reflecting telescope

Over the years since Newton's invention, astronomers have continued to make bigger and bigger telescopes with hopes of being able to see farther and farther into space. Telescopes on Earth, though, have one great problem. They must view the stars through thousands of meters of Earth's atmosphere. The atmosphere constantly moves and carries dust particles and water droplets in it. As a result, pictures taken of the heavens are often unclear.

The launch of the Hubble Space Telescope (HST) in 1990 at last gave astronomers a telescope that stays above Earth's atmosphere all of the time. The pictures it has taken of distant galaxies and our solar system are bright, clear, and beautiful.

240

SCIENCE BACKGROUND

Types of telescopes—There are at least five types of telescopes used to observe space. In addition to visual and radio telescopes, there are infrared, x-ray, and gamma ray telescopes. Each is designed to detect a specific type of wavelength.

SCIENCE MISCONCEPTIONS

Galileo did not invent the telescope. He was the first person to make scientific discoveries with the telescope. Galileo is said to have "reinvented" the telescope when his discoveries made it popular.

HISTORY

A precursor to the telescope was eyeglasses, called "spectacles." The same type of lenses used in telescopes were used for eyeglasses. Spectacles were worn as early as A.D 1300.

When telescopes were first introduced, many people thought they were just a hoax or an optical illusion.

Direct an ACTIVITY

Place convex and concave lenses in a learning center along with a newspaper article, and allow the students to look through the lenses to try to read the article. The student should observe how each lens produces a different size and quality image.

radio telescope

Refracting and reflecting telescopes are not the only instruments scientists use to study the stars. In addition to light waves, stars emit other kinds of waves, such as radio waves. **Radio telescopes** collect radio waves from space using a large concave-shaped disk. Radio telescopes can detect objects that do not give off enough light to be detected by other telescopes.

Spectroscopes

A **spectroscope** breaks down the light given off by a star into all its colors. It is similar to a prism, which breaks white light into the spectrum of color. The study of a star's color spectrum gives information about its temperature and composition. It also shows that all heavenly bodies, from dwarf stars up to the biggest galaxies, are moving. If an object is moving away from Earth, its colors' wavelengths become longer, and the colors shift more toward the red end of the spectrum. This action is called **redshift**. Astronomers have been amazed to learn that all the spectra outside our own galaxy show a definite redshift.

Science and the BIBLE

Many Creationists believe that most of the observed redshift is due to the stretching of space. The Bible says in Isaiah 40:22 that God "stretcheth out the heavens as a curtain, and spreadeth them out as a tent to dwell in." Only an omnipotent God can stretch the universe to bring glory to Himself.

 QUICK CHECK
1. What does *circumpolar* mean?
2. What is the difference between astronomy and astrology?
3. Why do scientists use spectroscopes when observing stars?

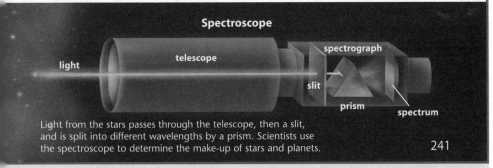

Spectroscope

Light from the stars passes through the telescope, then a slit, and is split into different wavelengths by a prism. Scientists use the spectroscope to determine the make-up of stars and planets.

241

Direct a DEMONSTRATION
Demonstrate separating colors of light

Materials: prism, flashlight, overhead projector, white paper

Shine light through a prism. You may do this with a flashlight by placing the prism on a lighted overhead projector or by holding the prism in a beam of sunlight near a window. Ask the students to describe what they see. Explain that a prism is similar to a spectroscope.

Try using flashlights with different types of bulbs, such as a halogen bulb, to see how light from different types of bulbs breaks into different spectrums.

Direct a DEMONSTRATION
Demonstrate that stretching affects how far apart objects are

Materials: balloon, permanent marker, measuring tape

Make two large dots on the uninflated balloon. Direct a student to measure the distance between the dots. Record the measurement.

Blow up the balloon. Direct another student to measure the distance between the dots. Record the measurement.

➤ What happened to the distance between the dots? *{It increased.}*

This is what many Creationists think is happening in the universe. As the universe stretches, the distances become greater and cause the redshift.

Discussion:
pages 240–41

➤ What advantage does a radio telescope have over other telescopes? *{It can detect objects that do not give off enough light to be detected by other telescopes.}*

➤ What is a *spectroscope*? *{an instrument that separates light into all of its colors}*

Point out the diagram *Spectroscope* on Student Text page 241. Discuss the path of light through the spectroscope.

➤ Why do astronomers study a star's color spectrum? *{Possible answers: to find out information about the star's temperature, composition, and movement}*

💡 Name the colors of the spectrum in order. *{red, orange, yellow, green, blue, indigo, violet}*

The order of the colors of the spectrum can be remembered easily through the acronym Roy G. Biv.

➤ What is redshift? *{If an object is moving away from Earth, its color spectrum has longer wavelengths, and the colors shift more toward the red end of the spectrum.}*

Refer the student to *Science and the Bible*.

➤ What do many Creationists believe causes the redshift? *{God's stretching of the heavens}*

Answers
1. revolves around the North Star
2. Astronomy is the scientific study of the stars. Astrology is the practice of trying to find guidance from the stars.
3. A spectroscope breaks light into colors, which helps scientists learn about the temperature and composition of stars.

Activity Manual
Reinforcement—page 155

Lesson 106

Student Text page 242

Objectives
- Make a model of a constellation
- Recognize and name some star groups and constellations

Materials
- dark umbrella
- self-stick dots or white chalk
- copies of TE pages A27–A28, *Pop Can Constellations*
- See Student Text page

Introduction

Prepare the umbrella by using the self-stick dots or chalk to mark the stars of constellations on the inside of the open umbrella.

▶ **Can we always see the same stars in the sky each night?** {no} **Why?** {Because Earth rotates around the Sun, Earth's position changes, giving it different viewpoints of the stars.}

Show the constellations inside the umbrella. With several students looking up at the "stars" under the umbrella, turn the umbrella slowly. Then hold the umbrella still, directing the students to circle slowly under the umbrella while looking at the "stars."

▶ **Which way do you think best demonstrates the way Earth and the stars move?** {Having the students move under the still umbrella best shows that Earth's view of the stars changes as it moves.}

▶ Learning the names and locations of constellations and stars will help you find them in the sky at different times of the year. This activity will help you learn to recognize some of the common star groups and constellations.

Purpose for reading:
Student Text page 242

The student should read the page before beginning the activity.

Assessment

Rubric—TE pages A50–A53

Select one of the prepared rubrics, or design a rubric to include your chosen criteria.

ACTIVITY: Pop Can Constellations

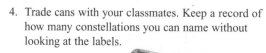

Process Skills
- Making and using models
- Observing

Sometimes the constellations in the night sky can be difficult to find. It is helpful to familiarize yourself with the constellations before trying to find them. This activity will give you an opportunity to make a representation of a constellation in order to help you learn what some constellations look like.

Procedure

1. Choose a constellation pattern available from your teacher.
2. Attach the pattern to the bottom of the can. Punch holes in the bottom of the soft drink can with the nails to make one of the constellations. Use the larger nail to make the larger holes on the pattern and the smaller nail to make the smaller holes.
3. Label the can with the name of the constellation it represents. Hold the can up to the light and look through the tab opening. Can you recognize the constellation?
4. Trade cans with your classmates. Keep a record of how many constellations you can name without looking at the labels.

Materials:
A clean, empty 12-ounce soft drink can
finishing nail
6-D common nail
hammer
blank label or masking tape
pen
constellation pattern

Conclusions
- What makes finding constellations difficult?

Follow-up
- Research to find out if, when, and where each constellation can be viewed from your home.

TEACHER HELPS

Although pop cans with small openings work best, this activity will also work with other cans or small plastic containers. If a plastic container is used, cover the outside of the container with dark paper to reduce the amount of light entering the container.

If you do not have time to include this activity in the unit, you may choose to make a set to place in a learning center and keep the constellation cans to reuse each year.

SCIENCE PROCESS SKILLS

Making a model

▶ **How will using the pop can constellations help you find constellations at night?** {Answers will vary, but elicit that repeatedly being able to see the shape of the stars without the distraction of other stars will help in identification.}

SCIENCE MISCONCEPTIONS

The Big Dipper and Little Dipper are not constellations. Because they each form a picture within a larger constellation, they are called *asterisms*.

Explorations: A Different Look

Student Text page 243

Lesson 107

All constellations appear to be approximately the same distance from Earth. However, in many constellations some stars are actually thousands of light-years farther from Earth than other stars.

In order to gain an appreciation for the varied distances of stars in a constellation, you will be constructing a three-dimensional model of a constellation.

What to Do

1. Use the star coordinate card your teacher gives you to plot the points for a given constellation on the graph paper.
2. Tape your plotted graph to the piece of cardboard or to the back of a foam meat tray.
3. Use one-inch squares of aluminum foil to make small balls at the end of 20 cm pieces of thread. You will need as many "stars" as there are points on your graph.
4. Use a large-eyed needle to poke holes in the cardboard or foam tray at each point plotted on the coordinate graph.
5. For each star, use the needle to pull the thread through the hole. Attach a piece of tape to the thread on the top side when the thread measures the correct length.
6. Repeat for the other stars.
7. Hold your constellation at eye level with the side labeled Earth pointing toward you.

Materials:
- cardboard (approximately 20 cm × 30 cm) or foam meat tray
- spool of thread
- aluminum foil
- centimeter ruler
- large-eyed needle
- tape
- star coordinates card
- graph paper

Objectives
- Make a model of a constellation
- Plot points on a graph
- Recognize the relative distances of stars

Materials
- picture of a mountain range
- copies of TE pages A29–A32, *Star Coordinates* and the graph paper
- See Student Text page

Introduction

Display the picture of the mountain range.

➤ Have you ever traveled where you could see mountains such as these in the distance?

➤ The mountains may have seemed close, but it may take hours or even more than a day to get to them. Mountains and other objects often appear closer than they really are.

➤ As you approach a range of mountains, you will often notice that not all of the mountains are next to each other as they first appeared. Some of the mountains are much farther away.

➤ This exploration will help you "see" how the different distances and sizes of the stars relate to form the "pictures" we view from Earth.

Purpose for reading:
Student Text page 243

The student should read the page before beginning the exploration.

Assessment

Rubric—TE pages A50–A53

Select one of the prepared rubrics, or design a rubric to include your chosen criteria.

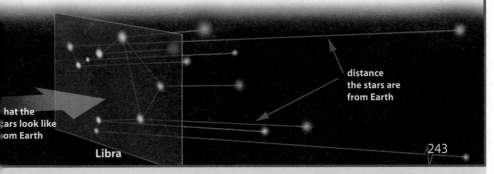
Libra — distance the stars are from Earth — what the stars look like from Earth

Teacher Helps

This activity requires a basic knowledge of coordinate points. You may need to review how to read coordinates and mark them on a graph.

➤ The first number in the parentheses is the x coordinate. The x numbers are along the bottom of the graph.

➤ The second number in the parentheses is the y coordinate. The y numbers are along the side of the graph.

➤ With your finger, find and mark the first x coordinate along the bottom of the graph.

➤ Follow the line upward until you reach the line that matches the y coordinate.

Some of the constellations are more difficult than others. Distribute the choices based on the skills of the student or group.

Encourage some of your more advanced groups to also research the absolute magnitude and/or the size of the stars in the constellation and adjust the sizes of the foil balls accordingly.

Lesson 108

Student Text pages 244–49
Activity Manual pages 156–58

Objectives

- Identify how many stars are in a binary star group and in a multiple star group
- Differentiate between an open star cluster and a globular cluster
- Identify our galaxy as the Milky Way
- Recognize that our galaxy is part of a cluster of galaxies called the Local Group
- Describe asteroids, meteoroids, meteors, meteorites, and comets

Vocabulary

binary system
multiple star group
open star cluster
globular cluster
galaxy
Local Group
asteroid
meteoroid
meteor
meteorite
comet

Materials

- star chart or picture of the night sky
- world or United States map

Introduction

Display the star chart or picture of the night sky.

➤ When you look at this picture, you see thousands of stars. Some of these points of light are not actually individual stars. One point of light may be a star group that has several stars in it.

➤ Today we are going to learn about galaxies and other types of star groups.

Purpose for reading:
Student Text pages 244–49

➤ What kind of galaxy is the Milky Way?
➤ What kinds of objects (besides planets and the Sun) occupy the solar system?

Star Groups

Very few stars travel through the universe alone. Most of them, about 85 percent, are members of a star family. The groups can be as large as galaxies with billions of stars or as small as a binary system made up of only two stars within a galaxy. The gravitational attraction of the stars on each other holds these groups together.

Small Groups of Stars

The smallest star group is a **binary system,** which contains only two stars. The two stars revolve around each other, held together by their gravitational pulls on each other. About half of all star groups are binary.

Beta Lyrai (LEER eye) is two stars. The stars are so close to each other, only 35 million kilometers (22 million miles) apart, that the gravity of each star pulls the other into an egg shape. They make one revolution around a center point every thirteen days.

Other small star groups have three or four stars and are called **multiple star groups.** Alpha Centauri (AL-fuh sen-TAWR-ee) is in a group of three stars. Like other star systems, multiple star groups remain together because of the attraction of their gravitational fields.

Alpha Centauri, a multiple star group, is part of the constellation Centaurus.

Science and HISTORY

In ancient times a famous binary star system was commonly used as an eye test. The middle "star" in the Big Dipper's handle is actually two stars, Alcor and Mizar. If a person could see both stars, he passed the eye test. Today most astronomers think these stars are only visual doubles. That is, they look close together but are actually too far apart to have much gravitational pull on each other. Interestingly, modern telescopes now show that both Mizar and Alcor are double stars, so there are at least four stars in the star system.

244

SCIENCE BACKGROUND

Visual doubles—Pairs of stars that appear to be binary but that are too far apart to have gravitational attraction may be called *visual doubles, visual pairs, optical doubles,* or *optical pairs.*

Mizar and Alcor—What was once thought to be the binary system of Mizar and Alcor may actually be six stars. When it was first discovered that Mizar was a binary pair, the two stars were named Mizar A and Mizar B. Through the use of spectroscopes, astronomers have determined that Alcor, Mizar A, and Mizar B are each binary stars. Binary stars such as these, which can be seen only with a spectroscope, are called *spectroscopic binaries.*

The prefix *bi-* means "two." A binary star system contains two stars.

Star charts may be found in encyclopedias and in other resources.

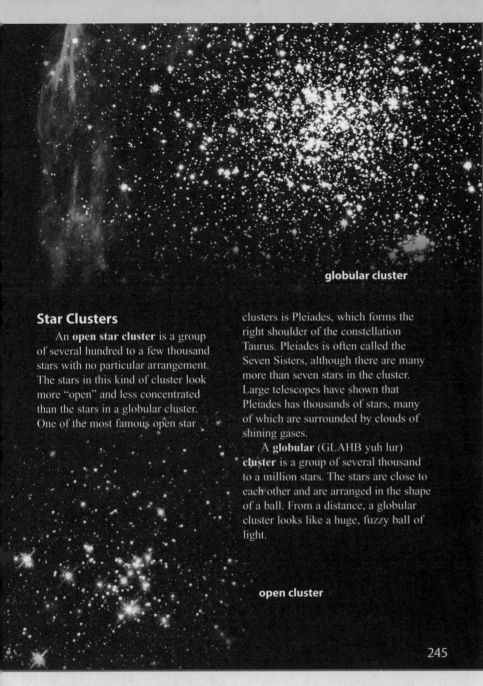

globular cluster

open cluster

Star Clusters

An **open star cluster** is a group of several hundred to a few thousand stars with no particular arrangement. The stars in this kind of cluster look more "open" and less concentrated than the stars in a globular cluster. One of the most famous open star clusters is Pleiades, which forms the right shoulder of the constellation Taurus. Pleiades is often called the Seven Sisters, although there are many more than seven stars in the cluster. Large telescopes have shown that Pleiades has thousands of stars, many of which are surrounded by clouds of shining gases.

A **globular** (GLAHB yuh lur) **cluster** is a group of several thousand to a million stars. The stars are close to each other and are arranged in the shape of a ball. From a distance, a globular cluster looks like a huge, fuzzy ball of light.

Discussion:
pages 244–45

➤ Which percentage of stars are part of a star family? *{85 percent}*

➤ What holds star groups together? *{gravitational attraction}*

➤ What is a *binary system*? *{It is the smallest star group, containing only two stars.}*

➤ Why do the stars in a binary group revolve around each other? *{Their gravitational pulls hold them together.}*

➤ What would happen if one star in a binary group became weaker? *{Possible answer: It would be drawn toward the stronger star, possibly causing it to "go nova."}*

➤ What kind of star group is *Alpha Centauri*? *{a multiple star group consisting of three stars}*

➤ How many stars make up a multiple star group? *{three or four}*

➤ What is an *open star cluster*? *{a group of several hundred to a few thousand stars with no particular arrangement}*

➤ What is a famous open star cluster? *{Pleiades}*

➤ What is a *globular cluster*? *{a group of several thousand to a million stars that are close to each other and are arranged in the shape of a ball.}*

➤ How are an open cluster and a globular cluster different? *{Possible answers: number of stars; arrangement of stars}*

Refer the student to *Science and History*.

➤ What stars were used as an eye test? *{Mizar and Alcor; the middle "star" in the Big Dipper's handle}*

➤ What are visual doubles? *{stars that appear to be together but are too far apart to have much gravitational pull on each other}*

Direct a DEMONSTRATION

Demonstrate binary stars

Instruct two students to stand facing each other. Instruct them to clasp left hands and right hands, forming an *X* between them. Instruct them to move in a circle around the center point made by their hands. The clasping of their hands is the "gravity" that keeps the two students "in orbit" around each other.

Observe the students as they move.

➤ Can you always see both students completely? *{no}*

➤ How does this compare to how we see binary stars? *{Possible answer: As binary stars orbit each other we may see both stars or only one at a time.}*

Discussion:
pages 246–47

▶ What is a *galaxy*? {*a huge star system that contains millions or billions of stars and covers many light-years of space*}

▶ What is the name of our galaxy? {*the Milky Way*}

▶ How large do astronomers think the Milky Way is? {*It measures about 100,000 light years and contains about 300 billion stars.*}

▶ What type of galaxy is the Milky Way? {*spiral*}

▶ Why do you think our galaxy got the name "Milky Way"? {*Possible answer: Through telescopes, the stars clustered together look milky white.*}

▶ What criteria do astronomers use to classify a galaxy? {*its shape and symmetry*}

 Refer the student to the pictures on Student Text pages 246–47.

▶ What are four kinds of galaxies? {*irregular, spiral, barred spiral, elliptical*}

💡 How is a barred spiral galaxy different from a spiral galaxy? {*A spiral galaxy looks similar to a wheel with at least two curved spokes. A barred spiral galaxy is a type of spiral galaxy that appears to have a straight bar through the center with the spokes coming from either end of the bar.*}

💡 Can a galaxy contain binary systems, constellations, and star clusters? {*yes*}

Galaxies

A **galaxy** is a huge star system that contains millions, or even billions, of stars and covers many light-years of space. Our own galaxy, the Milky Way, has about 300 billion stars. The distance across it is about 100,000 light-years.

Astronomers classify a galaxy according to its shape and symmetry. Our galaxy, the Milky Way, is a spiral galaxy.

irregular galaxy

spiral galaxy

barred spiral galaxy

246

 SCIENCE BACKGROUND

Our Sun in the Milky Way—The stars in a galaxy revolve around the galaxy's center point. For example, our Sun travels around the center of the Milky Way galaxy. The trip is a long one. Even though the Sun moves through space at 240 km (150 miles) per second, it would take it about 200 million years to complete one revolution. The time needed for a star to complete one revolution around the center of its galaxy is called a *galactic year*.

LANGUAGE Originally *galaxy* and *Milky Way* were interchangeable names for our star system. The Greek word for milk is *gala*. The name was used for our star system because of its milky appearance in the sky. According to the *Oxford English Dictionary*, Geoffrey Chaucer was one of the first people to use *galaxy* instead of Milky Way as a noun. It appeared in his poem "The House of Fame," written around 1384.

256   Lesson 108

elliptical galaxy

The Local Group

Our galaxy does not travel through space alone. More than thirty galaxies, including the Milky Way, form a cluster called the **Local Group.** These galaxies are our closest galactic neighbors. Altogether they take up an area in space three million light-years in diameter. The Milky Way and the Andromeda (an DRAHM ih duh) galaxies are two of the biggest in the Local Group.

Astronomers have discovered thousands of other galaxies. Most of these galaxies also occur in groups. One such cluster contains about 10,000 galaxies and makes our Local Group seem quite small in comparison.

Less than a century ago, though, astronomers were convinced that nothing existed beyond the boundary of the Milky Way. They thought our galaxy was a universe floating all alone in an enormous sea of empty space.

The expanse of our own galaxy is more than we can comprehend. Then we remember that there are billions of other galaxies, each holding millions and millions of stars spaced light-years apart. The universe is immense beyond our imagination. Yet God created all of it with a word and oversees it all with a glance. The more we learn about space, the more amazing we see our Creator's power to be and the more limited we see our own knowledge and abilities to be.

Discussion:
pages 246–47

➤ What is the *Local Group*? *{a cluster of galaxies, including the Milky Way, that travel together in space}*

➤ How does the size of the Milky Way compare to the sizes of other galaxies in the Local Group? *{It is one of the biggest galaxies in the Local Group.}*

➤ Which other large galaxy is part of the Local Group? *{Andromeda}*

➤ Is the Local Group the only group of galaxies that travel together? *{No; it is quite small in comparison to some groups of galaxies.}*

📖 Reiterate that God spoke the universe into existence.

➤ What can we learn about God by studying the universe? *{Accept any answers that indicate an understanding of God's omnipotence and majesty and/or man's limitations.}*

Direct an ACTIVITY

Help the student understand that sizes are relative. Instruct the student to start with his classroom and list progressively larger and larger locations. A possible list may include: classroom, school, city, county, state, country, continent, hemisphere, Earth, solar system, Milky Way, Local Group, universe. Emphasize the places discussed in this chapter.

Let each student make an individual list. Compare the lists of all the students. The goal is to include as many locations as possible.

Discussion:
pages 248–49

▶ **What are *asteroids*?** *{irregularly shaped pieces of rocks, metal, and dust}*

▶ **What is the largest known asteroid?** *{Ceres}*

▶ **Where are most asteroids in our solar system?** *{between Mars and Jupiter}*

💡 **Why do you think asteroids are sometimes called minor planets?** *{Possible answer: They orbit the Sun like planets.}*

Point out the pictures of the asteroids *Ida* and *Dactyl*.

▶ **What are some theories about how asteroids were formed?** *{Possible answers: Asteroids might be the remains of a planet that was destroyed in a collision. They might be leftover particles from the formation of our solar system. God created them from nothing.}*

📖 **Do we know how asteroids were formed?** *{no}* **What do we know about their formation?** *{that God formed asteroids}*

▶ **What theory about the formation of asteroids do we know is not true?** *{Any theory that does not recognize God as the Creator of the universe.}*

▶ **What causes a "shooting star"?** *{a meteoroid that enters Earth's atmosphere}*

▶ **Why does a meteor light up?** *{The friction caused by moving through the Earth's atmosphere causes heat and light.}*

▶ **Explain the relationships between a meteoroid, a meteor, and a meteorite.** *{A meteoroid is a chunk of metal or stone that heads toward our atmosphere. As the friction of our atmosphere causes the chunk to light up, we call it a meteor. If it lands on the Earth's surface, it is called a meteorite.}*

 Point out the location of Barringer Crater on a map.

▶ **Why has Barringer Crater been preserved while other craters on Earth have eroded?** *{Barringer Crater is in a desert, where there is little water erosion.}*

▶ **How did scientists determine that Barringer Crater was caused by a meteorite?** *{They examined other known meteorite sites and then compared the analyses of those sites to the soil samples from Barringer Crater.}*

Other Space Objects

Asteroids

Between Mars and Jupiter is an asteroid belt made of several thousand asteroids that orbit the Sun. **Asteroids** are irregularly shaped pieces of rock, metal, and dust and are sometimes called minor planets. Most asteroids are small, and some are as small as pebbles. Others are huge. The largest known asteroid, Ceres (SEER eez), has a diameter of about 1,000 km—approximately the distance from Detroit to Philadelphia.

Most asteroids orbit the Sun in a region between Mars and Jupiter. This asteroid belt is made of thousands of asteroids. Some astronomers think that the asteroids are the remains of a planet that was destroyed in a collision. Others think that asteroids are leftover particles from the formation of our solar system. No one can say for sure how God formed the asteroids, but we know

248

that any theory that does not recognize God as the Creator of the universe cannot be true.

Meteoroids

A **meteoroid** is a chunk of metal or stone that is moving toward Earth's atmosphere. We usually pay very little attention to meteoroids. However, sometimes a meteoroid enters Earth's atmosphere. Have you ever seen a "shooting star"? When the friction caused by a meteoroid's rapid movement through Earth's atmosphere causes it to light up, it is then called a **meteor**. Most meteors burn up in Earth's atmosphere, but a few impact Earth's surface. Those that hit Earth's surface are called **meteorites**.

Scientists believe they have found meteorite craters on Earth. One such crater is Barringer Crater in Arizona. This crater is nearly 1.6 km (1 mi) wide and 174 m (190 yd) deep. Since no one witnessed the formation of Barringer Crater, scientists tried to determine what caused the crater. They analyzed known meteorite impacts as well as the structure and soil of Barringer Crater. Scientists agree that the crater appears to

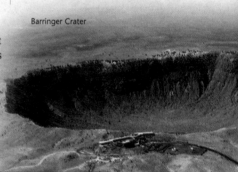
Barringer Crater

SCIENCE BACKGROUND

Asteroid satellite—Some asteroids are large enough to have other asteroids orbiting them.

Meteor shower—When a large number of meteors enters the atmosphere at about the same time and from the same area of space, we call it a *meteor shower*.

LANGUAGE

The word *asteroid* means "star-like."

comet

have been formed by a meteorite many years ago. Barringer Crater is one of the few craters on Earth that is still well preserved, probably because it is in a desert where there is little water erosion.

Comets

A **comet** is an icy chunk of frozen gases, water, and dust that orbits around the Sun over and over again. Some astronomers refer to comets as "dirty snowballs."

Comets have three parts: the *coma* and *nucleus* (which make up the head) and the *tail*. As a comet comes near the Sun in its orbit, the Sun melts some of the comet's ices, releasing the dust particles that trail behind the comet as its tail.

The time it takes a comet to orbit the Sun varies greatly. Some comets, called long-term comets, take thousands of years to orbit. Others, called short-term comets, take only a few years to return to Earth's view. The most famous short-term comet is Halley's Comet. It takes 76 years to make its journey around the Sun. Halley's Comet was last seen in 1986.

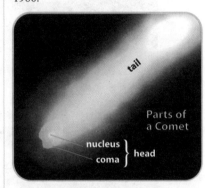
Parts of a Comet
tail
nucleus
coma } head

QUICK CHECK
1. How is a globular cluster different from an open star cluster?
2. What is the Local Group?
3. What is a "shooting star"?
4. What are the three parts of a comet?

249

Discussion:
pages 248–49

➤ What is a *comet*? *{a chunk of frozen gases, water, and dust that orbits the Sun}*

➤ What are the three parts of a comet? *{coma, nucleus, and tail}*

➤ Which two parts of a comet form the head? *{coma and nucleus}*

➤ What causes the tail of a comet to form? *{The Sun melts some of the comet's ices, releasing the dust particles that then trail behind it.}*

➤ What is the difference between a long-term comet and a short-term comet? *{A long-term comet takes thousands of years to orbit the Sun. A short-term comet takes only a few years to orbit the Sun.}*

💡 When is Halley's Comet expected to return to view from Earth? *{around the year 2062}*

Answers
1. A globular cluster is a ball-shaped group of several thousand to a million stars. An open cluster is a less concentrated group of several hundred to a thousand stars.
2. a group of galaxies traveling together, of which the Milky Way is a part
3. a meteor
4. coma, nucleus, and tail

Activity Manual
Bible Integration—page 156
This page explores some of the instances in which the Bible mentions stars.

Study Guide—pages 157–58
These pages review Lessons 105 and 108.

Assessment
Test Packet—Quiz 10-B
The quiz may be given any time after completion of this lesson.

HISTORY
Barringer Crater—The crater is named for Daniel Barringer, a mining engineer who led the research that concluded that the crater was made by a meteorite.

WRITING
Encourage the student to research and write about an Earth crater.

Barringer Crater—Arizona
Chicxulub—Yucatan Peninsula, Mexico
Aorounga—Chad, Africa
Wolf Creek—Australia
Manicouagan—Quebec, Canada

Some of the research may say that the craters were formed millions of years ago. Remind the student that according to the literal interpretation of the Creation account in the Bible, the Earth is not millions of years old.

Lesson 109

Activity Manual pages 159–60

Objectives
- Read and use a star chart
- Identify objects in the night sky
- Record observations

Materials
- star charts
- colored pencils or crayons
- transparency of TE page A33, *Stargazing*

Introduction

Display the various star charts.

➤ **Can a star chart always be used at any location on Earth?** {no} **Why?** {Possible answers: The stars seen from Earth change as Earth orbits the Sun. Most of the stars seen from the Northern Hemisphere are not seen from the Southern Hemisphere.}

Purpose for reading:
Activity Manual pages 159–60

The student should read all the pages before beginning the exploration.

Discussion:
Activity Manual pages 159–60

Display and mark the transparency, *Stargazing*, as Activity Manual page 159 is discussed.

➤ Using a star chart is similar to reading a road map.

➤ **Why are the stars different sizes?** {to show the magnitudes of the stars}

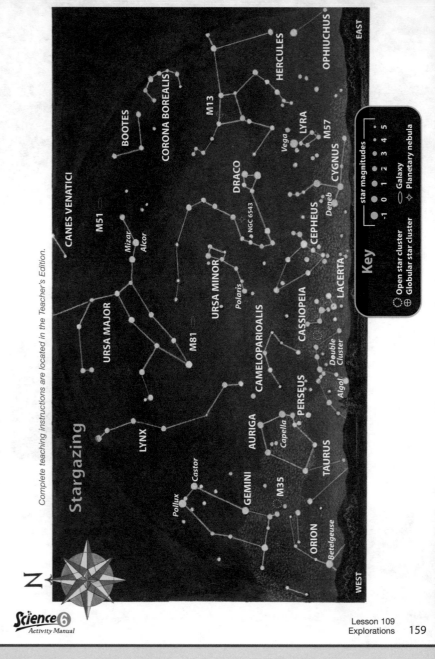

Star charts may be found in encyclopedias and in other resources.

The chart on Activity Manual page 159 shows the night sky for the Northern Hemisphere.

Encourage the student to find out whether there is a local astronomy group in your area. Many areas have amateur astronomers who meet regularly to look at the stars. They usually allow visitors to look through their telescopes.

Stargazing

Name _____

To be completed by _____ Date _____

Dear Parents,
   This activity is the culmination of your child's study of the stars. Although looking at pictures of the stars in a textbook is useful, it cannot compare to viewing the splendor of God's creation firsthand. The best conclusion to this chapter is for your child to observe the constellations in the evening sky. To get started, help him identify and locate a constellation. Then let him find the constellation patterns on his own.
   Please sign this letter to signify that your child has completed the assignment by the date above. Enjoy stargazing with your child!

Thank you for your help,
Your child's teacher

_____
Parent's signature

Suggestions for Stargazing
1. Wear layers of clothing to keep warm.
2. Find a place that is away from city lights.
3. Cover the end of a flashlight with red cellophane. The red light will not disturb your night vision like a white light can.
4. Use the star chart to help you find the constellations.
5. Lie down on a blanket or sit in a chair to steady your arms as you observe the stars with binoculars.

Write the answers.
1. What constellations did you find? *Answers will vary.*

2. List anything else you observed in the sky besides the stars and the Moon.
   Answers will vary.

3. What did you enjoy most about stargazing? *Answers will vary.*

4. Create your own constellation. On your own paper sketch it and name it.

Discussion:
Activity Manual pages 159–60

Draw the symbol for an open star cluster.
➤ **What does this symbol stand for?** *{open star cluster}*
➤ **With a yellow colored pencil, circle all the open star clusters.** *{next to Cassiopeia, M 35}*

Draw the symbol for a globular star cluster.
➤ **What does this symbol mean?** *{globular star cluster}*
➤ **With a purple colored pencil, circle the globular star cluster.** *{M 13}*

Draw the symbol for the planetary nebula.
➤ **What does this symbol stand for?** *{a planetary nebula}*
➤ **With a green colored pencil, circle the planetary nebulae.** *{NGC 6543, M 57}*

Choose several specific stars and direct the student to circle them with a white colored pencil. Examples include Mizar, Alcor, Pollux, Castor, Betelgeuse, Algol, Capella, Deneb, Vega, and Polaris.

Choose several constellations and direct the student to circle them with a blue colored pencil. Examples of constellations that are complete on this star chart are Lynx, Ursa Major, Bootes, Corona Borealis, Gemini, Auriga, Cameloparioalis, Cassiopeia, Cepheus, Ursa Minor, Draco, Lyra, and Hercules.

Continue asking questions about star magnitude, location of constellations to each other, etc., as time allows.

Direct the student to take Activity Manual pages 159–60 home and complete the stargazing activity.

Activity Manual
Explorations—page 159
This page allows students to experience reading a different kind of map.

Explorations—page 160
This page reinforces the star chapter with a stargazing activity.

Assessment
Rubric—TE pages A50–A53
Select one of the prepared rubrics, or design a rubric to include your chosen criteria.

Lesson 110

Student Text pages 250–51
Activity Manual pages 161–62

Objectives
- Measure mass and length
- Use a chart to record information
- Make and test predictions

Materials
- See Student Text page

Introduction

➤ Which space objects have a lot of craters? *{Possible answers: the Moon, Mercury}*

➤ Why does Earth not have many craters? *{The friction of Earth's atmosphere burns meteors up before they reach Earth.}*

➤ Craters have different sizes and depths. Why do you think some craters penetrate deeply into a space object while others leave a smaller print? *{Accept any answer.}*

➤ Today you will have the opportunity to make your own craters.

Purpose for reading:
Student Text pages 250–51
Activity Manual pages 161–62

The student should read all the pages before beginning the activity.

ACTIVITY: Crater Creations

Meteorites do not strike only Earth's surface. Several other planets and moons have many craters formed by meteorites hitting their surfaces. The size, shape, and composition of a meteorite affect the depth of the crater that it causes.

In this activity, you will make craters using balls of similar sizes but different masses.

Process Skills
- Hypothesizing
- Measuring
- Observing
- Recording data
- Identifying and controlling variables
- Communicating

Problem
How does the mass of a dropped object affect the depth of the crater it makes?

Procedure
1. Write a hypothesis in your Activity Manual, stating which object you think will make the largest crater at each height.
2. Write the names of the three objects in the table provided on the Activity Manual page. The name of each object should be written next to the color you will use to represent the object on the graph.
3. Weigh each object and record each object's mass on the table.
4. Lay out newspaper on the floor and place the foil pan on top of the newspaper. The newspaper should extend two or three feet beyond the sides of the pan.
5. Pour 3–4 inches of flour into the foil pan. Shake the pan gently to even the flour. Sprinkle a thin layer of chocolate milk mix on top of the flour. (The milk layer will allow you to see your craters better.) Once you begin making craters, do not bump or shake the pan. Disturbing the pan will destroy your craters.

Materials:
- three different round objects with similar diameter but different mass, such as a golf ball, Ping-Pong ball, and rubber ball
- newspaper
- deep foil pan or a dishpan
- 2–3 bags of flour
- powdered chocolate milk mix
- centimeter ruler
- mass scale
- 2 meter sticks or a tape measure
- 3 colored pens or pencils
- Activity Manual

250

Teacher Helps

Deep containers work best.

Rather than using powdered chocolate milk mix, you may use any colored powdered substance to coat the flour. Some possible substitutions include powdered tempera paint, powdered soft drink mix, or sand.

If you do not have a scale available to measure the mass of each of the objects, you can have the students hold the objects and rank each group of three objects from lightest to heaviest.

Although the graph indicates measuring the depth in centimeters, it may be necessary to change to millimeters if craters are very close to the same depth.

Because density is the actual determining factor in how deep a crater forms, it is very important to have balls as close to the same circumference as possible. This allows mass and density to be close to the same.

Science Background

Density or mass—Scientists would actually calculate and use density rather than mass when conducting experiments that are similar to this one. Mass is adequate for the student to use in this activity.

262 Lesson 110

6. Have your partner hold the meter stick so it touches the top of the flour. Drop each "meteorite" from 20 cm above a different area of the pan. Leave the objects in their craters until all are dropped.

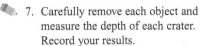7. Carefully remove each object and measure the depth of each crater. Record your results.

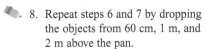8. Repeat steps 6 and 7 by dropping the objects from 60 cm, 1 m, and 2 m above the pan.

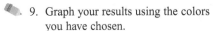9. Graph your results using the colors you have chosen.

Conclusions

- Was there a relationship between the mass of the object and the depth of its crater?
- Did the results change as the height variable changed?

Follow-up

- Compare the results of dropping a different set of objects, such as a baseball, an orange, and a tennis ball.
- Compare the depths of the craters of three objects that have the same mass but are different sizes.

251

Procedure

Observe the student's use of materials to ensure accuracy and consistency.

Point out the importance of keeping the measuring device as straight and tall as possible.

Remind the student to assign a different color to each object on the chart on Activity Manual page 161. The student should use these same colors to complete the graph on Activity Manual page 162.

Guide the student in writing numbers in the chosen increments along the left side of the graph, *Depth of craters*.

Conclusions

Provide time for the student to evaluate his hypothesis and answer the questions.

Activity Manual
Activity—pages 161–62

Assessment
Rubric—TE pages A50–A53

Select one of the prepared rubrics, or design a rubric to include your chosen criteria.

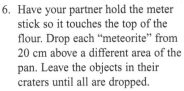

SCIENCE PROCESS SKILLS
Recording data

➤ What did you use to record your data? *{a chart}*

➤ Do you think that this is an effective way to record data on this activity? *{yes}* **Why?** *{Answers will vary, but elicit that the chart allows you to see all the information in one place.}*

➤ What is the advantage of making the graph? *{It is easier to see the relationships between the data collected from each object.}*

Chapter 10

Lesson 110 263

Lesson 111

Student Text page 252
Activity Manual pages 163–64

Objectives

- Recall concepts and terms from Chapter 10
- Apply knowledge to everyday situations

Introduction

Material for the Chapter 10 Test may be taken from Activity Manual pages 154, 157–58, and 163–64. You may review any or all of the material during the lesson.

You may choose to review Chapter 10 by playing "Constellations" or a game from the *Game Bank* on TE pages A56–A57.

Diving Deep into Science

Questions similar to these may appear on the test.

Answer the Questions

Information for Answer the Questions can be found on the following Student Text pages:

1. page 248
2. page 234
3. page 231

Solve the Problem

In order to solve the problem, the student must apply material he has learned. The answer for this Solve the Problem is based on the material on Student Text page 239. The student should attempt the problem independently. Answers will vary and may be discussed.

Activity Manual

Study Guide—pages 163–64

These pages review some of the information from Chapter 10.

Assessment

Test Packet—Chapter 10 Test

The introductory lesson for Chapter 11 has been shortened so that it may be taught following the Chapter 10 test.

Diving Deep into Science

Answer the Questions

1. Why are asteroids sometimes called minor planets?
   They are small, rocky objects that orbit the Sun as planets do.

2. What is the difference between an eclipsing variable star and a pulsating variable star?
   An eclipsing variable star is actually two stars that rotate around each other. Only the apparent magnitude changes. A pulsating variable star is only one star. Its absolute magnitude and its apparent magnitude change.

3. Why did the use of the Hubble Space Telescope cause astronomers to add to Hipparchus's scale?
   Because the Hubble Space Telescope is above the atmosphere, it is able to "see" fainter stars than telescopes on Earth can see. Astronomers continue to add to the scale as additional faint stars are seen.

Solve the Problem

You have a friend who tells you that when she's unsure about making a decision, she likes to look at a horoscope to get advice about what she should do. She claims to be a Christian, but she thinks that since the Bible was written long ago, it can't help her with the decisions of today. She asks you what you think about her ideas. What will you tell her?

Answers will vary, but emphasize that the Bible is the Word of God and that God is unchangeable. God tells us in His Word that the Bible should be "a lamp unto [our] feet and a light unto [our] path" (Ps.119:105). Christians should not look for guidance from sources that do not emphasize God's Word.

252

Review Game

Constellations

Divide the class into two teams. Choose two constellations with the same number of stars in each. Give each team a picture of one of the constellations. When a team member answers a review question correctly, he then draws a star to represent one of the stars in the constellation. Alternate questions between teams. The first team to draw all of the stars in its constellation wins. Bonus points may be given if the team can correctly identify the constellation. Use other constellations and continue playing as time allows.

Suggested constellations

5 stars: Lyra, Cassiopeia, Cancer

7 stars: Andromeda, Ursa Minor

8 stars: Lacerta, Lepus, Crater

These constellations are shown on TE page 260.

11

Solar System

GREAT & MIGHTY Things

In the 1500s Polish astronomer Copernicus published his theory about the order of the universe. He believed that the Sun was the center of the universe and that the planets revolved around the Sun. For this startling publication, Copernicus was considered a heretic. Most people believed that the planets, stars, and even the Sun revolved around Earth. Copernicus's idea was not new, but he was the first in about two thousand years to use mathematics to prove his idea. Copernicus's theory was debated by other scientists and leaders of the time. Only after many years did scientists such as Sir Isaac Newton and Johannes Kepler prove that although the Sun is not the center of the universe, it is indeed the center of our solar system. God's orderly pattern for the universe allows man to prove mathematically ideas that he can not prove experimentally.

253

Student Text page 253
Activity Manual page 165

Lesson 112

Objectives
- Demonstrate knowledge of concepts taught in Chapter 10
- Recognize that God's creation is orderly

Introduction

Note: This introductory lesson has been shortened so that it may be taught following the Chapter 10 test.

Can you imagine the devastation that would occur if we could not depend on the normal cycle of day and night? From the very beginning of Creation, God established "the evening and the morning" (Gen. 1:5). The entire universe maintains an orderliness that allows man to calculate where heavenly bodies will be located at any specific time. Even the mathematical systems that allow us to calculate these movements are a gift from God.

Chapter Outline
I. The Sun
   A. Parts of the Sun
   B. Solar Storms
II. The Seasons
III. The Planets
   A. The Inner Planets
   B. The Outer Planets
IV. Space Exploration
   A. Rockets
   B. The Space Shuttle
   C. Satellites
   D. Probes
   E. International Space Station

Activity Manual
Preview—page 165

The purpose of this Preview page is to generate student interest and determine prior student knowledge. The concepts are not intended to be taught in this lesson.

SCIENCE BACKGROUND

Aristarchus—Aristarchus (310–230 B.C.) was the first to calculate the sizes of and distances between the Sun, Earth, and the Moon. Although his measurements were not completely accurate, he correctly determined the relative sizes of the Sun, Earth, and the Moon. He also stated that Earth revolves around the Sun and that the Moon is closer to Earth than the Sun.

 BIBLE Just as the Sun is the center of our solar system, so should Christ be the center of a Christian's life. A godly Christian will center his daily thoughts, emotions, and actions on the One who rules in his heart as Savior. [BATs: 2b Servanthood; 3b Mind; 3c Emotional Control; 3d Body as a temple]

Lesson 113

Student Text pages 254–57
Activity Manual page 166

Objectives

- Identify the parts of the Sun
- Describe the characteristics of a solar storm
- Explain why Earth experiences seasons
- Understand that the Sun's gravitational pull keeps the planets in orbit

Vocabulary

photosphere	solar prominence
chromosphere	solar wind
corona	aurora
sunspot	revolution
faculae	axis
solar flare	rotation

Materials

- sunglasses
- sunscreen
- globe
- protractor

Introduction

Hold up the sunscreen and sunglasses for the students to see.

➤ **Why do we use these items?** *{to protect us from the effects of the Sun}*

➤ **How can the Sun that is so far away be harmful to us?** *{Answers may vary. Elicit that although the Sun is far away, it is very powerful.}*

➤ **The Sun, however, is also very important for our survival. Name some ways we depend on the Sun every day.** *{Possible answers: Plants need sunlight to survive, and people depend on plants for food. The Sun warms the Earth. The Sun gives us light.}*

Purpose for reading:
Student Text pages 254–57

➤ In what ways does the Sun affect life on Earth?
➤ What is an aurora?

Discussion:
pages 254–55

➤ **What are some ways the Sun affects life on Earth?** *{Possible answers: It provides heat and light. It affects Earth's climate and food supply. Its storms can disrupt our communication and navigation equipment. It can burn our skin and damage our eyes.}*

➤ **How far is the Sun from Earth?** *{150,000,000 km, or 93,000,000 mi}*

➤ **How long does it take the Sun's light to reach Earth?** *{less than 8½ minutes}*

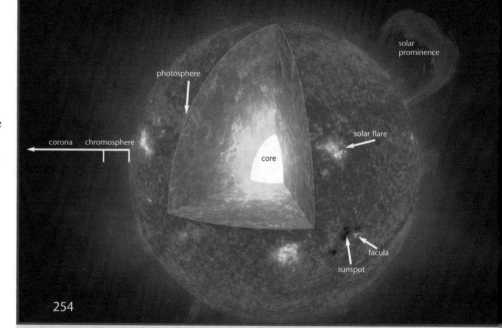

The Sun

Near the edge of the Milky Way shines an average-sized star that has far from average importance. This star, our Sun, is the center of our solar system. If God removed the Sun, there would be no life on Earth. Our heat, light, nourishment, and climate are all dependent on the Sun.

The Sun is the only star in the universe close enough for us to study. It is about 150,000,000 km (93,000,000 mi) from Earth. It takes less than 8½ minutes for light from the Sun to reach Earth. Even at this distance, though, the Sun is so powerful that it can burn your skin, and it can damage your eyes if you look directly at it.

Parts of the Sun

The surface of the Sun is called the **photosphere**. Because of Earth's great distance from the Sun, the photosphere appears smooth to us. However, it is actually bumpy and in constant motion. Gases move up from the interior of the Sun and create bulges on the surface like the surface of a pot of boiling water. Above the Sun's surface is its atmosphere, called the **chromosphere**. The **corona** is the outermost part of the Sun. Located above the Sun's chromosphere, the corona can be seen only during a solar eclipse or by special astronomical instruments. The corona is sometimes called "the crown" of the Sun.

SCIENCE BACKGROUND

Sunspots and solar flares—The number of sunspots that occur on the Sun appears to have a cycle of about eleven years.

A sunspot is actually many times larger than Earth.

Solar flares release energy that is greater than millions of atomic bombs.

The Sun affects many aspects of our life. But in heaven there is no need of the Sun, because the Son, Jesus Christ, is there. Just as we depend on the Sun for physical life, so we must depend on the Son for spiritual life. Jesus Christ's death and resurrection provide salvation to anyone who believes and trusts in Him. [BATs: 1a Understanding Jesus Christ; 1b Repentance and faith; Bible Promise: A. Liberty from Sin]

266 Lesson 113

Solar Storms

Just as Earth has storms, the Sun also has storms. These storms do not involve lightning or rain. The storms on the Sun are magnetic storms. They can affect life on Earth by disrupting communications satellites and GPS (Global Positioning System) navigation signals. The storms seem to be related to dark spots on the photosphere of the Sun, called **sunspots.** Astronomers believe that these spots look dark because they are cooler than the surrounding gases. Sunspots are usually accompanied by **faculae** (FAK yuh lee), bright clouds of gas on the photosphere.

Solar storms may also explode from the Sun's photosphere, creating **solar flares.** These flares become 20–30 times brighter than the rest of the Sun and then fade away in about one hour. An even more spectacular solar event is a **solar prominence.** Solar prominences are huge streams of gas that extend out past the Sun's chromosphere and into the Sun's corona. Unlike a solar flare, a solar prominence can last for days or even weeks.

Earth's magnetic field

aurora

A **solar wind** is made up of electrically charged particles from the Sun. Solar storms may cause an increase in the flow of these charged particles. The solar wind carries these particles from the Sun to Earth's atmosphere. Earth's magnetic field traps some of the particles and pulls them toward Earth's poles. As the particles collide with atoms and molecules in Earth's atmosphere, they emit energy in the form of beautiful colors of light. People near the North and South Poles are able to view this beautiful light show, called an **aurora** (uh RAWR uh). Near the North Pole an aurora is called *aurora borealis*, or the *northern lights*. An aurora that occurs near the South Pole is called *aurora australis*.

LANGUAGE

The surface of the Sun is called the *photosphere*. The Greek root *photo* means "light," and the word *sphere* refers to the visible circle of the Sun, which is the source of the Sun's light.

Aurora is the Latin word for *dawn*.

Borealis is the Latin word for *northern*.

Australis is the Latin word for *southern*.

➤ Why do you think the aurora near the South Pole is called the aurora australis? *{Possible answer: Australis means southern, and this aurora appears in the southern hemisphere.}*

MATH

The Sun is 150,000,000 km from Earth.

➤ If it takes 8.3 minutes for the Sun's light to reach Earth, how many km per minute is the light traveling? *{18,072,289.16 km/min}*

Calculate the km/min: 150,000,000 km ÷ 8.3 min = 18,072,289.16 km/min.

➤ How many km per second is the light traveling? *{301,204.82 km/sec}*

First calculate the number of seconds in 8.3 min: 8.3 min × 60 = 498 sec. Then calculate the km/sec: 150,000,000 km ÷ 498 sec = 301,204.82 km/sec

Round 301,204.82 km/sec to 300,000 km/sec. This is the generally accepted number for the speed of light. In the field of astronomy and space exploration, the speed of light is a very important concept.

Discussion:
pages 254–55

Point out each aspect of the Sun on the diagram on Student Text page 254 as it is discussed.

➤ What is the surface of the Sun called? *{the photosphere}*

➤ What is the Sun's atmosphere called? *{the chromosphere}*

➤ Where is the Sun's corona? *{It is the outermost part of the Sun located above the Sun's chromosphere.}*

➤ When can scientists see the corona of the Sun? *{during a total eclipse or when using special astronomical instruments}*

➤ What kind of storms does the Sun have? *{magnetic storms}*

➤ What are *sunspots*? *{dark spots on the Sun's surface}*

➤ What are the bright clouds of gas around sunspots called? *{faculae}*

➤ What is a *solar flare*? *{an explosion of a solar storm on the Sun's photosphere}*

➤ Which extends out farther and lasts longer—a solar flare or a solar prominence? *{a solar prominence}*

➤ What is a solar wind made of? *{electrically charged particles}*

➤ What is seen when these particles collide with atoms and molecules in Earth's atmosphere? *{an aurora}*

➤ What is an *aurora*? *{energy emitted in the form of beautiful colors}*

➤ What is another name for an aurora that occurs near the North Pole? *{aurora borealis, or the northern lights}*

💡 Why do auroras occur only near the Poles? *{Earth's magnetic field traps some of the particles from the solar wind and pulls them toward Earth at the Poles.}*

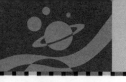

Discussion:
pages 256–57

➤ **How can the Sun hold the entire solar system in orbit around itself?** {Possible answers: The Sun's huge mass exerts a strong gravitational pull. God created the balance between the Sun's gravitational pull and the speed of each planet.}

➤ **Why are planets closer to the Sun at some times and farther away at other times?** {They have elliptical orbits.}

💡 **What shape is an ellipse?** {oval}

➤ **What makes one year for a planet?** {one complete trip around the Sun; a revolution around the Sun}

💡 **Is the revolution period (year) for each planet the same?** {no} **Why?** {The length of time it takes for each planet to go around the Sun is different.}

➤ **What is an *axis*?** {an imaginary line around which a planet rotates}

➤ **What is a complete rotation called?** {a day}

💡 **What is a hemisphere on Earth?** {half the Earth}

Point to the Northern and Southern Hemispheres on the globe.

➤ **What determines the seasons on Earth?** {the tilt of Earth during its revolution around the Sun}

Draw a vertical line. Mark a point near the center of the line. Place a protractor on the right side of the line and measure 23½° from the top. Draw a second line that intersects the first line through the center point at 23½°. This second line illustrates the tilt of the Earth.

The Seasons

The Sun contains over 98 percent of the mass of the entire solar system. Its huge mass exerts such a strong gravitational pull that it keeps the planets in orbit around it.

When God created the universe, He set everything in motion. The speed of each planet in our solar system balances the gravitational pull of the Sun. The planets stay in regular orbits because of this perfect balance. Each orbit that a planet makes around the Sun is called a **revolution**, or year. Because the planets' orbits are elliptical instead of circular, there are certain times in each planet's revolution when it is closer to the Sun.

Most areas of the earth have seasonal changes of temperature and light. A common misconception is that the seasons change depending on how close or far away from the Sun the earth is in its elliptical orbit. If that were true, people in the Northern and Southern Hemispheres would experience the same seasons at the same time. However, we know that they do not. So what does cause Earth's seasons?

Like the other planets, Earth rotates on an axis. The **axis** is an imaginary line around which a planet rotates. Each complete **rotation** around the axis is a day. Earth's axis, however, is not straight up and down. It is tilted 23½ degrees from the vertical. The four seasons on Earth are determined by Earth's tilt during its revolution around the Sun. As Earth travels around the Sun, sometimes the Northern Hemisphere tilts toward the Sun. At other times the Southern Hemisphere points toward the Sun. The hemisphere pointing toward the Sun receives the most direct sunlight and experiences summer. The day it receives the most direct sunlight is the longest day of the year in that hemisphere and is called the *summer solstice* (SOHL stihs). The *winter solstice* is the shortest day of the year for that hemisphere.

Between summer and winter are autumn and spring, which bring milder temperatures. During these seasons, neither hemisphere points directly toward or away from the Sun. Both hemispheres receive about the same amount of sunlight, and day and night are of about equal length in all parts of the world. The beginnings of these two seasons are called the *vernal equinox* (VUHR-nuhl EE-kwuh-NAHKS) and the *autumnal equinox* (aw-TUM-nuhl EE-kwuh-NAHKS). The word *equinox* comes from Latin. *Equi* means "equal," and *nox* means "night," so *equinox* refers to the equal length of day and night.

> **QUICK CHECK**
> 1. How far is the Sun from Earth? How long does it take the Sun's light to reach Earth?
> 2. What characteristic of Earth determines the seasons?
> 3. What is the difference between a planet's rotation and its revolution?

256

SCIENCE MISCONCEPTIONS
The seasons are not caused by how close to or far away the Earth is from the Sun.

TEACHER HELPS
Use a different color to draw the line showing Earth's tilt.

268   Lesson 113

The Seasons
(Northern Hemisphere)

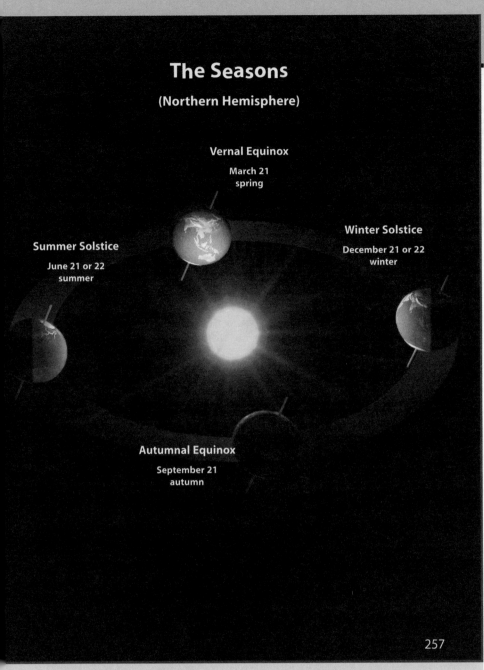

Discussion:
pages 256–57

Refer the students to *The Seasons* on Student Text page 257.

➤ When does a hemisphere receive the most direct sunlight and have the longest day? {at the summer solstice}

➤ What is the shortest day in a hemisphere? {the winter solstice}

➤ When the Northern Hemisphere is tilted toward the Sun, which season does it experience? {summer}

➤ Which season does the Southern Hemisphere experience while the Northern Hemisphere is experiencing summer? {winter} How can you tell? {It is not receiving the most direct sunlight.}

💡 How does the amount of daylight and darkness change as the Northern Hemisphere approaches its summer solstice? {The days become longer and the nights become shorter.}

💡 What happens at the South Pole during its summer solstice? {The Sun shines all day and all night, so there is no darkness.} What happens during its winter solstice? {There is no daylight.}

➤ What is the beginning of spring? {the vernal equinox} The beginning of autumn? {the autumnal equinox}

➤ What does *equinox* refer to? {an equal length of day and night}

➤ Why would the equator always experience warm temperatures? {It always receives direct sunlight.}

Answers
1. 150,000,000 km or 93,000,000 mi; less than 8½ minutes
2. the tilt of Earth as it revolves around the Sun
3. Rotation is the planet's spin on its axis. Revolution is the planet's orbit around the Sun.

Activity Manual
Reinforcement—page 166

Direct a DEMONSTRATION

Demonstrate seasons

Materials: lamp with its shade removed, globe

Position the lamp on a desk and darken the room. Hold the globe about 1 m from the lamp at the same level as the bulb. Walk around the lamp. Keep the tilt of the globe the same at all times. Stop the globe in the positions of the summer and winter solstices and of the vernal and autumnal equinoxes.

➤ Why is the temperature always hot at the equator? Why are the Poles always cold? {Because of Earth's tilt, the equator always receives direct light from the Sun, but the Poles do not receive direct light.}

Position the globe so the United States is experiencing summer.

➤ Which of the following countries experience summer at the same time as the United States? Argentina? {no} France? {yes} China? {yes} Australia? {no}

If desired, hold the globe close to the bulb long enough for the globe to feel warm where the light hits it the most.

Lesson 114

Student Text pages 258–61
Activity Manual pages 167–68

Objectives

- Describe similarities among the inner planets
- Explain how man has gradually learned about the planets
- Identify characteristics of Mercury, Venus, and Mars

Vocabulary

terrestrial

Materials

- picture of the solar system

Introduction

Display the picture of the solar system.

➤ Have you ever wondered what it would be like to visit one of the other planets in our solar system?

➤ If you were able to get there, could you survive?

➤ As we read today, you may want to imagine yourself visiting another planet.

Purpose for reading:
Student Text pages 258–61

➤ How are the inner planets similar?

➤ What has man used to learn about the planets?

Discussion:
pages 258–59

➤ Which planets are called the inner planets? *{Mercury, Venus, Earth, Mars}*

💡 What does *terrestrial* mean? *{Elicit that it means relating to Earth and its inhabitants.}*

➤ Why are the inner planets known as terrestrial planets? *{They are rocky, dense, and Earthlike in composition.}*

📖 Do you think it is interesting that Earth is the only inner planet that is not full of craters? Explain your answer. *{Accept any answer, but elicit that it is another evidence of God's design in setting Earth apart as the home for people.}* [Bible Promise: I. God as Master]

The Planets

Nine planets revolve around our Sun. We call Mercury, Venus, Earth, and Mars the inner planets because they are closest to the Sun. These four planets are also known as the **terrestrial** planets because they are rocky, dense, and earthlike in composition. These planets are small, solid, and relatively close together, and all except Earth are covered with craters. An observer on Earth can see Mercury, Venus, and Mars in the night sky without a telescope.

Between Mars and Jupiter is an area full of thousands of asteroids. This area is called an *asteroid belt*. Beyond the asteroid belt are the outer planets. Jupiter, Saturn, Uranus (YOOR uh nuhs), and Neptune are considered "gas giants" because their surfaces are made of gases. These planets are massive compared to the inner planets. They are also far away from each other. The distance between any two of the gas giants is greater than the distance between the Sun and Mars.

At the far reaches of the solar system lies Pluto. Pluto is not at all like the other outer planets. It is not a gas giant but rather a small solid-core planet.

First, man relied on his eyesight to observe the planets. Later, small telescopes helped him gather more information. But only in the last forty years or so has man developed technology that enables him to gain large amounts of new knowledge. With

the Sun — Mercury — Venus — Earth — Mars — asteroid belt — Jupiter

planet size to scale

258

SCIENCE BACKGROUND

Point of reference—Often a person uses a point, or frame, of reference to relate new information to something familiar. Since we have a good understanding of Earth measurements such as days and years, these Earth measurements are used as reference points to help us understand the rotations and orbits of the other planets.

 LANGUAGE Use a sentence to help the student remember the order of the planets. Use the first letter of each planet as the first letter of each word in the sentence. Write your own or use one of these.

My **v**ery **e**conomical **m**other **j**ust **s**natched **U**ncle **N**eil's **p**enny.

My **v**ery **e**nergetic **m**other **j**ust **s**wept **u**nder **N**eil's **p**late.

newer, more advanced telescopes on Earth and the Hubble Space Telescope in space, new ideas sometimes replace old ideas. But probes launched into space have given astronomers the best information. Some of these probes have flown by the planets at close range, and others have actually landed on the planets.

If the student completed the exploration "A Different Look" in Chapter 10, remind him that the distances of the foil stars were a scale model of the distances of real stars.

The sizes of the Sun and planets illustrated on Student Text pages 258–59 are to scale, but the distances between the planets are not to scale. In the diagram of the orbits, the distances are to scale, but the sizes of the Sun and planets are not to scale.

The word *terra* is Latin, meaning "land."

➤ Name other words that contain *terra*. {Possible answers: extraterrestrial, terra cotta, territory, terrain}

Discussion:
pages 258–59

➤ Which space bodies exist between Mars and Jupiter? {asteroids}

➤ What is this area called? {an asteroid belt}

➤ Which planets are called the outer planets? {Jupiter, Saturn, Uranus, Neptune, Pluto}

➤ Which planets are gas giants? {Jupiter, Saturn, Uranus, Neptune}

➤ How are the gas giants different from the other planets in our solar system? {Possible answers: They are much larger. They are gaseous rather than dense and rocky. They are far away from each other.}

➤ How is Pluto different from the other planets? {It is not a gas giant; rather, it is a small, solid-core planet.}

➤ Which advancements in technology have allowed man to learn more about the planets? {telescopes, space probes}

 Why would a probe likely give better information than telescopes? {Possible answers: Probes can get closer to planets. Probes can sometimes touch and collect materials from a planet's surface.}

Refer the student to the illustrations and diagrams on Student Text pages 258–59.

➤ What does "planet size to scale" mean? {Answers will vary, but elicit that in this illustration the relative sizes of the planets are accurate, although the pictured planets are much smaller.}

➤ Why is the orbit scale different from the size scale? {Answers will vary, but elicit that the orbit scale is much larger than the size scale.}

As time allows, continue asking questions about the relationships between the planets' sizes and their orbits as shown by the diagrams.

Discussion:
pages 260–61

💡 What influences the gravitational pull of an object? {its mass}

➤ How does Mercury's size compare with the sizes of the other planets? {It is the second smallest planet.}

➤ How does Mercury's size compare with the size of Earth? {It is 1/3 the size of Earth.}

➤ Why can't Mercury hold an atmosphere? {Mercury has a weak gravitational field because of its small mass.}

💡 Why do you think Mercury has so many craters? {Possible answers: It has no atmosphere to burn meteors before they hit the planet. The gravity of the Sun may pull more space objects toward Mercury.}

➤ Why does Mercury have the shortest year in our solar system? {It is close to the Sun, so its orbit is smaller than those of the other planets.}

➤ Why was Venus called the "Morning Star" and also the "Evening Star"? {People thought it was actually two different stars that shone brightly in the morning and in the evening.}

➤ Why is Venus considered Earth's twin? {It is almost the same size as Earth. It is the closest planet to Earth. Both planets are similar distances from the Sun.}

➤ In which ways is Venus a hostile environment for humans? {Possible answers: The temperatures are too hot for humans to survive. The density of the air would crush humans.}

Retrograde means "moving backward."

➤ What does it mean that Venus has "retrograde rotation?" {The planet rotates from east to west, which is the opposite of Earth's rotation.}

💡 Why did scientists have a hard time gathering data about Venus? {Its thick cloud cover prevented observation.}

➤ How did scientists eventually find out more about Venus? {Space probes such as *Mariner 2*, *Magellan*, and *Venera* penetrated the cloud cover.}

272   Lesson 114

Mercury

The Inner Planets

Mercury: the planet closest to the Sun

Mercury is the planet closest to the Sun. It can often be seen near the horizon before sunrise and after sunset. Mercury is the second smallest planet, about ⅓ the size of Earth. Because of its small size, Mercury has a weak gravitational field and therefore cannot hold an atmosphere.

The temperature on Mercury varies greatly. It reaches more than 400°C (800°F) on the side facing the Sun and drops to as low as –170°C (–300°F) on the side facing away from the Sun.

Mercury has the shortest year in the solar system. It takes 88 Earth days for Mercury to revolve around the Sun. However, it has a very long "day." It takes 59 Earth days for Mercury to rotate on its axis one time.

During 1974 and 1975 cameras onboard the probe *Mariner 10* took the first clear pictures of Mercury. These pictures revealed a barren world scarred by craters, similar to Earth's Moon.

Venus

260

Venus: the Evening Star

Venus is the brightest object in the morning and evening sky because its thick cloud covering reflects sunlight well. Centuries ago observers thought that Venus was actually two separate stars, and they called it "The Morning Star" and "The Evening Star."

Venus is sometimes referred to as Earth's twin. The two planets are almost the same size and are similar distances from the Sun. Venus is also the closest planet to Earth. However, the two planets are actually very different.

Venus would not be a friendly place for human visitors. Its atmosphere is 96 percent carbon dioxide. Carbon dioxide traps the Sun's heat, similar to the way in which a car with its windows rolled up on a hot day traps heat. In fact, even though Mercury is closer to the Sun, the thick cloud that covers Venus causes it to be hotter than Mercury. Temperatures on Venus reach 450°C (900°F)—hot enough to melt lead! Venus also has an atmosphere so dense that it would crush a person in just a few seconds.

Unlike most of the planets, Venus has a retrograde rotation. It rotates from

MATH — Draw a Venn diagram with two intersecting ovals. Direct the student to label one outer section Earth and the other outer section Venus. List differences of Earth and Venus in each corresponding section. List any similarities of Earth and Venus in the center, intersecting section. Direct the student to make another Venn diagram to compare and contrast Earth and Mars.

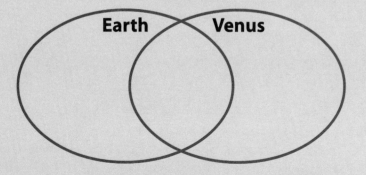

east to west instead of west to east as Earth does. Venus's period of rotation and revolution are almost the same. A year on Venus is equal to 224.7 Earth days, and a day on Venus is equal to 243 Earth days.

Little was known about Venus's atmosphere and surface until spacecraft penetrated its thick cloud cover. But since Venus is so close to Earth, many space probes, such as *Mariner 2, Magellan,* and *Venera,* have visited the planet.

Mars: the red planet

Earth is the third planet from the Sun. The fourth planet, Mars, is the last of the inner planets. Though it is smaller than Earth, Mars is the planet most like Earth. Mars has a tilt similar to Earth's, so it also experiences seasonal changes. Its polar ice caps grow and shrink depending on the season. Because of its thin atmosphere and distance from the Sun, Mars is very cold, having an average temperature of −63°C (−81°F). But on a sunny summer day it might reach 30°C (86°F). Unlike Earth's nitrogen and oxygen atmosphere, Mars's atmosphere is mainly carbon dioxide.

One of the brightest objects in the night sky, Mars is visibly red. Iron oxide, or rust, in its soil causes its rusty color. Although Mars is dry like a desert, it has some land features that cause some scientists to believe Mars might have had liquid water at one time. Currently scientists suspect the presence of liquid and frozen water under the surface of Mars.

A day on Mars is 24.5 Earth hours—almost the same length as an Earth day. A year on Mars, however, is about 687 Earth days, almost twice as long as an Earth year.

Because Mars is more like our own planet than any other, the idea that life might exist on the planet has fascinated people for centuries. In 1997 the *Mars Pathfinder* landed on Mars. A remote-controlled rover named *Sojourner* took pictures of the planet and gathered soil and rock samples. *Sojourner* found no evidence of any type of life, past or present, on the planet.

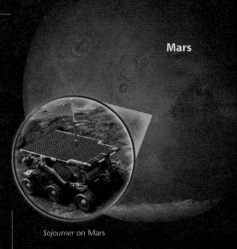

Mars

Sojourner on Mars

1. Name the planets, starting closest to the Sun and moving outward.
2. How is Mars like Earth?

Percival Lowell, an amateur astronomer who lived from 1855–1916, spent much of his life trying to prove that there was intelligent life on Mars. Through a telescope, he observed lines on the planet and interpreted them to be canals built by intelligent life. He further believed that dark patches on the planet were areas of vegetation and lighter patches were deserts. Lowell published several books and convinced many people that life did indeed exist on Mars.

Lowell's theory became so widespread that it became the theme of books and other forms of entertainment. In 1938 actor Orson Welles presented a radio drama that included a news report that Martians had invaded Earth. Many people who heard the drama thought it was a real news report. Panic spread quickly, and many people locked themselves into their houses or fled for cover.

Discussion:
pages 260–61

➤ Which planet is the fourth from the Sun? *{Mars}*

➤ How is Mars like Earth? *{Possible answers: It has a similar tilt, experiences seasons, has an atmosphere, and has days of about the same length.}*

➤ What does the growing and shrinking of polar ice caps tell us about Mars? *{Mars experiences seasons.}*

➤ What causes Mars to be cold? *{its thin atmosphere and distance from the Sun}*

💡 Could you survive the temperature on Mars? *{Answers will vary. Elicit that the temperature on Mars is sometimes comfortable, but not usually.}*

💡 Could you live by breathing the air on Mars? Why? *{No; the air on Mars is mostly carbon dioxide. Humans need oxygen to live.}*

➤ Why is Mars red? *{Iron oxide in the planet's soil causes the planet to be red.}*

➤ What has caused men to think that Mars may have had liquid water? *{the land features}*

💡 What kind of land features do you think might give scientists that idea? *{Accept any reasonable answers. Guide the student to recall information about weathering and erosion from Chapter 2.}*

➤ The length of Mars' day is almost the same as Earth's. How is its year different? *{A year on Mars is almost twice as long as a year on Earth.}*

💡 Why would Mars have a longer year? *{It is farther from the Sun.}*

➤ Why have people wondered whether life might exist on Mars? *{Mars is the planet most like Earth.}*

Answers

1. Mercury, Venus, Earth, Mars, Jupiter, Saturn, Uranus, Neptune, Pluto

2. Possible answers: similar tilt; experiences seasons; has an atmosphere; days are about the same length

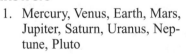

Activity Manual
Study Guide—pages 167–68
These pages review Lessons 113–14.

Assessment
Test Packet—Quiz 11-A
The quiz may be given any time after completion of this lesson.

Lesson 115

Student Text pages 262–65

Objectives

- Explain some ways God made Earth unique
- Describe why the same side of the Moon always faces Earth
- Give details about the *Apollo 11* mission
- Describe the causes of solar and lunar eclipses

Vocabulary

satellite
solar eclipse
totality
lunar eclipse

Introduction

➤ When did astronauts first land on the Moon? *{1969}*

➤ Who was the first man to walk on the Moon? *{Neil Armstrong}*

➤ This lesson will answer these and other questions about Earth and the Moon.

Purpose for reading:
Student Text pages 262–65

➤ Which characteristics of the Moon are different from those of Earth?

➤ In what positions do the Earth, Moon, and Sun have to be to produce a solar eclipse or a lunar eclipse?

Discussion:
pages 262–63

➤ In what ways is Earth unique? *{It is the only planet where man can survive in his natural state. The temperatures are neither too hot nor too cold for humans to survive. It is the only planet where water exists in liquid form on the surface.}*

📖 What can we learn about God by observing the uniqueness of Earth? *{Answers will vary, but elicit that Earth demonstrates God's care for us in providing Earth for our use.}*

➤ How often does Earth rotate? *{once every 24 hours}*

➤ Why is this rotation important? *{It allows the atmosphere to receive adequate heating and cooling to provide an overall moderate climate.}*

Earth

Earth: home sweet home

Earth also is an inner planet, situated between Venus and Mars in its orbit around the Sun. Of all the planets, God made Earth unique. We know that God created the other planets too, but Earth is where He put man. Earth is special because it is the only planet on which man can survive in his natural state.

God has placed Earth at the perfect distance from the Sun, neither so close that it is too hot nor so far that it is too cold. The moderate temperature allows water to exist as a liquid. Water is essential for all life and shows another way God has planned for man's needs.

Earth rotates once every 24 hours. This rotation allows the atmosphere and surface of Earth to receive adequate heating and cooling each day to maintain Earth's overall moderate climate. Earth orbits the Sun every 365¼ days. For three of every four years, our calendars show only 365 days. Every fourth year we add an extra day, February 29, to catch the calendar up with Earth's actual revolution. We call this year with an extra day a *leap year*.

Gravity holds Earth's atmosphere in place. The atmosphere helps maintain the warmth from the Sun. It also filters out the Sun's harmful rays and protects us from meteors that otherwise might crash onto Earth's surface.

Evolutionists believe that Earth came into being by chance. But as we observe the intricate processes that occur only on Earth, we can see that Earth's marvelous design points to an all-powerful Creator.

The Moon

Earth has one natural satellite, the Moon. A **satellite** is any object that revolves around another body in space. Because the Moon has no atmosphere, there are no sounds, no clouds, no rain, and no colors in the sky. Lack of an atmosphere also means that the Moon has no protection from charged particles from the Sun or from meteorites. Therefore, unlike Earth, the Moon is full of craters. Most meteors burn up in Earth's atmosphere without hitting Earth's surface, but the Moon has no protection from them.

the Moon

262

SCIENCE BACKGROUND

Natural satellites—Earth has one natural satellite, and it was given the name *Moon*. Many planets have one or more natural satellites orbiting around them. The word *moon* has come to be used in place of the words *natural satellite*. Astronomers often use these words interchangeably.

The Moon rotates once on its axis as it makes one revolution around Earth. Consequently, the same side of the Moon always faces Earth. Until man began sending spacecraft to the Moon, observers wondered whether the back side would be different from the front side. When astronauts finally circled the Moon, they discovered that the back side of the Moon is similar to the side we see.

The Moon does not give off its own light. Instead, it reflects the light from the Sun. The Moon appears to change shape as different areas of its surface are lighted by the Sun. We call this the changing phases of the Moon. The Moon takes about 29½ days to pass from one new moon to the next.

Have you ever heard of someone being called "loony"? The word *loony* is a shortened form of the word *lunatic*, which comes from *luna*, the Latin word for moon. The phases of the Moon were believed to influence people's behavior, so some people who acted strangely during a full moon were called lunatics.

Phases of the Moon as seen from Earth

Discussion:
pages 262–63

➤ How long is Earth's revolution? {365¼ days}

➤ What is a *leap year*? {every fourth year, in which an extra day is added to catch the calendar up with Earth's revolution}

➤ What are the benefits of Earth's atmosphere? {Earth's atmosphere helps maintain the Sun's warmth, filters the Sun's harmful rays, and protects Earth from meteorites.}

➤ What is a *satellite*? {any object that orbits around another object}

💡 Why was space travel necessary to determine what is on the backside of the Moon? {The Moon's rotation and revolution are the same, so the same side of the Moon always faces Earth.}

➤ Why does the Moon appear light? {It reflects light from the Sun.}

📖 Read Matthew 5:16. How does the Moon's relationship to the Sun picture a Christian's relationship to Christ? {The Moon reflects light from the Sun. It does not give off light of its own. In the same way, Christians are to reflect God's light so that God receives glory.} [BAT: 7b Exaltation of Christ]

➤ In what ways does the absence of an atmosphere make the Moon different from Earth? {Possible answers: Lack of an atmosphere on the Moon means that the Moon has no sound, no clouds, no rain, no colors in the sky, and no protection from meteorites and harmful rays from the Sun.}

➤ How long does it take for the Moon to pass from one new moon to the next? {29½ days}

➤ What causes the phases of the Moon? {Different areas of the Moon are lighted by the Sun at different times.}

💡 Look at the *Phases of the Moon* diagram. Give the meaning of the following words. (Possible answers given) *Waning* {becoming smaller} *Gibbous* {more than half of the Moon is showing} *Crescent* {less than half of the Moon is showing}

Refer the student to *Fantastic Facts*.

➤ Where does the word *loony* come from? {It is a shortened form of *lunatic*, which is from *luna*, the Latin word for moon.}

➤ Why were people sometimes called lunatics? {It was believed that the phases of the Moon affected people's behavior, so if someone acted strangely during a full moon, he was called a lunatic.}

Direct a Demonstration
Demonstrate why the same side of the Moon always faces Earth

Materials: globe

Place a globe or another object representing Earth in the center of the room. Direct a student to use his body to represent the Moon. Tell him to walk around "Earth" but keep the front of his body toward "Earth" at all times. Point out that during his revolution around Earth, he also rotates once in relationship to the room. His revolution around Earth and his rotation occurred at the same rate, which kept the same side of the Moon facing Earth at all times.

 Read Psalm 104:13–14 and 16–24. Direct the student to list the different things on Earth that God designed. Discuss the purpose of each. [BAT: 3a Self-concept; Bible Promise: I. God as Master]

Discussion:
pages 264–65

- **Which President issued a challenge to put a man on the Moon?** {President John F. Kennedy}
- **What name was given to the project to put an American on the Moon?** {Project Apollo}
- **In what year was the first Moon landing?** {1969}
- **Who were the astronauts on board *Apollo 11*?** {Neil Armstrong, Buzz Aldrin, and Michael Collins}

Refer the student to the picture at the top of Student Text page 264.

- **What was the *lunar landing module*?** {the spacecraft that actually landed on the Moon}

The command module orbited the Moon while Neil Armstrong and Buzz Aldrin were on the Moon.

- 💡 **Why did the astronauts need protective space suits?** {Possible answers: The Moon has no air for men to breathe. The lack of an atmosphere means there is no protection from the harmful rays of the Sun.}
- **What now-famous statement did Neil Armstrong make as he stepped onto the Moon?** {"That's one small step for man, one giant leap for mankind."}
- 💡 **What did Armstrong mean by that statement?** {Accept any answer, but elicit that his first step on the Moon demonstrated that man was making great strides in science by successfully putting a man on the Moon.}
- **How many more Moon landings were made by the United States?** {six}
- 💡 **Why do you think astronauts brought back the rocks, pebbles, dust, and sand?** {Answers will vary. Elicit that scientists wanted to analyze objects from the Moon.}

Apollo 11 lunar module

Project Apollo

In May of 1961, President John F. Kennedy issued a challenge to the American people to put a man on the Moon before the end of that decade. Project Apollo was begun to accomplish that mission. Eight years later, the United States was ready to send a man to the Moon on the *Apollo 11* mission.

On July 16, 1969, Neil Armstrong, Edwin E. ("Buzz") Aldrin, and Michael Collins left the Earth aboard *Apollo 11*. Neil Armstrong commanded the mission. The pilot of the lunar landing module was Buzz Aldrin. Michael Collins piloted the command module.

The three-stage rocket took three days to reach the Moon. As the lunar module approached the Moon, Armstrong looked for a safe landing spot. When the lunar module was nearly out of fuel, Aldrin called out how many seconds of fuel were left. Armstrong finally found a landing spot, and they landed with fuel left for only twenty seconds.

The astronauts waited for the moon dust to settle before they opened the hatch. Clad in a protective spacesuit, Neil Armstrong went down the ladder first. As he stepped from the landing pad onto the Moon's surface, he said, "That's one small step for man, one giant leap for mankind." Buzz Aldrin soon joined his fellow crewman.

Armstrong and Aldrin had only about two hours to complete their work on the Moon, but they took time to set up cameras so TV viewers on Earth could watch. They also set up an American flag. Aldrin stood at attention beside it while Armstrong took his picture. The men collected two cases of moon rocks and dust and set up several experiments. They left a plaque that read

"Here men from the planet Earth first set foot upon the Moon, July 1969, A.D. We came in peace for all mankind."

The United States made six Moon landings between 1969 and 1972. Ten more American astronauts walked on the Moon. Altogether they spent 79 hours working outside the landing craft. They brought back 382 kg (842 lb) of moon rocks, pebbles, dust, and sand.

Astronauts from *Apollo 11* — Michael Collins, Neil Armstrong, "Buzz" Aldrin

WRITING Direct the student to interview someone who was alive during the first Moon landing. The student should write down the information from the interview to share with others. Some information the student could gather is as follows: memories of the first Moon landing, including the person's age at the time, where he was, if he watched the landing live on television, and how he felt about men landing on the Moon. Other information the student may gather from the interview could include how the world was at that time and changes the person has seen since then.

TECHNOLOGY Each of the three crewmen of the *Apollo 11* mission had specific jobs to do. The lunar module was the craft that landed on the Moon. Both Armstrong and Aldrin left the lunar module to walk on the Moon and perform the other tasks. Collins remained in orbit around the Moon in the command and service modules. At the completion of the mission, all three men landed on Earth inside the command module.

solar eclipse

Eclipses

A **solar eclipse** is a spectacular and rare event that occurs when the Moon passes between Earth and the Sun and casts its shadow on Earth. When we view it from Earth, the Moon appears to be exactly the same size as the Sun. In reality, the Sun is many times larger than the Moon or Earth, but since the Moon is so much closer to Earth, the Moon and the Sun appear to be the same size. During a solar eclipse, the Moon's circle covers the Sun completely, leaving only the Sun's corona visible. Some astronomers call it a "remarkable coincidence" that the Moon when viewed from Earth appears to be exactly the same size as the Sun, creating this amazing phenomenon. Of course, Christians know that it is no coincidence. Although they may not understand why, Christians recognize that this event is part of the handiwork of God and that it declares His glory.

The phase of an eclipse when the Moon appears to cover the Sun completely is called a **totality**. Because the Moon is relatively small, the area of Earth that will witness a total covering of the sun during an eclipse is about a one-hundred-mile area. The rest of the Earth will see only a partial eclipse. Witnessing a totality is a very rare and special treat.

A lunar eclipse is more common than a solar eclipse. About every six months a lunar eclipse can be seen somewhere in the world. A **lunar eclipse** occurs when the Moon passes through the shadow of the Earth. When the Moon is in totality, it reflects beautiful colors, such as violet or apricot.

1. Why do we always see the same side of the Moon from Earth?
2. Which Apollo mission was the first to land on the Moon?
3. What causes a solar eclipse?

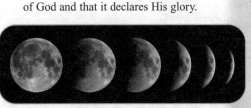

lunar eclipse

Direct an ACTIVITY

Materials: penny for each student

Illustrate how the Moon appears to be larger than the Sun during a solar eclipse. Instruct each student to close one eye and hold the penny a few centimeters in front of his open eye. Tell him to look around to find an object that appears the same size as the penny and then to open both eyes and walk to the object to compare the actual size of the object with the actual size of the penny.

LANGUAGE Relate the name of each eclipse with what is viewed from Earth. *Solar* means "sun." In a solar eclipse, the Sun is covered. *Lunar* means "moon." In a lunar eclipse, the Moon is covered.

Discussion:
pages 264–65

➤ What causes a solar eclipse? *{The Moon passes between Earth and the Sun, causing the Moon to block the Sun.}*

➤ Why can the Moon, which is so small, completely hide the Sun, which is so large? *{The Moon is closer to Earth, so it appears larger than the Sun.}*

➤ Which part of the Sun is seen around the Moon during a solar eclipse? *{the corona}*

➤ What is a *totality*? *{when the Moon covers the Sun completely during a solar eclipse; when Earth's shadow covers the Moon completely during a lunar eclipse}*

➤ How much of Earth will see a total covering of the Sun during a solar eclipse? *{usually about a one-hundred mile area}*

➤ What causes a lunar eclipse? *{The Moon passes through the shadow of Earth.}*

➤ Which is more common—a solar eclipse or a lunar eclipse? *{a lunar eclipse}*

➤ How often do lunar eclipses occur? *{about once every six months somewhere in the world}*

Discuss times when the student may have seen an eclipse.

Answers
1. The Moon rotates once on its axis as it makes one revolution around Earth.
2. *Apollo 11*
3. The Moon passes between Earth and the Sun and blocks the Sun's light.

Lesson 116

Student Text pages 266–67
Activity Manual pages 169–70

Objectives

- Construct a solar oven that will melt a marshmallow
- Infer the relationship between materials used and results

Materials

- See Student Text page

Introduction

Note: Two lesson days are allotted for this activity. On the first day, introduce the activity, set guidelines and a due date for the solar ovens, and begin planning. The second lesson day may occur at a later time when you are ready to test the solar ovens.

➤ **What are some things you have used to open the tight clear plastic wrapper of an item such as a CD or a toy?** *{Possible answers: fingernails, scissors, knives, keys, pens}*

➤ Objects are sometimes used in ways they were not originally intended when people creatively try to solve problems.

Purpose for reading:
Student Text pages 266–67
Activity Manual pages 169–70

The student should read all the pages before beginning the activity.

Discussion:
Student Text pages 266–67

➤ **How do you think the scientists felt when faced with this seemingly impossible task?** *{Accept any answers.}*

📖 Emphasize that scientists often rely on the basic laws of science when trying to find solutions to problems. They can depend on these laws because of the orderliness and reliability of God's creation. [Bible Promise: I. God as Master]

ACTIVITY: Spare Parts Solar Oven

Process Skills
- Observing
- Inferring
- Identifying variables
- Recording data

A Successful Failure

"A successful failure" is the way Jim Lovell described the *Apollo 13* mission to the moon. In 1970 Jim Lovell, Jack Swigert, and Fred Haise were headed to the Moon onboard the *Apollo 13* spacecraft. Several days into their flight an accident occurred that damaged the command module and depleted much of its electricity, oxygen, water, and heat. The astronauts were forced to take refuge in the attached lunar module.

Unfortunately, the equipment in the lunar module for filtering the air was designed for only two men. It was not able to filter enough air for three men, and carbon dioxide began building up in the lunar module. The equipment in the command module was usable, but it did not fit the hook-ups in the lunar module. Without a creative solution to the filtering problem, the crew could not make it back to Earth alive.

The task seemed almost impossible, but the scientists at mission control on Earth were not about to give up. They collected objects that were identical to those the astronauts had on the spacecraft. Using items such as plastic bags, cardboard, and lots of duct tape, the scientists found a solution. The instructions for the device that would save the astronauts' lives were radioed to the spacecraft. The ingenious solution saved the astronauts' lives, and they returned safely to Earth.

Something from Nothing

Thankfully, your task is not to solve a life-threatening problem. Your task is to create a solar oven that will successfully melt a marshmallow. Like the Apollo scientists, though, you will be limited in what objects you can use. You may use only the items your teacher makes available to you.

266

TEACHER HELPS

The ovens should reach temperatures of 200–225°F. The student can try heating other foods in the solar ovens.

This activity can be completed outside at any temperature. However, the activity takes less time when the temperature is 60°F or higher.

These types of boxes work well for this activity: pizza boxes, shoeboxes, and copier paper boxes. Local businesses may be willing to donate boxes.

You could make s'mores with the melted marshmallows.

Certain science principles will help your student:

- Black absorbs light energy, and white reflects light energy.

- The direction and angle of the Sun will affect the temperature inside the oven.
- A closed oven will have less air circulation and will be hotter.

BIBLE

God has given us everything we need to do His will in our lives.

➤ **Does anything that happens in your life surprise God?** *{Elicit that God knows all things (Job 34:21).}*

God controls all things. He has promised to guide those who trust in Him (Ps. 146:5–10).

Problem
How can I create a solar oven that will melt a marshmallow?

Procedure
 1. Draw your solar oven design in your Activity Manual.

 2. List the materials you will use.

 3. Write an explanation of why you chose your design and how it will help your solar oven to collect heat.

4. Construct your solar oven. Be sure to leave a door or hole to insert the marshmallow if the oven is enclosed.

5. Take your solar oven outside, place a marshmallow inside, and observe.

 6. Record your observations.

Materials:
cardboard box
marshmallows
watch or clock
Any of the following:
  aluminum foil
  aluminum pie plate
  plastic wrap
  cardboard or card stock
  black and white construction paper
  black and white paint
  paper towels
  newspaper
  black trash bag
  craft sticks
  string
  scissors or craft knife
  tape
  glue
  paper fasteners or paper clips
  paintbrushes

Conclusions
- Was there a feature of your design that seemed to cause the solar oven to heat well?
- Was there a feature that kept the oven from working?

Follow-up
- Make improvements to your oven and test it again.
- Try heating other foods in your solar oven.

Procedure
Provide time for the student to construct his solar oven. Remind the student to leave an opening for inserting the marshmallow.

Place the oven outside in a sunny spot. Allow the oven to preheat for at least 30 minutes before inserting the marshmallow.

Insert a marshmallow into the oven. Record the amount of time that it takes for the marshmallow to melt. Record any other observations made while the oven is being tested.

Conclusions
Provide time for the student to answer the questions.

Discuss which design elements were most effective for trapping solar heat.

Activity Manual
Activity—pages 169–70

Assessment
Rubric—TE pages A50–A53
Select one of the prepared rubrics, or design a rubric to include your chosen criteria.

SCIENCE PROCESS SKILLS
Inferring
➤ Why would someone choose to use black for the solar oven? *{Answers will vary. Elicit that black items tend to get hot faster and hold heat better.}*

➤ Why would someone use plastic wrap and foil? *{Answers will vary. Elicit that it might work like a greenhouse.}*

➤ What can you infer from seeing the results of the solar ovens? *{Answers will vary.}*

 WRITING Direct the student to do further research on the *Apollo 13* mission and write a short play involving four characters—the three astronauts and a voice from mission control. The play should include dialogue for the four characters as the event may have occurred. Allow time for the play to be read aloud.

 ART The Apollo astronauts used plastic bags, cardboard, hoses, and duct tape to make a useful device. Give the student the same objects, and encourage him to construct his own useful invention.

Lesson 117

Student Text pages 268–71
Activity Manual pages 171–72

Objectives

- Identify characteristics of each of the outer planets
- Recognize that the *Voyager* probes have explored the outer planets

Materials

- ruler

Introduction

➤ Describe the worst storm you have experienced. *{Answers will vary.}*

➤ What are the names of some of the most severe types of storms on Earth? *{Possible answers: hurricane, typhoon, tsunami, blizzard}*

➤ What are some of the conditions of these storms? *{Possible answers: high winds, rain, extreme temperatures}*

➤ The other planets also have weather, and some even experience storms.

Purpose for reading:
Student Text pages 268–71

➤ How do the outer planets' days and years compare to those of Earth?

➤ Which features qualify an object in space to be considered a planet?

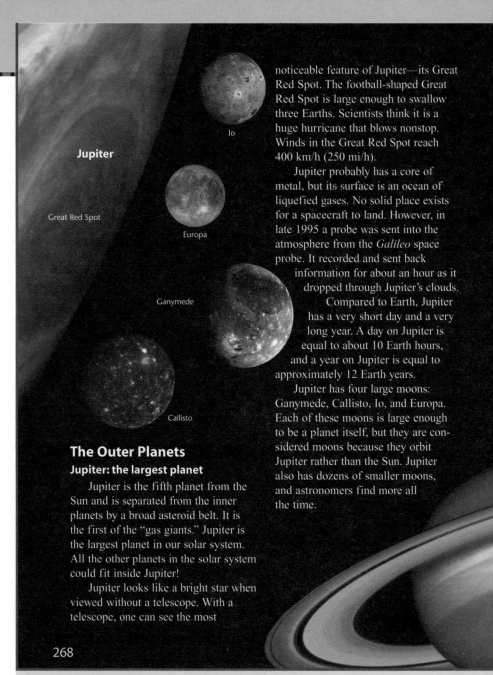

The Outer Planets
Jupiter: the largest planet

Jupiter is the fifth planet from the Sun and is separated from the inner planets by a broad asteroid belt. It is the first of the "gas giants." Jupiter is the largest planet in our solar system. All the other planets in the solar system could fit inside Jupiter!

Jupiter looks like a bright star when viewed without a telescope. With a telescope, one can see the most noticeable feature of Jupiter—its Great Red Spot. The football-shaped Great Red Spot is large enough to swallow three Earths. Scientists think it is a huge hurricane that blows nonstop. Winds in the Great Red Spot reach 400 km/h (250 mi/h).

Jupiter probably has a core of metal, but its surface is an ocean of liquefied gases. No solid place exists for a spacecraft to land. However, in late 1995 a probe was sent into the atmosphere from the *Galileo* space probe. It recorded and sent back information for about an hour as it dropped through Jupiter's clouds.

Compared to Earth, Jupiter has a very short day and a very long year. A day on Jupiter is equal to about 10 Earth hours, and a year on Jupiter is equal to approximately 12 Earth years.

Jupiter has four large moons: Ganymede, Callisto, Io, and Europa. Each of these moons is large enough to be a planet itself, but they are considered moons because they orbit Jupiter rather than the Sun. Jupiter also has dozens of smaller moons, and astronomers find more all the time.

268

SCIENCE BACKGROUND

Saturn's rings—When Galileo first discovered the rings of Saturn, he thought that they looked like cup handles on both sides of the planet.

Saturn: the ringed planet

Saturn, the sixth planet away from the Sun, is the second largest planet in our solar system. It too is a gas giant, and its core and surface appear to be similar to Jupiter's.

Saturn is known as "the ringed planet," because its rings are brighter and larger than the rings of any other planet. Although Saturn's rings look like a solid band, they are actually made up of many small, frozen particles that reflect the Sun's light. Saturn rotates once about every 11 Earth hours. Its revolution, however, takes almost 30 Earth years.

Saturn has more known moons than any other planet in our solar system. Eighteen named moons orbit Saturn, but at least twelve more have been observed but not named.

Uranus: the planet that rotates sideways

Uranus is the seventh planet in our solar system. Even with a good telescope, Uranus appears as only a faint blue-green disk in the sky. The space probe *Voyager 2* traveled for eight years before it reached Uranus. The probe revealed a blue-green planet that rotates on its side. Uranus rotates from west to east like most other planets, but the rotation appears to be from the bottom to the top since the planet is tipped over so far. Its rotation takes approximately 17 Earth hours.

Uranus is so far away from the Sun that its side facing away from the Sun is not much colder than the side facing the Sun. It takes Uranus 84 Earth years to orbit the Sun. Its curious tilt, however, makes for interesting "days." Each pole spends 21 Earth years in endless daylight and another 21 Earth years in total darkness.

Like other gas giants, Uranus has a liquid surface. If you were visiting Uranus, you would not be able to breathe because the atmosphere is poisonous methane gas. This poisonous atmosphere gives the planet its bluish color.

Uranus

Saturn

269

Direct a DEMONSTRATION

Demonstrate how small particles can appear to form a solid

Materials: flashlight, flour

Saturn's rings look like a solid band, but they are actually made up of smaller particles that reflect the Sun's light. To illustrate this, shine a flashlight in a darkened room. Take a pinch of flour and sprinkle it above the rays of the flashlight. As the flour enters the rays of the flashlight, it appears to be solid when seen from a distance.

LANGUAGE

Most of the moons in our solar system were named after classic mythological characters, but the names of Uranus's moons come from characters in Shakespeare's writings—Juliet, Prospero, Puck, and Rosalind.

Direct the student to research the names of several moons in the solar system. He should find out who or what each moon was named after and whether there was a reason for that particular name.

Discussion:
pages 268–69

▶ Jupiter is what kind of planet? {a gas giant}

▶ How does the size of Jupiter compare to the sizes of the other planets in our solar system? {It is the largest planet. All of the other planets could fit inside Jupiter.}

▶ What do some scientists think the Great Red Spot on Jupiter is? {a huge hurricane}

▶ What is the surface of Jupiter? {an ocean of liquified gases}

▶ Why does this surface make Jupiter a difficult planet to explore? {There is no place for a spacecraft to land.}

▶ How many moons does Jupiter have? {four large moons and dozens of smaller moons}

▶ Jupiter's moon Ganymede is larger than Mercury. Why is it considered a moon rather than a planet? {It orbits Jupiter rather than the Sun.}

▶ How large is Saturn compared to the other planets? {It is the second largest planet.}

▶ Is Saturn the only planet with rings? {no}

▶ Why is Saturn considered "the ringed planet"? {Its rings are brighter and larger than the rings of any other planet.}

▶ What are Saturn's rings made of? {many small, frozen particles}

▶ Which spacecraft reached Uranus? {the *Voyager 2* probe}

▶ How is Uranus different from all the other planets? {It is tipped over and rotates on its side.}

Hold a ruler in a vertical position and rotate it so that the top of the ruler moves downward 98°. This is the angle of the axis of Uranus.

▶ What gives Uranus its bluish color? {its methane gas atmosphere}

Chapter 11   Lesson 117   281

Discussion:
pages 270–71

➤ How does the weather on Neptune compare to the weather on other planets in our solar system? {It has the most violent weather.}

➤ What feature does Neptune have that makes it similar to Jupiter? {Neptune's Great Dark Spot is similar to Jupiter's Great Red Spot.}

➤ What is the coldest place known in our solar system? {Neptune's moon Triton}

💡 What are some reasons that humans would not survive naturally on the gas giants? {Possible answers: The atmospheres are poisonous. The surfaces are not solid. The temperatures are too cold.}

➤ What do scientists think Pluto's surface is made up of? {rock and ice}

💡 Why does Pluto's orbit sometimes bring it closer to the Sun than Neptune? {Because Pluto's orbit is narrower than Neptune's, it sometimes goes inside Neptune's orbit.}

Neptune: the blue planet

Neptune, the last of the gas giants, is a dark and unfriendly world to humans. Neptune has the most violent weather in the solar system. The winds on Neptune reach 2000 km/h (1,240 mi/h)—ten times faster than the winds of a hurricane! Astronomers have observed that Neptune's storms appear as large dark spots. One such spot was named Neptune's Great Dark Spot after Jupiter's Great Red Spot.

If you were visiting Neptune, you would not be able to see the Sun, stars, or moons through the planet's thick cloud cover of methane gas. Neptune rotates in about 16 Earth hours and takes 165 Earth years to orbit the Sun.

Neptune has eight known moons. Its largest moon, Triton, is the coldest place known in the solar system. The surface temperature of Triton is –235°C (–391°F).

Pluto: the smallest planet

Pluto, the most remote planet in our solar system, is also the smallest and coldest of the planets. Scientists believe that rock and ice make up Pluto's surface.

Pluto is unique because sometimes it is the planet farthest away from the Sun and sometimes it is not. Pluto's orbit is elongated, or stretched out. So for 20 years of its 248 Earth-year orbit, Pluto is closer to the Sun than Neptune. Pluto's period of rotation, or day, is 6.4 Earth days. Like Venus, Pluto has a retrograde rotation.

Some astronomers suggest that Pluto is not a true planet but rather a

270

Astronomers discovered Neptune using mathematics. Uranus's orbit was so erratic that scientists predicted that another planet was affecting it. Through mathematical calculations, astronomers figured out where Neptune should be located and found the planet. Pluto was discovered in the same way.

➤ Why would the existence of Neptune affect the orbit of Uranus? {Elicit that the gravitational pull of Neptune may cause Uranus's orbit to be erratic.}

satellite lost from Neptune. In order to be classified as a planet, an object must meet two criteria: it must be large enough for its own gravity to keep it in the shape of a sphere, and it must orbit the Sun rather than another planet. Pluto meets both requirements.

Some people consider Pluto to be a double planet. Its moon, Charon (KAH ruhn), is half its size and is relatively close to the planet. Before the Hubble Space Telescope provided clearer images of the planet, astronomers could not tell Charon and Pluto apart. Charon orbits Pluto once every 6 days and 9 hours. Pluto's rotation on its axis takes only slightly longer.

Voyagers 1 and 2

Much of the information we have about Jupiter, Saturn, and the other outer planets comes from the *Voyager* probes launched in 1977. *Voyager 1* visited Jupiter and Saturn and sent back large amounts of new information. *Voyager 2* went by Jupiter and Saturn but then continued on to visit Uranus and Neptune. Both probes are out of our solar system now, and they continue to travel through space, close to thirty years later! Scientists never expected the probes to last that long. Scientists learned a lot about the gas giants during the *Voyager* flybys, and they hope to learn more about our galaxy as the probes continue to travel. Each probe has gold-plated phonograph records on it with sounds and pictures from Earth. Scientists wanted the record to explain where Earth is in our solar system and what humans are like, in case the probe should be intercepted by alien life.

1. Why can a spacecraft not land on Jupiter?
2. Which planet rotates on its side?
3. Which planet in our solar system has the most violent weather?
4. What are the names of the probes that explored the gas giants?

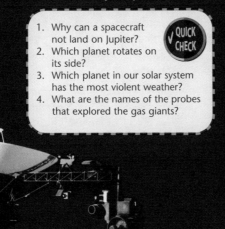

Voyager 1

271

WRITING Direct the student to write a paragraph congratulating scientists on the outstanding performances of *Voyager 1* and *Voyager 2*. Tell him to include some of each probe's accomplishments.

Discussion:
pages 270–71

➤ What are two criteria for an object to be classified as a planet? *{It must be large enough that its own gravity keeps it in the shape of a sphere, and it must orbit the Sun rather than another planet.}*

➤ Does Pluto meet these criteria? *{yes}*

💡 Why do some people think of Pluto as a double planet? *{Its satellite, Charon, is close to it and is almost half its size.}*

➤ Why is Charon considered a satellite of Pluto rather than a separate planet? *{Charon orbits Pluto rather than the Sun.}*

➤ Even with the Hubble Space telescope, Pluto is so far away that it is difficult to get clear pictures with our current technology.

➤ Which space probes have visited the gas giants and sent information to Earth? *{Voyager 1 and Voyager 2}*

➤ Which planets did *Voyager 1* visit? *{Jupiter and Saturn}*

➤ Which planets did *Voyager 2* visit? *{Jupiter, Saturn, Uranus, Neptune}*

Answers

1. The surface is liquified gases. It is not solid.
2. Uranus
3. Neptune
4. *Voyager 1* and *Voyager 2*

Activity Manual
Reinforcement—pages 171–72

Chapter 11   Lesson 117

Lesson 118

Activity Manual pages 173–74

Objectives

- Construct a scale model of the solar system
- Gain a greater understanding of the vastness of our solar system

Materials

- softball
- sand
- 2 poppy seeds
- mustard seed
- 2 dried peas
- 2 BBs
- measuring wheel

Introduction

God has created a vast universe. Our solar system is only a small part of it, but even our solar system is so huge that it is difficult for us to imagine. Humans are very tiny in relation to the universe, yet God still loves us.

Read Psalm 8:3–4.

➤ What is another word for *heavens* in verse 3? {universe}

➤ Who created the universe? {God}

➤ God is powerful enough to have created everything, yet these verses say He is mindful toward us. What do you think *mindful* means? {Elicit that *mindful* means that God thinks about and cares for us.}

➤ The Son of Man in verse 4 is Jesus Christ. Why did Jesus come to Earth to visit man? {Elicit that Christ came to Earth to provide salvation.}

➤ To give us an idea of how very small humans are in this vast solar system, we will make a model of our solar system.

Purpose for reading:
Activity Manual pages 173–74

The student should read all the pages before beginning the exploration.

Solar Walk

Name _____

A. The chart lists the relative size of each planet and its distance from the Sun. To calculate the distance between each planet, subtract the previous planet's distance from the Sun (Column 3) from the current planet's distance from the Sun (Column 3). Use the information in the last column to take a solar walk that shows a scale model of the solar system.

Example: Venus's distance from the Sun − Mercury's distance from the Sun
6.7 − 3.6 = 3.1

Planet	Relative size of planet	Relative distance from the Sun in meters	Relative distance from the previous planet in meters
the Sun	softball	—	—
Mercury	grain of sand	3.6	—
Venus	poppy seed	6.7	6.7 − 3.6 = 3.1
Earth	poppy seed	9.3	9.3 − 6.7 = 2.6
Mars	mustard seed	14.0	14.0 − 9.3 = 4.7
Jupiter	pea	48.0	34
Saturn	smaller pea	89.0	41
Uranus	BB	180.0	91
Neptune	BB	230.0	50
Pluto	grain of sand	370.0	140

B. After returning from your solar walk, answer the questions.

1. Did the size of the solar system surprise you? Why or why not?
   Answers will vary.

2. Why would it be impossible to fit a scale model using the same relative size and distance in a classroom? Answers will vary but should include that either the planets would be too small to see or the distance between them would be too large to fit in the classroom.

> **Psalm 8:3–4**
> When I consider thy heavens, the work of thy fingers, the moon and the stars, which thou hast ordained; What is man, that thou art mindful of him?

TEACHER HELPS

An area about 375 meters (425 yards) long is needed to complete this activity. If that much space is not available, include as many planets as you can in the space that you have.

You might be able to borrow a measuring wheel from someone in one of these occupations: builder, realtor, engineer, carpet installer, policeman, or surveyor.

If you do not have access to a measuring wheel, use a prepared piece of string with the measurements marked with tape.

Place a bright flag in the grass at each planet so you can look back and see the distances easily.

Since many of the "planets" are small, place them in paper cups if you want to leave them at their markers.

SCIENCE MISCONCEPTIONS

Solar system diagrams and models show the planets in a row. Planets are never lined up like this at one end of the solar system. These models are actually comparing the orbital paths of the planets. Very rarely will the actual distance between any two objects in the solar system match the chart.

WRITING
If you choose not to do "Travel Brochure" in Lesson 120, you may want to assign groups of students to research each of the planets. Stop at each planet on your walk and allow the students to present their research.

Procedure

In order to keep the planets in correct proportion to one another, the distances and sizes of the planets must be reduced at the same rate. These calculations have already been made. The student needs to determine the relative distances between the planets before starting the space walk. He is given each planet's relative distance from the Sun.

Guide the student in calculating the distances between the planets. Subtract the previous planet's distance from the Sun from the current planet's own distance from the Sun.

Use a measuring wheel to make accurate measurements. Instruct the student to walk slightly ahead of you and approximate the distance to the next planet by counting his strides, estimating a meter distance with each stride. Stop at each planet to look at the object being used to represent it. Compare the distances of the "planets" from the "Sun" and the distances between the "planets."

Activity Manual

Explorations—page 173

Study Guide—page 174
This page reviews Lessons 115 and 117.

Assessment

Rubric—TE pages A50–A53
Select one of the prepared rubrics, or design a rubric to include your chosen criteria.

Direct a DEMONSTRATION
Demonstrate the relative sizes of the planets

Materials: assorted balls or round objects

Construct a model of the relative sizes of the planets inside the classroom. The following objects may be used: Mercury—a small bead; Venus—a large marble; Earth—a large marble; Mars—a small marble; Jupiter—a basketball; Saturn—a soccer ball; Uranus—a softball; Neptune—a softball; Pluto—a tiny bead. Using this scale, the Sun is 2.4 meters in diameter. You could paint the Sun on a large sheet of paper and hang it on a wall.

Direct a DEMONSTRATION
Demonstrate the relative distances of the planets' orbits.

Materials: masking tape, meter stick

To make a classroom representation of the distances between the planets, place a 4-meter piece of masking tape on the floor. Measure and mark the planet distances from the chart using centimeters instead of meters. At this scale, the planets would be too small to model.

Chapter 11   Lesson 118   285

Lesson 119

Student Text pages 272–75
Activity Manual pages 175–76

Objectives
- Explain how a rocket uses thrust to launch
- Define Newton's third law of motion
- Distinguish between a satellite and a probe

Vocabulary
probe dehydrated
International Space Station

Materials
- bicycle helmet
- package of dehydrated food (e.g., packaged noodle mix, powdered drink mix)
- small electronic device
- model of a three-stage rocket (optional)

Introduction

Display the helmet, food, and electronic device.

➤ **What do these items have in common?** *{Accept any answer.}*

The students are not likely to recognize the connection between the items shown and space exploration. After they have offered several guesses, continue the discussion.

➤ **Each of these items was developed as a result of space exploration. The foam used in a bicycle helmet was developed for use in aircraft seats. The process of dehydrating food produced food that was lightweight and took less space. Electronic devices were miniaturized to fit inside a spacecraft.**

Purpose for reading:
Student Text pages 272–75

➤ How has the last forty-five years of space travel and technology affected how we live and what we know about the solar system?

➤ How is a space probe different from a satellite?

Space Exploration

Rockets

The development of rockets has been essential to space exploration. The Chinese were probably the first people to invent rockets, about 1,000 years ago. Their rockets used gunpowder and were used as weapons. In 1926, American Robert Goddard launched the first liquid-fueled rocket. Though his rocket went only 12.5 m (41 ft) high and traveled only about 97 km/h (60 mi/h), it signaled the beginning of modern rocket science. By 1942, Germany had produced the V-2, a long-range rocket used as a weapon during World War II.

In 1955 Wernher von Braun (VAIR-nuhr vahn BROWN), the German who developed the V-2, began working for the United States and led a team to develop rockets for space travel. To overcome Earth's gravity and establish an orbit around Earth, a rocket must reach a speed of 27,350 km/h (17,500 mi/h). To travel beyond Earth's orbit requires even greater speed.

Rockets work in a way similar to jet engines. Hot gases being pushed very quickly out of a nozzle, or small opening, in the rocket's engines create thrust, or forward force. Newton's third law of motion says that for every action there is an equal and opposite reaction. As the hot gases are pushed out, the gases push against the rocket. You probably have demonstrated this principle using a balloon. If you blow up a balloon and then let it go, the balloon sails away in an opposite direction from the air moving out of the balloon.

Saturn 1

balloon moves opposite way

air comes out one way

SCIENCE BACKGROUND

Rocketry—Rocketry actually involves all of Newton's laws of motion. At take-off, a tremendous force is necessary for the spacecraft to overcome inertia (Newton's first law). The greater the mass of the spacecraft is, the more force is needed to lift it off the launch pad (Newton's second law). As stated in the Student Text, lift-off is also an example of Newton's third law. The laws of motion are discussed in Chapter 9, *Motion and Machines*.

 LANGUAGE In Latin *astro* means "star," and *naut* means "sailor" or "ship." An *astronaut* is a "star sailor."

➤ **What other words can you think of that contain the roots *astro* or *naut*?** *{Possible answers: astroturf, astronomy, cosmonaut, nautical}*

 SCIENTIST Although Wernher von Braun developed the V-2 rocket, which was used against London in World War II, his dream was that rockets would be able to leave Earth's orbit. This dream caused him to be arrested by the Gestapo in Nazi Germany.

After he was released, he arranged the surrender of his rocket team to the Americans.

286 Lesson 119

Science and History

Though at times space travel seems commonplace and ordinary, it is never without risks. During a training mission in 1967, an Apollo spacecraft on the ground caught fire and killed three astronauts. *Apollo 13* was almost lost to drift in space because of an explosion on the way to the Moon. In 1986 the space shuttle *Challenger* exploded soon after liftoff. Seven crew members were killed. The entire shuttle fleet was grounded while changes were made to a thin rubber ring that had failed and caused the explosion. After a few years the shuttle fleet again began making frequent trips into space. In 2003 the space shuttle *Columbia* broke apart while reentering Earth's atmosphere. Another seven crew members were lost. Again the shuttle fleet was grounded while scientists tried to determine the cause of the accident and to take appropriate action to avoid its reoccurrence. Despite the tremendous list of things that could go wrong, the space program has had remarkably few accidents. But every time a life is lost, we are reminded that reaching for new frontiers is always dangerous and that life is brief (James 4:14).

The Space Shuttle

Early space missions to the Moon used huge three-stage rockets to send people into space. Each rocket could be used only one time. Then in 1981 the United States launched the first space shuttle, a reusable space vehicle. Though it launches much like a rocket, it returns to Earth like a glider airplane. A shuttle can carry up to eight crew members. Sometimes these crew members are not regular astronauts. They are specialists who work with the shuttle's cargo or with experiments that are part of the mission. The shuttle's main task is to transport equipment, but it also serves as the "bus" for astronauts traveling to and from the International Space Station. In fact, the shuttle program's complete name is the Space Transportation System (STS). Scientists are working to create a new type of shuttle that would both take off and return to Earth as an airplane, thus improving economy and safety.

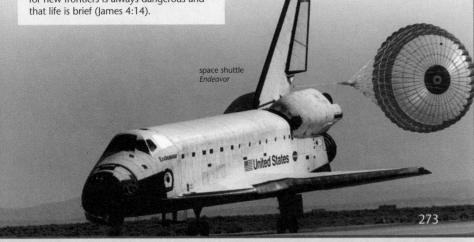

space shuttle *Endeavor*

Three-stage *Saturn* series rockets were used by NASA for the Apollo missions to the Moon. Each rocket was a series of sections stacked in the order of their use. After a section performed its function, it separated from the remaining rocket. The first, or bottom, stage ignited to provide the thrust necessary to lift the rocket off the ground. The ignition of the second stage boosted the rocket into position for orbiting the earth. The third stage then ignited to give the rocket the velocity to maintain the orbit.

After the orbits were complete, the third stage sent the rocket toward the Moon. The command and service modules then aligned with the lunar module to prepare for the third stage to separate.

To return to Earth, the service module provided the thrust needed to leave the Moon's orbit. The only part of the huge rocket to return to Earth was the small, cone-shaped command module that had left Earth at the top of the rocket.

Read James 4:14.
➤ Are we guaranteed to have safety more than the astronauts? {no}
➤ Are young people guaranteed to live until tomorrow any more than adults? {no}
➤ What should our response be to the shortness and uncertainty of life? {We should be sure of our salvation, and we should live each day for God's glory.}

Discussion:
pages 272–73

> When were rockets probably first used and by whom? {1,000 years ago; Chinese}
➤ Who is credited with starting modern rocket science? {an American, Robert Goddard}
➤ What did he do? {launched the first liquid-fueled rocket}
➤ What German led a team to develop rockets for space travel? {Wernher von Braun}
💡 What challenges must a scientist plan for a rocket to overcome in order for it to go into space successfully? {Possible answers: Earth's gravity, atmosphere, and orbit}
➤ What does Newton's third law of motion say? {For every action there is an equal and opposite reaction.}
➤ How does Newton's third law relate to rockets? {As hot gases leave the rocket, they push the rocket in the opposite direction.}
➤ What kind of rockets were used to go to the Moon? {huge three-stage rockets}

Refer to the *Saturn 1* rocket on Student Text page 272. The *Saturn* rocket series were the rockets used to go to the Moon.

If available, show the stages of the rocket model.

💡 How is the space shuttle different from a regular rocket? {Possible answers: It is reusable. It returns to Earth like a glider airplane. It can carry more cargo and a larger crew.}

💡 Why was the shuttle program named the Space Transportation System? {The shuttle was designed to carry equipment and astronauts to and from the International Space Station.}

Discuss space shuttle lift-offs and landings that the student may have seen.

Refer the student to *Science and History*.

➤ What are some examples of space travel accidents? {Apollo spacecraft that caught on fire; Apollo 13, Challenger, Columbia}

💡 Why is space travel so dangerous? {People traveling in space cannot easily leave a damaged spacecraft and survive. The spacecraft supplies what is needed for the crew to live and return to Earth. Because of this, even small failures can have disastrous results.}

💡 Do you think the benefits of space travel are worth the risks involved? {Answers will vary.}

Discussion:
pages 274–75

➤ **What is a *satellite*?** *{an object that orbits another object in space}*

➤ **What was the first artificial satellite to orbit Earth?** *{Sputnik I}* **Who launched it and when?** *{Soviet Union on October 4, 1957}*

➤ **How big was *Sputnik*?** *{about the size of a basketball; 83 kg}*

➤ **How did the launch of that satellite affect the world?** *{It unofficially began the "space race."}*

➤ **Why were the United States and the Soviet Union so concerned about the "space race"?** *{The countries were not friendly, and each was afraid the other would gain control of space.}*

➤ **Is there still a "space race" today?** *{No, the United States and Russia sometimes cooperate in space exploration.}*

The Soviet Union no longer exists. It has split into many smaller countries, of which Russia is the largest.

➤ **Which kinds of satellites are orbiting Earth today?** *{Possible answers: the Hubble Space Telescope; satellites for tracking weather, spying, communication, and distance education}*

💡 **Why do you think satellites look like stars?** *{Answers will vary.}*

💡 **Name some things that your family uses that rely on signals from satellites.** *{Possible answers: satellite television, GPS, some radios, satellite phones}*

➤ **What is a *probe*?** *{a research spacecraft sent beyond Earth's orbit}*

💡 **How is a probe different from a satellite?** *{A probe does not orbit the Earth like a satellite does. Probes travel to other parts of the solar system.}*

➤ **Why would a probe use a robotic arm?** *{to gather and analyze samples}*

Refer the student to *Science and Math*.

➤ **What happened to the *Mars Climate Orbiter*?** *{It probably burned up in Mars's atmosphere or crashed.}*

➤ **Why were the scientists' measurements off?** *{the two teams used different measuring systems}*

MightySat, a 320 kg (705 lb) US Air Force/Phillips Laboratory satellite

Satellites

A satellite is an object that orbits another object in space. In addition to Earth's natural satellite, the Moon, many man-made satellites orbit Earth. The first artificial satellite ever to orbit the Earth was *Sputnik I*. Launched by the Soviet Union (U.S.S.R.) on October 4, 1957, *Sputnik* was about the size of a basketball and weighed 83 kg (183 lb). It took only 98 minutes to orbit the Earth. The launch of *Sputnik* unofficially began what was called the "space race." Because of the political tension between the Soviet Union and the United States, each country tried to keep the other from gaining control of the realm of space. Today, Russia and the United States sometimes cooperate in space exploration.

Satellites have many purposes. The Hubble Space Telescope is a satellite of Earth that is used for space exploration. Other satellites are used for communication. Satellites enable you to talk to someone on the other side of the world or to watch a live news report from a distant country. Satellites are also used for tracking the weather, for international spying, and even for distance education.

You can usually see several satellites in the night sky. Some look like shooting stars, except that they move more slowly and stay in sight longer. Thousands of satellites orbit the Earth every day.

Probes

Probes are research spacecraft sent beyond Earth's orbit. They do not have astronauts aboard, so they can travel longer and farther. Probes can relay

In 1999 NASA lost the *Mars Climate Orbiter* because of a simple mathematical error. One team of scientists did calculations using the English system while the other team did calculations using the metric system. As a result their measurements were off, and the probe probably either burned up in Mars's atmosphere or crashed.

274

SCIENCE BACKGROUND

Space animals—Dogs and monkeys went to space before men. A dog, Laika, was the first animal to be sent into space. It was sent to space by the Russians on *Sputnik II* one month after *Sputnik I* was launched.

Measurement systems—The United States is one of the few countries that still uses the English system of measurement. Metric measurements are most often used for science research so that data is easily interpreted by anyone who uses it.

WRITING

Some catastrophes, such as plane crashes, have occurred because of mix-ups between using liters or gallons. Direct the student to write a persuasive paragraph in support of or against changing the United States' official measuring system to the metric system.

Do you wish you could grow a few inches? While in space, an astronaut's vertebrae spread apart because of the lack of gravity, and the astronaut "grows" one to two inches. When the astronaut returns to Earth, though, his height returns to normal.

images of planets. Some use robotic arms and other instruments to collect and analyze samples from other planets. *Voyagers 1* and *2*, which visited the outer planets, and the *Mars Global Surveyor* are probes.

International Space Station

The **International Space Station** (ISS) is a facility that orbits Earth and is maintained and used by astronauts from sixteen different countries. Astronauts began building the space station in 1995. They have continued to add onto it while living in it and conducting experiments that will benefit Earth and further our knowledge of space. The astronauts are also learning what effects living in space over a long period of time has on humans.

Living in space is much different from living on Earth. The weightlessness in space affects how the astronauts eat, sleep, work, and exercise. Astronauts cannot eat anything that produces crumbs because the crumbs would float away.

Most of the food is **dehydrated**, meaning it has had the water removed. Dehydrated food is easier to store and takes up less space than regular food. Drinking out of a cup in space would cause problems too, because the water would float out of the cup. Astronauts keep their drinks sealed and drink them through straws.

Sleeping in space is also different. Some astronauts float around freely while they sleep. Others choose to strap themselves down in sleeping bags that are secured to the wall. Astronauts have to exercise every day. Without the force of gravity pulling on their muscles and causing them to work harder, their muscles can deteriorate.

1. What does Newton's third law of motion say? How is this demonstrated by rockets?
2. Why is the shuttle called a "transportation" system?
3. What is a probe?

artist's concept of future space station

275

TECHNOLOGY

Some people question the value of space exploration. However, there are multitudes of things we use everyday that are the results of space technology. The miniaturization of computers is probably one of the most obvious. The metal used in braces was developed for antennas in space. Medical technology has been particularly advanced due to space research. The list of practical uses of space technology is lengthy and is continually growing.

HISTORY

Early in 2004 scientists successfully put two probes on the surface of Mars, the Mars Exploration Rover *Spirit* and the Mars Exploration Rover *Opportunity*. The rovers were designed to investigate and send back to Earth extensive data about the surface of Mars.

Discussion:
pages 274–75

➤ **What is the International Space Station (ISS)?** *{a facility that orbits Earth and is maintained and used by different countries}*

➤ **What is the purpose of the ISS?** *{to conduct experiments where there is a lack of gravity in order to develop new materials that will benefit Earth and to study the effects of living in space on humans}*

➤ **What does *dehydrated* mean?** *{the water has been removed}*

➤ **What are some differences in living conditions that astronauts face while living in space?** *{Possible answers: They must eat food that does not produce crumbs, drink only from closed containers with straws, float or secure themselves while sleeping, and exercise to keep their muscles toned.}*

➤ **How do astronauts try to lessen the effects of zero gravity on their muscles?** *{They exercise regularly.}*

Refer the student to *Fantastic Facts*.

➤ **Why does an astronaut get taller in space?** *{The vertebrae spread apart because of the lack of gravity.}*

💡 **What are *vertebrae*?** *{the bones that form the backbone, or spine}*

💡 **Why does the astronaut return to normal height when he returns to Earth?** *{Earth's gravity pulls the vertebrae back into place.}*

Answers

1. For every action, there is an equal and opposite reaction. As gases push out toward the ground, the rocket moves upwards in the opposite direction.
2. Space shuttles transport equipment and people to and from the International Space Station and Earth.
3. a research spacecraft sent beyond Earth's orbit

Activity Manual
Reinforcement—page 175

Bible Integration—page 176

This page highlights the Christlike qualities in the life of Johannes Kepler.

Assessment
Test Packet—Quiz 11-B

The quiz may be given at any time after the completion of this lesson.

Lesson 120

Activity Manual pages 177–78

Objectives
- Design a travel brochure for a planet
- Collect data

Materials
- encyclopedias and other resource books about planets
- variety of real travel brochures

Introduction

➤ If you had the opportunity to plan your family's vacation, where might you go?

➤ How could you learn about vacation places that you have never seen or heard about?
{Possible answers: advertisements in newspapers or magazines; commercials; travel brochures; word of mouth}

➤ Today you will design a travel brochure that describes a fictional vacation to a planet in our solar system.

Purpose for reading:
Activity Manual pages 177–78

➤ What kinds of information do you need to include in your travel brochure?

The student should read all the pages before beginning the exploration.

Travel Brochure

Name_____

A. The year is 2050. You are a graphic designer for a travel agency on Earth. You have received this memo from your boss.

1500 Moonbeam Way
Space Station, FL 34000
(987) 555-1654

Jack McAllister, Head of Marketing
Final Frontier Travel Agency
February 6, 2050

To: All brochure designers

Thank you for your fine work in designing brochures for the Lunar Vacation. We have been selling many vacations to the Moon. I'm sure that our talented designers deserve much of the credit for the attractive brochures. I am confident that you are ready for this next project.

We are planning eight new and exciting vacations. There will be a vacation to each of the planets in our solar system. Choose one of the planets and design a travel brochure for that planet. Please make your brochure attractive and informative. For the benefit of our clients, please include the following information:

1. the weather on the planet
2. landforms to visit
3. the history of the planet's exploration
4. survival items for travelers
5. any additional information that you find helpful

Thank you for your fine work for Final Frontier Travel Agency.

Sincerely,

Jack McAllister
Jack McAllister

The key to the success and interest in this activity is the teacher. Be enthusiastic. Make the project light-hearted but focused.

Many planets will have limited information. Allow the students to make up many features and create unusual solutions to traveling needs. The value of this activity is in doing research and in exercising creativity within boundaries.

If students are working in groups, assign each student to research and write one paragraph of the brochure. They could share the responsibility for the design and art elements.

A traveler will need oxygen, food, water, housing, transportation, protection from heat, harmful radiation, and meteor showers, etc. The types of survival items will vary depending on the environment of each planet.

What to do

B. Follow the directions given by your boss, Jack McAllister, in the memo. A guide has been provided for your research about your planet. Make your brochure using your notes. Be prepared to explain the information you include in your brochure.

1. the weather on the planet

2. landforms to visit

3. the history of the planet's exploration

4. survival items for travelers

5. any additional information that you find helpful

Discussion:
Activity Manual pages 177–78

Display the sample brochures.

➤ **What do all the brochures have in common?** *{Elicit that they each try to get the reader to visit somewhere or do something.}*

➤ **What characteristics make some brochures more interesting than others?** *{Possible answers: pictures; types of details included and how they are written; size of the brochures}*

Ask questions as needed to encourage students' ideas.

➤ **What are the weather conditions on your planet during the day and during the night?**

➤ **How long is a day on your planet?**

➤ **What important sites can be seen on your planet?**

➤ **Which fictional space explorers visited your planet? When did they visit it? What was the name of their spacecraft? What important tests and experiments did they perform while visiting your planet?**

➤ **What special survival items will you supply for the travelers? What will they need to bring?**

➤ **What is the most unusual fact about your planet?**

Activity Manual
Explorations—pages 177–78

Assessment
Rubric—TE pages A50–A53

Select one of the prepared rubrics, or design a rubric to include your chosen criteria.

Lesson 121

Student Text pages 276–77
Activity Manual pages 179–80

Objectives

- Hypothesize how design affects the performance of a balloon rocket
- Construct a balloon rocket
- Demonstrate an understanding of Newton's third law of motion

Materials

- See Student Text page

Introduction

➤ Rockets have been used for special purposes for centuries. They have been used for weapons, fireworks, and space flight. In this activity you will have the opportunity to construct a rocket and learn about thrust, the force that enables rockets to move.

Purpose for reading:
Student Text pages 276–77
Activity Manual pages 179–80

➤ Do you think you can design a balloon rocket that will fly a long distance?

The student should read all the pages before beginning the activity.

ACTIVITY: Rocket Race

Rockets need thrust to escape Earth's atmosphere. A balloon can demonstrate thrust. As a balloon deflates, it is propelled in the opposite direction of the flowing air. In this activity you will design and test a "rocket" that uses a balloon for thrust.

Problem
How can I make a balloon rocket propel a long distance?

Procedure
Your balloon rocket must include
- a clothespin to close the balloon.
- a piece of drinking straw attached along the top of your rocket.

1. Draw and label a diagram of your rocket to use as your hypothesis. You may construct a multi-stage rocket, but it must be designed so the second balloon starts deflating after the first stage has deflated.
2. List the materials you will use.
3. Build your rocket according to your diagram. Blow up your balloon to the desired size. Twist and secure the end with a clothespin. Add the design elements that you think will make your rocket go farther. Remember that the force propelling your rocket must come from the air escaping the balloon.
4. Cut the drinking straw to whatever size you choose. Securely attach the straw to the upper side of your rocket. Thread the fishing line through the straw.
5. Hold or attach the fishing line so that it is stretched tightly between two people or fixed points. At the appointed time, release the clothespin from your rocket.

Process Skills
- Hypothesizing
- Measuring
- Making and using models
- Observing
- Inferring
- Recording data

Materials:
balloons of various sizes and shapes
clothespin, clip-style
drinking straw
tape
glue
card stock or construction paper
foam cup
ten-meter fishing line
meter stick or tape measure
Activity Manual

276

TEACHER HELPS

Reducing the size of the balloon opening with a rubber band may reduce the airflow, causing the rocket to go a longer distance. This may be helpful if a large balloon is used.

Try to launch the rocket within one hour after it is inflated. The opening may not open easily if left twisted and closed for a longer period of time.

SCIENCE BACKGROUND

This activity correlates with Chapter 9, *Motion and Machines*, in which Newton's third law of motion is discussed.

SCIENCE PROCESS SKILLS

Hypothesizing

➤ What do you use as a hypothesis in this activity? *{the labeled diagram}*

➤ Is this a true hypothesis? *{No; a true hypothesis is a statement.}*

➤ Could you write a statement to replace your diagram? *{yes}*

Allow several students to try writing hypothesis statements. Guide them as they define conditions and make predictions. The statements may be lengthy.

➤ Do you think it is easier to draw a diagram or write a statement for this hypothesis? *{Answers will vary.}*

➤ Diagrams are sometimes included with a hypothesis.

6. Measure and record the distance traveled by your rocket.

Conclusions
- Was your hypothesis (design) effective?
- What features of your rocket helped it go farther?

Follow-up
- Change the design to make a balloon rocket go farther.
- Launch several rockets at the same time along different pieces of fishing line. Compare the design of each rocket to its speed and the distance it traveled.

Procedure
Attach the fishing line to two points in a room, hallway, or outdoors. You may also choose to have two students hold the ends of the fishing line.

With a piece of tape, mark the starting point on the fishing line.

If individual rockets are made, you may launch several rockets at the same time on different pieces of fishing line.

Conclusions
After the activity discuss which design elements were most effective in making the rocket propel further. There are several factors that may have contributed to the success of the rocket: 1) the size of the balloon; 2) the velocity of the escaping air; 3) the shape of the balloon; 4) the amount of air in the balloon. The student should conclude that the most effective factor was the thrust that resulted from the amount of air that was able to escape from the balloon.

As time permits, allow the student to make changes to his rocket and test it again based on the knowledge he gained from testing it the first time.

Activity Manual
Activity—pages 179–80

Assessment
Rubric—TE pages A50–A53
Select one of the prepared rubrics, or design a rubric to include your chosen criteria.

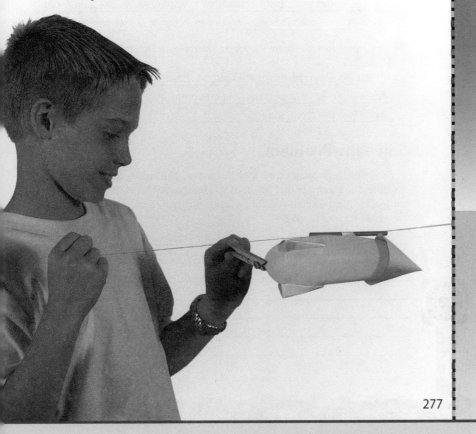

Direct an Activity

Materials: 2 straight balloons (identified as A and B below), clothespin, 2 rubber bands, 15 cm x 30 cm strip of stiff paper, tape, drinking straw

Make a multi-stage rocket by inflating balloon A completely and closing it with a clothespin. Place 2 rubber bands around the balloon at 5 and 8 cm from the rounded end.

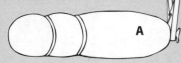

Inflate balloon B about two-thirds full, and push the air toward the rounded end to make the end near the opening (neck) longer. Twist the neck and insert the neck under both rubber bands on balloon A so that balloon B extends from the rounded end of balloon A. Adjust as needed to keep air from leaking.

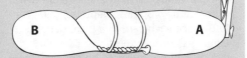

Wrap and tape the paper strip around both balloons so the rubber bands are covered and the balloons are aligned.

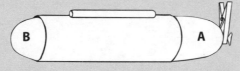

Attach the straw and launch as a single rocket.

Lesson 122

Student Text page 278
Activity Manual pages 181–82

Objectives

- Recall concepts and terms from Chapter 11
- Apply knowledge to everyday situations

Introduction

Material for the Chapter 11 Test will be taken from Activity Manual pages 167–68, 174, and 181–82. You may review any or all of the material during the lesson.

You may choose to review Chapter 11 by playing "The Solar System" or a game from the *Game Bank* on TE pages A56–A57.

Diving Deep into Science

Questions similar to these may appear on the test.

Answer the Questions

Information relating to the questions can be found on the following Student Text pages:

1. page 262
2. page 273
3. pages 256–57

Solve the Problem

In order to solve the problem, the student must apply material he has learned. The answer for this Solve the Problem is based on the material on Student Text page 275. The student should attempt the problem independently. Answers will vary and may be discussed.

Activity Manual

Study Guide—pages 181–82

These pages review some of the information in Chapter 11.

Assessment

Test Packet—Chapter 11 Test

The introductory lesson for Chapter 12 has been shortened so that it may be taught following the Chapter 11 test.

Diving Deep into Science

Answer the Questions

1. Why do astronauts need to wear protective suits on the Moon?
   Accept any reasonable answer, but elicit that the Moon has no atmosphere. Therefore, the Moon has no breathable air. The suits provide oxygen for the astronauts and protect them from the Sun's harmful rays and extreme temperatures.

2. Space shuttles have many advantages. What is one disadvantage of shuttles?
   Accept any reasonable answer, but elicit that repeated usage causes additional safety issues that a single-use rocket would not have. Also, the shuttles cost more than small, single-use rockets.

3. Why would there be more sea ice in Antarctica in October than there is in March?
   In the Southern Hemisphere October comes after winter, when the sea ice is probably greatest. March comes after summer, when there would be the least sea ice.

Solve the Problem

Your family is going on a hiking and overnight camping trip. All the equipment and supplies that you need will have to be carried in your backpacks. What invention(s) for the space program could help you minimize the weight of your supplies?

Accept any reasonable answer. The dehydrated food used in the space program would lighten the load. Small computer devices are helpful also, and they are an indirect result of the space program.

278

Review Game

The Solar System

Prepare a set of eleven cards labeled Sun, Mercury, Venus, Earth, Mars, asteroid belt, Jupiter, Saturn, Uranus, Neptune, and Pluto. Place each card in a plain dark envelope. Ask review questions. Each time a student answers correctly, he chooses an envelope, but he does not look inside the envelope.

After all of the envelopes are chosen, instruct the students with envelopes to come to the front of the room. At a signal, each opens his envelope. The students then arrange themselves from left to right, starting with the Sun and moving outward. Once they are finished, allow the rest of the class to decide if they are in the correct order. Instruct the students to replace the cards in the envelopes. Shuffle the envelopes and repeat as time allows.

Student Text pages 279–81
Activity Manual page 183

Lesson 123

Objectives
- Demonstrate knowledge of concepts taught in chapter 11
- Understand the interrelationships of science concepts
- Recognize that man's inferences are sometimes faulty

Unit Introduction

Note: This introductory lesson has been shortened so that it may be taught following the Chapter 11 test.

In Unit 5 students will learn about how God uses factors such as DNA to create unique offspring.

Chapter 12 discusses God's plan for reproduction. Both plant and animal sexual reproduction and asexual reproduction are presented.

Chapter 13 explains how an organism inherits traits from parents.

Both chapters use correct terminology but are tempered with an understanding of the sensitive nature of the topics.

➤ **Look through Unit 5. What are the topics of the chapters we will study in this unit?** *{Possible answers: plant and animal reproduction, genetics, heredity}*

➤ **How do the chapters relate to each other?** *{Answers will vary, but elicit that living organisms reproduce and possess characteristics inherited from their parents.}*

Unit 5: God's Continuing Plan

Teacher Helps
Additional information is available on the BJU Press website.
www.bjup.com/resources

Unit 5 — Lesson 123 — 295

Project Idea

The project idea presented at the beginning of each unit is designed to incorporate concepts of each chapter as well as information gathered from other resources. You may choose to use the project as a culminating activity at the end of the unit or as an ongoing activity while the chapters are taught.

Unit 5—Fish Traits

Guide the student in setting up an aquarium that contains several types of the same species of fish. Choose fish that have distinctive markings, coloring, or fin shapes. Discuss the similarities and differences between the fish. The student could research the dominant and recessive traits of the fish. Some fish hatcheries will also provide fish eggs, such as salmon or trout, which can be raised in the classroom and then released into their natural habitats. If animals cannot be in the classroom, allow the student to set up a pea plant experiment similar to Mendel's pea plants.

Technology Lesson

A correlated lesson for Unit 5 is provided on TE pages A12–A13. This lesson may be taught with this unit or with the other Technology Lessons at the end of the course.

Bulletin Board

Note: This suggested bulletin board idea may be modified for use in a learning center.

Write Psalm 139:14*a* in the center of the board. Direct the student to collect pictures of his family or pictures of other families that share obvious physical traits, such as widow's peaks, curly hair, face shape, etc. Place the pictures on the bulletin board and discuss some of the traits evident in the pictures. Remind the student that although families share common traits, God has created each person as a unique individual.

Variation: Use Genesis 1:25 instead of Psalm 139:14. Collect pictures of pets and other animals and their offspring, such as a picture of a cat and kittens, or a picture of a baby giraffe and its mother.

God planned for living things to reproduce after their own kind. In Chapter 12 you will learn about some of the ways that plants and animals produce offspring.

What do pea plants have to do with your ear lobes? In Chapter 13 you will learn about how an Austrian monk discovered that some traits are dominant over others.

You have probably seen a milk mustache, but have you ever seen a "pollen mustache"? In Chapter 12 you will find out that bees are not the only animals that help pollinate plants.

280

God's Marvelous Creation

Psalm 139:14a
"I will praise thee; for I am fearfully and wonderfully made."

296 Lesson 123

12 Plant and Animal Reproduction

GREAT & MIGHTY Things

A pile of dirty rags and wheat kernels turns into full-grown mice. Frogs come from slimy mud. Wormlike maggots develop from rotting meat. It may seem strange, but many people used to believe these things. It was not until the 1600s that spontaneous generation, the belief that life comes from nonliving things, was questioned. Francesco Redi, an Italian doctor, proved that rotting meat does not create maggots. Through his experiments, he found that the maggots actually came from eggs that flies laid on the meat. God told us in Genesis 1 that plants and animals reproduce after "their own kind." Man took thousands of years to reaffirm what God had already told him.

281

SCIENTIST Francesco Redi, an Italian physician and poet, was among the first to show that spontaneous generation was not true. In 1668 he performed a series of tests that showed that maggots come from the eggs of flies rather than from rotting meat. In his experiments he controlled the variable of meat's exposure to flies. Through this structured method he was able to show the relationship of flies and maggots and whether or not meat was involved.

Introduction

God created living organisms and commanded them to be fruitful and multiply. Even the plants were commanded to bring forth seed (Gen. 1:11). In each case, God said that the organism would bring forth "its kind" or another organism similar to itself.

Man, however, forgot God's Word. Because man could not see the processes of reproduction, he developed some strange ideas about how living organisms began.

Chapter Outline

I. Plant Reproduction
   A. Seeds in a Fruit
   B. Seeds in Cones
   C. Spores

II. Animal Reproduction
   A. Placental Gestation
   B. Marsupial Gestation
   C. Eggs
   D. Parental Care

III. Asexual Reproduction
   A. Binary Fission
   B. Budding
   C. Fragmentation
   D. Regeneration
   E. Vegetative Reproduction

Activity Manual

Preview—page 183

The purpose of this Preview page is to generate student interest and determine prior student knowledge. The concepts are not intended to be taught in this lesson.

Chapter 12   Lesson 123   297

Lesson 124

Student Text pages 282–85
Activity Manual page 184

Objectives
- Explain the purpose for each part of a flower
- Differentiate between pollination and fertilization
- Explain how scientists classify fruits
- Describe the process of germination

Vocabulary
sepal
stamen
anther
pistil
ovary
ovule
style
stigma
cross-pollination
self-pollination
fertilization
zygote
embryo
fruit
germinate
cotyledon
seed coat

Materials
- Assorted fruits, such as tomatoes, squash, acorns, maple tree seeds, or plums (Include some that the student would not normally classify as fruit.)
- standard measuring spoons

Introduction
Show students the collection of fruit.
➤ **Sort these items into categories.** {Possible answers: fruit, vegetables, seeds}

After each fruit is placed into a category, explain that scientists would classify all of these as fruit.

Purpose for reading:
Student Text pages 282–85
➤ How is a flower pollinated?
➤ How would a scientist classify a fruit?

Discussion:
pages 282–83
➤ **What is an *angiosperm*?** {a plant that produces seeds enclosed in a fruit}
➤ **What is the main purpose of flowers?** {to produce seeds}
➤ **Why else might God have designed flowers?** {Possible answers: Flowers provide food for some of God's creatures. God provided their beauty for our enjoyment.} [Bible Promise: H. God as Father]

From the very beginning of Creation, God knew that man's sin would bring death to the earth. God already had a plan for man's spiritual birth through the death and resurrection of Jesus Christ, His Son. God also planned for plants, animals, and humans to reproduce and fill the earth (Gen. 1:22, 28).

Plant Reproduction

In Genesis 1:11–13, the Bible tells us that God created plants on the third day of Creation. Just by His word, all the grasses, trees, shrubs, and bushes came into being. Genesis 1:29–30 also tells us that God planned for the plants to be food for mankind and for the animals. To fulfill God's plan each plant must have a way to reproduce, or make a copy of itself. God's design for most plants is to reproduce through either seeds or spores.

Seeds in a Fruit
Angiosperms are plants that produce seeds enclosed in a fruit. These seed-filled fruits develop from pollinated flowers. Though plant flowers vary in size, color, and odor, each flower has the same function—to produce seeds.

Parts of a flower
The **sepals** (SEE puhlz) of a flower protect a developing flower bud by enclosing the bud until it is ready to open. Usually the sepals are green, but in some plants they look just like the petals of the flower. The **stamen** (STAY muhn), the male part of the flower, usually has a thin stalk called a filament. The knoblike structure, or **anther** (AN thur), at the top of the filament produces the pollen.

In the center of the flower is the **pistil** (PIS tuhl), the female part of the flower. The pistil has three main parts. The bottom of the pistil is called the **ovary** (OH vuh ree), and it has one or more **ovules** (OH vyoolz), or places where the eggs are produced. A long, slender stalk called a **style** connects the ovary to the top of the pistil. The **stigma** (STIG muh) is the sticky tip of the pistil. It traps the pollen grains that fall from the anthers or that are carried into the flower by insects or other animals.

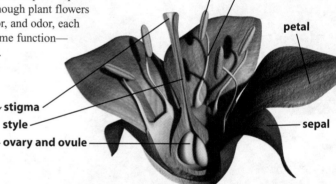

282

SCIENCE BACKGROUND

Flowers and pollination—The color, shape, and markings of flower petals, as well as the flower's scent, serve to attract the insects or animals that help pollinate that plant. Bees cannot see the color red and are usually attracted to blue, purple, or yellow flowers. Butterflies are often attracted to orange, yellow, pink, or blue flowers. Red or orange flowers usually attract birds, and bats and moths that pollinate at night are attracted to white or cream-colored flowers.

Flowers pollinated by bats often have a strong, musky odor. Other flowers may have an odor that is unpleasant to people but is very attractive to the beetles and flies that pollinate those plants. The markings and color patterns of the flower help the animals or insects locate the nectar.

Tepals—Sepals are called *tepals* when the sepals and petals are similar in color and shape.

Carpels—Some flowers have more than one pistil. These pistils are often fused together into a compound pistil. The individual pistils of a compound pistil are often referred to as *carpels*.

298 Lesson 124

hummingbird

In order to make just one pound of honey, honeybees fly more than 55,000 miles and visit about 2,000,000 flowers. A bee is able to visit 50–100 flowers on each collection trip. In her lifetime, the average worker bee makes only ½ of a teaspoon of honey.

Discussion:
pages 282–83

 Point out each part of the flower as it is discussed.

- ➤ **What is the purpose of the sepals?** *{to protect the developing flower bud}*
- ➤ **What is the *stamen*?** *{the male part of the flower}*
- ➤ **What parts of the flower make up the stamen?** *{the filament and anther}*
- ➤ **What produces the pollen?** *{the anther}*
- ➤ **What is the female part of the flower?** *{the pistil}*
- ➤ **What parts of the flower make up the pistil?** *{ovary, style, stigma}*
- ➤ **Why is the stigma sticky?** *{The stigma is sticky in order to capture grains of pollen when the flower is pollinated.}*
- ➤ **What are the *ovules*?** *{the places where eggs are produced}*
- ➤ **How are flower petals important to the flower?** *{The petals attract the insects and animals that pollinate the flower.}*
- ➤ **What kinds of plants are usually pollinated by wind?** *{plants that have small, inconspicuous flowers}*
- ➤ **How do insects pollinate flowers?** *{As an insect visits a flower, pollen collects on its body. Another flower is pollinated when some of the pollen falls off of the insect and onto the stigma of the flower.}*
- ➤ **What is necessary for cross-pollination to occur?** *{The pollen must be transferred to another plant.}*
- ➤ **When does self-pollination occur?** *{when pollen is transferred from the anther of a flower to the stigma of the same flower or to another flower on the same plant}*
- ➤ **What happens to a flower after it is pollinated?** *{A pollen tube grows into an ovule. A male sperm cell travels down the pollen tube and unites with a female egg cell.}*
- ➤ **What is the difference between pollination and fertilization?** *{Pollination involves the transfer of pollen from flower to flower. Fertilization is the joining of an egg cell and a sperm cell to produce a zygote.}*

Refer the student to *Fantastic Facts*.

- ➤ **How many flowers do honeybees have to visit to make one pound of honey?** *{2 million}*
- ➤ **How much honey does one average bee make in a lifetime?** *{1/2 of a teaspoon}*

Show the student a ½ teaspoon.

Pollination and fertilization

A flower is pollinated when a grain of pollen lands on its stigma. This pollination can happen in many different ways. Plants pollinated by the wind usually have small and inconspicuous flowers. In other flowers, the color, shape, and fragrance of the petals attract insects, birds, and other animals. Bees and butterflies pollinate plants as they search for nectar. Some plants are even pollinated by bats or rodents.

As the insect enters the flower, pollen from the anthers collects on the insect's head, back, or legs. When the insect visits another flower, some of that pollen falls onto the stigma and pollinates this other flower. **Cross-pollination** happens when pollen is transferred from the anther of one flower to the stigma of a flower on another plant. **Self-pollination** occurs when the pollen is transferred from the anther to the stigma of the same flower or to another flower on the same plant.

After the pollen lands on the stigma, a tiny pollen tube grows and reaches down into the ovule of the pistil. **Fertilization** (FUHR tl ih ZAY shun) occurs when a male sperm cell inside the pollen travels down the pollen tube and unites with a female egg cell. This fertilized egg is called a **zygote** (ZY gote). Reproduction that involves both a male and a female cell is called *sexual reproduction*.

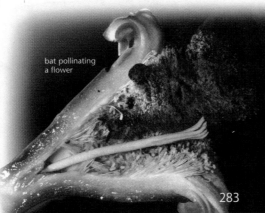
bat pollinating a flower

SCIENCE MISCONCEPTIONS

Pollination—Insects are only one group of animals that pollinate flowers. Birds and small animals, such as bats and the Australian honey possum, also pollinate flowers.

Fruit—Be sure that students understand that scientists use the term *fruit* to describe the part of any flower (the ovary) that contains the seeds.

 LANGUAGE The term *angiosperm* comes from two Greek words meaning "seed in a vessel" (ovary). This term is introduced in Chapter 6, *Plant Classification*.

Discussion:
pages 284–85

- What happens to a flower after fertilization? {*The petals fall off, and the zygote begins to grow.*}
- What is another name for the tiny new plant that develops from the zygote? {*embryo*}
- What does the ovule become? {*the seed coat*}
- What part of the flower will become the fruit? {*the ovary*}
- 💡 What is the purpose of a fruit? {*to contain and protect the seeds*}
- 💡 How is the way we commonly use the word *fruit* different from the scientific definition of *fruit*? {*We commonly think of fruit as something that is eaten. The scientific definition means the protective covering around a seed, whether or not the covering is edible.*}
- In which two categories do scientists usually group fruits? {*dry and fleshy*}
- What are some common dry fruits? {*Possible answers: nuts, corn, grains, seeds from some shrubs, trees, and grasses*}
- Give some examples of fruits that have seeds embedded in the outside of their flesh. {*strawberries and pineapples*}
- Give some examples of fruits that have several seeds inside a core that is surrounded by a fleshy layer. {*Possible answers: apples, pears*}
- Give some examples of fruits that do not have a core, are mostly fleshy, and have seeds that are scattered throughout the flesh. {*Possible answers: oranges, watermelons, cucumbers*}
- Why do some fruits have many seeds and other fruits have only one? {*Fruits with many seeds develop from flowers that have more than one ovule in the pistil. Fruits with one seed develop from flowers that have only one ovule in the pistil.*}

Types of fruit

After fertilization, the petals of the flower fall off, and the zygote begins to grow. The zygote develops into an **embryo** (EM bree oh), a tiny new plant. The ovule that surrounds it becomes the seed coat, or outer covering of the seed. The ovary also grows larger and eventually develops into the fruit of the plant.

When you think of fruit, you probably think of edible fruits such as apples, oranges, peaches, or pears. However, not all fruits are edible. The protective covering around a maple tree seed or a mistletoe berry would not be good for humans to eat, but these are still fruits. A **fruit** is simply the part of the plant that contains the seeds. Fruits that have multiple seeds develop from flowers that have more than one ovule in the pistil.

Scientists often classify fruits as either dry or fleshy. *Dry fruits* include nuts, corn, and other grains, as well as seeds from some shrubs, trees, and grasses. Some *fleshy fruits,* such as strawberries and pineapples, have many seeds embedded in the outside of their flesh. Apples and pears have several seeds inside a core that is surrounded by a fleshy outer layer. Other fruits, such as oranges, watermelons, and cucumbers, do not have a core. Their seeds are scattered throughout their flesh. Fruits that have only one seed, such as peaches, develop from flowers that have only one ovule in the pistil. This one seed is usually enclosed in a hard pit or stone.

284

SCIENCE BACKGROUND

Name of fruit by seed type—Fruits such as watermelons, oranges, and grapes that have seeds embedded in their flesh are called *berries*. Many fruits that we would normally consider berries, such as strawberries and blackberries, are actually *compound fruits*. Fleshy fruits with a single seed, such as cherries and peaches, are called *drupes*. *Pomes* are fleshy fruits, such as apples, that have more than one seed enclosed in a core.

Seed leaves—Cotyledons, or seed leaves, do not usually look like the true leaves of the plant. For example, the cotyledons of a carrot resemble tiny straight blades of grass, but the leaves of a carrot are circular with ruffly edges.

Parts of a Seed

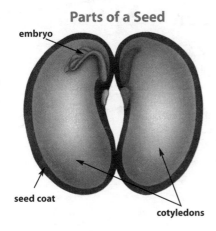

Germination

Seeds cannot **germinate** (JUR muh NATE), or sprout, without the right conditions. Most seeds need water, light, and the proper temperature and type of soil to be able to sprout. If conditions are not right, seeds may lie dormant, or inactive, for long periods of time. When conditions are suitable, the seeds begin to germinate.

Seeds have three parts: an embryo, stored food, and a seed coat. The embryo is the young plant that has developed from the zygote. It has the beginnings of roots, stems, and leaves. The seed has food stored inside one or two **cotyledons**, or special seed leaves. The **seed coat** is the outer covering that protects the embryo and food. It also helps to keep the seed from drying out.

Germination begins when the seed absorbs water and swells. The seed coat splits open, and the embryo begins to grow. The root starts to grow first, and then the stem grows. As the stem grows, the cotyledons stay attached to it so that the new plant can use the stored food. When the first leaves start to grow and the plant is capable of making all of its own food, the cotyledons drop off.

1. Identify the male and female parts of a flower.
2. Define fruit.
3. Explain the difference between pollination, fertilization, and germination.

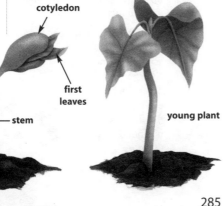

Direct a Demonstration
Identify the parts of a seed

Materials: lima bean seeds

Open a lima bean seed and show the student the embryo and cotyledons. Identify the parts of the seed. If dry lima bean seeds are used, they will open more easily if they have been soaked in water overnight.

BIBLE
The Bible compares the hearing of the Word of God to seed that is sown. In the Bible example some seed fell on stony ground and could not grow. Some seed fell where birds could find it and eat it. Some seed fell among thorns and was choked out. But some fell on good ground, grew, and yielded a crop. Just as seeds require the right conditions for growth, the Word of God can take root and yield fruit in the life of a person whose heart is prepared to obey God. [BATs: 6a Bible study; 6c Spirit-filled]

Discussion:
pages 284–85

➤ What is another word for *sprout*? {germinate}

➤ What basic conditions do most seeds need for germination? {light, water, right temperature, and soil}

💡 Why do you think some plants produce many seeds? {Possible answers: Not all of the seeds produced will land in a place where conditions are suitable for germination. Some seeds may not reach maturity.}

➤ What are the three parts of a seed? {embryo, stored food (cotyledon), and seed coat}

➤ What are *cotyledons*? {special seed leaves where food is stored for the plant embryo}

➤ Why does a seed need cotyledons? {Cotyledons provide food for young plants until the plants are able to make food on their own.}

➤ What is the purpose of the seed coat? {to protect the embryo and food inside the seed and to help prevent the seed from drying out}

➤ Which part of a seed always germinates first? {the root}

💡 How do germination, pollination, and fertilization work together as a cycle for plant reproduction? {In germination, the seed sprouts, and a young plant develops. As the plant matures, flowers are produced. Insects and animals pollinate the flowers so that fertilization can take place. After fertilization, the seed develops.}

Answers

1. The female part of a flower is called a pistil. The male part is called a stamen.
2. A fruit is the part of the plant that contains the seeds.
3. Pollination involves the transfer of the pollen (containing sperm cells) from the stamen to the egg cells in the pistil. Fertilization occurs when the sperm cell unites with the egg cell. Germination occurs when the seed formed through fertilization sprouts and begins to grow.

Activity Manual
Reinforcement—page 184

Lesson 125

Student Text pages 286–87
Activity Manual pages 185–86

Objectives
- Measure the parts of a flower
- Identify the parts of a flower

Materials
- See Student Text page

Introduction

Note: Because this activity is designed for the student to observe and identify the parts of a flower, it does not include a problem or hypothesis.

Purpose for reading:
Student Text pages 286–87
Activity Manual pages 185–86

The student should read all the pages before beginning the activity.

ACTIVITY: Flower Dissection

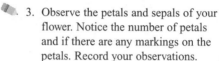

Honeybees have three simple eyes that detect light and darkness and two compound eyes. Working together, these eyes provide a very detailed picture of the honeybee's environment. Honeybees cannot see the color red, but they can see some ultraviolet colors that humans cannot see. Many of the flowers that honeybees pollinate reflect these ultraviolet colors. Because of their job as pollinators, honeybees get an up-close view of a flower. In this activity, you will be getting a "bee's-eye view" of the inside of a flower.

Procedure

1. Write the type of flower you are observing in the materials list in your Activity Manual.
2. Lay a piece of black paper on your desk for a work surface.
3. Observe the petals and sepals of your flower. Notice the number of petals and if there are any markings on the petals. Record your observations.
4. Measure the length of the flower from the base of the sepal to the tips of the petals. Record your measurement.
5. Carefully remove the petals and the sepals. Observe the stamens of your flower. After recording your observations, carefully remove the stamens without damaging the other parts of the flower.

Materials:
- large flower
- magnifying glass
- small knife or other cutting tool
- sheet of black paper
- centimeter ruler
- toothpick
- Activity Manual

286

SCIENCE BACKGROUND

Classifying flowers—Flowers are classified as either complete or incomplete. Complete flowers have petals, pistils, sepals, and stamens. Incomplete flowers are missing one or more of these parts.

Flowers are further classified as perfect or imperfect. Perfect flowers have both stamens and pistils, while imperfect flowers have either stamens or pistils but not both.

Both pecans and corn have imperfect flowers, with both male and female flowers on the same plant. Asparagus, holly, ginkgo, and pistachio have separate male and female plants. In order to bear fruit, these plants must grow close enough to each other for pollination to occur.

TEACHER HELPS

Not all flowers have separate stamens. The stamens of some flowers are fused to the pistil and may be harder to dissect. Flowers that work well with this dissection include lilies, daffodils, snapdragons, and tulips.

Have extra flowers on hand in case of accidents.

Be aware of any students who may have severe reactions to pollen.

302   Lesson 125

6. Study one of the stamens under your magnifying glass. Draw a picture of the stamen and label the anther and filament.
7. Measure and record the length of each stamen.
8. Brush one of the stamens gently over the black paper so that some pollen grains are visible on the paper. Use your magnifying glass to observe the pollen grains.
9. Measure the length of the pistil, observing the stigma, style, and ovary. Record your measurement. See if the sticky tip of the stigma is able to pick up any pollen grains from the black paper. Examine the pistil with your magnifying glass and draw a picture of it.
10. Lay the pistil on your paper. Carefully cut the widest part of the pistil in half from top to bottom. Separate the two parts and observe the inside of the ovary. Record your observations.

Conclusions
- Why is it important to follow a specific procedure when dissecting a flower?
- How could inaccurate measurements affect the results of this activity?

Follow-up
- Try dissecting other flowers.

287

SCIENCE PROCESS SKILLS

Measuring/Using Numbers—Discuss the accuracy of the measurements used in the activity.

➤ **What are some problems that may cause inaccurate measurement?**
{Possible answers: not starting at the ends of the pieces; misreading the ruler; not accurately removing the pistil or stamen}

➤ **Which measurement is more accurate, centimeters or millimeters?**
{millimeters}

Procedure
Guide the student through each step of the dissection. Examine and discuss each flower part before going on to the next step.

 Direct the student to use the diagram on Student Text page 282 to help with identifying the flower parts.

➤ **How does the length of a stamen compare to the length of the pistil?**

➤ **Does your flower have all the parts pictured in the diagram on page 282?**

➤ **What color is the pollen of your flower?**

Direct the student to use the toothpick to move the plant parts as they are examined.

Adjust the procedures if your flowers do not have all the parts shown on Student Text page 282.

Demonstrate procedures as necessary. In step 10, you may choose to cut the ovary rather than have the student cut it.

Conclusions
If several of the same type of flowers are used, calculate the average size of each flower part measured. The student can then compare his flower to the average.

Prepare a chart for recording flower measurements.

 Studying details of God's organisms helps us appreciate God's wonderful designs.

Activity Manual
Activity—pages 185–86

Assessment
Rubric—TE pages A50–A53
Select one of the prepared rubrics, or design a rubric to include your chosen criteria.

Lesson 126

Student Text pages 288–91
Activity Manual pages 187–88

Objectives

- Explain how conifers reproduce
- Compare and contrast seeds and spores
- Identify some organisms that reproduce by spores

Vocabulary

spore   fruiting body

Materials

- fruit such as an apple or an orange that has seeds
- knife

Introduction

Slice the apple or other fruit and display the seeds inside.

➤ What do we call plants that produce seeds enclosed in a fruit? {angiosperms}

➤ The seeds you will study today are not protected by a fruit. These seeds do not have a protective covering, and they often develop inside cones.

Purpose for reading:
Student Text pages 288–91

➤ How are the seeds of conifers different from the seeds of fruit-bearing plants?
➤ How do plants without seeds reproduce?

Seeds in Cones

Some seeds are not protected by fruits. Instead, they develop inside cones or are sometimes protected by a fleshy seed coat that looks like a berry. Plants that have seeds like these are conifers, some of the most common gymnosperms.

Conifers are usually evergreen trees or shrubs. They can range in height from 8 cm (3.1 in.) to over 91 m (299 ft) tall. Many conifers, such as pines, have woody cones. However, some conifers, such as junipers, have softer, fleshier cones.

Conifers usually produce male and female cones on the same tree. The male cones usually grow in the lower branches of the tree or out at the tips of the branches. They are generally smaller and softer than the female cones. These cones produce tiny grains of pollen that contain the sperm cells. The pollen grains often overflow the scales on the cone. Though the pollen is dispersed by the wind, not all of it reaches the female cones on other conifers. You may have noticed some of this pollen as fine yellow dust on the ground or even on your car in the spring. After the male cones lose their pollen, they usually disintegrate and fall off the tree.

Each species of conifer has a uniquely textured pollen grain. Some pollen grains have smooth surfaces, but others have rough and bumpy surfaces. Even though the pollen of many different trees might be in the air at any one time, each conifer can be pollinated only by pollen from its own species.

male cones

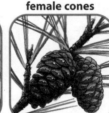
female cones

TEACHER HELPS

You may choose to have samples of male and female pinecones and fern fronds with visible spore cases to show during the discussion. Refer to Student Text page 288 for help in identifying cones.

SCIENCE MISCONCEPTIONS

Students may confuse the seeds of yews and junipers with fruits. Remind them that since yews and junipers are conifers, the berrylike flesh is the seed coat and not a fruit. If you were to open a juniper berry, you would not find a seed inside it.

Life Cycle of a Conifer

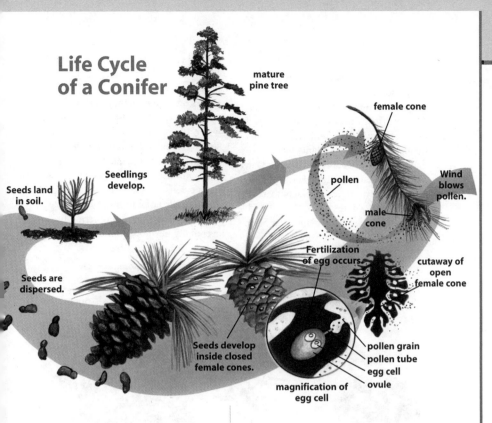

Female cones often grow in the upper branches of the tree. For some pine trees, the female cones are small and green or brown. Other conifers have purple or reddish cones. Before pollination, these cones have their scales slightly open. Each cone has at least one ovule at the base of each scale. The wind blows the pollen into the slightly opened scales of the female cones. After the pollen enters, the scales close. The pollen grain begins to slowly grow a long tube down toward the egg cell in the ovule. The sperm cell then travels down the tube and fertilizes the egg cell, forming a zygote.

After fertilization, the ovule develops into a seed. The zygote becomes the embryo of the seed. The rest of the ovule develops into the seed coat and the stored food for the seed. As the seed matures, the female cone grows larger. The scales of the cone open, and the fully developed seed is carried away by the wind.

289

Discussion:
pages 288–89

💡 **How are gymnosperm seeds different from angiosperm seeds?** {Angiosperm seeds develop in a fruit. Gymnosperm seeds do not have a protective covering. They develop in cones or have a fleshy seed coat.}

▶ **How are male cones different from female cones?** {Male cones are often found in the lower branches of the tree or out at the tips of the branches. They are usually smaller and softer than female cones.}

▶ **How is the pollen from the male cones scattered?** {by the wind}

💡 **Why do you think conifers produce such large quantities of pollen each spring?** {Not all the pollen gets into the female cones.}

▶ **Why do pollen grains from different conifers have different textures?** {Each type of cone can be pollinated only by pollen from its own species.}

▶ **How are conifers pollinated?** {The wind blows the pollen into the slightly open scales of a female cone. The scales close, and the pollen grain begins to grow a long tube down toward the egg cell.}

▶ **What happens to the cone as the seed develops?** {The cone grows larger.}

▶ **How are the seeds dispersed?** {by the wind}

📖 Refer the student to *Life Cycle of a Conifer* on Student Text page 289. Discuss the steps shown in the diagram.

Direct a DEMONSTRATION
Examine types of cones

Materials: assorted male and female cones of various ages

Display the cones and discuss the similarities and differences among them.

▶ **What difference do you see between young male and female cones?** {Possible answer: If the cones are from the same tree the male cones would be smaller and softer.}

▶ **What differences do you see between young and mature female cones?** {Possible answer: The scales of young cones are closed and the scales of mature cones are open.}

Break a scale off the mature female cone.

▶ The seeds of gymnosperms, such as this conifer, develop inside the cone on each scale that is pollinated. These seeds have only a seed coat. They do not have a fruit.

Discussion:
pages 290–91

➤ **How is a spore different from a seed?** {A spore is smaller than a seed, is made of only one cell, and does not have any stored food.}

➤ **What are two kinds of plants that reproduce by spores?** {ferns and mosses}

➤ **Where do spores develop on a fern?** {in tiny spore cases on the undersides of the fronds}

➤ **What kinds of conditions are best for a fern spore to germinate?** {Fern spores develop best in shaded, moist areas.}

➤ **How are spores dispersed?** {by wind and water}

Each spore case of a fern can contain thousands of spores.

➤ **How is the heart-shaped plant that develops from a spore related to an adult fern?** {This plant produces the male and female cells necessary for fertilization. After fertilization, the zygote develops into an adult fern.}

💡 **Ferns produce and release thousands of spores. Why does a plant need to release so many spores?** {Not all the spores land in a place suitable for germination.}

Refer the students to *Life Cycle of Ferns* on Student Text page 290. Discuss each step of the diagram.

Spores

Although many plants reproduce by seeds, some plants reproduce by spores. A **spore** is much smaller than a seed and consists of only one cell. Spores do not have any stored food available for the new plant to use. Plants that reproduce by spores usually need more water than seed-bearing plants. Ferns and mosses are two types of plants that reproduce by spores.

Ferns are vascular plants that have underground stems as well as underground roots. Spores develop in tiny spore cases on the undersides of the fronds, or leaves, of the fern. When the spores are released, they are carried away from the parent plant by wind or water. If a spore lands in moist, shaded soil, it begins developing into a small, flat, heart-shaped plant. This small plant produces both the male and female cells for the fern. The male and female cells unite to form the zygote. An adult spore-producing fern will eventually develop from this zygote.

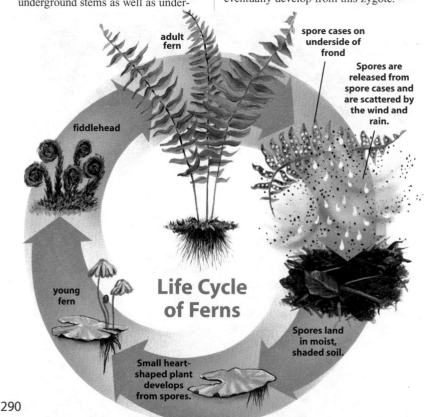

Life Cycle of Ferns

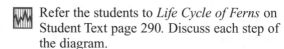

Fern gametophyte—the part of the fern life cycle that includes the germination and fertilization of the fern; A young fern plant at this stage of the life cycle is very small and inconspicuous, often much smaller than a straight pin. The heart-shaped growth gradually dissolves as the mature plant develops.

Fern sporophyte—the part of the fern life cycle that includes the fiddleheads and mature fronds that produce spores; This is the stage most often seen and recognized as a fern plant.

Moss gametophyte—This first stage of the moss life cycle includes the lush part of the plant most often seen.

Moss sporophyte—In this second stage of the moss life cycle, the plant develops a sporophyte at the end of a slender stalk, called a *seta*.

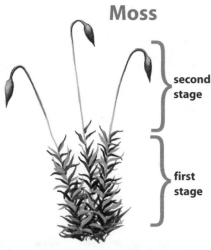

Moss

second stage

first stage

Moss is another plant that reproduces by spores. Like ferns, mosses undergo two stages of reproduction. In the first stage, small, fuzzy, stemlike structures grow, often looking like a soft, green carpet. These plants produce the male and female cells. After fertilization, a slender stalk grows up out of the moss plant and produces the spores that will become new moss.

Molds, yeast, mushrooms, and toadstools are not plants. They are fungi. However, they also can reproduce by spores. Most fungi produce **fruiting bodies,** structures that contain their spores. Fruiting bodies differ in appearance from one type of fungus to another. The cap of a mushroom is its fruiting body, and spores are released from tiny openings underneath the cap.

Most bread and fruit molds, however, send up little clublike structures that contain their spores. Other fungi, such as yeast and truffles, form their spores in tiny sacs.

The methods of forming and releasing spores vary from one kind of fungus to another, but when the fruiting body opens, the spores scatter. Spores that land near food, warmth, and moisture germinate and develop into new fungi. Spores are able to travel great distances and are capable of remaining in the air for years. Perhaps you know someone who is allergic to mold spores. That person may need a special filter in his home to capture the spores and keep them from germinating.

QUICK CHECK
1. How are conifers pollinated?
2. How is a spore different from a seed?
3. What structures in fungi contain the spores?

mold on bread

Direct a Demonstration

Demonstrate how to make spore prints

Materials: large mushroom cap, white paper, container big enough to cover the mushroom

Lay the mushroom cap on the sheet of white paper with the gill side down.

Cover the cap with the container and leave it in a place where it will not be disturbed for two days.

Carefully remove the container and cap. Examine the spore print with a hand lens.

Discussion:
pages 290–91

➤ **How are the reproduction methods of mosses and ferns similar?** {Possible answers: Both reproduce through two stages. Both have spores.}

➤ **What are the two stages of moss reproduction?** {The fuzzy green growth of moss produces both the male and female cells. After fertilization, a slender stalk that contains spores grows up out of the moss plant. These spores are then released to form new moss.}

➤ **Molds and mushrooms used to be classified as plants. In what kingdom do scientists now classify them?** {fungi}

➤ **How do fungi reproduce?** {by spores}

➤ **What are *fruiting bodies*?** {fungus structures that contain the spores}

➤ **What are three examples of fruiting bodies?** {Possible answers: the cap of a mushroom, clublike structures on molds, and tiny sacs on yeast and truffles}

➤ **What conditions must be met for fungus spores to germinate?** {food, warmth, and moisture}

Answers

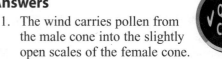

1. The wind carries pollen from the male cone into the slightly open scales of the female cone.
2. A spore is smaller than a seed and consists of only one cell. It also does not have any stored food.
3. fruiting bodies

Activity Manual

Study Guide—pages 187–88
These pages review Lessons 124 and 126.

Assessment

Test Packet—Quiz 12-A
The quiz may be given any time after completion of this lesson.

Lesson 127

Student Text pages 292–95
Activity Manual page 189

Objectives
- Recognize that animals begin as a single cell
- Compare and contrast placental and marsupial gestation
- Differentiate between the different types of eggs
- Explain why some animals lay many eggs

Vocabulary
gestation
marsupial mammal
placental mammal

Materials
- picture of a mother cat or other animal with several babies that look different from the mother

Introduction

Display the picture of the mother cat and kittens.

➤ Why do you think the kittens do not look like their mother?

📖 In today's lesson you will discover how God has created each individual unique and different.

Purpose for reading:
Student Text pages 292–95

➤ What are three different ways that baby animals develop?

➤ How are amphibian eggs different from reptile eggs?

Animal Reproduction

Kittens, pups, fawns, fry, and tadpoles all are names we give to baby animals. Each of these babies will grow and develop into a different adult animal. However, in one way, all of these animal babies are similar. After God created each animal, He declared that each would reproduce after its own kind (Gen. 1:25). Each animal begins its life as a single cell. That single cell divides and grows, eventually becoming an adult animal.

This beginning cell is formed when a male sperm cell joins with a female egg cell to form a zygote. Each of these reproductive cells has only half of the number of chromosomes found in an ordinary body cell. When the egg and sperm cells unite, fertilization occurs. The newly formed zygote receives half of its chromosomes from each parent. This division ensures that the baby animal will have the same number of chromosomes as its parents. It also allows the new animal to be a unique individual, a mixture of both parents and not an exact copy of either one.

After the zygote cell divides for the first time, it is called an embryo. The embryo's cells divide again and continue to grow through the process of mitosis. The embryo continues to develop either inside or outside the mother's body. Some animals develop completely inside their mothers' bodies. Other animals develop in a pouch outside their mothers' bodies or in eggs. The period of time between fertilization and the birth of the animal is called **gestation.**

292

Students may need to review mitosis, meiosis, and chromosomes from Chapter 4, *Cells and Classification*.

Mitosis—the process of cell division in which the cell makes an exact copy of itself; The chromosomes duplicate once, and the cell divides once.

Meiosis—the two-stage process of cell division in which the chromosomes duplicate once, but the cell divides twice; Each new cell has half the chromosomes of the parent cell.

Chromosomes—the genetic structure of a cell that contains the DNA, or chemical code, that gives the cell directions about what it should do

Chromosomes are mentioned again in Chapter 13, *Heredity and Genetics*.

SCIENCE MISCONCEPTIONS

Emphasize that although humans are often classified biologically as a type of mammal, man is not an animal, because he is created in the image of God. God created man to have dominion over the animals.

The placenta is not the fluid that surrounds the embryo. It is a separate tissue that connects the amniotic sac (the fluid that surrounds the embryo) and the mother's blood supply.

308 Lesson 127

puppies

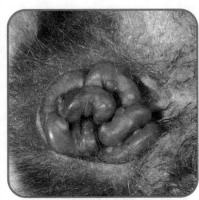

baby opossums

Placental Gestation

Mammals whose young develop inside the mother's body are called **placental** (pluh SEN tuhl) **mammals.** Inside the mother, a fluid-filled sac surrounds the embryo, or developing baby. The embryo receives food and oxygen from the mother's blood through a placenta. The placenta allows the embryo to receive nourishment from the mother until it is completely developed. The placenta also empties the embryo's wastes into the mother's blood.

The length of gestation is different for each animal. A mouse embryo develops in about twenty days. However, an elephant takes almost two years to develop. When placental mammal babies are born, they are no longer dependent on their mothers for life. Though they are often helpless and need parental care, their body systems can function independently.

Marsupial Gestation

Some mammals carry their young in their bodies for only a short time. These **marsupial** (mar SOO pee uhl) **mammals** have a pouch on the mother's body, where their young finish developing. The gestation period for marsupials is very short. For example, baby opossums are born after only thirteen days. Newborn marsupials are also very tiny. Just one teaspoon can hold about twenty baby opossums! Even the largest marsupials are less than 2.5 cm (1 in.) long when they are born.

The newborn marsupial uses its front legs to crawl to its mother's pouch. Some marsupials have deep and roomy pouches. Other pouches are just loose flaps of skin. Some pouches are lined with fur, and others are not. All pouches protect the babies and provide milk for them as they continue to grow.

Discussion:
pages 292–93

➤ What common beginning do all animals have? *{a single cell}*

➤ What is unique about reproductive cells? *{They have only half the number of chromosomes of a regular body cell.}*

➤ How do reproductive cells allow a baby animal to be different from its parents? *{Each parent gives half of the necessary chromosomes for the baby animal. The baby has characteristics of both parents, but it is not exactly like either.}*

➤ What is the difference between an embryo and a zygote? *{A zygote is the fertilized egg cell, or the beginning cell. After the cell divides for the first time, it is called an embryo.}*

➤ What is *gestation*? *{the length of time between fertilization and the birth of the animal}*

➤ How does the embryo receive food and oxygen? *{from its mother's blood through the placenta}*

➤ What happens to the embryo's wastes? *{The placenta empties them back into the mother's blood.}*

➤ Have you ever noticed the thin membranes inside an eggshell? *{Answers will vary.}*

A placenta functions in a similar manner as the egg membranes. It allows nutrients and oxygen to pass through it for the developing embryo, but it also acts as a filter to help keep the embryo's blood separate from the mother's blood.

➤ What is unique about a mother marsupial mammal's body? *{It has a pouch.}*

➤ How does a newborn marsupial get to its mother's pouch? *{After birth it uses its front legs to crawl to the pouch.}*

💡 Why is the gestation time for marsupial mammals so short? *{Marsupial young are born before they are fully developed. They finish their development in their mother's pouch.}*

➤ What does the marsupial's pouch provide for the young? *{protection and milk}*

Discussion:
pages 294–95

▶ **What are some of the purposes for eggs?** *{to provide protection, nutrients, food, and waste removal for the developing embryo}*

📖 **What are some possible reasons God designed eggs of different shapes?** *{Possible answers: to fit the types and locations of various nests}*

▶ **Where are most amphibian and fish eggs laid?** *{in water}*

▶ **How are most eggs laid in water protected?** *{They are usually covered in a clear, jellylike fluid.}*

▶ **What do these eggs usually look like?** *{transparent with a dark spot where the embryo is developing}*

Discuss any times that students may have seen eggs in water.

▶ **What causes most eggs laid in freshwater rivers to sink?** *{They are sticky, so they pick up grains of sand that make them heavier and cause them to sink.}*

💡 **Why would eggs laid in fresh water need to sink?** *{Possible answer: Freshwater eggs need to sink so that they do not float with the current into salt water where they cannot survive.}*

💡 **Why is this not a problem for saltwater eggs?** *{Possible answer: Saltwater eggs will not float into an environment unsuited for their development.}*

▶ **Where are most saltwater eggs located?** *{at or near the surface of the water}*

▶ **What are some characteristics of eggs laid on land?** *{They usually have either leathery or brittle shells.}*

💡 **What kind of shells do birds have?** *{brittle}*

Allow students to share experiences they have had with eggs and nests.

▶ **Why are reptile eggs often white but birds' eggs are usually colored?** *{Reptile eggs are usually buried, but the birds' eggs need coloring for camouflage.}*

📖 Point out God's perfect design in the way that He has designed eggs to suit the environment and the instincts of the parents. Even though we may sometimes interpret animal parents' caring for their young as a loving or tender emotion, animals feed and protect their young through instinct. For example, the cuckoo lays her eggs in the nests of other birds. Young cuckoos are then cared for because the bright orange color of their open mouths acts as a stimulus for the foster parent birds to feed them.

Lesson 127

Eggs

Numerous other animals develop inside eggs that have been laid on land or in water. All birds lay eggs, as do many species of fish, amphibians, and reptiles. Eggs vary in size, shape, texture, and appearance from animal to animal. But all eggs provide protection, nutrients, food, and waste removal for the developing animal.

robin eggs

frog eggs

The shape of the egg often varies to suit the type of nesting place. The guillemot (GILL uh MAHT), a bird that usually lives near rocky seashores, lays her eggs on the bare rock of narrow cliff ledges. Her eggs are tapered to help prevent them from falling off the ledge. Other birds lay wedge-shaped eggs. Since this shape allows the eggs to lie closer together in the nest, it is easier for the parent bird to keep each egg at an even temperature. Reptile eggs are usually round or oval.

Most amphibians and fish lay their eggs in water. Eggs laid in water are often covered in a clear, jellylike fluid that protects them in the water. These eggs are usually transparent with a dark spot where the embryo is developing. Eggs laid in freshwater rivers are usually sticky. They pick up grains of sand, which make the eggs heavier and cause them to sink. Eggs that are laid in salt water often float on or near the surface of the water.

Land eggs usually have either leathery or brittle shells. Some reptiles have soft, leathery eggs, but other reptiles, such as crocodiles, have hard-shelled eggs. Reptile eggs are white, and the reptile mothers usually bury or cover their eggs. Other eggs, especially bird eggs, are often camouflaged to match their nesting environments. These colorings and markings help protect the eggs from predators.

SCIENCE BACKGROUND

Placental mammal—The term *placental mammal* is technically inaccurate since marsupial mammals use a placenta too. However, the marsupial mammal's placenta is more like an egg yolk, and it nourishes the young only until they are ready to crawl to the mother's pouch.

midwife toad with eggs

Parental Care

Some animals lay thousands of eggs, and other animals lay only one or two. Most fish and reptiles lay their eggs and then leave, not returning to their eggs. A few fish, like salmon, die after the eggs are laid. When these eggs hatch, the young are already able to take care of themselves. However, many of these young, as well as many of the eggs, will become food for other animals.

Some animals, though, remain to guard and care for their eggs. Some species of fish take care of the eggs by fanning them with their fins to prevent silt from settling on them. Several species of frogs and toads carry their eggs around with them on their backs. Some even carry their eggs in their vocal sacs! Most parent birds share the responsibility of guarding their eggs and raising their young.

Species that do not provide any parental care often lay large numbers of eggs at a time. Usually species that provide more parental care lay fewer eggs. But lack of parental care is not the only reason that some animals lay many eggs. Why would animals that are lower on the food chain need to lay more eggs? Species lower on the food chain need to lay more eggs to help keep their populations balanced.

1. How is marsupial gestation different from placental gestation?
2. What do all eggs provide for a developing animal?
3. Why do some animals generally lay more eggs than others?

295

The student should research and write a paragraph about how an animal cares for its young.

Suggested animals:

- beaver
- Darwin frog
- elephant
- emperor penguin
- kangaroo
- koala
- lion
- platypus
- spiny anteater
- wolf

Discussion:
pages 294–95

💡 Do you think most animals that lay eggs provide parental care? {Answers will vary, but elicit that most do not.}

➤ What must be true of young animals that receive no parental care? {They must be able to care for themselves as soon as they are hatched.}

➤ What happens to many eggs and newly hatched young? {They become food for other animals.}

➤ Why do some fish fan their eggs with their fins? {to keep silt from settling on them}

➤ What are other ways that some animals care for their eggs? {Possible answers: Some frogs and toads carry their eggs on their backs, and some carry their eggs in their vocal sacs. Parent birds guard their eggs.}

➤ Why do animals that provide little parental care often lay so many eggs? {Without parental protection fewer of the eggs and young survive to adulthood.}

💡 What is the food chain? {Possible answer: The series of events in which organisms prey on one another.}

💡 At what position do you think animals that lay large numbers of eggs are on the food chain? {lower position} Why? {because more eggs and young become food for other animals}

Answers

1. Possible answers: Marsupial mammals are born underdeveloped and finish developing in their mother's pouch. Placental mammals develop inside the mother's body.

2. Eggs provide protection, nutrients, food, and waste removal for the developing animal.

3. Possible answers: Generally, animals that do not provide parental care lay more eggs. Animals that are lower on the food chain lay more eggs.

Activity Manual
Reinforcement—page 189

Lesson 128 — Activity Manual page 190

Objectives

- Recognize the value that God places on life
- Recognize that God provides eternal life

Introduction

Note: This lesson is designed as an interactive Bible study using Activity Manual page 190. You may choose to omit the discussion and guide the student through completion of the page alone.

The discussion focuses on the value God places on life and His omniscient design of each of us before birth. The lesson ends with an emphasis on salvation for eternal life.

The terms *abortion* and *euthanasia* do not appear in the discussion, but you may choose to incorporate your own thoughts as appropriate for your student.

➤ People have different opinions about when life begins and when it should end.

➤ In this lesson we will look at the Bible to see what God says about life and death.

Discussion:
Activity Manual page 190

1. Read Genesis 1:27.
   ➤ How was man created different from the animals? *{God made man in His own image.}*

2. Read Genesis 4:8–12.
   ➤ What did Cain do to Abel? *{Cain murdered Abel.}*
   ➤ How did God respond to Cain's actions? *{God judged Cain and punished him.}*

3. Read Exodus 20:13.
   ➤ What does this commandment say? *{We are not to kill anyone.}*
   ➤ Think about Genesis 4:8–12 and Exodus 20:13. Do you think God values the life of each person? *{yes}*

   Exodus 21:12 states that the punishment for murder is death for the murderer.

What Value Does God Place on Life?

Name _____

A. Read the verses and complete the statements about God's value on life.

- **Genesis 1:27** — 1. Man is a special creation, made in God's own __image__.
- **Genesis 4:8–12** — 2. After Cain killed his brother Abel, God __cursed__ him.
- **Exodus 20:13** — 3. One of the Ten Commandments says, "Thou shalt not __kill or murder__."
- **Isaiah 44:2** — 4. God made me and formed me in the __womb__.
- **Jeremiah 1:5** — 5. God knew who Jeremiah would be and what he would do __before__ Jeremiah was even born.
- **Exodus 4:10–12** — 6. Regardless of man's imperfections, or disabilities, God made each person for a specific purpose. He made the dumb (mute), the __deaf__, the seeing, and the __blind__.
- **Job 12:9–10** — 7. God controls the soul of every living thing and the __breath__ of all mankind.
- **Matthew 10:30–31** — 8. God knows all about me. He knows the __number__ of hairs on my head, and my life is valuable to Him.
- **Job 33:4** — 9. God made me, and His __breath__ gave me life.
- **Psalm 139:13–14** — 10. The Bible tells us to praise God, for we are fearfully and wonderfully made. Marvelous are His __works__.

B. Match the description and the verse.

- __E__ 11. Life and death are controlled by the Lord.
- __A__ 12. Physical life is a gift from God.
- __C__ 13. Death is a result of sin.
- __B__ 14. Every man has an appointed time to die.
- __D__ 15. God provides eternal life through Jesus Christ.

A. Acts 17:25
B. Hebrews 9:27
C. Romans 5:12
D. Romans 6:23
E. I Samuel 2:6

190 Lesson 128
Bible Integration

SCIENCE BACKGROUND

Abortion—Abortion is the act of killing a baby before it is born.

Euthanasia—The term *euthanasia* is used to describe the act of ending a person's life because of illness, old age, or physical disability. Euthanasia is practiced and accepted in many countries. The guidelines for using euthanasia vary from country to country. Some countries allow it only for people over the age of eighteen, but some countries have proposed legislation to allow euthanasia for disabled individuals under the age of eighteen.

Assisted suicide—As of 2003 in the United States, Oregon was the only state to allow assisted suicide. However, this does not mean that euthanasia and assisted suicide are not issues for other states. Some states are currently hearing court cases about terminating the lives of spouses who have suffered strokes, are in comas, or have other long-term health problems.

Discussion:
Activity Manual page 190

4. Read Isaiah 44:2.
   - Who formed each of us before we were born? *{God}*
   - We are each designed and formed by God. He created us with exactly the hair color, eye color, height, and abilities that He wanted for us.

5. Read Jeremiah 1:5.
   - God formed the prophet Jeremiah before he was born. What else did God plan for Jeremiah before he was born? *{God planned that Jeremiah would be a prophet.}*
   - What attribute of God tells you that God has a plan for your life? *{God is omniscient, or all-knowing.}*
   - God has a specific plan for each person's life. Each person is born for a purpose.

6. Read Exodus 4:10–12.
   - What does Exodus 4:11 say about people born with disabilities? *{God made them.}*
   - God created Moses exactly as He needed for Moses to lead the Israelites. God creates each person with the abilities that he needs to fulfill God's purpose for his life.

7. Read Job 12:9–10.
   - Do we have control of whether we live and breathe? *{no}* Who does? *{God}*

8. Read Matthew 10:30–31.
   - What does Matthew 10:30–31 tell us about God? *{God knows exactly how many hairs we have on our heads, and our lives are valuable to Him.}*

9. Read Job 33:4.
   - Who gives us life? *{God}*

10. Read Psalm 139:13–14.
    - According to Psalm 139:13–14, why should we praise God? *{We are fearfully and wonderfully made.}*

Psalm 139:16–18 continues to tell us how the writer marvels over God's knowledge and plan of the details of each life before birth and God's continued thought and care to sustain life.

Guide the student in matching the statements and references. After the matching is completed, discuss the statements and verses in order.

11. Discuss I Samuel 2:6. Life and death are controlled by the Lord.
    - Who controls life and death? *{the Lord}*

12. Discuss Acts 17:25. Physical life is a gift from God.
    - What does God provide for all people? *{life, breath, and all things necessary to live}*

Other verses that support this truth include James 4:13–15 and Luke 12:16–21.

13. Discuss Romans 5:12. Death is a result of sin.
    - How did death enter into the world? *{by the sin of one man, Adam}*
    - Have all people sinned? *{yes}*
    - What is the punishment for sin? *{death}*
    - Do you think Romans 5:12 is talking about physical death, spiritual death, or both? *{both}*

14. Discuss Hebrews 9:27. Every man has an appointed time to die.
    - What does Hebrews 9:27 say happens after a person dies? *{judgment}*

15. Discuss Romans 6:23. God provides eternal life through Jesus Christ.
    - The payment for sin is death. What does God provide for each person? *{the gift of eternal life}*

    Not only did God provide each of us with physical life, but He also gives us the free gift of eternal life. After we die, we will continue to live eternally in either heaven or hell. When Jesus died on the cross, He took our punishment for our sins. Christ's death paid the penalty for sin. Although He died and was buried, Christ rose again, victorious over sin and death. Everyone who has accepted Christ's death as the payment for his sins will live forever in heaven. This gift of eternal life available through Jesus Christ is far more important than physical life. Have you chosen to accept God's free gift?

Activity Manual
Bible Integration—page 190

Lesson 129

Student Text pages 296–99
Activity Manual pages 191–94

Objectives

- Identify some methods of asexual reproduction
- Set up an experiment to observe and compare the rate of growth of a plant cutting and a seed

Vocabulary

asexual reproduction
binary fission
budding
regeneration
fragmentation
vegetative reproduction

Materials

- potato with sprouts

Introduction

Note: Included with this two-page text lesson is an activity. The activity, *It's a Race,* will not take long for the student to set up. However, he will need to spend at least one week observing his carrots. The setup can be done at the beginning or at the end of this lesson.

➤ Have you ever seen potatoes that have sprouted or tried to grow an African violet from a leaf? *{Answers will vary.}*

Display the potato for the student to examine.

➤ What are the white things growing on the potato? *{sprouts, roots}*

➤ Sometimes pieces of a plant can grow into a new plant in unusual ways. Today's lesson will explain another type of reproduction.

Purpose for reading:
Student Text pages 296–97

➤ What are some different methods of asexual reproduction?

Discussion:
pages 296–97

➤ What do we call reproduction that involves only one parent? *{asexual reproduction}*

➤ How is an organism that is a result of asexual reproduction different from one that is the result of sexual reproduction? *{It is identical to its parent instead of being a genetic mixture of two parents.}*

➤ What is *binary fission*? *{An organism splits in half to form two separate organisms.}*

Asexual Reproduction

Most animals and plants come from two parents. Each new organism is unique, a mixture of both parents' characteristics. However, some organisms can reproduce from only one parent. These new organisms are identical to the parent. The process of reproducing from only one parent is called **asexual reproduction**.

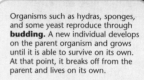

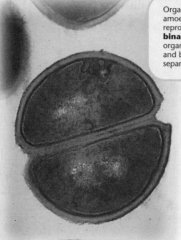

Organisms such as amoebas and bacteria reproduce through **binary fission**. An organism splits in half and becomes two separate organisms.

Organisms such as hydras, sponges, and some yeast reproduce through **budding**. A new individual develops on the parent organism and grows until it is able to survive on its own. At that point, it breaks off from the parent and lives on its own.

A few organisms can reproduce through **regeneration**. For example, if the arm of a sea star is broken off and it includes a piece of the center of the sea star, then that arm can grow to become a new sea star.

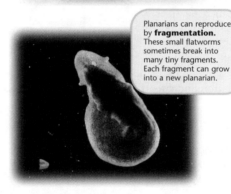

Planarians can reproduce by **fragmentation**. These small flatworms sometimes break into many tiny fragments. Each fragment can grow into a new planarian.

296

SCIENCE BACKGROUND

Asexual reproduction—All methods of asexual reproduction involve mitosis rather than meiosis.

Layering—Another method of asexual reproduction is called *layering*. Some trees and bushes, such as the forsythia, may have branches that bend over and touch the ground. If left undisturbed, a new plant develops from where the branch touches the ground.

Stem reproduction—There are many forms of stem reproduction, including bulbs (daffodils, tulips), corms (gladioluses, crocuses), rhizomes (irises, cattails), stolons (strawberries), and tubers (potatoes, artichokes). Unlike other potatoes, the sweet potato tuber is actually underground storage in a root rather than in a stem.

Vegetative reproduction—Plant asexual reproduction is called vegetative reproduction.

Root cuttings—Root cuttings from radishes, carrots, and sweet potatoes will sprout new greenery. The root usually does not regrow.

Stem and leaf cuttings—Stem and leaf cuttings will root faster if a rooting compound is applied before they are placed in water.

Sometimes part of a plant that usually is not involved in reproduction is able to develop into a new plant. This is called **vegetative reproduction.**

Spider plants grow new plants at the end of the stalks of the parent plant. These plants will develop into mature plants if they touch soil.

Some plants, like pineapples, can grow a new plant from the crown of the fruit.

Plants such as African violets and ivy are able to grow from *cuttings*. If a cut stem is placed in water, new roots will grow. The new roots can develop into a new plant.

Strawberries and many grasses often grow runners, or creeping stems called *stolons*. These stems grow across the soil, occasionally putting down roots and developing new plants.

Potatoes reproduce from underground stems called *tubers*. These tubers are stems that store food for the new plant.

✓ QUICK CHECK
1. How is an organism that is a product of asexual reproduction different from one that is the product of sexual reproduction?
2. Name five methods of asexual reproduction.

297

SCIENCE MISCONCEPTIONS

Be sure that students understand that when the arm of a sea star is broken off, that arm will regenerate into a new sea star only if it contains some of the center part of the original sea star. The original sea star can grow a new arm to replace the one that was broken off. This is also sometimes called regeneration, but it is not a form of asexual reproduction.

Discussion:
pages 296–97

➤ **What are some organisms that reproduce through budding?** *{Possible answers: hydra, sponges, some yeast}*

➤ **What is a difference between budding and fragmentation?** *{Possible answer: An organism that reproduces through budding does not separate from the parent until it is able to live on its own. If an organism that reproduces through fragmentation is broken into pieces, each piece develops into a new organism.}*

➤ **What must be included for an arm of a sea star to regenerate into a new sea star?** *{The arm must include a piece of the center of the sea star.}*

➤ **Sometimes part of a plant that is not usually involved with reproduction develops into a new plant. What is this called?** *{vegetative reproduction}*

➤ **What are some types of vegetative reproduction?** *{Possible answers: cuttings; tubers; stolons; new spider plants growing at the end of the stalks of the parent plant; a new pineapple plant growing from the crown of the fruit}*

➤ **How do plants reproduce through cuttings?** *{Possible answer: When stem or leaf cuttings are placed in water, the cuttings grow new roots. The cuttings can then be transplanted into soil to develop into new plants.}*

💡 **Potato plants can reproduce both sexually and asexually. Why might a potato plant or other organism need to be able to reproduce in both ways?** *{Possible answer: When conditions are not right for seeds to develop, the plant can reproduce asexually.}*

Answers

1. The organism is identical to its parent instead of being a genetic mixture of two parents.
2. binary fission, budding, fragmentation, regeneration, and vegetative reproduction

Activity Manual
Reinforcement—page 191

Study Guide—page 192

This page reviews Lessons 127 and 129.

Assessment
Test Packet—Quiz 12-B

The quiz may be given any time after completion of this lesson.

Chapter 12   Lesson 129   **315**

Purpose for reading:
Student Text pages 298–99
Activity Manual pages 193–94

The student should read all the pages before beginning the activity.

Note: This activity is included as part of Lesson 129. The observation time for the activity is at least one week.

Materials

- See Student Text page
- copy of TE page A34, *Plant City Times*, for each student

➤ News reporters desire to be the first to have their breaking stories in print. They use observation and note-taking skills to provide the facts that will interest their readers. After you finish covering the big race, use information from your observations to write an exciting, yet informative, article for the front page of the *Plant City Times*.

ACTIVITY: It's a Race!

You are the newest reporter for the *Plant City Times*. It's your job to cover the big race—the race between a seed and a cutting. Public interest is high, so be sure that you provide all the details!

Problem
Which will grow greenery six centimeters high first—the carrot top or the carrot seed?

Procedure

1. Formulate your hypothesis. Under equal conditions, which plant do you think will reach six centimeters first? Record your hypothesis in your Activity Manual.

2. Fill each container with soil. Position the carrot top in one of the containers so that most of the top is under the soil. Be sure to leave part of the top above the soil. Add enough water to keep the soil damp.

3. Plant the carrot seeds about one centimeter deep in the soil. Add enough water so the soil is damp.

Materials:
two cups or containers
potting soil
one carrot top, greenery removed
two carrot seeds
water
centimeter ruler
Activity Manual

298

Use a carrot that still has the greenery attached. Twist the greenery off of the carrot. Then cut the carrot root 1.5 to 2 centimeters from the base of the top. Carrot tops can be cut and prepared the night before and kept refrigerated.

Tall, clear containers provide more depth for the carrot seeds' roots to grow. You may need to add gravel or put holes into the bottom of the containers for drainage.

Some other root vegetables, such as radishes, can be used.

SCIENCE MISCONCEPTIONS

Since the carrot top is a root cutting, the student should not expect his carrot top to grow a new carrot root. It should, however, grow new stems and leaves.

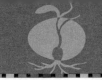

4. Place the containers in a sunny place where they will not be disturbed or knocked over.

5. Observe the containers at the same time each day. Record your observations. When new greenery is visible, record the number of stems and leaves. Measure and record the length of the longest stem. Add water as needed to keep the soil moist.

6. Keep measuring each plant until one of them has greenery that is six centimeters long.

7. Write a newspaper article about the race. Include a catchy headline and all the important information.

Conclusions

- Was your hypothesis correct? Why or why not?
- Gently remove the plants from the soil. How are the roots different?

Follow-up

- Try the activity with a different root vegetable. Compare the results.
- Plant the seeds of several different vegetables and graph their rates of growth.

299

Procedure

Demonstrate how to plant the carrot top and the carrot seeds. Guide the student through Steps 2 and 3.

Allow the student to select a sunny spot in the classroom for his container. Set a time for the plants to be observed each day. If plants will not be observed over the weekend, be sure that the cuttings and seeds have enough water to last through the weekend. (The cuttings will require more water than the seeds.)

When green growth is visible, remind the student to handle his cuttings gently as he measures the greenery. Emphasize the importance of accurate measurements.

Conclusions

Provide time for the student to evaluate his hypotheses and answer the questions.

Activity Manual
Activity—pages 193–94

Assessment
Rubric—TE pages A50–A53

Select one of the prepared rubrics, or design a rubric to include your chosen criteria.

SCIENCE PROCESS SKILLS

Observing—Discuss the importance of careful observation.

➤ **What information are you being asked to observe?** *{number of stems visible and number of leaves visible}*

➤ **Do you think that you will be able to observe new information each day?** *{Answers will vary. Elicit that it may take several days before any changes can be seen.}*

➤ **Why do you think it is important to observe the plants daily even if there is no new information to record?** *{In order to maintain consistent, reliable records, observations must be done in a systematic way.}*

Chapter 12

Lesson 129 317

Lesson 130

Student Text Page 300
Activity Manual pages 195–96

Objectives

- Recall concepts and terms from Chapter 12
- Apply knowledge to everyday situations

Introduction

Material for the Chapter 12 Test will be taken from Activity Manual pages 187–88, 192, and 195–96. You may review any or all of the material during the lesson.

You may choose to review Chapter 12 by playing "Busy Pollinators" or a game from the *Game Bank* on TE pages A56–A57.

Diving Deep into Science

Questions similar to these may appear on the test.

Answer the Questions

Information relating to the questions can be found on the following Student Text pages:

1. page 283
2. page 284
3. page 297

Solve the Problem

In order to solve the problem, the student must apply material he has learned. The answer for this Solve the Problem is based on the material on Student Text pages 283–84. The student should attempt the problem independently. Answers will vary and may be discussed.

Activity Manual

Study Guide—pages 195–96

These pages review some of the information from Chapter 12.

Assessment

Test Packet—Chapter 12 Test

The introductory lesson for Chapter 13 has been shortened so that it may be taught following the Chapter 12 test.

Diving Deep into Science

Answer the Questions

1. How does the way that bees and flowers need each other show God's design?
   Answers will vary, but elicit that God designed the flowers to attract bees and provide nectar for them. The bees benefit from the nectar and then help to pollinate the plants.

2. Why is an acorn an example of a fruit?
   Accept any reasonable answer, but elicit that "fruit" means the part of the plant that contains the seeds. An acorn contains the seeds of an oak tree.

3. Crabgrass is a kind of grass that produces stolons. Why is it so difficult to get rid of crabgrass that has taken over a yard?
   The crabgrass stolons take hold of the soil in many places and produce roots. Therefore, the crabgrass has many roots in many places and is hard to pull up.

Solve the Problem

Your friend Brent planted some peach trees in his yard late last year. This year, just as the trees were starting to bloom, the weather turned cold, and there was a hard freeze. Most of the blossoms on the trees turned brown and fell off. Brent told you yesterday that he cannot figure out why his peach trees have little fruit. Can you explain why?

If the blossoms fell before they could be pollinated and fertilized, no fruit will develop.

Review Game

Busy Pollinators

Draw two large flowers for display. Divide the class into two teams. The teams are bees gathering pollen and pollinating flowers. Ask review questions, alternating between teams. A tally mark is made on the team's flower for each correct answer. The team with the most tallies at the end of the game wins.

Variation: Prepare paper flowers. Each time a team member answers a review question correctly, he receives a paper flower. The team with the most flowers wins.

13
Heredity and Genetics

GREAT & MIGHTY Things

Suppose you heard that if you learned to play the piano well, all of your children would also be able to play the piano. That might not be surprising unless you also heard they would never need to take a lesson or practice. That theory, called *pangenesis,* was a popular theory of Charles Darwin. Darwin proposed that an attribute an organism acquires during its lifetime could be passed on to its offspring. For example, a giraffe that stretches its neck toward the leaves high in a tree will have offspring with longer necks.

The pangenesis theory was popular with those who held to the theory of evolution. It provided a means for one organism to gradually change into another organism. However, the pangenesis theory had no facts to support it. God's design for living organisms allows them to inherit exactly the traits they need to bring glory to Him.

Student Text page 301
Activity Manual page 197

Lesson 131

Objectives
- Demonstrate knowledge of concepts taught in Chapter 12
- Recognize that a parent's acquired abilities are not part of inherited traits

Introduction

Note: This introductory lesson has been shortened so that it may be taught following the Chapter 12 test.

When man chooses not to believe God's record of Creation, he has difficulty explaining how organisms of one species change into organisms of another species. Over the years, man has tried to explain his belief with many different theories.

➤ What was the *pangenesis theory*? {the theory that said offspring inherit traits that a parent acquired during its lifetime}

➤ Why was this theory popular with evolutionists? {It provided an explanation for organisms' gradually changing to other organisms.}

➤ What have you observed about this idea? Can someone play the piano without practicing just because he or she has a parent who can play? {no}

Chapter Outline
I. Heredity
II. Genetics
   A. DNA: The Double Helix
   B. Father of Genetics
   C. Dominant and Recessive Genes
   D. Genetic Disorders and Diseases
   E. Genetic Engineering

Activity Manual
Preview—page 197

The purpose of this Preview page is to generate student interest and determine prior student knowledge. The concepts are not intended to be taught in this lesson.

Lesson 132

Student Text pages 302–4
Activity Manual pages 198–200

Objectives

- Explain how chromosomes, DNA, and genes are related.
- Identify some learned and inherited traits
- Take a survey of a sampling group
- Graph recorded survey results

Vocabulary

trait
gene
DNA
heredity
inherited

Introduction

Note: Included with this two-page text lesson is an activity. The activity, *It's All in the Genes,* should be completed as part of the lesson.

➤ In Chapter 4 we learned about the parts of a cell. Which part of the cell has the coded instructions that tell the cell what to do? *{DNA or chromosomes}*

➤ Not only does DNA tell a cell what to do, it also determines characteristics that you as an individual will have, such as what you will look like and what abilities you have.

Purpose for reading:
Student Text pages 302–3

➤ What is *heredity*?
➤ Are all of your characteristics inherited?
➤ What are some examples of traits that are determined by your genes?
➤ How are learned traits different from inherited traits?

Heredity

Is your hair curly or straight? Do you have freckles? Can you roll your tongue? What color are your eyes? These are just some of the **traits,** or characteristics, that you have inherited from your parents. These traits are controlled by **genes,** small pieces of DNA found in the *chromosomes* in your cells.

You are a unique individual. God designed you and knew all about you even before you were born (Ps. 139:14, 16). Because of the process of meiosis, you inherited twenty-three chromosomes from your father and twenty-three from your mother. Unless you have an identical twin, no one else has the same combination of chromosomes and genes that you have. Your chromosomes contain **DNA** (deoxyribonucleic [dee AHK see RYE boh noo KLAY ihk] acid), the chemical code that tells your cells what to do. Your genes are small sections of DNA that determine your different traits.

Genes control many visible traits, such as the length of your eyelashes, the color of your hair, and whether you are a boy or a girl. However, genes also control traits that we cannot see. Your genes determine the shape of your red blood cells, your blood type, and whether or not you are colorblind.

Heredity is the passing of traits from parents to offspring. You have **inherited,** or received, traits from your parents. But not all of your traits are permanent and unchangeable. Your genes determine some of your traits, such as eye color and freckles. You have no control over these traits. However, other traits are determined by your genes but influenced by your environment and health habits. A person may have inherited the genes for tallness, but if he is malnourished, he

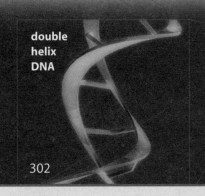

double helix DNA

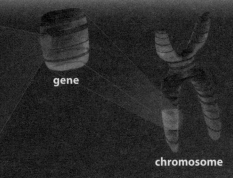

gene

chromosome

302

Be sensitive to students who are adopted or who do not live with their birth parents. Assignments that include finding out about inherited family traits may be difficult for these students to do.

The following definitions from Chapters 4 and 12 may need to be reviewed.

Chromosomes—tight bundles of DNA found in the nucleus of cells

Meiosis—the two-stage process of cell division in which the chromosomes duplicate once, but the cell divides twice; Each new cell has half of the chromosomes of the parent cell.

Mitosis—the process of cell division in which a cell makes an exact copy of itself; The chromosomes duplicate once, and the cell divides once.

The terms *heredity* and *inherit* come from the same root as *heir* and *heritage*.

320   Lesson 132

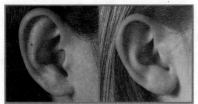
attached ear lobe   unattached ear lobe

right thumb dominance   left thumb dominance

bent thumb   straight thumb

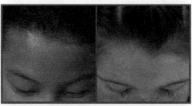
straight hairline   widow's peak

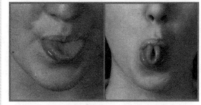
cannot roll tongue   can roll tongue

will not grow to his full height. Your musical and academic abilities, height, weight, and blood pressure are just some of the traits that your environment and habits can influence.

Other traits are learned rather than inherited. For example, you did not inherit the ability to speak in a certain language. You learned the language of the people around you. Scientists are not sure how much of a trait is related to genetic inheritance and how much is influenced by environment, habits, and things that we learn.

Most traits are controlled by many genes, so it is difficult to determine which genes you may have. For example, eye color is controlled by multiple genes, so blue eyes come in many shades, not just one.

Other traits, however, you either have or you do not. Fold your hands together. Which thumb is on top? Unless you make a conscious effort to change, the same thumb will almost always be on top. Above are some other traits that are easy to identify.

Quick Check
1. How are genes, DNA, and chromosomes related?
2. What is heredity?

Direct a Demonstration

Demonstrate surveying sampling groups

Materials: tally chart with three choices—hot breakfast, cold breakfast, and no breakfast

Choose five students as the sampling group. Use tally marks to record the usual breakfast choice of each student on the chart.

- Who was the sampling group? *{the five students}*
- Is this a large sampling group or a small sampling group? *{a small sampling group}*
- How many in the sampling group usually have a hot breakfast? *{Answers will vary.}*
- How many in the sampling group usually have a cold breakfast? *{Answers will vary.}*
- How many in the sampling group usually do not eat breakfast? *{Answers will vary.}*

Repeat the activity using five different students and add their responses to the previous results.

- How big is the sampling group now? *{ten students}*
- Which sampling group probably gives a better representation of breakfast-eating habits—the five students or the ten students? *{the ten students}* Why? *{The greater the sampling group is, the more accurate the representation of the whole is.}*

Discussion:
pages 302–3

- Which verses tell us that God knew all about us before we were even born? *{Possible answer: Psalm 139:14–16}*

Read aloud Psalm 139:14–16.

- Why should these verses give us great comfort? *{They tell us that God made us and knows all about us.}*
- What are *traits*? *{characteristics}*
- What controls inherited traits? *{genes}*
- What are *genes*? *{small pieces of DNA found in the chromosomes of cells}*
- What is *DNA*? *{the chemical code that tells the cells what to do}*
- DNA controls what the cells do. What should control what a Christian does? *{Possible answers: the Holy Spirit and the Bible}* [BATs: 6a Bible study; 6c Spirit-filled]
- What are some traits determined by genes? *{Possible answers: length of eyelashes, hair color, shape of red blood cells, blood type, colorblindness}*
- Which word describes the passing of traits from parents to offspring? *{heredity}*
- How are your traits determined? *{Possible answers: heredity or genes, environment, and health habits}*
- Language is an example of a learned trait. How is a learned trait different from an inherited trait? *{People do not inherit the ability to speak in a certain language. Language is learned from other people.}*
- Other than heredity, what can affect your musical or academic abilities? *{Possible answers: the instruction you receive; your study and practice habits; your overall health}*

Answers
1. Genes are small sections of DNA. DNA is the chemical code that tells the cells what to do. Chromosomes contain DNA.
2. Heredity is the passing of traits from parents to offspring.

Activity Manual
Bible Integration—page 198
This page traces the earthly genealogy of Jesus.

Purpose for reading:
Student Text page 304
Activity Manual pages 199–200

The student should read all the pages before beginning the activity.

Note: This activity is included as part of Lesson 132. You may choose to do the Direct a Demonstration on TE page 321 before the student begins the activity.

Materials

- See Student Text page

➤ Today you will take a survey using a sampling group to determine the occurrence of certain traits. A sampling group is the group that is surveyed.

Procedure

Refer the student to Student Text page 303 for pictures of the traits. Remind each student to record tally marks while completing the survey.

After the survey is completed, guide the student in labeling and completing the bar graphs.

Conclusions

➤ Did one of the traits in each pair occur more often than the others? If so, which one?

Activity Manual
Activity—pages 199–200

Assessment
Rubric—TE pages A50–A53

Select one of the prepared rubrics, or design a rubric to include your chosen criteria.

 It's All in the Genes

Process Skills
- Collecting data
- Interpreting data
- Communicating

Kayla was a frustrated girl. She could not roll her tongue. All of her friends could do it, and Kayla really did try. Even though she practiced and practiced, she couldn't get her tongue to roll. It just flopped in her mouth and refused to stand up in a nice, neat roll. Surely something was wrong with her!

Perhaps you have felt like Kayla. Although you try, you cannot get your tongue to roll. But now you know that rolling your tongue is not a matter of practice. It's a matter of genes.

For this activity, you will take a survey to find out how many people have specific genetic traits. The people that you survey will represent a *sampling group*. When taking a survey, the larger your sampling group is, the more accurate your data, or gathered information, is likely to be.

Procedure

Materials: Activity Manual

1. Look at the traits that your teacher gives you. Survey a minimum of fifteen people to determine which of the traits they have. If you are unsure of what to look for, you may refer to page 303 of your text.
2. Record your findings in your Activity Manual.
3. Prepare a bar graph to show your data.

Conclusions

- Compare your findings with those of others. In each pair of traits, did one trait show up more frequently than the other?

Follow-up

- Figure out the ratio between the two corresponding traits shown on your bar graph.
- Survey another fifteen people. See if the ratio changes as your sampling group increases.

Ears (Example)

| Attached earlobes | |||| | |
| Unattached earlobes | |||| |||| |

304

 Teacher Helps

In this lesson, help the student realize that one trait tends to occur more often than the corresponding trait. Lesson 135 discusses dominant and recessive traits.

Because of the relatively small sampling group, you may actually find that your sampling will give an inaccurate representation of the dominant trait. If that occurs, emphasize to the students that small samplings sometimes give faulty information.

Science Process Skills

Recording data—Recording data allows others to determine what information you found, as well as the size of the sampling group used. The sampling group size may have a large impact on the validity of the collected information.

Science Background

The following traits are examples of some common dominant traits.

bent pinky
curly hair
dimples (many variations possible)
freckles (many variations possible)
hair on fingers
left thumb dominance when hands are folded
right-handedness
straight thumb (cannot bend backwards)
unattached earlobes
widow's peak

Genetics

DNA: The Double Helix

How can your genes determine your hairline and the shape of your ears? Each of your genes contains a section of the DNA found in your chromosomes. This DNA contains all of the instructions for your cells. Each time a cell divides through mitosis, the DNA duplicates itself so that each new cell will have a copy of your DNA pattern.

Structure of DNA

For many years scientists studied DNA, trying to determine its shape and structure. Then in 1953, after seeing an x-ray photograph of DNA, James Watson and Francis Crick announced that they had discovered "the secret of life." Their model of DNA was shaped like a twisted spiral ladder. Because of its shape, they called it a *double helix*.

Sugar and phosphate molecules form the sides of this ladder. The rungs are formed with the four basic molecules of DNA, called *bases*. These four bases are similar to a four-letter alphabet for DNA. Even before Watson and Crick discovered the shape of the DNA molecule, scientists learned that only certain bases would fit together. Base A fits only with base T, and base G fits only with base C. The order in which the bases are arranged creates the code, or pattern, for each gene.

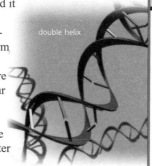

double helix

Meet the SCIENTIST: ROSALIND FRANKLIN

Rosalind Franklin

Rosalind Franklin (1920–1958), a British scientist working at King's College in London, was the first person to photograph a DNA molecule. Her colleague Maurice Wilkins showed the photograph and some of her work to James Watson and Francis Crick. Her research enabled them to conclusively identify the twisted spiral structure of DNA. She supported their model of DNA with other research that she had done. Franklin continued to study DNA, but she also researched plant viruses and the structure of the live poliovirus. She died from cancer in 1958, four years before Wilkins, Crick, and Watson received the Nobel Prize for their DNA discoveries.

James Watson

Francis Crick

305

SCIENCE BACKGROUND

Names of bases—The base molecules for DNA are adenine (A), thymine (T), guanine (G), and cytosine (C).

DNA replication—When DNA replicates itself, it "unzips" the ladder. New bases floating in the nucleus pair up with the bases on each side of the ladder, creating two copies of the same DNA.

HISTORY

The Nobel Prize is awarded yearly to men and women who make outstanding advances in chemistry, economics, literature, medicine, peace, and physics. It is named for the Swedish scientist Alfred Bernard Nobel (1833–96), who is known for inventing dynamite. In his will, Nobel stated that much of his fortune should be used to award the Nobel Prizes. The first Nobel Prizes were awarded in 1901.

SCIENTIST

Rosalind Franklin and Maurice Wilkins both studied DNA at King's College in London, but they did not get along with each other. Some people believe that Maurice Wilkins showed Franklin's DNA photograph and research to James Watson and Francis Crick without her permission. At that time, women scientists did not get the same respect and benefits as their male colleagues. Many people wonder whether the Nobel Prize would have been shared with Rosalind Franklin as well if she had been alive.

Chapter 13

Student Text Pages 305–6
Activity Manual pages 201–4

Lesson 133

Objectives
- Identify the structure of a DNA molecule
- Recognize James Watson and Francis Crick as those who identified DNA structure
- Make a model of a DNA molecule
- Identify ways DNA testing is used

Materials
- transparency of TE page A35, *Coded Instructions*

Introduction

Display the transparency *Coded Instructions*.

➤ What is the name of this code? {the Morse code}
➤ How is it useful? {Possible answer: to send telegraph messages}

Direct the student to decode the message.

➤ What is the message from the code? {DNA is the chemical code that tells cells what to do.}

Purpose for reading:
Student Text pages 305–6

➤ What does a molecule of DNA look like?
➤ What are ways that DNA testing is used?

Discussion:
page 305

➤ How do new cells have the same pattern of DNA as old cells? {Each time a cell divides, the DNA replicates itself so that each new cell has a copy of the organism's complete DNA.}
➤ Which two scientists received credit for being the first to discover the structure of DNA? {James Watson and Francis Crick}

Refer the student to *Meet the Scientist*.

➤ Whose research and photograph helped Watson and Crick with their discovery? {Rosalind Franklin's}
➤ The word *helix* means "a coil, or something that is twisted." Why do you think a DNA molecule is described as a *double helix*? {Answers may vary but should include that DNA looks like a twisted ladder or two coils twisted together.}
➤ What kinds of molecules make up the structure of DNA? {The sides of the "ladder" are made up of sugar and phosphate molecules. The "rungs" are the four molecules called bases—A, T, G, and C.}
➤ Which base fits with base T? {base A}
➤ Which base fits with base C? {base G}

Lesson 133   323

Discussion:
page 306

➤ **Why can DNA patterns be used to identify organisms?** *{Every organism has a different DNA pattern.}*

💡 **Would the DNA in your blood cells be the same as the DNA in the saliva in your mouth? Why?** *(Yes; your DNA is the same in all the cells of your body.)*

💡 **Why would the U.S. government say that it might never have another unknown soldier?** *{Technology has improved so much that DNA can be used to identify people who have died.}*

➤ **What is another use of DNA testing?** *{to help solve crimes}*

📊 **Compare the DNA samples on page 306. Which suspect committed the crime? How do you know?** *{suspect B; His DNA sample matches the sample produced as evidence.}*

➤ **How are fingerprints and DNA similar?** *{Possible answer: Everyone, with the exception of identical twins, has a different DNA pattern, just as everyone, including identical twins, has different fingerprints.}*

Answers

1. the four basic molecules of DNA
2. Possible answers: to identify servicemen killed in action; to aid police in identifying a criminal

Activity Manual

Reinforcement—pages 201–2

Reinforce the student's knowledge about the structure of DNA by making a model from chenille wires. Compare models and discuss the variety of color sequences. Relate the variety seen in the students' models to the innumerable combinations of real DNA. This activity uses inch rather than centimeter measurements.

Study Guide—pages 203–4

These pages review Lessons 132 and 133.

Assessment

Test Packet—Quiz 13-A

The quiz may be given any time after completion of this lesson.

324   Lesson 133

Patterns of DNA

Even with only four "letters," many different patterns of DNA are possible. Every organism has a different DNA pattern, even within the same species. However, within the organism, every cell has the same DNA pattern, no matter what the job of the cell is. The DNA pattern of your blood is the same as the DNA pattern of your skin. Just like fingerprints, no one, except identical twins, has the same DNA pattern as another person.

Have you ever heard of DNA testing? Scientists and investigators take samples of DNA and compare them. They can use these DNA samples to help identify soldiers who were killed in action. In every war there have been servicemen killed who could not be identified. The Tomb of the Unknown Soldier is a memorial in the United States that honors American servicemen whose remains could not be identified. But with the increased use of DNA testing, government officials have remarked that America may never have another unknown soldier.

DNA testing is also used to help solve many crimes. Crime scene investigators can use samples of hair, skin, and blood cells from the scene of a crime to help identify the criminal. Machines analyze the DNA and show the DNA as a pattern of bands somewhat similar to a barcode on a product. Detectives can then compare the DNA pattern with DNA from suspects in the case.

1. What are DNA bases?
2. What is one way that DNA testing is used?

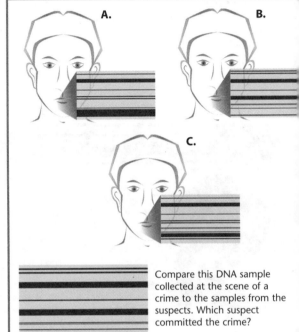

Compare this DNA sample collected at the scene of a crime to the samples from the suspects. Which suspect committed the crime?

306

Teacher Helps: When doing Activity Manual pages 201–2, you may choose to assign the students to work in pairs.

Encourage each student to design a DNA molecule that is different from the other DNA molecules.

SCIENCE BACKGROUND

DNA of the "Unknown Soldier"—In 1998, DNA testing was done on the Vietnam War soldier buried at the Tomb of the Unknown Soldier in Washington, D.C. He was later identified as 1st Lt. Michael J. Blassie. This DNA testing was actually mitochondrial DNA testing.

Genetic DNA is found in the chromosomes, which are in the nucleus of a cell. This DNA is different for each individual, since it is a mixture received from the father and the mother. Mitochondrial DNA is found in the mitochondria, which are located in the cytoplasm of a cell. Because mitochondrial DNA is inherited from the mother only, siblings with the same mother will have similar mitochondrial DNA. This DNA is passed on to other generations through the female side of the family only.

Explorations: DNA Extraction

Student Text page 307
Activity Manual pages 205–6

Lesson 134

DNA molecules are the building blocks of all living organisms. DNA is the mysterious substance that makes a plant a rose instead of a daisy or an animal a cat instead of a dog. It is also responsible for giving you the skin color, hair color, and eye color that you have.

Most DNA molecules are part of a cell's nucleus. Though the molecules are very small, you can perform an activity that will allow you to see the strands of DNA molecules.

What to do

1. Measure 15 mL of the wheat germ into a clear container.
2. Pour 45 mL of water into the same container.
3. Stir thoroughly. Add 8 mL of detergent and stir occasionally for 5 minutes.
4. Add 4 mL of meat tenderizer and stir occasionally for 5 minutes.
5. Tip the container slightly and gently pour 45 mL of alcohol along the side of the container. The alcohol on top should form a separate layer.
6. Carefully set the container upright. Allow the solution to sit for at least 10 minutes. Observe the white, stringy substance that moves into the alcohol layer. Use a toothpick to lift one of the strings up. This is a DNA molecule.
7. Refer to your Activity Manual for additional information about extracting DNA molecules.

Materials:
- raw wheat germ
- water
- liquid detergent
- meat tenderizer
- rubbing alcohol
- toothpicks, wooden skewers, or craft sticks
- metric measuring spoons
- clear plastic containers
- Activity Manual

Objectives
- Extract DNA from organic matter

Materials
- See Student Text page

Introduction

➤ Are DNA molecules large or small? *{small}*

➤ How do you know? *{Possible answer: DNA molecules are parts of cells, which are very small.}*

➤ Since DNA molecules stick together easily, we can do an activity that will allow us to actually see strands of DNA molecules.

Purpose for reading:
Student Text page 307

The student should read the page before beginning the exploration.

Procedure

Follow the directions on Student Text page 307 to extract strands of DNA. After completing the extraction, discuss the questions on Activity Manual pages 205–6.

Activity Manual

Explorations—pages 205–6

Read and discuss these pages after completing the activity.

Assessment

Rubric—TE pages A50–A53

Select one of the prepared rubrics, or design a rubric to include your chosen criteria.

 Raw wheat germ is available in health food stores and in other stores that carry organic flours in bulk.

 God made each person unique. He made us to bring Him honor and glory. [Bible Promise: I. God as Master]

You may choose to use other organic substances as described in FAQ #1 on Activity Manual page 205.

Some liquid dish and laundry detergents work better than others. Test your brand to make sure that it gives the desired result.

Chapter 13

Lesson 135

Student Text pages 308–11
Activity Manual page 207

Objectives

- Describe some of Mendel's experimental procedures
- Explain some of Mendel's conclusions
- Recognize the difference between dominant genes and recessive genes

Vocabulary

purebred plant
P generation
hybrid
dominant trait
recessive trait
phenotype
genotype
codominant
incomplete dominance

Introduction

➤ Suppose you are a scientist who has been researching and studying a topic for several years. You have spent much time and effort on your experiments and have carefully recorded your results. You are almost ready to publish your results when you discover that an unknown scientist did similar experiments years earlier and made the same discovery that you made. Would you acknowledge the other scientist's work or ignore it?

📖 The Bible tells us not to lie. [BAT: 4c Honesty]

➤ Today's lesson includes a very similar situation.

Purpose for reading:
Student Text pages 308–11

➤ Who is known as the Father of Genetics?
➤ How are dominant and recessive genes different?

Discussion:
pages 308–9

➤ Why did Gregor Mendel begin studying peas? *{He wanted to find out how traits were passed from generation to generation.}*

➤ Why are peas a good type of plant to study? *{Possible answers: Peas grow quickly, making it possible for Mendel to study several hundred generations. Peas also have traits that can be traced easily.}*

➤ What is the difference between purebred plants and hybrid plants? *{Purebred plants will continue to show the same trait(s) for many generations when pollinated naturally. A hybrid plant is a mixture of two purebred plants with different forms of the same trait.}*

Father of Genetics

Genetics (juh NET ihks) is the study of how traits are inherited. The idea that genes determine many physical traits began with Gregor Johann Mendel, an Austrian monk and scientist. The son of a farmer, Mendel became a monk in order to continue his education. He was in charge of the monastery gardens and was also a substitute teacher at a school nearby.

Mendel studied peas for eight years, wanting to discover how traits were passed on from generation to generation. Pea plants grow quickly, so he was able to study several hundred generations of plants. The plants also have traits that are easy to trace because they appear only in one of two forms. For example, a pea plant has either yellow seeds or green seeds. It is either a tall plant or a short plant and has either white flowers or purple flowers.

Mendel began his experiments with purebred plants. **Purebred plants** are plants that show the same trait for many generations when pollinated naturally. Pea plants usually self-pollinate, so Mendel cross-pollinated tall pea plants with short pea plants. He took pollen from the stamens of one plant and added it to the pistil of another plant. He then removed the stamens from the second plant. He referred to this pair of plants as the parent generation, or **P generation**. Much to his surprise, all of the new pea plants were tall. The trait for shortness seemed to have disappeared. Mendel hypothesized that the shortness trait was still there but was hidden. These new plants

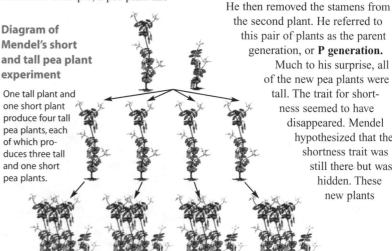

Diagram of Mendel's short and tall pea plant experiment

One tall plant and one short plant produce four tall pea plants, each of which produces three tall and one short pea plants.

308

SCIENCE MISCONCEPTIONS

The fact that a trait is dominant does not mean that the majority of people possess that trait. A dominant trait simply means that this trait will be expressed if it is present in a person's genes. The majority of people, especially when surveying only a small sample, such as in Lesson 132, may actually have the recessive trait instead.

 SCIENTIST

Gregor Mendel (1822–1884) loved studying and learning, but he had great test anxiety. He could not become a full-time teacher because he failed his teacher's examinations twice.

Mendel's younger sister gave him her dowry so that he could continue studying and preparing for university work. Her money was not enough to cover expenses, so Mendel eventually joined the Augustinian monastery that emphasized teaching and research. In gratitude for his sister's unselfish gift to him, Mendel later helped to raise and support her three sons.

were **hybrids,** plants produced by crossing purebred parent plants that each have a different form of the same trait.

To test his hypothesis, Mendel allowed the tall hybrid plants to self-pollinate. The next generation included both tall and short plants. For every three tall plants, there was one short plant—a ratio of three to one.

Mendel continued testing his plants while keeping detailed records of the results. In another experiment, he crossed plants with round seeds with plants that had wrinkled seeds. Only the round seeds appeared in the hybrid generation. The trait for wrinkled seeds was hidden. He also experimented with the flower color, seed color, and pod shape.

Mendel concluded that offspring inherit traits in pairs of factors, receiving one factor from each parent. He also realized that some traits were hidden in some generations but reappeared in following generations. This led to the idea of dominant and recessive traits. A **dominant trait** is the characteristic that is shown in the hybrid generation. The **recessive trait** is hidden in the hybrid generation and appears in later generations only when no dominant factor is inherited.

Mendel's discoveries about heredity contradicted the theories of his time. He presented his work to other scientists, and in 1866 he published a report about his discoveries. However, most people ignored his work. In 1900, three other scientists who had done similar experiments read Mendel's report. They gave Mendel the credit for discovering that traits are passed from generation to generation. Because of this, Mendel is now known as the Father of Genetics.

	Seed shape	Seed color	Pod color	Pod shape	Plant height	Flower color
Dominant	round	yellow	green	full	tall	purple
Recessive	wrinkled	green	yellow	flat	short	white

309

The term *genetics* comes from the Greek word *genesis*, meaning "origin" or "beginning."

The term *dominant* comes from the Latin word *dominus*, meaning "lord" or "rule."

The term *recessive* comes from the Latin word *recedere*, meaning "to retreat" or "go back."

Discussion:
pages 308–9

▶ **Why did Mendel cross-pollinate his pea plants?** {Pea plants are usually self-pollinated. By cross-pollinating, Mendel could combine plants with different purebred traits.}

💡 **Why do you think Mendel removed the stamens from the second plant after he had added pollen to its pistil?** {to prevent the plant from self-pollinating}

▶ **What did Mendel call the generation of plants that he cross-pollinated?** {the parent generation, or P generation}

📊 Refer the student to the *Diagram of Mendel's short and tall pea plant experiment* on Student Text page 308.

▶ **What trait did the hybrids show after Mendel crossed tall purebred plants with short purebred plants?** {The hybrids were all tall.}

▶ **What did Mendel hypothesize about the trait for shortness?** {He thought that it was still there but just hidden.}

▶ **How did Mendel find out that the hybrids had a trait for shortness?** {He allowed them to self-pollinate.}

▶ **Was his hypothesis correct?** {yes}

▶ **What was the ratio of tall plants to short plants? Explain.** {3:1; Mendel noticed that he had three tall plants for each short plant.}

💡 **What did Mendel do that lets us review his research?** {He kept detailed records.}

▶ **What were Mendel's conclusions?** {Possible answers: Traits are inherited in pairs of factors, with each parent giving the offspring one factor. Some traits are hidden in one generation but reappear in following generations.}

💡 **Why do you think most people ignored Mendel's work when he presented his results?** {Possible answers: His findings contradicted the current theories of his time. He worked in a monastery and was not working with other scientists.}

▶ **Why is Mendel now known as the Father of Genetics?** {The scientists who found Mendel's work gave him the credit for discovering that traits are passed from generation to generation.}

📊 Use the chart on Student Text page 309 to discuss Mendel's findings.

▶ **Which pod color is dominant?** {green}

▶ **Which flower color would probably not appear in the hybrid generation?** {white}

Discussion:
pages 310–11

➤ Why do we usually have two genes for each trait? *{because we received one set of genes from each parent}*

➤ What is the difference between dominant and recessive genes? *{Dominant genes will be expressed if they are present. A recessive gene will be hidden if a dominant gene is present. A recessive gene will be expressed only if the person has two recessive genes for that trait.}*

➤ What term is used to describe a trait that is seen? *{expressed}*

➤ If a person has one gene for a bent thumb and one gene for a straight thumb, why will the person have a straight thumb? *{because the gene for a straight thumb is dominant}*

➤ What genes must a person have if he has a bent thumb? *{He must have two recessive genes for a bent thumb.}*

💡 A person who has hair on his fingers has at least one dominant gene for that trait. If a person did not have any hair on his fingers, what would you know about that person's genes for this trait? *{He inherited two recessive genes for this trait.}* Why? *{because the recessive trait can be shown only if no dominant genes are present}*

➤ What is the appearance of an organism called? *{phenotype}*

➤ What is the genetic arrangement of an organism called? *{genotype}*

💡 *Genotype* is a compound word that combines a form of which word with *type*? *{gene}*

➤ Can you tell the genotype of an organism by observing its phenotype? Why? *{No; recessive traits in a genotype may not appear in the phenotype.}*

🎬 Look at the pea plants pictured on page 310. What is the same for the first two plants—the phenotype or the genotype? *{phenotype}*

➤ What genotype might a tall pea plant have? *{It could have two dominant genes for tallness, or it could have one dominant gene and one recessive gene.}*

Purebred tall plant: two genes for tallness

Hybrid tall plant: one gene for shortness and one gene for tallness

Purebred short plant: two genes for shortness

Dominant and Recessive Genes

Today we know that what Mendel called factors are actually genes. Because you received one set of genes from each parent, you have two genes for each trait. Genes for a certain trait, such as the shape of your thumb, are in the same place on each chromosome. A straight thumb is a dominant trait. If a person has one gene for a straight thumb and one gene for a bent thumb, the person will have a straight thumb. The gene for the dominant trait, called the dominant gene, will always be *expressed*, or shown, if it is present in a person's chromosomes.

The gene for a bent thumb is a recessive gene. It is hidden, or masked, when a dominant gene is present. A person can have a bent thumb only if he has inherited two recessive genes for thumb shape.

In Mendel's experiments, all of the pea plants in the first generation were tall. Since tallness is dominant for pea plants, each plant was tall even if it had one gene for tallness and one gene for shortness. The plant would be short only if it received two genes for shortness.

You cannot tell by its appearance whether a plant has two genes for tallness or if it has one gene for shortness and one for tallness. The physical appearance of an organism is called its **phenotype** (FEE nuh TIPE). The genetic combination, or arrangement of genes within the organism, is its **genotype** (JEN uh TIPE).

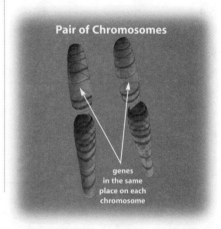

Pair of Chromosomes

genes in the same place on each chromosome

310

SCIENCE BACKGROUND

Forms of genes—Some genes have different forms, or *alleles*. A person will have only two genes for a trait, even though many different forms of the gene exist in the population. An example of this is a person's blood type.

LANGUAGE The prefix *pheno-* means "to show or display." Another word with this prefix is *phenomenon*. A phenomenon is a fact or occurrence, often unusual, that is perceived through the senses.

328 Lesson 135

expressed. For example, some cattle have both red and white hairs. The gene for red hair and the gene for white hair are codominant. Both genes are expressed instead of one being recessive, or hidden.

At other times some genes blend together. This is called **incomplete dominance.** Red snapdragon flowers crossed with white snapdragons produce pink snapdragons. When crossed, the hybrid pink flowers produce some red, some pink, and some white flowers.

Mendel tested the color of pea plant flowers by crossing a purebred purple-flowered plant with a purebred white-flowered plant. The dominant trait, purple, masked the recessive trait of white flowers in the hybrid generation. However, the purple hybrid flowers produced some white flowers.

The phenotypes of the purebred purple flowers and the hybrids were the same. Both sets of plants showed purple flowers. However, their genotypes were different. The purebred flowers had two dominant genes for purple color. The purple hybrids, though, had one dominant gene for purple and one recessive gene for white.

Sometimes genes are not just dominant and recessive. Some genes are **codominant,** with both genes being

Snapdragons

first generation — red, white
second generation — pink

QUICK CHECK
1. What did Mendel conclude from his research on pea plants?
2. What is the difference between a phenotype and a genotype?

311

Discussion:
pages 310–11

➤ When Mendel crossed purebred purple-flowered pea plants with purebred white-flowered pea plants, why did all the new plants end up with purple flowers? *{Purple is the dominant trait for pea plant flower color.}*

➤ How were the genotypes of the purple purebred flowers and the purple hybrids different? *{The purple purebred flowers had two dominant genes for purple color. The purple hybrid flowers had one dominant gene for purple color and one recessive gene for white color.}*

➤ Are genes always strictly dominant and recessive? *{no}*

➤ Which term is used when both genes are expressed instead of one being dominant and the other recessive? *{codominant}*

➤ What is an example of codominance? *{Possible answer: cattle that have both red and white hairs}*

➤ What is *incomplete dominance?* *{when genes seem to blend together in one generation}*

➤ Which type of plant usually shows incomplete dominance? *{snapdragons}*

➤ Which color is produced when red snapdragons and white snapdragons are crossed? *{pink}*

➤ Which color is produced when pink hybrid flowers are crossed? *{white, red, and pink}*

Answers

1. Mendel concluded that traits are inherited in pairs of factors, with each parent giving the offspring one factor. He also found that some traits are hidden in one generation but reappear in following generations.

2. The phenotype is simply the organism's physical appearance, but the genotype is the arrangement of genes in the organism.

Activity Manual
Reinforcement—page 207

Lesson 136

Student Text pages 312–15
Activity Manual pages 208–10

Objectives

- Predict genetic probability using a Punnett square
- Interpret a pedigree chart
- Identify some sex-linked traits

Vocabulary

pedigree   sex-linked trait

Materials

- family tree

Introduction

Display a family tree.

▶ What is the purpose of a family tree? {to trace one's ancestors}

▶ In today's lesson, you will be learning about something similar to a family tree.

Purpose for reading:
Student Text pages 312–15

▶ Why are Punnett squares useful?

▶ What is a *pedigree*?

Discussion:
pages 312–13

▶ What are *Punnett squares*? {charts that show possible genetic outcomes}

▶ Why did Reginald Punnett use these squares? {to make Mendel's charts easier to understand}

▶ A parent has two genes for each trait. How many genes for each trait will a parent pass on to a child? {one}

▶ What do the letters on the outside of a Punnett square represent? {each parent's genotype for that trait}

▶ What does a capital letter represent? {the dominant gene}

▶ What does a lowercase letter represent? {the recessive gene}

Punnett squares

When Mendel's research was rediscovered, scientists studied his results carefully. Mendel had written some of his ideas about possible genetic combinations in chart form. Reginald Punnett (PUHN net), an English geneticist, was especially interested in Mendel's charts. Punnett used squares to make Mendel's charts easier to understand. Punnett squares show the genetic possibilities of a particular trait that can result for the offspring of a specific set of parents.

To use a Punnett square, geneticists write one parent's genotype at the top of the Punnett square. The other parent's genotype is written at the left edge of the Punnett square. Geneticists use letters to represent the genes. An uppercase letter represents the dominant gene, and a lowercase letter represents the recessive gene. Although each parent has two genes for a trait, a parent can give only one gene for that trait to his or her offspring.

Geneticists take one gene from the top and one gene from the side to fill in the small boxes in the Punnett square. The four boxes show the possible genotypes for the offspring of the parents. For example, when a purebred tall pea plant (TT) is crossed with a purebred short pea plant (tt), all of the offspring will have the phenotype, or appearance, of tallness. The boxes in the Punnett square show that each offspring has a dominant gene.

However, if two hybrids are crossed, the results are very different. The Punnett square shows that there is the possibility that one plant will have two recessive genes. That plant will be the only short pea plant, although three of the offspring carry the recessive gene. Notice in the Punnett square that the dominant gene is listed first in the boxes no matter which parent it comes from.

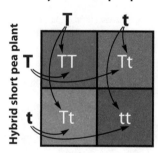

312

 Any letter can be used to represent genes for a Punnett square. However, it is usually better to use a letter that can be easily distinguished between its upper case and lower case.

Each of these Punnett squares shows four possible genetic combinations for one offspring. The squares do not show that the parents will have four offspring. For example, if a cat has four kittens, she may not have a kitten with each of the genetic combinations represented. Some kittens may have the same combination. Some of the combinations may not appear in any of the kittens.

 Punnett squares were named for the English geneticist, Reginald Punnett (1875–1967). In 1905 he wrote the first genetics textbook. In 1910 he became the first professor of genetics at Cambridge University. He researched worms, sweet peas, maize, and poultry.

330 Lesson 136

Punnett squares also show the probability of a certain outcome. For example, what would be the probability of parents whose hairlines form widow's peaks having a child with a straight hairline? Notice that one of the parents has a recessive gene for a straight hairline. Use an uppercase *W* to represent the dominant gene of a widow's peak. The lowercase *w* represents the recessive gene of a straight hairline.

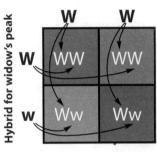

Purebred for widow's peak

There is no possibility that a child of this couple would have a straight hairline. All the children would carry the dominant gene for a widow's peak and, therefore, would show that trait. However, unlike the offspring of the two hybrid pea plants, the probability is that only two of these offspring, or 50 percent, would carry the recessive gene to the next generation.

Several different genes control the color and length of a cat's fur, but the gene for short hair is dominant. What would be the probability of producing a longhaired kitten if one parent has short hair with a recessive gene for long hair and the other parent is longhaired? *H* represents the dominant gene for short hair, and *h* represents the recessive gene for long hair. The kitten would have a 50 percent probability of having long hair.

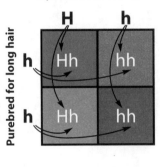

Hybrid for short hair

Direct a Demonstration

Practice Punnett squares

Having hair on the fingers is a dominant trait. Using H for fingers with hair and h for fingers without hair, complete several Punnett squares with the class. Possible parent combinations include Hh and hh, HH and hh, HH and Hh, or Hh and Hh.

	H	h
h	Hh	hh
h	Hh	hh

	H	H
h	Hh	Hh
h	Hh	Hh

	H	H
H	HH	HH
h	Hh	Hh

	H	h
H	HH	Hh
h	Hh	hh

Discussion:
pages 312–13

Refer the student to the Punnett squares pictured on Student Text pages 312–13. The discussion refers to these squares as Charts 1 and 2 on page 312 and Charts 3 and 4 on page 313.

➤ **Look at Chart 1. How many offspring shown in the Punnett square carry a dominant gene?** *{all of them}*

💡 **Would these plants be tall or short?** *{tall}* **Why?** *{If both dominant and recessive genes are present, the dominant gene will be expressed in the phenotype.}*

➤ **Look at Chart 2. Which letters represent the parents' genes in this Punnett square?** *{Tt and Tt}*

💡 **What do these letters tell us about each parent's genotype?** *{Possible answers: Both parents are hybrid pea plants. They each have one dominant gene and one recessive gene for height.}*

➤ **What is the probability that the offspring will show the dominant trait?** *{3 out of 4, or 75%}*

➤ **What is the probability that one of these new plants will be short?** *{1 out of 4, or 25%}*

➤ **Look at Chart 3. Which letters represent the parents' genes in this Punnett square?** *{WW and Ww}*

➤ **What is the probability that the offspring will show the dominant trait of having a widow's peak hairline?** *{4 out of 4, or 100%}*

➤ **How are the probable offspring in this Punnett square different from those in Chart 2?** *{Possible answer: Three offspring in Chart 2 carry the recessive gene. Only two offspring in Chart 3 carry the recessive gene.}*

💡 **Why is there a difference?** *{In Chart 2, both parents have both the dominant gene and the recessive gene. In Chart 3, one parent has two dominant genes and the other has one dominant gene and one recessive gene.}*

➤ **Look at Chart 4. Which letters represent the parents' genes in this Punnett square?** *{Hh and hh}*

➤ **What is the probability that a kitten will have the recessive trait of long hair?** *{2 out of 4, or 50%}*

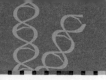

Discussion:
pages 314–15

➤ What is a *pedigree*? {*a chart that traces a trait through a family tree*}

➤ What is the difference between the circles and squares on a pedigree? {*Circles represent females, and squares represent males.*}

▨ Look at the *Pedigree for Tongue Rolling* on page 314. Why are some symbols shaded in this pedigree? {*Shaded symbols show the people who possess the recessive trait. On this pedigree they show those who cannot roll their tongues.*}

➤ What are the names of Leon and Beverly's children {*Anna, David, Paul, and Amy*}

➤ How does this pedigree show the relationship of the parents in one generation to their children in the next generation? {*with vertical lines*}

💡 What would be two possible genotypes for the father, Leon, in this family tree? {*He could have two dominant genes (RR) for tongue rolling or one dominant gene and one recessive gene (Rr).*}

➤ Since the mother, Beverly, cannot roll her tongue, what must be true about her genotype? {*She must have two recessive genes (rr).*}

➤ How many of Leon and Beverly's children can roll their tongues? {*3*}

💡 Since their son Paul cannot roll his tongue, is Leon's genotype two dominant genes or one dominant gene and one recessive gene? {*one dominant gene and one recessive gene (Rr)*} How do you know this? {*Paul must have two recessive genes (rr) for the recessive trait to be expressed.*}

Continue asking questions as time allows.

📖 A pedigree shows how children's traits are related to their parents' traits. When a person accepts Christ as Savior, he becomes a child of God. His traits should show the world that God is his Father. [Bible Promise: D. Identified in Christ]

Pedigrees

You probably have heard of pedigreed dogs or cats. The ancestors of these animals are recorded for many generations. This is so traits can be traced back through the generations. A **pedigree** is similar to a family tree, but instead of tracing people, it traces a particular trait. By using lines and symbols on a chart we can demonstrate how dominant and recessive traits show up in each generation.

For example, the following pedigree traces the trait of tongue rolling through three generations. The circles on the pedigree indicate females, and the squares represent males. The horizontal lines signify marriage. Vertical lines connect the parents to their children. Shaded symbols show that the person cannot roll his tongue. Symbols that are not shaded show that the person possesses the dominant trait, tongue rolling. The family members who can roll their tongues have either two dominant genes or one dominant gene and one recessive gene for this trait.

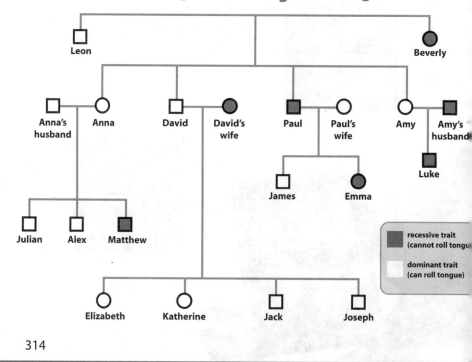

314

SCIENCE BACKGROUND

Carrier—A person who has a gene that is not expressed is called a *carrier*. In sex-linked traits, the mother is often a carrier.

X and Y chromosomes—The sex chromosomes are identified as X and Y. A female has two X chromosomes, and a male has one X chromosome and one Y chromosome. Genes for the sex-linked traits are found on these chromosomes.

 When identifying possible genotypes from a pedigree, it may be helpful to use a Punnett square.

From mother to son

Some traits are passed from mothers to sons. Although the daughters of the family may inherit a gene for that trait, the trait is usually visible in the sons only. Traits like these are called **sex-linked traits**.

One of the most common sex-linked traits is colorblindness. People who are colorblind usually have trouble distinguishing between red and green. In some instances, people with the trait for colorblindness also have difficulty with blue and yellow. In severe cases, people cannot distinguish any colors. These people see everything in shades of black and white.

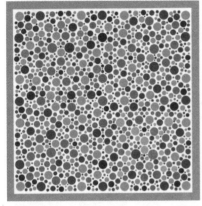

Colorblindness test—People with normal color vision see two colored symbols, an X and an O, among the gray dots. People with an inherited color vision deficiency see only an X, only an O, or neither symbol. This is for demonstration only. Photo courtesy of Jay Neitz.

Another sex-linked trait is hemophilia (HEE muh FILL ee uh), an illness that prevents a person's blood

Science and History

Queen Victoria, ruler of England from 1837 to 1901, was a carrier for hemophilia. One of her sons had hemophilia, and some of her daughters were carriers for the disease. As her daughters married, the trait spread to other European royal families. One of her best known descendants was the son of her granddaughter, the Empress Alexandra of Russia. Alexis Romanov, heir to the Russian throne, had hemophilia. He and his parents and sisters were murdered during the Russian revolution.

from clotting properly. Proteins in the blood that help to stop bleeding are either missing or not working properly. Falls and bumps often bring great danger by causing internal bleeding. The disease cannot be cured, but today with proper medical care people with hemophilia can live healthy lives.

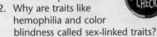

1. Why do geneticists use Punnett squares?
2. Why are traits like hemophilia and color blindness called sex-linked traits?

Science Misconceptions

Blindness and colorblindness are not the same. People with colorblindness are not blind. They just cannot see all colors.

Discussion:
pages 314–15

➤ What are *sex-linked traits*? {traits that are passed from mothers to sons}

➤ What are two of these traits? {colorblindness and hemophilia}

➤ Which two colors do most colorblind people have trouble distinguishing? {red and green}

💡 What are some things that might be difficult for someone who is colorblind? {Possible answers: matching clothes, coloring a picture, or driving}

➤ What is *hemophilia*? {an illness that prevents a person's blood from clotting properly}

💡 Is hemophilia a contagious disease? Explain. {No; a person cannot catch it from someone. It is inherited.}

Refer the student to *Science and History*.

➤ Which sex-linked trait was passed through Queen Victoria's family? {hemophilia}

💡 Some of Queen Victoria's daughters were described as carriers for the disease of hemophilia. What does the term *carrier* mean? {The daughters had the gene for hemophilia but did not suffer from the disease.}

💡 Sometimes when you fill out a medical questionnaire, it will ask for medical information about your parents and grandparents. Why do you think doctors ask questions about parents and grandparents on a questionnaire about a child? {Since some medical problems are hereditary, it would be helpful to know whether there are any family traits that might affect the child.}

Answers

1. Punnett squares show the genetic possibilities that could result for the offspring and the probability of a certain outcome.

2. These traits are passed from mothers to sons. Daughters do not usually inherit these traits, but they may be carriers of them.

Activity Manual
Reinforcement—pages 208–9

Study Guide—page 210
This page reviews Lessons 135 and 136.

Lesson 137

Student Text pages 316–17
Activity Manual pages 211–12

Objectives

- Use Punnett squares to predict genotypes
- Construct paper pets based on predicted genotypes

Materials

- See Student Text page
- TE pages A36–A37, *Parental Genotype Cards*, reproduced and cut apart (enough for 2 cards per student)

Introduction

▶ Why do puppies or kittens not look exactly like their parents? *{Since offspring receive genes from each parent, each puppy or kitten is a genetic mixture of both parents.}*

▶ Today you will use genetic information of two parents to determine the possible genotypes of their offspring.

Purpose for reading:
Student Text pages 316–17
Activity Manual pages 211–12

The student should read all of the pages before beginning the activity.

ACTIVITY

Paper Pet Genetics

Sometimes when a baby is born you will hear people comment on his heredity. They may say things like, "He has a nose just like his father's," or "He's definitely got the Tucker ears." No baby will have all the traits of one parent. He will have traits from both parents.

As the child grows, more family traits become evident. He may be athletic like his mother and tall like his father. Maybe his hair will be curly like his grandfather's hair, but instead of being brown, it might be red like his grandmother's hair.

For this activity, you will be given two genotypes to use as "parents" for your paper pet. Each parent will have genes for four traits. Based on the parental genotypes, you must construct the faces of three paper pets that could be the offspring of those "parents." Each paper pet must be unique, having its own genotype.

Process Skills
- Making and using models
- Inferring
- Interpreting data
- Communicating

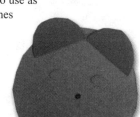

Procedure

1. Look at the two parental genotypes provided by your teacher. Use the chart in your Activity Manual to determine whether each trait of your "parents" is dominant or recessive.

2. Begin with the genes for color. Prepare a Punnett square using the parental genotypes for color. Choose a color for your first offspring based on the results in your square.

3. Complete a Punnett square for face shape. Cut the shape of the face based on the results in your square. The face should be at least 10 cm wide.

4. Complete a Punnett square for eye shape.

Materials:
blue, green, yellow, and orange construction paper
compass or large circle pattern
centimeter ruler
scissors
glue
crayons or markers
parental genotype card
Activity Manual

316

Review and discuss the following terms as necessary.

genotype—the arrangement of genes in an organism; its genetic combination

phenotype—the physical appearance of an organism

Punnett square—a chart that shows the genetic possibilities of a certain trait that can result for the offspring of a specific set of parents

dominant gene—a gene that will always be expressed if it is present in an organism

recessive gene—a gene that will not be expressed in an organism if any dominant genes are present

You may choose to have the student complete all four Punnett squares before handling the other materials.

Each parental genotype card represents one parent.

The student may find it helpful to color the pictures on his cards to reflect the phenotypes.

The size of the paper pets will vary, depending on student preference.

Use yellow paper and cut the shape of the eyes based on the results in your square.

5. Complete a Punnett square for ear shape. Use orange paper and cut the shape of the ears based on the results in your square.

6. Draw any remaining facial details you like.

7. Use the Punnett squares you have completed to construct two more paper pet offspring. Remember that each paper pet must be different in some way.

8. Present your paper pets to the class. Be prepared to explain the traits you used.

Conclusions
- What would happen to the possible offspring if you added another trait?

Follow-up
- Choose a single trait and show the pedigree for it using your "parents" and offspring.
- Choose a "mate" for each offspring, and continue the pedigree for another generation.

SCIENCE PROCESS SKILLS
Making and using models—Scientists use models to teach concepts about things that are too large or too small to be seen easily.

➤ **How are models used in this activity?** *{Possible answer: Each pet shows how dominant and recessive traits show up in offspring.}*

➤ **Are your pets accurate models?** *{not really}* **Why?** *{Possible answers: Few traits are controlled by only a single gene. Real organisms have more than four genes.}*

BIBLE No two people are exactly alike. Each person was created unique by God. [BAT: 3a Self-concept]

Procedure:
Distribute pairs of *Parental Genotype Cards*. Help the student determine which traits are dominant and which are recessive.

Demonstrate procedures as necessary. The student may need to complete a practice Punnett square before completing those relating to his pets.

➤ Look at the *Genetic Information* chart on Activity Manual page 211. Which face color is dominant? *{blue}*

➤ Face color is represented with the letter B. How will we show whether face color is dominant? *{B}* recessive? *{b}*

Since Punnett squares show possible genetic outcomes, many different answers are possible. However, the choices of color and shape should be based on one of the outcomes shown in the Punnett square.

Allow students to show their paper pets to the class and explain why they chose specific traits for them.

Conclusions
➤ In what ways do your pets resemble their parents?

➤ Do any of your pets look exactly the same as pets from other parents? Why? *{probably not; The genotypes of the parents are different.}*

➤ Would there be greater or less variety in the offspring if you added another trait? *{greater}*

Activity Manual
Activity—pages 211–12

Assessment
Rubric—TE pages A50–A53
Select one of the prepared rubrics, or design a rubric to include your chosen criteria.

Chapter 13

Lesson 137 335

Lesson 138

Student Text pages 318–21
Activity Manual page 213

Objectives

- Identify and discuss some common genetic diseases and disorders
- Explain why genetic diseases are not easy to cure
- Name some examples of genetic engineering

Vocabulary

sickle cell anemia   Down syndrome
cystic fibrosis   genetic engineering

Materials

- road map

Introduction

Display the road map.

➤ Why do we use maps? *{Possible answer: to find the location of a place}*

➤ When scientists figure out the DNA pattern of an organism, the pattern is called a *DNA map*. Why do you think it is called a map? *{Possible answer: The DNA map shows where the genes in the DNA are located.}*

➤ A DNA map may be used to identify individual organisms, but in this lesson we will discover other ways DNA mapping is used.

Purpose for reading:
Student Text pages 318–21

➤ How are genetic diseases different from contagious diseases?

➤ What is *genetic engineering*?

Genetic Disorders and Diseases

Some diseases, such as sickle cell anemia and cystic fibrosis, are inherited. These diseases are not contagious, so they cannot be spread from person to person. Instead, they are genetically passed from parent to child. Sometimes the gene that causes the disorder or disease is recessive. That means that in order for a child to inherit the disease, he must receive a recessive gene from each parent. However, a single dominant gene can also cause some genetic disorders. Scientists have identified the genes that cause many of these inherited diseases. Although much research has been done, many of these diseases have no known cure.

Sickle cell anemia

All people have both red and white blood cells. Normal red blood cells are round and flexible. They can bend and move easily through narrow blood vessels in the body. A person who has **sickle cell anemia** has some red blood cells that are hard and curved, like a farmer's sickle. These sickle-shaped blood cells can get stuck in the body's blood vessels. When blood vessels are blocked, oxygen cannot get to all parts of the body. This causes pain in the place where the blood vessel is blocked. It also may cause other health problems if the blood vessels remain blocked for too long.

Sickle cells are very fragile and can break apart easily. This causes the person to have *anemia,* or not enough red blood cells. Without enough red blood cells, a person's body does not get the oxygen it needs. The person often feels tired and gets infections easily. A person with sickle cell anemia needs to have plenty of fluids and should avoid things that decrease oxygen, such as smoking. This disease is most often found in people of African descent and in people from countries around the Mediterranean Sea. Blood tests can determine whether or not a person has inherited this disease.

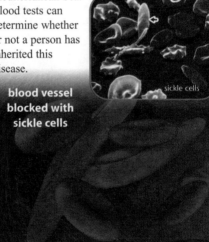

normal red blood cells

normal blood vessel

blood vessel blocked with sickle cells

sickle cells

318

SCIENCE BACKGROUND

Sickle cell anemia—The part of the red blood cell that picks up oxygen and moves it throughout the body is the hemoglobin molecule. When sickle cell anemia is present, the hemoglobin molecules harden into rods and clump together. This makes the blood cells sickle shaped.

The only known cure for sickle cell anemia is a bone marrow transplant. However, usually only a brother or sister who does not have the disease is a close enough match to donate bone marrow. The surgery can also add complications with other medicines or increase the risk of other health problems.

 TEACHER HELPS

Genetic diseases are not contagious. It is important to be kind and courteous toward everyone, including those who have genetic disorders or diseases. Be sensitive to students who may have genetic diseases. Some students may wish to share information about their diseases, but others may prefer that the rest of the class not know about their situations.

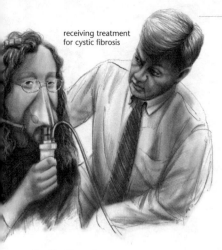
receiving treatment for cystic fibrosis

Cystic fibrosis

Cystic fibrosis (SIS-tic fye-BROH-sis) is a genetic disease that affects the lungs and digestive system. It is found primarily in people of European descent. A person with cystic fibrosis has mucus that is thicker than normal. This thick mucus clogs the lungs and air passages, increasing the chances of infection. Someone who has cystic fibrosis may cough often and may tire quickly.

This disease can also affect the digestive system. It prevents food from being fully digested, so the body does not receive enough nutrients. People with cystic fibrosis often take medicines with their meals to help their bodies absorb more nutrients. No cure for this disease has been found, but new treatments and medicines help people with cystic fibrosis live longer, healthier lives.

Down syndrome

A person who has the genetic disorder **Down syndrome** often has an extra chromosome. Most people receive 23 chromosomes from each parent—46 in all. Sometimes, though, one of the chromosomes makes an extra copy of itself, giving the person 47 chromosomes. This extra chromosome may cause developmental disabilities, such as delayed motor and language skills. People with Down syndrome do not all show the same symptoms, but many have hearing and vision problems, learning disabilities, or heart problems. With early training, many people with Down syndrome live productive lives.

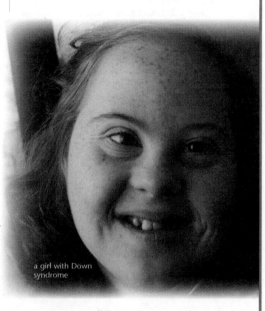
a girl with Down syndrome

Science Background

Down syndrome—Down syndrome is not always caused by an extra chromosome. Some rare cases of Down syndrome are caused by other chromosomal irregularities. Some people with Down syndrome may have some cells with the normal count of 46 chromosomes and other cells that have 47 chromosomes.

Huntington's disease—Huntington's disease is another genetic disorder that attacks the nervous system and the brain. It gradually takes away a person's ability to walk, talk, think, and reason. Most of the time, the disease does not begin until adulthood. Unlike many genetic disorders, Huntington's disease is caused by one dominant gene.

WRITING Many genetic diseases and disorders are named for the doctor or scientist who first diagnosed or identified the disease. The student could research Down syndrome, Huntington's disease, or another disease to determine how the disease got its name.

Discussion:
pages 318–19

➤ What are some genetic diseases? {Possible answers: sickle cell anemia, cystic fibrosis, Down syndrome}

➤ Why are these diseases called *genetic diseases*? {They are inherited. They are not contagious diseases.}

➤ What is different about the red blood cells of a person with sickle cell anemia? {Some of the red blood cells are hard and curved like sickles instead of being round and flexible.}

➤ What happens when a person's body does not have enough red blood cells? {The person has anemia.}

➤ Which two parts of the body are most often affected by cystic fibrosis? {the lungs and the digestive system}

💡 Recessive genes cause both cystic fibrosis and sickle cell anemia. In order for a child to inherit one of these diseases, would he need to inherit a recessive gene from one or both parents? {both} Why? {Recessive genes can be expressed only if no dominant gene is present for that trait.}

➤ How many chromosomes do most people have? {46}

➤ How many chromosomes does a person with Down syndrome have? {47}

➤ What kinds of difficulties might a person with Down syndrome have? {Possible answers: delayed motor and language skills; vision and hearing problems; learning disabilities; heart problems}

📖 A child born with one of these genetic disorders is not a mistake. God does not make mistakes. He has planned every detail of each person's life and has created each of us for His purpose and for His glory. He gave us exactly the abilities and characteristics that we need in order to fulfill His purpose for our lives. [Bible Promise: I. God as Master]

Discussion:
pages 320–21

➤ How many genetic diseases have been identified and named? *{over 5,000}*

➤ Why is it difficult to find a specific gene? *{There are many genes of different sizes.}*

➤ What is involved in genetic engineering? *{Genetic engineering involves changing a gene or moving some of one organism's genes into another organism.}*

➤ What are some reasons for genetic engineering? *{Possible answers: to treat genetic diseases; to cause a plant to require less water; to have a plant produce its own insecticide; to make food more nutritious}*

➤ How are some genetic diseases treated? *{through gene therapy}*

➤ What can be dangerous about gene therapy? *{Possible answer: If a gene gets put back in the wrong place, new problems could happen.}*

➤ What are some other ways that genetic research has helped people with genetic diseases? *{Possible answers: bacteria producing insulin for diabetics; new treatments for hemophilia and burn patients}*

Refer the student to *Fantastic Facts*.

➤ Why did scientists make a fly that had so many eyes? *{They wanted to find out which genes controlled the development of certain body parts. They also wanted to find ways to help solve human visual problems.}*

📖 Man's natural curiosity is a gift from God. That curiosity often results in new ideas that help others.

💡 Besides doctors, what other professionals might use genetic information? *{Accept any reasonable answers. Possible answers are farmers, animal breeders, and horticulturists.}*

Genetic Engineering

More than 5,000 genetic disorders and diseases have been identified and named. Doctors and scientists know that these diseases occur when certain genes are not working correctly or are missing. However, a scientist cannot quickly fix a gene or add a missing gene. Before a scientist can change a gene, he has to find the one that is not working correctly. The DNA packed inside a person's chromosomes is divided into many different genes. Scientists estimate that there are 30,000–40,000 genes of various sizes. Remember that DNA is made up of a sequence of four bases. Some genes have fewer than 10,000 base pairs, while others have more than two million.

Genetic engineering involves changing a gene or moving some of one organism's genes into another organism. Changing a gene can be risky. With so many genes of different sizes, it is hard to get the gene back into exactly the right place on the chromosome. Putting a gene into the wrong place can result in many new problems. However, genetic engineering is done for many different reasons.

One type of genetic engineering uses gene therapy to treat some genetic diseases. Doctors can substitute a healthy gene for one that is not working properly. Even though this does not cure the disease, it has helped many cystic fibrosis patients. Also, scientists have discovered a way to add a gene to bacteria that makes the bacteria produce insulin for diabetics. New treatments for hemophilia and burn patients have also benefited from genetic research.

Some scientists wanted to know which genes controlled how large body parts, such as legs, livers, and eyes, formed. They experimented with fruit flies and found a gene that controls eye development. Using this gene, they created a fly that had fourteen eyes on its wings, antennae, and legs. They hope to use this discovery to help solve human vision problems.

SCIENCE BACKGROUND

Genetic engineering—Genetic engineering is a topic that can generate strong feelings and pose ethical questions. Some feel that new discoveries will help end world hunger and will help people live healthier lives. Others are concerned about the long-term safety and results of these new discoveries. Concerns also arise over the access and use of the genetic records of individuals.

Cloning—Cloning is a form of genetic engineering. Its many definitions have created misunderstandings. Plants grown from cuttings, such as African violets, are actually clones. They are genetically identical to the parent plant. Identical twins could also be considered clones, since they are genetically identical to each other. An excellent resource article written from a Christian scientific perspective on cloning is available in the *Teacher's Resource Guide to Current Events for Christian Schools* published in 2000 by BJU Press.

 TECHNOLOGY One of the most recent genetic engineering projects added a fluorescence gene to zebra fish. These fluorescent fish have been used to help detect pollutants and are also sold as aquarium fish.

Any new technology or discovery will have positive and negative aspects. Genetic engineering is an example of this. Wise discernment and judgment are needed to determine if the benefits of using the technology outweigh the potential problems.

Other scientists have studied plants. Some scientists try to find ways to change the genes of a plant so that the plant will require less water. These plants, then, could be grown during droughts. Others have invented a cotton plant that produces its own insecticide. To do this, scientists inserted a gene that makes the plant produce poisons to kill the insect pests. Farmers who plant this type of cotton can use fewer chemical pesticides on their crops.

Sometimes the genes added by scientists are from plants that are very different from each other. For instance, by adding genes from bacteria and a daffodil to rice, scientists can make a type of rice that helps the human body produce more vitamin A. Scientists even found that adding a wheat gene to corn made corn plants taste bad to the insects that would usually eat them.

Genetic engineering has the potential to be both beneficial and harmful, depending on how it is used. Some people think that money for genetic engineering is often spent on unnecessary research. Others are concerned that new changes might result in unexpected problems. For example, plants that grow their own pesticides might also kill some beneficial insects.

Many new scientific discoveries have been made recently. What scientists are learning about the microscopic things in our universe truly is amazing. However, it is most important to remember our God who created this world and all things in it. None of these discoveries have surprised Him—He planned them all.

Romans 11:33 "O the depth of the riches both of the wisdom and knowledge of God! how unsearchable are his judgments, and his ways past finding out!"

1. How are genetic diseases different from other diseases?
2. What are some examples of genetic engineering?

genetic engineering

SCIENCE BACKGROUND

Testing on animals—Many people oppose planting test crops and testing new things on animals. However, new discoveries cannot be considered safe and beneficial without testing of some sort. Things need to be tested to determine if they are safe to use, but often the public does not want to test them until they are proven safe.

Discussion:
pages 320–21

➤ **Why would it be useful for a plant to need less water?** {The plant could be grown during droughts.}

💡 **What would be another benefit of a plant's needing less water?** {Possible answer: It could be grown in dry places, such as deserts, that do not receive much water.}

➤ **How did scientists get a cotton plant to grow its own insecticide?** {They inserted a gene that made the plant produce poisons that would kill insect pests.}

➤ **How would this be beneficial?** {Farmers who plant this type of cotton can use fewer chemical pesticides.}

➤ **What was added to a type of rice that would help the human body produce more Vitamin A?** {genes from bacteria and from a daffodil}

➤ **Why did scientists add a wheat gene to some corn plants?** {to make the corn taste bad to the insects that would normally eat and destroy it}

➤ **What are some concerns that some people have about genetic engineering?** {Possible answers: money being spent on unnecessary research; unexpected problems}

📖 **Do any of these new discoveries surprise God?** {no} **Why not?** {God planned them all as part of His creation.}

📖 God gave man dominion over the earth to use its resources wisely. Some people debate whether different uses for genetic engineering are wise uses of God's resources.

Answers

1. Genetic diseases are inherited from one's parents. They are not contagious diseases.

2. Possible answers: gene therapy; using bacteria to produce insulin; changing the genes of a plant to make it require less water; producing cotton that grows its own insecticide

Activity Manual
Reinforcement—page 213

Assessment
Test Packet—Quiz 13-B
This quiz may be given any time after completion of this lesson.

Lesson 139

Student Text page 322
Activity Manual pages 214–16

Objectives
- Recall concepts and terms from Chapter 13
- Apply knowledge to everyday situations

Introduction

Material for the Chapter 13 Test will be taken from Activity Manual pages 203–4, 210, and 214–16. You may review any or all of the material during the lesson.

You may choose to review Chapter 13 by playing "Sketch a Face" or a game from the *Game Bank* on TE pages A56–A57.

Diving Deep into Science

Questions similar to these may appear on the test.

Answer the Questions
Information relating to the questions can be found on the following Student Text pages:

1. page 306
2. pages 308–11
3. pages 308–11

Solve the Problem
In order to solve the problem, the student must apply material that he has learned. The answer for this Solve the Problem is based on the material on Student Text pages 310–11. The student should attempt the problem independently. Answers will vary and may be discussed.

Activity Manual

Study Guide—pages 214–16
These pages review some of the information from Chapter 13.

Assessment

Test Packet—Chapter 13 Test
The introductory lesson for Chapter 13 has been shortened so that it may be taught following the Chapter 13 test.

Diving Deep into Science

Answer the Questions

1. A series of crimes has been committed. The criminal's hair was found at one place and some blood was found at another place. How can crime investigators determine whether the same person committed the crimes?
   Answers will vary, but elicit that they could test the DNA of the hair and the blood. Since all cells in a person's body have the same DNA, both the hair and the blood would show the same DNA pattern if they were from the same person.

2. Why can you not tell whether a plant is purebred or a hybrid by observing only one generation?
   The dominant trait will express itself in both a purebred and a hybrid. You can identify a purebred only after multiple generations have shown no recessive trait.

3. Why would it take fewer generations to determine the genotype of a plant that self-pollinated than it would the genotype of a plant that was cross-pollinated?
   The genotype of the self-pollinating plant would be the same each time. It would not introduce any new genetic traits. It would simply express the traits that were already part of its genetic makeup.

Solve the Problem

Juanita really enjoys the pink snapdragons in her yard. One year she decided to try to pollinate two of the plants herself to get the plants to produce seeds. She saved the seeds and planted them the next year. What a disappointment! Instead of all pink snapdragons, she had red, white, and pink flowers. Can you think of a reason Juanita did not get all pink snapdragons?

Juanita's pink snapdragons are hybrids caused by incomplete dominance. When the pink hybrid snapdragon was crossed with another pink hybrid snapdragon, the purebred traits of red and white and the pink hybrid traits resulted.

322

Review Game
Sketch a Face

Prepare a set of 20 gene cards. Make 2 cards for each of the following dominant traits: square face, round eyes, triangle ears, round nose, and curved-line mouth. Make 2 cards for each of the following recessive traits: round face, oval eyes, semicircle ears, square nose, and straight line mouth. You may choose to use different colored inks to label the dominant and recessive gene cards or to display a chart that identifies the dominant and recessive traits.

Divide the class into two teams. Shuffle the cards and place them face down. When a student answers a review question correctly, he draws one of the gene cards. Whenever a team has two gene cards for the same trait, the team should determine the phenotype of that facial feature and draw it on the board. If at least one of the cards is a dominant gene, then the facial feature should be the dominant trait. After a team member has drawn that facial feature, place the gene cards in a discard pile. Any other gene cards drawn by that team for the same trait should also be placed in the discard pile. Shuffle the discard pile and add those cards back to the other cards as needed. The first team to draw a complete face wins.

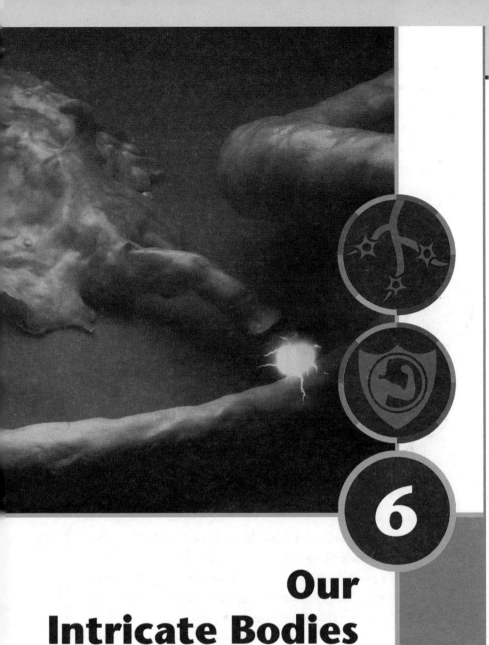

Unit 6: Our Intricate Bodies

Student Text pages 323–25
Activity Manual page 217

Lesson 140

Objectives

- Demonstrate knowledge of concepts taught in Chapter 13
- Recognize the interrelationship of science concepts
- Recognize that man's inferences are sometimes inaccurate

Unit Introduction

Note: This introductory lesson has been shortened so that it may be taught following the Chapter 13 test.

In Unit 6 the student will see the marvels of the human body. No study could be a greater reminder of God's wonderful care for us. The nervous system discussed in Chapter 14 and the immune system discussed in Chapter 15 are two of the most complex systems of the body.

The nervous system, endocrine system (also discussed in Chapter 14), and immune system provide most of the control and protection of the other systems of the body.

➤ **Look through Unit 6. What kinds of topics do you think you will be studying in this unit?** *{Possible answers: nervous system, endocrine system, diseases, pathogens, immune system}*

➤ **Why do you think these chapters are organized into the same unit?** *{Answers will vary, but elicit that they are both about the human body.}*

➤ **What does the word *intricate* mean?** *{Possible answers: detailed or complex}*

 Why would we call our bodies intricate? *{Answers will vary, but elicit that God made our bodies with very complex systems that all work together for our benefit.}*

Teacher Helps: Additional information is available on the BJU Press website. www.bjup.com/resources

Project Idea

The project idea presented at the beginning of each unit is designed to incorporate concepts of each chapter as well as information gathered from other resources. You may choose to use the project as a culminating activity at the end of the unit or as an ongoing activity while the chapters are taught.

Unit 6—A Look Inside

Technology continues to miniaturize. Direct the student to write about a microscopic voyage inside the body. The student should describe the parts of the nervous system, endocrine system, and immune system in a fictionalized context. For example, the nerves could be referred to as power lines, hormones as messengers, macrophages as policemen, and antibodies as ammunition.

Technology Lesson

A correlated lesson for Unit 6 is provided on TE pages A14–A15. This lesson may be taught with this unit or with the other Technology Lessons at the end of the course.

Bulletin board

Note: This suggested bulletin board idea may be modified for use in a learning center.

Place a silhouette of a person's head in the center of the board. Draw the parts of the brain on colored paper and attach them to the silhouette.

Write interesting brain facts and trivia on note cards or on sentence strips and attach them to the bulletin board. You may choose to write the facts in question and answer form. Discuss the facts with the student. Look through resources for interesting brain facts or trivia, such as the following:

The average human brain is 140 mm wide. The left side of the brain has 186 million more neurons than the right side of the brain. The optic nerve contains 1,200,000 nerve fibers. The human brain has about 3 million miles of axons. The human brain contains about 100 billion neurons.

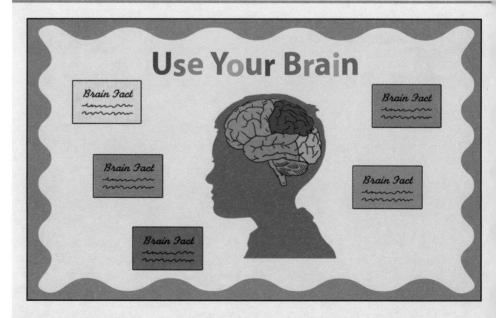

342   Lesson 140

Chapter 14: Nervous System

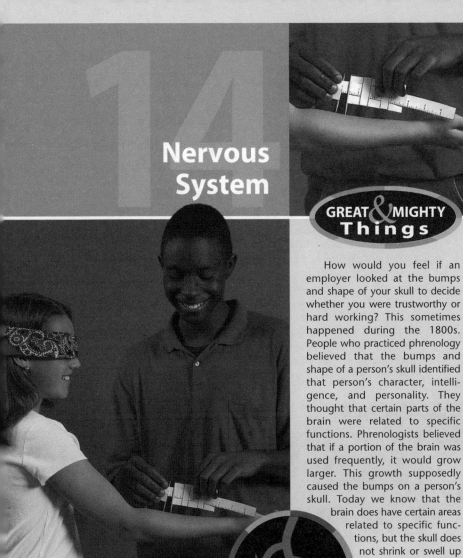

GREAT & MIGHTY Things

How would you feel if an employer looked at the bumps and shape of your skull to decide whether you were trustworthy or hard working? This sometimes happened during the 1800s. People who practiced phrenology believed that the bumps and shape of a person's skull identified that person's character, intelligence, and personality. They thought that certain parts of the brain were related to specific functions. Phrenologists believed that if a portion of the brain was used frequently, it would grow larger. This growth supposedly caused the bumps on a person's skull. Today we know that the brain does have certain areas related to specific functions, but the skull does not shrink or swell up based on how often parts of the brain are used. God designed our brains to control all the functions of our bodies.

Introduction

Choose a volunteer that will allow another student to feel his head for shapes and bumps. Ask the following question of the student doing the examining:

➤ **Did you feel any bumps or unusual shapes on the skull?** {Answers will vary.}

Having bumps on one's skull is normal. Ask this question of the student being examined:

➤ **How would you feel if your math grade was based on the size and shape of the part of your skull that was thought to be related to math?** {Answers will vary.}

➤ **Some people used to believe this.**

Chapter Outline

I. Structure of the Nervous System
   A. The Central Nervous System
   B. The Peripheral Nervous System

II. Interactions with the Nervous System
   A. The Five Senses
   B. Memory
   C. Sleep and the Nervous System
   D. The Endocrine System
   E. Disorders and Drugs

Activity Manual

Preview—page 217

The purpose of this Preview page is to generate student interest and determine prior student knowledge. The concepts are not intended to be taught in this lesson.

God is the Creator of all. He designed the brain to control all of the functions of the human body. He planned and prepared each detail in a way that will bring honor and glory to Him. [Bible Promises: I. God as Master]

Each one of us is uniquely created in God's image. He gives each individual special talents and abilities. [BAT: 3a Self-concept]

Lesson 141

Student Text pages 326–29
Activity Manual page 218

Objectives

- Identify the two main parts of the nervous system
- Describe the parts of the central nervous system
- List the four lobes of the cerebrum
- Differentiate among the functions of the three parts of the brain

Vocabulary

central nervous system	lobes
peripheral nervous system	cerebellum
brain	brain stem
cerebrum	spinal cord

Materials

- bike, football, or motorcycle helmet

Introduction

Display the helmet.

➤ **What is the purpose of a helmet?** {head protection}

Place the helmet on a student.

➤ **Let's use the helmet to picture the head. If the head of the student inside the helmet represents the brain, what does the hard outer shell of the helmet represent?** {the skull}

➤ **The padding inside the helmet also represents a part of the body that protects the brain inside the skull. Today we will learn about this and other parts of the brain.**

Purpose for reading:
Student Text pages 326–29

➤ Which part of the brain helps you identify sounds as speech, music, or noise?

➤ How is your spinal cord different from your spinal column?

Each day people use their senses—seeing, hearing, tasting, smelling, and touching—to observe God's world. But none of the information gathered by the senses would be of any value without a way to understand it. God designed a complicated network to gather and process, or interpret, information. This network is called the nervous system.

Even the most complex computer network cannot compare to the human nervous system. Just imagine all that is happening in your body while you read this paragraph! Your eyes gather information, and your ears hear sounds. Your hands touch the book. But besides all this, your nervous system keeps your heart beating and your lungs breathing without your even having to think about it. Your skin feels the temperature of the room and your body stays balanced in your seat, all because of your nervous system. But that is only the beginning of the nervous system's responsibilities.

Many parts of the nervous system work together to allow a person to go white-water rafting.

326

Structure of the Nervous System

The nervous system is divided into two main parts. The **central nervous system** consists of the brain and the spinal cord. This part of the nervous system makes decisions and controls the body's actions. The **peripheral** (puh RIF ur ul) **nervous system** consists of millions of nerve cells that communicate with the central nervous system about what goes on in and around the body.

The Central Nervous System
Brain

The **brain** acts as the command center for the body. Thousands of pieces of data are transmitted to and from the brain every second. The brain organizes and interprets this information and tells the body how to respond. It not only controls actions and speech but also influences emotions. The brain is protected both by the skull and by *cerebrospinal* (SEHR uh broh SPY nuhl) *fluid*. This fluid acts like a cushion and shock absorber for the brain and the spinal cord.

You might expect something as hard working as your brain to be quite large. Actually, the brain weighs only about 1.4 kg (3 lb)! It is shaped like a large, wrinkly walnut. The brain has

---

SCIENCE BACKGROUND

Lobes of the brain—The cerebrum actually includes a fifth lobe, the *insula*. The insula is involved with the automatic functions of the brain stem, the sense of smell, and other functions such as taste and digestion. It is located within the Sylvian fissure that separates the temporal lobes from the frontal and parietal lobes. It is usually not visible in drawings of the brain. Sometimes it is called the island of Reil because it was first described by Johann Christian Reil in 1809. *Insula* is the Latin name for "island."

SCIENCE BACKGROUND

Cerebrospinal fluid—Cerebrospinal fluid is made in the choroid plexus, blood vessels that line certain areas in the brain. This fluid is a clear, water-like liquid made from substances in the blood. It includes glucose (sugar), protein, and white blood cells. The fluid is continuously made, circulated, and reabsorbed.

Headaches may occur whenever the level of cerebrospinal fluid is low, such as after a doctor has taken a sample of the fluid to test for disease. Too much cerebrospinal fluid can also cause headaches, hydrocephalus, and other problems.

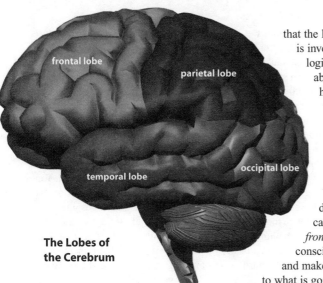

The Lobes of the Cerebrum

three distinct parts: the cerebrum (SEHR uh bruhm), the cerebellum (SEHR uh BEL uhm), and the brain stem. Each part has different functions, but all three parts work together to allow you to live and interact with your environment.

The largest part of the brain is the **cerebrum,** which means "brain." The cerebrum takes up most of the space inside the skull. It can be divided into two halves, the left hemisphere and the right hemisphere. The left hemisphere controls the right side of the body, and the right hemisphere controls the left side of the body. Many scientists think that the left hemisphere is involved with math, logic, and speech abilities. The right hemisphere seems to control creativity and musical and artistic abilities.

The cerebrum also can be divided into distinct areas called **lobes.** The *frontal lobe* controls conscious movement and makes a person alert to what is going on around him. It is the center of reasoning and decision-making and also influences personality. The *parietal* (puh RY uh tuhl) *lobe* interprets pain, touch, and temperature, as well as some tastes and pressure on the skin. Another part, the *temporal* (TEM pur uhl) *lobe,* deals with hearing, speech, and memory. This lobe helps classify sounds as speech, music, or noise. The *occipital* (ahk SIP uh tuhl) *lobe* stores information about what a person sees. This lobe receives messages from the eyes and interprets those messages. God designed each part of the cerebrum to help people understand and appreciate the world that He created.

Discussion:
pages 326–27

▶ Which part of your body gathers and interprets information? {*the nervous system*}

▶ What are the two main parts of the nervous system? {*the central nervous system and the peripheral nervous system*}

▶ Which parts of the body make up the central nervous system? {*the brain and spinal cord*}

▶ What is the job of the central nervous system? {*to make decisions and control the body's actions*}

▶ Which part of the central nervous system organizes and interprets information? {*the brain*}

▶ What protects the brain from injury? {*the skull and cerebrospinal fluid*}

▶ About how big is your brain? {*1.4 kilograms, or 3 pounds*}

▶ What does the brain look like? {*a large, wrinkled walnut*}

▶ What are the three parts of the brain? {*cerebrum, cerebellum, and brain stem*}

▶ What are two different ways that the cerebrum can be divided? {*hemispheres and lobes*}

▶ Which hemisphere do scientists think controls math, logic, and speech abilities? {*the left hemisphere*}

💡 Why might someone who is very artistic or musical be referred to as a "right-brained" person? {*Some scientists think that the right hemisphere controls artistic and creative abilities.*}

💡 When a person has a stroke, a part of the brain is damaged because of lack of oxygen. Often there is weakness on one side of the body. If a person's right side is affected, which side of the brain was probably damaged? {*the left*}

📖 Refer the student to the diagram *The Lobes of the Cerebrum* on Student Text page 327.

▶ What are some functions of the frontal lobe? {*It controls conscious movement, makes a person alert to what is going on around him, is the center of reasoning and decision making, and influences personality.*}

▶ Which lobe interprets pain, touch, and temperature? {*the parietal lobe*}

▶ How is the function of the temporal lobe different from the function of the occipital lobe? {*The temporal lobe deals with hearing, speech, and memory. The occipital lobe receives messages from the eyes.*}

Direct a DEMONSTRATION
Demonstrate how cerebrospinal fluid protects the brain

Materials: two raw eggs, two plastic jars with screw-on lids, water

Place one egg inside a jar and close the lid tightly. The egg represents the brain, and the jar represents the skull. Direct the student to shake the jar vigorously.

▶ What happened to the "brain" while it was shaken? {*It became broken and damaged.*}

Fill the other jar with water. Place the second egg in the water. Tightly close the jar.

▶ What do you think will happen when you shake the jar this time? {*Answers will vary.*}

Direct the student to shake the jar vigorously.

▶ What kept the "brain" from breaking? {*the water*}

▶ What part of our body does the water represent? {*the cerebrospinal fluid*}

Discussion:
pages 328–29

 Refer the student to the diagram *Parts of the Brain* on Student Text page 328.

➤ Where is the cerebellum located? {underneath the cerebrum}

➤ How does its size compare to that of the cerebrum? {The cerebellum is much smaller.}

➤ What is the purpose of the cerebellum? {It receives orders from the frontal lobes and sends messages to muscles throughout the body in order to accomplish tasks.}

➤ Which aspects of movement does the cerebellum control? {the speed and force at which a person moves}

💡 When a person is learning to roller-skate, which part of his brain will tell his muscles what to do? {the cerebrum}

💡 After a person has learned how to skate, which part of his brain remembers how to do it? {the cerebellum}

➤ Why might a person with an injured cerebellum have difficulty with eating, talking, or walking? {because the cerebellum helps to control balance and muscle coordination}

➤ Which part of the brain connects the brain to the spinal cord? {the brain stem}

➤ Name some actions controlled by part of the brain stem. {breathing, heartbeat, blood pressure, swallowing, digestion}

➤ Why are these actions called involuntary activities? {You do not have to think about them in order for them to happen.}

📖 Why do you think God has designed our bodies to do some actions automatically? {Answers will vary. Elicit that we would never be able to sleep if we had to consciously remember to breathe or swallow.}

The next part of the brain is the **cerebellum,** which means "little brain." The cerebellum is located underneath the cerebrum and is much smaller. It receives orders from the frontal lobes and sends messages to muscles throughout the body in order to accomplish tasks. The cerebellum does not decide when or where a person should move, but it does control the speed and force with which he moves.

Whenever you learn a new activity, such as bike riding, the cerebrum directs your muscles. Once the activity has been learned, the cerebellum takes over. It remembers how to do that task. The cerebellum also helps to control balance and muscle coordination. If this part of the brain is damaged, a person may have difficulty with motor skills such as eating, talking, or walking.

The final part of the brain is the **brain stem.** The brain stem is located below the cerebrum and in front of the cerebellum. It connects the brain to the spinal cord. Part of the brain stem also controls the functions necessary for life, such as breathing, heartbeat, blood pressure, swallowing, and digestion. These are involuntary activities. You do not have to think about them to make them happen. God has designed our brains to operate some functions automatically. Think of how hard it would be if you had to remember to breathe, make your heart beat, and digest your food all at the same time.

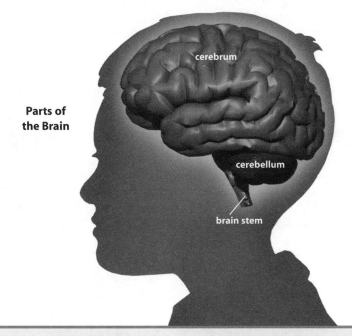

Parts of the Brain

328

 The brain gives directions to the muscles, helping each part of the body perform its function. In a similar way, the Bible provides directions for Christians to follow. Studying the Bible helps believers to become familiar with these instructions and to grow spiritually. [BAT: 6a Bible study]

SCIENCE MISCONCEPTIONS

Although certain parts of the brain perform certain tasks, in many cases if there is an injury, other parts of the brain can be trained to take over some of the functions that are impaired.

346   Lesson 141

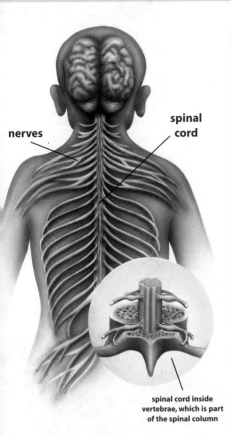

nerves

spinal cord

spinal cord inside vertebrae, which is part of the spinal column

Spinal cord

Can you feel the bumpy bone that goes down the center of your back? That backbone, your *spinal column*, protects your spinal cord. The spinal cord is inside the tunnel made by the *vertebrae*, or bones, in your spinal column. It is surrounded by cerebrospinal fluid and covered by three membranes. These membranes act like filters, protecting the spinal cord from any harmful substances that may be in the blood stream.

God created the **spinal cord** to be the main pathway of information connecting the brain to the rest of the body. The spinal cord is a column of nerve fibers about as thick as one of your fingers. It will be about 43–45 cm (17–18 in.) long when you are an adult. Usually the spinal cord ends at a person's waist.

The spinal cord is divided into thirty-one sections. Each section has pairs of nerves that branch out from between the vertebrae in the spinal column. These nerves continue to branch out and reach all parts of the body. Nerves connect with every spot of skin as well as with each organ and muscle.

The central nervous system is a very important part of the body. Injuries to the brain or spinal column can result in problems such as blindness, paralysis, and loss of speech or movement. Sometimes an injury to the central nervous system can be fatal. That is one reason it is important to wear the proper protective equipment for sports and other athletic activities.

> **QUICK CHECK**
> 1. What functions does the central nervous system have?
> 2. What are the two main parts of the central nervous system?
> 3. What are the three parts of the brain?

329

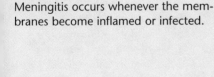

Meninges—The three membranes, or meninges, that protect the spinal cord also cover and protect the brain. The *pia mater* is the inside layer, closest to the central nervous system. The middle membrane is the *arachnoid mater,* and the outside membrane is called the *dura mater.* Meningitis occurs whenever the membranes become inflamed or infected.

Discussion:
pages 328–29

➤ Where is your spinal column located? {down the center of your back}

➤ What is another name for the spinal column? {backbone}

➤ What bones in your spinal column make up the tunnel for your spinal cord? {vertebrae}

➤ Other than your spinal column, what else protects your spinal cord? {cerebrospinal fluid and three membranes}

➤ What is your spinal cord? {a column of nerve fibers}

Emphasize to the student that the spinal cord is the nerve fibers and the spinal column is the bones that protect the spinal cord.

➤ How thick is your spinal cord? {about as thick as your finger}

➤ How long will your spinal cord be when you are an adult? {43–45 centimeters or 17–18 inches}

➤ Into how many sections is your spinal cord divided? {31}

💡 Why do you think the spinal cord is called "the main pathway of information"? {The spinal cord is connected to the brain, and nerves branch out of the spinal cord to all parts of the body.}

💡 Some people say that the spinal cord looks like a tree. Why do you think they say this? {Possible answer: The spinal cord resembles the trunk, with nerves as the branches.}

💡 When people talk about the spine, are they referring to the bones or to the nerves? {Elicit that usually they are referring to the bones.}

➤ What are some things that could result from an injury to the brain or spinal cord? {blindness, paralysis, loss of speech or movement, death}

Answers
1. The central nervous system makes decisions and controls the body's actions.
2. the brain and the spinal cord
3. cerebrum, cerebellum, and brain stem

Activity Manual
Reinforcement—page 218

Lesson 142

Student Text pages 330-33
Activity Manual pages 219-20

Objectives

- Identify the parts of a neuron
- Explain how neurons send messages
- Compare the two parts of the peripheral nervous system
- Describe how a reflex occurs

Vocabulary

neuron
sensory neuron
motor neuron
dendrite
impulse
axon
synapse
reflex

Materials

- stopwatch or watch with a second hand

Introduction

➤ What are some different ways that we can send messages to our friends? {Possible answers: letters, e-mail, phone calls, instant messaging}

➤ Why do we send those messages? {to stay in contact with our friends; to share experiences}

➤ Today we will be studying how the body sends messages to and from the brain.

➤ Why do you think it is important for the brain to stay in contact with the parts of the body? {Possible answer: The brain is responsible for controlling the body parts, so it must be in contact with them.}

Purpose for reading:
Student Text pages 330-33

➤ Which part of the peripheral nervous system helps your body adjust to its external environment?

➤ What is a *reflex*?

Discussion:
pages 330-31

➤ What are *neurons*? {nerve cells}

➤ How are neurons similar to other cells in your body? {They each have a cell body with a nucleus, chromosomes, and DNA.}

➤ What is unique about neurons when compared to other cells? {They can communicate with each other.}

The Peripheral Nervous System

Neurons

When you stub your toe, how do you know that it hurts? Thousands of tiny nerve cells send a message up to your brain. The brain interprets the incoming message as pain, and you become aware that your toe is hurting. The nerve cells are called **neurons** (NUHR ahnz). In some ways neurons are similar to the other cells in your body. Each has a cell body with a nucleus, chromosomes, and DNA. However, neurons have the unique ability to communicate with each other. The neurons located outside the brain and spinal cord make up the peripheral nervous system.

The shape and size of a neuron depend on its function and location in the body. Some neurons are **sensory neurons.** They carry messages to the brain. Others are **motor neurons,** sending messages from the brain and spinal cord to the muscles. Neurons can live for a long time, longer than most cells. However, most neurons that die are not replaced.

The **dendrite** (DEN dryt) receives the electrical **impulse,** or message, from another neuron. The dendrite passes that message to the cell body. The cell body passes the message to the **axon** (AK sahn), which sends the impulse on to the next neuron. Although a neuron usually has only one axon, it can have many dendrites. Some nerve cells have as many as 10,000 dendrites! The nerve fibers in your body are actually bundles of axons and dendrites from many neurons. Your body has over ten billion long and microscopically thin nerve cells. Some of the longest neurons have axons more than a meter long.

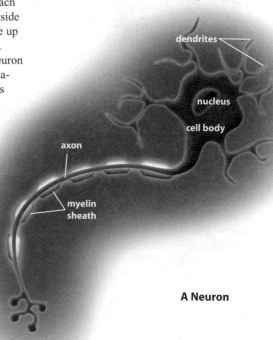

A Neuron

330

SCIENCE BACKGROUND

Cells—The parts of cells and their functions within cells are discussed in Chapter 4, *Cells and Classification.*

How neurons send messages—The electrical impulse is created by the movement of ions such as sodium and potassium through the neuron's cell membrane.

 LANGUAGE

The word *neurotransmitter* is made up of the prefix *neuro,* meaning "nerve," and *transmit,* meaning "to send."

Direct an ACTIVITY

Materials: pony beads, plastic lacing, copy of TE pages A38-A39, *Bead Neuron*

Direct the student to make a bead neuron. Use the bead neuron as a model to review and reinforce the parts of a neuron. The student may find it helpful to color the beads on the instructions the same colors as the beads he is using. You may choose to use a bead neuron instead of the ruler in the Demonstration on TE page 349.

Even though your body contains billions of neurons, the neurons do not touch each other in order to send messages to and from the brain. If you touch a paper clip to your finger, the paper clip presses against the dendrites in your skin. The pressure you feel travels as an electrical impulse through the dendrites, to the cell bodies, and then to the axons.

Between each neuron is a little gap called a **synapse** (SIN aps). As the electrical impulse arrives at the synapse, the electricity causes chemicals called *neurotransmitters* (NUHR oh TRANZ miht uhrz) to be released. These chemicals cross the synapse and carry the message on to the next sensory neuron. The impulse continues from neuron to neuron, sometimes as fast as 120 m (400 ft) per second. This impulse could travel the length of a football field in less than one second.

When the sensation reaches your brain, the brain interprets it and lets your finger know that it is experiencing pressure. By the time you feel the pressure of the paper clip, the message has already traveled to the brain, been interpreted, and traveled back through the motor neurons to tell your finger to move. The amount of electricity involved in sending the nerve message to the brain is about one-tenth of a volt.

God gave some axons a protective covering called a *myelin* (MYE uh lin) *sheath*. This extra insulation helps the neuron send messages faster. Because myelin sheaths are white, areas of the nervous system with myelin sheaths are called white matter. Areas of the central nervous system where the neurons do not have myelin sheaths are sometimes called gray matter.

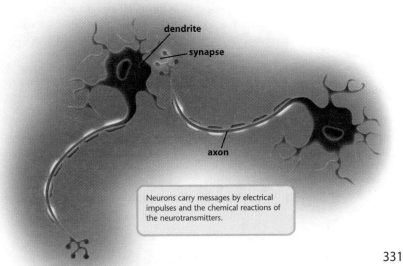

Neurons carry messages by electrical impulses and the chemical reactions of the neurotransmitters.

331

Discussion:
pages 330–31

- Does the peripheral nervous system include the neurons outside the brain and spinal cord or inside the brain and spinal cord? {outside}
- What determines the shape and size of a neuron? {its function and location in the body}
- How are sensory neurons different from motor neurons? {Sensory neurons carry messages to the brain. Motor neurons carry messages from the brain to the muscles.}
- We know that most cells are replaced when they die. How are neurons different from this? {Most neurons are not replaced when they die.}
- 💡 Why might a person who has had a severe cut on a finger or toe lose feeling in that part of the body? {Possible answer: The neurons there might have been injured and not replaced.}

Refer the student to the diagram *A Neuron* on Student Text page 330 as each part is discussed.

- What are *impulses*? {the electrical messages sent from neuron to neuron}
- Which part of a neuron receives the impulse? {the dendrite}
- Which part of a neuron sends the impulse to the next neuron? {the axon}
- What are *nerve fibers*? {bundles of axons and dendrites from many neurons}
- What is a *synapse*? {the little gap between each neuron}
- How do the neurons send impulses? {The impulse is received by the dendrites, sent to the cell body, and then sent to the axon. At the axon, chemicals are released that carry the message across the synapse to the dendrites of the next neuron.}
- What are *neurotransmitters*? {chemicals that cross the synapse and carry the impulse on to the next neuron}
- What protective covering helps neurons send messages faster? {the myelin sheath}
- How is white matter different from gray matter? {White matter consists of neurons that have myelin sheaths. Gray matter consists of neurons without myelin sheaths.}

Direct a DEMONSTRATION

Materials: 6 or more students, ruler

Direct the students to stand with their arms outstretched at their sides. There should be a small space between the fingertips of each student, so that the students cannot touch each other. The students represent neurons, and the ruler represents an impulse. Start the impulse by placing one end of the ruler in the hand of the first student.

- Which part of the neuron is represented by the hand that receives the ruler (impulse)? {dendrite}

The student passes the ruler over his head to his other hand.

- The impulse moves from the dendrite to which part of the neuron? {axon}

The student passes the ruler to the next person without touching hands.

- What is the name of the gap between the axon of one neuron and the dendrites of the next? {synapse}

Continue passing the ruler and reviewing how neurons send impulses.

Discussion:
pages 332–33

▶ What are the two parts of the peripheral nervous system? {somatic nervous system and autonomic nervous system}

▶ Which part helps the body adjust to its external environment? {somatic nervous system}

▶ Which muscles are controlled by the somatic nervous system? {the skeletal muscles}

💡 How are the skeletal muscles different from muscles controlled by the autonomic nervous system, such as the heart? {Most skeletal muscle movement is voluntary. Other muscles, such as the heart, are involuntary muscles. You cannot control them.}

▶ What is the purpose of the autonomic nervous system? {It regulates the body's internal environment.}

Check a student's heart rate for 30 seconds as he sits still in a chair. Record his heart rate. (It will probably be easiest if he finds his pulse on his neck and then counts the number of heartbeats for the 30 seconds.) Direct the same student to run in place or do jumping jacks for one minute. Check his heart rate again and record the number.

💡 Why are these two numbers different? {The first number shows how fast the heart was beating while the student was still and resting. The second number shows how fast the heart was beating after exercise.}

▶ Did the student control his heart rate, or did it change automatically? {It changed automatically.}

▶ Which part of the peripheral nervous system regulates your heartbeat? {autonomic nervous system}

▶ What are some other activities controlled by the autonomic nervous system? {Possible answers: temperature, blood pressure, breathing, digestion}

▶ Which kind of neuron is used by the autonomic nervous system? {motor neuron}

▶ How often is your autonomic nervous system functioning? {continuously}

Somatic nervous system

The peripheral nervous system can be separated into two parts. One part, the *somatic* (soh MAT ihk) *nervous system*, controls your skeletal muscle movements. The somatic nervous system helps your body adjust to its external environment. The sensory neurons gather information about things in your environment and send that information to your central nervous system. The central nervous system then sends messages to your muscles, making them contract and relax as you move.

Autonomic nervous system

The peripheral nervous system also helps to regulate your internal environment. This part of the peripheral nervous system is sometimes called the *autonomic nervous system* because it controls involuntary activities. Usually you do not have conscious control over these activities.

For example, the autonomic nervous system controls your heart rate. Stress or fear can cause your heart to beat faster. But when you are resting or digesting food, the autonomic nervous system slows your heartbeat. The autonomic nervous system also helps your body maintain a constant temperature. When you are cold, your body starts to shiver. If you get too warm, your body releases perspiration through its pores. All of these reactions happen automatically.

The autonomic nervous system also regulates your blood pressure, breathing, digestion, and many other bodily functions. If you had to think about each one of these activities in order for it to occur, you would probably not be able to do anything else! The autonomic nervous system uses motor neurons to keep your body running smoothly. It works continuously, even when you are sleeping.

332

SCIENCE BACKGROUND

Autonomic nervous system—The autonomic nervous system is divided into the sympathetic nervous system and the parasympathetic nervous system. During stressful times, the sympathetic nervous system prepares the body to respond quickly by increasing the heart rate, blood pressure, etc. At times of rest, such as digestion and relaxation, the parasympathetic nervous system decreases the heart rate and blood pressure.

Shivering—When the body is chilled, the sympathetic nervous system causes the body to shiver. This shivering helps warm the body because of the increased muscle activity.

Yawning—Even though a yawn seems quite "contagious" at times, it is not considered to be a reflex. People used to think that yawns occurred when the lungs did not have enough oxygen or when someone was bored. Although yawns may occur at these times, they are not caused by lack of oxygen in the lungs or by boredom. Scientists are not certain what causes yawning, but some believe that a low oxygen level in part of the hypothalamus triggers yawns.

Photic sneezing—Some people tend to sneeze when exposed to a bright light such as sunlight. This genetic reflex is called *photic sneezing*.

Reflexes

Sometimes your body responds to a situation before your brain makes a conscious decision. For example, if you touch a hot stove accidentally, the electrical impulse immediately begins to travel from neuron to neuron until it reaches the spinal cord. Before the impulse passes on to the brain, an automatic message is sent back to your hand, telling your muscles to move your fingers away from the hot stove.

While this is happening, the impulse continues to the brain. Your brain interprets the message, and you realize that you are touching something hot and should move your hand. Since the whole process happens so quickly, you do not notice the time difference. From the time you touch the stove until you move your hand and your brain registers what has happened, less than one-thousandth of a second has passed.

How fast is a sneeze? This reflex usually happens whenever the lining of the nose becomes irritated by foreign particles. Scientists have found that a person can sneeze about as fast as a baseball pitcher can pitch a fastball. Some sneezes have been as fast as 150 km/h (about 100 mi/h). Sneezes can be quite powerful because they involve not only the nose, but also muscles in the chest, abdomen, face, throat, and eyelids. Sneeze droplets can travel as far as 1.5 m (5 ft) before they settle.

Even though this seems to happen all at once, you actually pull your hand back before your brain tells you that there is pain. This is called a reflex. A **reflex** is an action that happens before the brain has time to think about the action. A reflex is hard—sometimes impossible—to control.

QUICK CHECK
1. What is the difference between sensory and motor neurons?
2. What is the purpose of the autonomic nervous system?
3. How are reflexes different from other muscle movements?

333

Discussion:
pages 332–33

💡 Why might your body need to respond to a situation before your brain decides on the action? {Possible answer: to protect itself from danger}

▶ Which part of the central nervous system would send an automatic message telling your hand to move away from a hot stove? {the spinal cord}

▶ Does this happen before or after the brain recognizes that your hand is touching something hot? {before}

▶ What do we call an action that happens before the brain can think about it? {reflex}

💡 Can you tell that your hand moves before you think about the stove's being hot? {no} Why? {The body reacts so quickly that the movements and thoughts seem to happen at once.}

▶ What are some other reflexes? {Possible answers: sneezing, blinking, knee reflexes}

📖 Reflexes are an example of how God designed the body with built-in ways to protect itself.

Refer the student to *Fantastic Facts*.

▶ What causes most sneezes? {the lining of the nose being irritated by foreign particles}

▶ How fast can a sneeze be? {150 kilometers per hour or 100 miles per hour; about as fast as a pitcher's fastball}

▶ Which parts of the body are involved in a sneeze? {nose, chest muscles, abdomen, face, throat, and eyelids}

Answers
1. Sensory neurons carry messages from the senses to the brain. Motor neurons carry messages from the brain to the muscles.
2. to control involuntary activities
3. The brain controls all muscle movements except reflexes. Reflexes happen automatically before the brain has time to think about the action.

Activity Manual
Reinforcement—page 219

Study Guide—page 220
This page reviews Lessons 141 and 142.

Assessment
Test Packet—Quiz 14-A
The quiz may be given any time after completion of this lesson.

This diagram is provided to help you see the organization of the nervous system.

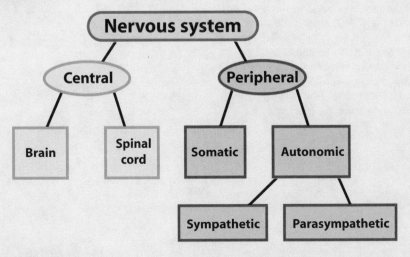

Lesson 143

Student Text pages 334–35
Activity Manual pages 221–22

Objectives

- Explore variables that affect reaction time

Materials

- See Student Text page

Introduction

➤ Have you ever had a 150 km/h (100 mi/hour) fastball pitched to you?

➤ If you had, were you able to hit it?

➤ Why can some baseball players hit balls this fast, but you cannot? {Possible answers: The baseball player is stronger, faster, or has practiced more.}

➤ Why do some softball teams use orange balls for practice? {Elicit that orange balls are easier to see.}

➤ Do you think orange balls help the softball player react more quickly to hitting the ball? {Accept any answer.}

➤ In today's activity, you will be testing your reaction time and determining ways to improve it.

Purpose for reading:
Student Text pages 334–35
Activity Manual pages 221–22

The student should read all the pages before beginning the activity.

Reaction Time

Your body continually reacts to your environment. Most of these reactions, such as your body temperature, are automatic. However, you can change the speed of some reactions. Many factors can affect your body's ability to react quickly. In this activity, you will test variables that can either increase or decrease your reaction time.

Problem

How does changing a variable affect my reaction time?

Procedure

1. Have a partner hold the top of the white strip of poster board. Hold your thumb and index finger on either side of the bottom of the strip without touching it.

2. When your partner lets go of the strip, try to catch it between your fingers as quickly as you can. Mark the place on the strip where your thumb and finger caught it. Measure the distance from the bottom of the strip to the mark. Record the measurement in your Activity Manual.

3. Repeat two times. Each time, mark with a different color or symbol. Average your measurements.

4. Change places with your partner and repeat.

5. Choose a color of poster board that you think will improve your reaction time. Test your time again using the colored strip of poster board. Record and average your measurements.

Process Skills
- Predicting
- Measuring
- Inferring
- Identifying and controlling variables
- Recording and interpreting data

Materials:
assorted 3 cm × 30 cm strips of poster board, one white and three different colors
centimeter ruler
additional items as needed
Activity Manual

334

 Reactions are "actions that we do quickly with minimal thought." They are generally a result of inherited or practiced traits. Our reactions to others are also a result of our inherited personalities and our practiced traits. A Christian's reactions should be loving, kind, helpful, honest, and friendly. As Christlike reactions are practiced, the person is more likely to react properly to others. [BATs: 3c Emotional Control; 4b Purity; 4e Honesty; 5b Giving; 5d Communication; 5e Friendliness]

6. Compare the results of the two tests. Explain how changing the variable of color affected your reaction time.
7. List three other variables that might change your reaction times.
8. Predict whether each change will increase or decrease your reaction time.
9. Test each of your variables and record the results.

Conclusions
- Were your predictions correct?
- Which changes increased your reaction time?

Follow-up
- List everyday activities that benefit from an increased reaction time.

335

Procedure
Provide time for the student to test his reaction time using both strips of poster board. After recording measurements, discuss how changing the color affected the reaction time.

➤ Did you notice a difference in your reaction times when you used the colored strip instead of the white strip of poster board?
➤ Did the colored strip increase or decrease your reaction time?
➤ What are some things other than the color of the strip that might affect your reaction time?
➤ Would these variables be more likely to increase or decrease your reaction time?

Allow the student to test additional variables as time allows.

Conclusions
Provide time for the student to evaluate the results and answer the questions.

Activity Manual
Activity—pages 221–22

Assessment
Rubric—TE pages A50–A53
Select one of the prepared rubrics, or design a rubric to include your chosen criteria.

SCIENCE PROCESS SKILLS
Identifying and controlling variables

➤ What was the main variable that you changed? *{the color of the strip}*
➤ What are other variables that you could not control? *{Possible answers: individual reaction times; benefit of practice; how the person felt}*
➤ Could these variables have been controlled? *{probably not}*
➤ What are some ways the effects of these uncontrolled variables could be minimized? *{Possible answers: do the tests over several days and average the results; add more people to the sampling group}*

➤ Which characteristic of the strip was changed for the second test? *{the color}*
➤ What other variables were changed in later tests? *{Answers will vary based on the student's choices.}*
➤ When you changed variables, were other variables created that you could not control? *{Answers will vary.}*
➤ Why is it important for scientists to identify and control as many variables as possible? *{Possible answer: Uncontrolled variables may give inaccurate results.}*

Chapter 14

Lesson 143 353

Lesson 144

Student Text pages 336–39
Activity Manual page 223

Objectives

- Recognize how the five senses interact with the nervous system
- Read diagrams for information
- Identify the nerves associated with hearing, sight, and smell
- Explain how the different senses communicate with the brain

Introduction

➤ What would it be like if you could not use some of your five senses?

➤ How would the lack of senses affect the way that you gather information? {Answers will vary.}

➤ The lack of which sense would probably most affect a scientist's conclusions? {sight} Why? {Most conclusions are drawn from observations that are based on sight.}

📖 Today we will be studying the five senses and seeing how they interact with the nervous system to keep us aware of God's world.

Purpose for reading:
Student Text pages 336–39

➤ Which part of the ear changes vibrations into nerve impulses?

➤ In which part of your skin is your sense of touch located?

Discussion:
pages 336–37

➤ Which system of our body helps all five senses function? {the nervous system}

➤ What are sound waves caused by? {vibrations}

 Refer the student to *The Ear* diagram on Student Text page 336.

➤ Which part of the ear vibrates as it receives sounds? {the eardrum}

➤ Which part of the ear changes the vibrations into nerve impulses? {the cochlea}

➤ Which three small bones transfer vibrations from the eardrum to the cochlea? {anvil, hammer, and stirrup}

➤ Which nerve transmits nerve impulses from the ear to the brain? {the auditory nerve}

➤ Which part of your body interprets the sound vibrations received by your ear? {the brain}

Interactions with the Nervous System

The Five Senses

Our five senses help us to be aware of the world around us. Without these senses, we would not be able to understand or appreciate God's creation. But our senses only gather information. The interaction of the senses and the central nervous system allows us to interpret the sensory information, or stimuli, that is gathered. All five senses can function only with the help of the nervous system.

Hearing

Sound waves, caused by vibrations, are funneled into the ear canal by the outer ear. The vibrations continue to move through the middle ear and inner ear, where the cochlea changes them into nerve impulses. Finally, the impulses reach the brain, which interprets them to let you know what sounds you are hearing. Without your brain to interpret the sounds, your ear would still receive sound waves, but the vibrations of your ear would have no meaning to you.

The Ear

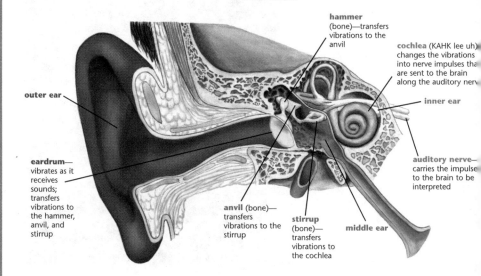

336

 Cochlear implants—Cochlear implants can help some deaf people convert sound waves into electric currents that stimulate the auditory nerve to send impulses to the brain. A cochlear implant functions similarly to the hair cells in the cochlea that normally change sound waves into the electrical impulses. The benefits of a cochlear implant vary from person to person, but these implants have helped many deaf people discern sounds that had been unknown to them previously.

Hearing and locating sounds—Play a game of Marco Polo within an established boundary. The person who is "It" must either have his eyes closed or be blindfolded. As he seeks for the other players, he calls out, "Marco." The other players move around within the boundaries and answer, "Polo." The person who is "It" must listen to the replies and move to tag another player. The player tagged will become the new "It."

The Eye

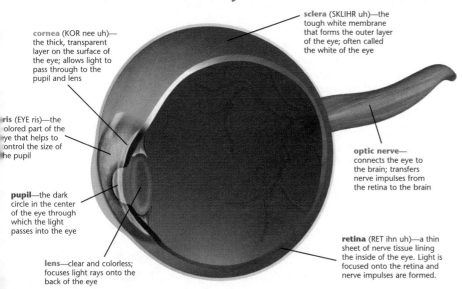

cornea (KOR nee uh)—the thick, transparent layer on the surface of the eye; allows light to pass through to the pupil and lens

iris (EYE ris)—the colored part of the eye that helps to control the size of the pupil

pupil—the dark circle in the center of the eye through which the light passes into the eye

lens—clear and colorless; focuses light rays onto the back of the eye

sclera (SKLIHR uh)—the tough white membrane that forms the outer layer of the eye; often called the white of the eye

optic nerve—connects the eye to the brain; transfers nerve impulses from the retina to the brain

retina (RET ihn uh)—a thin sheet of nerve tissue lining the inside of the eye. Light is focused onto the retina and nerve impulses are formed.

Sight

Without light you would not be able to see anything. When light bounces off objects, the parts of the eye work together to allow the brain to see an image. However, images received by the brain are upside down. The brain flips the images over and recognizes what you are seeing.

We speak of the eye as seeing. However, the eye only provides the means for the brain to receive sensory information. Only as the sensory receptor neurons in the retina collect information and send it along for the brain to interpret can we actually see.

337

Discussion:
pages 336–37

Refer the student to *The Eye* diagram on Student Text page 337.

➤ What must bounce off an object in order for you to see that object? {light}

➤ What is the purpose of the lens of the eye? {It focuses light rays onto the back of the eye.}

➤ In which part of the eye are nerve impulses formed? {the retina}

➤ What does an image look like on the retina? {The image is upside down.}

💡 What parts of the eye does light go through before the image is formed on the retina? {cornea, pupil, lens}

➤ Which nerve sends messages from the eye to the brain? {the optic nerve}

➤ Can your eye see without the brain's being involved? Why? {No; the eye only gathers the sensory information and sends it to the brain. The brain interprets the information and identifies what is seen.}

📖 Read Matthew 13:13–15.

➤ Our eyes and ears receive information that is not understood until our brains interpret the impulse. In the same way, we can hear and see things that God has for us to learn, but not understand them without the guidance and help of the Holy Spirit. [BAT: 6a Spirit-filled]

Direct a Demonstration

Demonstrate how the eye sees things upside down

Materials: glass fish bowl filled with water, black poster board, white poster board, candle

Make a small pencil hole in the middle of the black poster board. Stand the black poster board against one side of the bowl. Lean the white poster board against the other side of the bowl. Light the candle. Position the candle so that the light is shining through the hole in the black poster board. Darken the room. Adjust the white poster board until an image of the flame appears on it.

➤ Which part of the eye does the black poster board represent? {the pupil}

➤ Which part of the model represents the eyeball? {the bowl filled with water}

➤ Which part of the eye does the white poster board represent? {the retina}

➤ How does the image appear? {upside down}

➤ How is this similar to the eye? {The curved eyeball causes the light entering the eye to bend, inverting the image as it forms on the retina.}

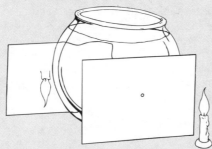

Chapter 14   Lesson 144

Discussion:
pages 338–39

Refer to the diagram on Student Text page 338 as each topic is discussed.

➤ Where are some of your taste buds located? *{inside bumps on your tongue}*

➤ Where are the sensory receptors for taste located? *{inside the taste buds}*

➤ Which one of your other senses is directly linked to your sense of taste? *{the sense of smell}*

➤ What type of cell detects odor particles that have entered the nose? *{olfactory receptor cells}*

➤ Which nerve sends impulses from the nose to the brain? *{the olfactory nerve}*

➤ How is your sense of touch different from the other four senses? *{It is the only sense located all over the body.}*

➤ Which part of your skin contains the nerve receptors for your sense of touch? *{the inner layer, or the dermis}*

➤ What types of things are detected by the sensory receptors in your skin? *{Possible answers: movement, pressure, temperature changes, pain}*

💡 People often wish that they did not feel the pain of a headache, a broken bone, or other injury. We often take medicine to lessen the amount of pain felt. Can you think of a reason why it is important that we feel pain? *{Possible answer: If we did not feel any pain, we might injure ourselves more severely.}*

Taste

Taste buds help us recognize different tastes and flavors. If you look inside your mouth, you will see lots of tiny bumps, called *papillae* (puh PILL ee) all over your tongue. Those bumps are *not* taste buds. The taste buds are located inside the bumps. Some bumps contain only a few taste buds, but others have more than one hundred taste buds.

Inside the taste buds are sensory receptors. The receptors receive the taste and send it along to the brain to be interpreted. The brain then decides what the taste is. However, your sense of taste is directly affected by your sense of smell. If you have a bad head cold that keeps you from being able to smell, you will probably notice that your food is not as tasty. Some foods have almost no taste without the sense of smell. This is why some people hold their noses when they take bad-tasting medicine.

Smell

The air contains many different odor particles. As you breathe in, air enters into the nasal cavity. Inside the nasal cavity, the odor particles first pass through a thick layer of mucus. Then they float up to the top of the nasal cavity.

Olfactory (ohl FAK tuh ree) *receptor cells* detect the particles and send impulses to the olfactory nerve. The olfactory nerve sends the impulses to the brain. The brain interprets the message and identifies the smell.

Touch

You use your sense of touch to feel things. Each of the other four senses is located in just one certain place. But the sense of touch is located all over your body. The outer skin, or *epidermis* (EP ih DUHR mihs), however, is not responsible for your sense of touch. There are no nerve receptors in the epidermis.

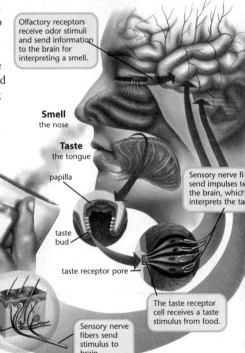

338

SCIENCE BACKGROUND

Tasting and smelling—The senses of taste and smell are closely linked and are often referred to as the chemical senses. Adults have about 10,000 taste buds located on the tongue, cheek, palate, and throat. The taste receptors respond to chemicals dissolved in the saliva. Humans have about 10 million olfactory receptors that respond to odor chemicals dissolved in the mucus layer.

Taste buds are usually replaced about every 10–14 days, and olfactory receptors are replaced about every 4–8 weeks. These receptor cells are the only sensory cells that the body replaces regularly. However, the number of taste buds declines as people age.

Scientists have found that people can recognize at least 5,000 different tastes and more than 10,000 different odors. Many memories are often associated with specific smells and tastes.

Science and the Bible

The book of Genesis tells us that Isaac was blind in his old age. When his son Jacob came to him pretending to be Esau, Isaac's senses deceived him. Isaac thought he smelled Esau because Jacob wore Esau's clothes. Jacob's hands, which had been covered with hairy goatskin, felt like Esau's hairy hands. Even the meat had been prepared to taste like the meat Esau usually brought. Although Isaac recognized the voice of Jacob, his other senses deceived him, and he gave Jacob the blessing meant for Esau.

Your sense of touch originates in the *dermis*, or inner layer of skin. The dermis is filled with tiny sensory receptors. Some of these receptors detect movement and pressure. Others recognize temperature changes or detect pain. These sensory receptors send messages to the brain about what you touch. The brain processes the information and sends messages back, letting you know how things feel.

You might have noticed the pressure of your watch when you first put the watch on. In just a short while, though, you do not even feel your watch on your arm. When your brain constantly receives the same pressure signals from your skin, it becomes used to the pressure. God designed your senses to adapt to the environment around you to keep from being overloaded with stimuli. Have you ever noticed a certain odor, such as air freshener, when you entered a room? After you have been in the room for several minutes, though, the odor is not as noticeable.

Your body has many different types of receptors that send specific messages to the brain. These receptors allow the brain and senses to work together, keeping us aware of the world around us. However, we cannot completely trust our senses. Sometimes the information we gather is inaccurate. For example, optical illusions can confuse our sight perception. Only one source of information—God's Word—is completely accurate and trustworthy.

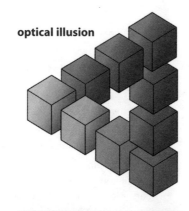

optical illusion

> 1. Which nerves are associated with sight and hearing?
> 2. Which other sense is closely associated with your sense of taste?
> 3. How is your sense of touch different from your other senses?

✓ QUICK CHECK

339

Direct an Activity

Materials: two different brands of soft drinks, popcorn, or potato chips

Remove or hide the labels so that only the person directing the challenge knows which product is which. Allow the student to predict which brand he will prefer. Record the predictions. Provide time for the student to taste each sample. Record his sample preferences. Reveal which brand is which, and compare the predictions with the results.

Discussion:
pages 338–39

➤ **Why do you not feel the pressure of your watch after you have worn it for a while?** *{The sense of touch adapts to the pressure because the brain has become used to the pressure.}*

➤ **Why do our senses adapt to the environment around us?** *{Possible answer: to keep us from being overloaded with stimuli}*

➤ **Is the information received by our senses always accurate?** *{no}* **Give an example of a time when the information received may not be accurate.** *{Possible answer: optical illusions}*

📖 **What is the only completely accurate and trustworthy source of information?** *{God's Word}* [BAT: 8b Faith in the power of the Word of God]

📖 Refer the student to *Science and the Bible*.

➤ **Which of Isaac's senses was impaired?** *{his eyesight}*

➤ **Which of his senses were working?** *{hearing, smell, touch, taste}*

➤ **Isaac's sense of smell was working accurately, but it still deceived him. Why?** *{Isaac could smell that the clothing was Esau's, but he did not realize that Jacob was wearing it.}*

➤ **What other senses of Isaac's were deceived?** *{touch and taste}* **Explain.** *{Jacob's arms were covered with skins to feel like Esau's. The meat was prepared to taste the way Esau would have cooked it.}*

Answers

1. The optic nerve is associated with sight. The auditory nerve is associated with hearing.
2. the sense of smell
3. It is the only sense located all over your body.

Activity Manual
Reinforcement—page 223

Chapter 14   Lesson 144   357

Lesson 145

Student Text pages 340–41
Activity Manual page 224

Objectives

- Predict and identify areas of the body that are the most sensitive to touch

Materials

- See Student Text page
- copy of TE page A40, *Touch Tester* for each group

Introduction

➤ Have you ever felt a fly crawling on your arm?

➤ The sensory receptor cells in your arm registered the touch of the fly and transmitted impulses to your brain about the fly's movement. Sensory receptors are located all over your body, but some areas of your body have more receptors than other areas have. In this activity you will determine which place on your body is the most sensitive to touch.

Purpose for reading:
Student Text pages 340–41
Activity Manual page 224

The student should read all the pages before beginning the activity.

Touch Tester

ACTIVITY

The nerve endings in the skin contain neurons that send messages to the brain about the things we touch. God made some areas of our bodies more sensitive than other areas by giving them more neurons. In this activity you will test and compare the sensitivity of different places on your body.

Process Skills
- Predicting
- Measuring
- Inferring
- Recording data

Problem

Which place on your body—the arm, finger, palm, or neck—is most sensitive to touch?

Procedures

Note: This activity uses English rather than metric measurements.

1. Assemble the Touch Tester that your teacher gives you.
2. Predict the sensitivity of the areas of your body listed in your Activity Manual. Number from 1–4, with 1 being the place on your body you think will be the most sensitive.
3. With your eyes closed or blindfolded, have your partner begin testing the areas of your body listed.
4. To use the touch tester, begin with both toothpicks together at the 0 mark. Gently press the toothpicks on the skin. Determine the number of toothpicks that are felt. If only one toothpick is felt, slide the toothpick to the next mark and test the skin again. Continue sliding the toothpick and retesting until both toothpicks are felt.
5. Record the distance between the toothpicks.

Materials:
scissors
2 toothpicks
tape
blindfold (optional)
touch tester
Activity Manual

340

TEACHER HELPS

When assembling the touch tester, make sure that the placement of the tape does not interfere with the movement of the centimeter strip.

This activity works best when done in groups of two.

6. Repeat steps 3–5 to test the sensitivity of each place on your body.

 7. Number the places again based on your measurements. Write 1 next to the place with the smallest measurement.

Conclusions
- Were your predictions correct?
- Think how you use each part of the body that you tested. Why do you think God made some places on your body more sensitive than other places?

Follow-up
- Research to find how the distance between the toothpicks compares with the number of neurons in each area of skin.
- Test other areas of your body, such as your face, ear, the top of your foot, and the bottom of your foot.

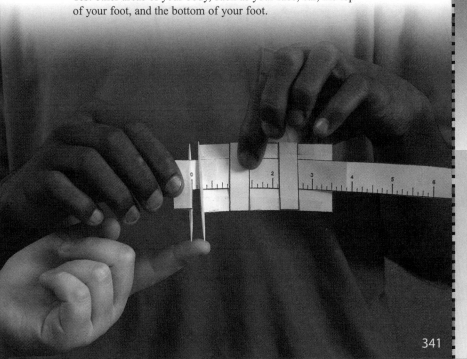

Science Process Skills

Predicting—Predicting is forming an idea of a future event based on previous knowledge.

➤ On what did you base your predictions? *{Answers will vary, but elicit that the student's predictions were based on previous knowledge of his own body's sensitivities.}*

➤ After you tested the first place on your body, did you still think your predictions would be accurate? Why? *{Answers will vary.}*

➤ Suppose you were asked to predict something for which you have no background knowledge. Would you be making a good or reasonable prediction? Why? *{No; it would be a guess since it is not based on observation or knowledge.}*

➤ What could you do to make your prediction more valid? *{Possible answers: conduct research; make observations of similar situations}*

➤ Why do you think scientists research their topics and the research results of other scientists extensively before planning and making their own predictions about their own experiments? *{Since predicting implies background knowledge, scientists must gain this knowledge before they can anticipate a result.}*

Procedure
Reproduce the touch tester pattern. Provide time for the student to assemble the touch tester.

Remind the student to predict his sensitivity before he begins testing.

Conclusions
Provide time for the student to evaluate the results and answer the questions.

Activity Manual
Activity—page 224

Assessment
Rubric—TE pages A50–A53
Select one of the prepared rubrics, or design a rubric to include your chosen criteria.

Lesson 146

Student Text pages 342–45
Activity Manual pages 225–28

Objectives
- Differentiate between short-term memory and long-term memory
- Identify two categories of long-term memory
- Describe some characteristics of REM sleep

Vocabulary
memory
short-term memory
long-term memory

Materials
- prepared list of 10 two-digit numbers

Introduction

Display the list of numbers and give the student one minute to study the numbers. Remove the list and direct the student to write what he remembers. Check the student's list for accuracy. Direct the student to turn over his list and set it aside. Partway through the lesson, direct the student again to write what is remembered from the list without looking at it. Check the list for accuracy. The results will be used in the discussion for Student Text page 343.

Purpose for reading:
Student Text pages 342–45

➤ What section of the brain seems to be necessary for developing long-term memories?
➤ What is REM sleep?

Discussion:
pages 342–43

➤ What is *memory?* {the ability to remember or recall something}

📖 Why do you think God told the Israelites to remember what He had done for them? {Possible answers: to give glory to Him; to remind them of their need of God; to give courage when things were difficult} [BATs: 7b Praise; 7e Humility; 8d Courage]

➤ How do scientists think memories are formed? {by neurons communicating with each other}

➤ How is short-term memory different from long-term memory? {Short-term memory is stored only temporarily. Long-term memory is more permanent.}

➤ What are some examples of short-term memories? {Possible answers: grocery list, phone number, score of a basketball game}

Memory

Memory plays a very important role in our daily lives. Suppose you woke up one morning and could not remember your name, how to tie your shoes, or where you put your homework assignment. You would have a difficult time functioning that day. But God has given your brain the ability to store and retrieve information. In fact, God tells us to remember. The ability to remember is called **memory.** In Deuteronomy 8:2 God told the Israelites to remember how He brought them out of Egypt and protected them during their forty years of wandering.

What kinds of things do you remember? You remember sounds, smells, things that you saw and, hopefully, things that you have studied. The entire brain is involved in making and keeping memories. Scientists think that memories are formed by neurons communicating with each other. They identify a memory as a specific pathway from neuron to neuron. The brain continually makes new connections between neurons as you learn and process new information.

Some memories are kept longer than other memories. **Short-term memory** stores information only temporarily. This information might be a phone number, a grocery list, or the

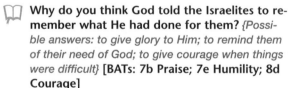

SHORT-TERM MEMORY   LONG-TERM MEMORY

number of points scored in a basketball game. Usually when you look up a phone number, your short-term memory remembers the number only for a little while. However, the more often you see or hear something, the longer you will remember it.

Long-term memory can store information for months, years, or even for a lifetime. These memories can be a mixture of sensory information as well as facts and experiences. Some scientists think that emotions play a part in transferring information to long-term memory. If something you experience was unpleasant, you want to remember to avoid it. On the other hand, if the experience was comforting or pleasurable, you usually want to repeat it.

342

SCIENCE BACKGROUND

Memories—The brain classifies memories in many different ways, sometimes associating a memory with one of the senses. Certain sounds, smells, or sights can sometimes bring back a specific memory quickly. The more times a memory is reviewed and stored in the brain, the more new neural pathways form, making it easier to access the information again. An infant has many more neurons in his brain than he will have in adulthood. Neurons that are not used or stimulated to form pathways often die or cease functioning.

SCIENCE MISCONCEPTIONS

When an item stored in long-term memory is "forgotten," that memory is not necessarily gone. It may be that the person is not accessing that memory through the same neural pathway that stored the memory.

360   Lesson 146

Scientists do not know exactly how some information in short-term memory transfers into long-term memory. However, an area inside the temporal lobe, called the *hippocampus* (HIP uh KAM puhs), seems to be necessary for making new long-term memories. Scientists and doctors have found that if damage occurs to this area of the brain, people can still remember old long-term memories but cannot store any new long-term memories. Their memories of current events or new people and places are held only in short-term memory.

Long-term memories can also be described as *declarative* or *procedural*. Declarative memories involve any knowledge that requires you to recall specific facts. These facts include your friend's birthday, vocabulary words, and definitions. Procedural memories include remembering how to ride a bike, play a violin, and paint a picture.

Scientists are not exactly sure how we learn. But scientists do know that different people learn in different ways and that the ability to learn changes as people grow older. Though some things are easier to learn as children, other information requires more maturity to understand. We know that God never intends for us to stop learning. We should say with David in Psalm 143:10, "Teach me to do thy will; for thou art my God: thy spirit is good; lead me into the land of uprightness."

SCIENCE BACKGROUND

Memory loss—*Amnesia* is a general term used to describe memory loss. A person may experience amnesia if the brain is injured. Depending on the cause, amnesia may be temporary or permanent. This memory loss could be caused by trauma, diseases, infections, drug and alcohol abuse, or by reduced blood flow to the brain.

Discussion:
pages 342–43

➤ What are some examples of long-term memories? {Possible answers: music or sports skills, names of parents and best friends, major accomplishments}

➤ Which area in the brain seems to be responsible for transferring short-term memories into long-term memory? {the hippocampus}

➤ What can happen when the hippocampus is damaged? {New long-term memories cannot be made. Old long-term memories are remembered, but all new experiences are stored in short-term memory only.}

➤ What are two different types of long-term memories? {declarative and procedural}

➤ How are declarative and procedural memories different? {Declarative memories include remembering specific facts. Procedural memories include remembering how to perform certain skills.}

💡 Why would remembering how to perform a skill be called a procedural memory? {A procedure is a way to do or perform something.}

➤ Do all people learn the same way? {no}

💡 What are some ways people learn or receive information to be remembered? {Possible answers: reading, hearing, doing}

➤ What happens to the ability to learn as a person gets older? {It changes.}

📖 Why do you think God may allow learning abilities to change with age? {Accept reasonable answers. Elicit that the types of things we need to learn and remember change.}

➤ Do you think God intended for anyone to get old enough that he would no longer need to learn new things? {no}

➤ The Bible reminds us that a wise man will continue to hear instruction and gain knowledge (Job 32:4–9; Prov. 1:5; 8:33).

Discuss the results of the tests performed in the Introduction.

➤ Did your results change from the first test to the second test? {probably yes}

➤ Which type of memory did you use as you took the first test? {short-term}

➤ Which type of memory did you use as you took the second test? {long-term, or declarative}

💡 How could you have studied the list differently if you had known you were going to have the second test? {Answers will vary.}

Discussion:
pages 344–45

➤ **What does your nervous system do while you sleep and your body rests?** {The nervous system keeps the autonomic functions going and appears to do some sensory housekeeping.}

💡 **What are some things you do to help your parents with housekeeping?** {Possible answers: put things in their proper places; clean; throw away unusable things}

💡 **Do you think you use all the sensory information that comes into your body each day?** {no} **Why?** {Possible answer: Much of the information is unimportant.}

💡 **What does *sensory housekeeping* mean?** {Accept reasonable answers. Elicit that our brains also put information into their proper places and clear out unnecessary information.}

💡 **What are three different types of sleep stages?** {light sleep, deep sleep, and REM sleep}

➤ **Which stage of sleep does a person experience first?** {light sleep}

➤ **Does your heart rate increase or decrease as you begin to fall asleep?** {decrease}

➤ **Which part of the nervous system regulates your heart rate and breathing as you sleep?** {the autonomic nervous system}

➤ **Why is it difficult to wake someone up out of deep sleep?** {The body is very relaxed.}

➤ **What are some characteristics of REM sleep?** {Possible answers: The brain is very active. The eyes move quickly back and forth. Some muscles twitch.}

➤ **Why is REM sleep important to the body?** {Possible answers: It helps the brain develop and helps the brain sort through and organize information received throughout the day.}

➤ **In which stage of sleep do most dreams occur?** {REM}

💡 **What characteristics of REM sleep might you observe in a sleeping animal?** {Possible answers: muscles in ears and paws twitching; eyes moving back and forth under the eyelids}

➤ **What is an *electroencephalograph* (EEG)?** {a machine used to study how the brain works}

➤ **What does an EEG measure?** {electrical impulses of neurons in the brain}

💡 **Why can electrical impulses be used to measure the brain's functions?** {Neurons communicate with each other through electrical impulses.}

Sleep and the Nervous System

When you sleep your body rests, but your nervous system remains very active. Not only does the nervous system maintain the autonomic functions such as breathing and heartbeat, it appears to do some sensory housekeeping as well.

Scientists have identified several different stages of sleep. The brain remains active throughout all the stages of sleep, but its level of activity changes. When you first fall asleep, your autonomic nervous system slows down your heart rate and breathing. Your body prepares to rest, and you enter the first stages of light sleep. Later, in deep sleep, your body becomes very relaxed. Waking up out of deep sleep can be quite difficult.

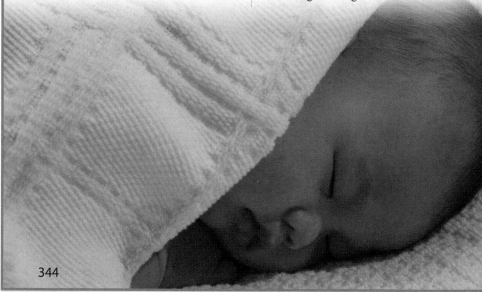

344

People go back and forth between periods of light sleep and deep sleep throughout the night. Scientists have found that there is also another stage of sleep where the brain is very active. The eyes move back and forth quickly, even though the eyelids are closed. Often a person's muscles begin to twitch, and the brain seems to be as active as if the person were awake. This stage of sleep is called Rapid Eye Movement, or *REM* sleep. REM sleep occurs only after the body has gone through periods of light and deep sleep.

REM sleep is very important to our bodies. Some scientists think that this stage of sleep helps our brains to develop. Infants usually spend about 50 percent of their sleep in REM sleep. Scientists also think that REM sleep may be one of the ways that our brains sort through and organize all of the

SCIENCE BACKGROUND

Sleep medicines—Some sleep medicines used to treat insomnia or other health problems cause the body to skip REM sleep and spend more time in deep sleep. When the person is no longer using the medicine, the body sometimes tries to catch up on missed REM sleep, thus causing more problems. Each stage of sleep is important for the body's health.

information received throughout the day. Most dreams occur during REM sleep. A person who can remember details about a dream probably woke up during REM sleep.

Scientists use a machine called an *electroencephalograph* (ih LEK troh en SEF uh luh graf), or EEG, to study how the brain works while people sleep. The EEG measures the electrical impulses produced by the neurons in the brain. By studying sleep patterns, scientists have noticed that both the quality and the quantity of sleep are important for a person's health. Failing to get adequate rest can affect every area of a person's life. Most scientists also believe that getting a good night's sleep improves a person's memory.

Children usually need about nine to ten hours of sleep each night. Adults require a little less—about seven to eight hours each night. While we are sleeping, our bodies rest and can work on other functions, such as repairing cuts and bruises. Sleep also gives children's bodies time to grow.

Our brains filter sounds while we sleep so that familiar noises do not bother us. The brain can also allow the body to rest while it stays alert to certain noises. Just ask any mother how long it takes for her to awake to her child's cry at night. The brain's multiple abilities show evidence of the wonderful Creator.

a child connected to an EEG

Quick Check
1. Give an example of a short-term memory.
2. How is declarative memory different from procedural memory? How are the two similar?
3. Why do scientists think that REM sleep is important for the body?

Discussion:
pages 344–45

➤ **How many hours of sleep do children need?** *{nine to ten}*

➤ **What are some benefits of getting a good night's sleep?** *{Possible answers: health, improved memory; The body works on other functions such as repairing cuts and bruises. Children's bodies grow.}*

💡 **Why might children need more sleep time than adults do?** *{Possible answers: A child's body is growing. Children have more sensory information to "housekeep."}*

💡 **Why does your body need more sleep when you are sick or injured?** *{The body must repair itself. During sleep, many body functions are slowed down and are not attended to, so the body's energy can be more concentrated on healing.}*

💡 **Why can someone who lives near a railroad track sleep soundly as a train passes by, but an overnight guest might be awakened by the sound of the train?** *{The brain filters out familiar noises while we sleep. The person who lives there is used to the sound of the train, but his guest is not.}*

📖 Emphasize the amazing skills that God designed our brains to do.

Answers

1. Possible answers: phone number, grocery list, basketball score
2. Declarative memory involves the recall of specific facts. Procedural memory involves remembering how to do a certain skill. Both are long-term memories.
3. REM sleep helps the brain develop and may also be one of the ways the brain sorts through and organizes all the information received throughout the day.

Activity Manual
Reinforcement—page 225

Bible Integration—page 226
This page discusses things God tells Christians to remember.

Expansion—pages 227–28

Chapter 14

Lesson 146

Lesson 147

Student Text pages 346–49
Activity Manual pages 229–30

Objectives

- Compare the nervous system and the endocrine system
- Identify the function of some glands in the endocrine system
- Identify some common nervous system disorders
- Recognize some of the problems resulting from drug abuse

Vocabulary

hormone
endocrine gland
target cell
hypothalamus
pituitary gland
epilepsy
multiple sclerosis
Parkinson's disease
Alzheimer's disease

Materials

- padlock with correct key or key to the classroom door
- several keys that will not unlock the padlock or the door

Introduction

Distribute the keys and direct the students to try to unlock the lock.

➤ **Did every key work in the lock? Why?** {no; Not all the keys were made to fit that lock.}

➤ Only the key made for that lock will unlock it. Today we will be learning about some chemical messages used by our bodies. These messages and the cells that they affect are somewhat like locks and keys. The chemical messages can affect or "unlock" only certain cells.

Purpose for reading:
Student Text pages 346–49

➤ What is the hypothalamus?
➤ What are some nervous system disorders?

Discussion:
pages 346–47

➤ **What are two ways that the endocrine system differs from the nervous system?** {It uses chemical messengers instead of electrical impulses, and it works more slowly than the nervous system.}

➤ **What are hormones?** {the chemical messengers used by the endocrine system}

➤ **Where are hormones produced and released?** {in endocrine glands}

The Endocrine System

The nervous system is directly connected to another system of the body, the endocrine (EN duh krin) system. Together these systems control all the functions of the human body.

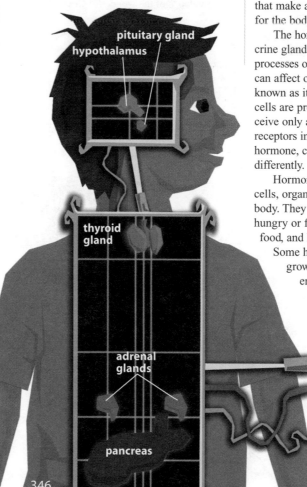

The endocrine system works more slowly than the nervous system. Instead of electrical impulses, the endocrine system uses chemical messengers called **hormones**. Most of the body's hormones are produced in the **endocrine glands**, special groups of cells that make and release the hormones for the body.

The hormones released by the endocrine glands speed up or slow down the processes of certain cells. Each hormone can affect only a specific group of cells, known as its **target cells**. These target cells are programmed by genes to receive only a certain hormone. Special receptors in the target cell bind to the hormone, causing the cell to function differently.

Hormones influence almost all the cells, organs, and functions of your body. They regulate whether you feel hungry or full, how your body uses food, and how you handle stress. Some hormones control your growth process, body temperature, and even your sleep.

The **hypothalamus** (HY poh THAL uh muhs) is a group of special cells near the base of the brain. Neurons in the hypothalamus help regulate the pituitary (pih TOO ih TEHR ee) gland, which is located just underneath it. Even

SCIENCE BACKGROUND

Hormone production—Many other body organs, including the kidney and stomach, also secrete hormones. The kidney produces a hormone that stimulates red blood cell development in bone marrow. The stomach produces a hormone that increases the amount of hydrochloric acid in the stomach. Another hormone produced by the stomach causes the constriction of stomach muscles.

Pancreas—The pancreas makes some hormones that are used to help digest food. The insulin secreted by the pancreas causes cells to absorb sugar from the blood. If the pancreas does not produce enough insulin, the blood will contain too much sugar, resulting in *hyperglycemia*, also known as diabetes. If the pancreas produces too much insulin, the cells absorb too much sugar, resulting in low amounts of sugar in the blood. This is known as *hypoglycemia*.

364 Lesson 147

though it is only about the size of a pea, the **pituitary gland** is very important for your body. Sometimes called the master gland, the pituitary gland produces hormones that control other glands in the endocrine system. This gland also produces a growth hormone that helps bones grow and develop.

Many times the amount of hormones released also depends on the circumstances in a person's life. Your body has two *adrenal* (uh DREE nuhl) *glands* located on top of your kidneys. These glands help your body respond to stressful or dangerous situations. They release a substance called *adrenaline* (uh DREN uh lin), which increases your blood pressure and heart rate during stress. You may have heard of someone who showed great strength and endurance in a dangerous situation. This strength was possible because of the hormones released by the adrenal glands. These hormones also increase a person's heart rate and cause a person to tremble when he is nervous or scared.

The *pancreas* is located near the stomach, and it releases the hormone insulin. *Insulin* helps to control the amount of sugar in the bloodstream. If a person's pancreas does not make enough insulin, a disease called diabetes could develop.

Your *thyroid gland* is located in your neck, just below your voice box. This gland is shaped like a butterfly. It controls how your body uses food to make energy. It also influences your body's growth and development. If the

Gigantism is caused by an overproduction of the growth hormone.

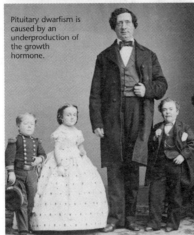
Pituitary dwarfism is caused by an underproduction of the growth hormone.

thyroid releases too many hormones, a person may become nervous or hyperactive or may lose too much weight. When the gland does not release enough of the hormones, the person often gains weight and feels tired all the time. The pituitary gland and the hypothalamus control these glands and many others.

347

In the past, people who exhibited dwarfism or gigantism were sometimes part of circus or carnival sideshows. Being on exhibit was often their only means of income.

Charles S. Stratton, known as Tom Thumb, traveled with P. T. Barnum. Tom Thumb was only 33 inches tall. He was a great attraction throughout the U.S. and Europe in the 1800s.

As an adult, Ella Ewing was 8 feet 4.5 inches tall. She worked as part of several museum displays and traveled with the Barnum & Bailey circus. Ella is thought to have been the tallest woman, but she is not recorded in the *Guinness Book of World Records*.

The tallest person recorded by the *Guinness Book of World Records* is Robert Wadlow (1918–1940) of Alton, Illinois. By age 10, Robert was 6 feet, 5 inches tall and wore size 17½ shoes. His growth was caused by an overactive pituitary gland. When he died in 1940, he was 8 feet, 11.1 inches tall. He did not work for a circus or carnival.

The tall man in the top photo on Student Text page 347 is Robert Wadlow. In the bottom photo, Tom Thumb appears on the far right.

Discussion:
pages 346–47

➤ **What are two ways that hormones can affect cells?** {They speed up or slow down certain processes.}

➤ **Are all the cells in the body affected by each hormone?** {no} **Why?** {Each hormone can affect only a certain group of cells, its target cells.}

➤ **What programs these target cells to receive only a certain hormone?** {genes}

➤ **What are some life processes regulated by hormones?** {Possible answers: hunger; how the body uses food; how the body handles stress; growth; body temperature; sleep; weight}

➤ **What group of cells near the base of the brain connects the nervous system with the endocrine system?** {hypothalamus}

➤ **What does the hypothalamus do?** {helps regulate the pituitary gland}

➤ **Why is the pituitary gland sometimes called the master gland?** {It makes hormones that control the other glands in the endocrine system.}

➤ **What is another special hormone produced by the pituitary gland?** {the growth hormone}

➤ **Which glands help the body respond to dangerous or stressful situations?** {the adrenal glands}

➤ **What hormone is released during times of stress?** {adrenaline}

➤ **Which hormone does the pancreas produce to regulate the amount of sugar in the bloodstream?** {insulin}

➤ **What disease can occur when the pancreas does not produce enough insulin?** {diabetes}

➤ **Which gland is shaped like a butterfly?** {thyroid}

➤ **What does the thyroid gland control?** {how the body uses food}

Refer the student to the pictures on Student Text page 347.

➤ **What causes gigantism?** {overproduction of the growth hormone}

➤ **What condition is caused when the body does not produce enough growth hormone?** {pituitary dwarfism}

Discussion:
pages 348–49

- Why do our bodies not always work in the ways God designed them to function? *{Answers will vary, but elicit that sin affects all of God's creation, including our bodies.}*

➤ What causes a disorder? *{The body fails to function as it should.}*

➤ Some disorders are called *diseases*. What is different between these "diseases" and real diseases such as chickenpox or colds? *{These "diseases" are not contagious. They are malfunctions, or disorders, of the body. Chickenpox and colds are contagious.}*

➤ Which nervous system disorder happens when neurons in the brain send their impulses too quickly and at irregular rates? *{epilepsy}*

➤ Many different things can cause a person to have a seizure. What is noticeable about the seizures of someone with epilepsy? *{The seizures are repeated and usually have a regular pattern.}*

➤ Which part of neurons does the disease multiple sclerosis destroy? *{the myelin sheath covering the axon}*

➤ What occurs in places where the myelin sheaths are destroyed? *{Nerve impulses are stopped and cannot continue to move.}*

💡 Why might a person with Parkinson's disease have difficulty eating or writing? *{Parkinson's disease causes damage to brain cells that control movement.}*

➤ What is the most noticeable symptom of Parkinson's disease? *{muscle tremors}*

➤ What do Parkinson's disease and Alzheimer's disease have in common? *{Both diseases cause damage to cells in the brain.}*

➤ How are they different? *{Parkinson's disease damages cells that control movement, but Alzheimer's disease affects thinking processes.}*

💡 What are some possible causes of nervous system diseases? *{Possible answers: heredity; head and back injuries; improper development of the nervous system before birth; drug abuse; unhealthful habits}*

➤ How do drugs such as cocaine and marijuana affect the body? *{They change the way neurons in the brain send and receive information.}*

➤ Why might a person addicted to a drug such as cocaine keep taking more of the drug? *{The drug changes the way the neurons normally work and causes his body to want more of the drug even though it is harming him.}*

366 Lesson 147

Disorders and Drugs

Both the nervous system and the endocrine system are extremely important to our bodies' health. God designed our bodies in a wonderful and marvelous way. The Bible says that we are "fearfully and wonderfully made" (Ps. 139:14). However, because of man's sin, our bodies do not always work in the ways God designed them to function.

Sometimes disorders are called diseases. But you cannot catch these diseases. These disorders occur when the body fails to function as it should. Sometimes doctors can treat the symptoms of a disorder. But often they do not know the causes or the cures for the disorders.

Epilepsy (EP uh LEP see), often called seizure disorder, occurs when the neurons in the brain send their electrical impulses too quickly and at an irregular rate. Other conditions besides epilepsy can cause seizures. However, a person with epilepsy has repeated seizures, usually of a similar pattern. Doctors can prescribe medicine that helps to control the seizures, but they have not discovered a cure.

Another disease, **multiple sclerosis**, destroys the myelin coating that covers the axon in some neurons. This causes the neurons to "short-circuit" so that the impulses cannot keep moving along. The symptoms that a person with this disease has depend on the location of the damaged nerves. People with multiple sclerosis may experience muscle weakness, paralysis, or loss of vision.

Parkinson's disease and Alzheimer's disease are two diseases of the nervous system that occur mainly in elderly people. **Parkinson's disease** causes damage to certain brain cells that control movement. This disease can cause a person's head, arms, and hands to tremble. A person who has Parkinson's disease may have trouble keeping his balance and doing simple tasks such as eating.

Alzheimer's (ALTS hy muhrz) **disease** also destroys brain cells, but in a different way. This disease affects thinking processes. At first, a person with Alzheimer's disease usually has trouble with short-term memories. Later, the person may lose the ability to learn new information or to reason. A person with this disease may not be able to recognize family members and friends. Many people with this disease also suffer from depression and anxiety.

Scientists and doctors are searching for cures for these and other nervous system diseases. Some nervous system problems may be inherited or may happen because of head or back injuries. For some people, the nervous system does not develop properly before birth. Sometimes drug abuse or unhealthful habits cause or intensify nervous system disorders.

348

SCIENCE BACKGROUND

Disease and age—Although Parkinson's disease and Alzheimer's disease occur mainly in older people, there are cases of both diseases occurring in younger people. Alzheimer's disease usually does not begin until after age 65, but it may begin as early as age 40. Parkinson's disease usually does not affect people younger than age 30, but it can develop at any age.

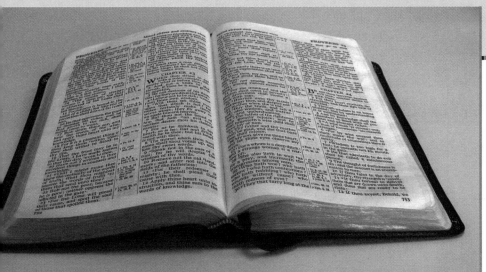

When you think of drug abuse, you probably think first of illegal drugs. Drugs such as cocaine and marijuana are harmful to the body. These types of drugs change the way that neurons in the brain send and receive information. Drugs affect the nervous system in many different ways, and some drugs can be addictive. By altering how the neurons work, some drugs make a person's body want to have more of the drug. The person then continues taking the drug even though it actually harms him and could be fatal.

But not all drug abuse happens with drugs such as cocaine, marijuana, or heroin. Some athletes take additional hormones called steroids to make themselves stronger or more muscular. However, with the improved performance may come dangerous side effects, such as seizures and heart attacks. Even legal drugs such as cold and fever medicines can be abused. Any time a person uses a medicine in excess or in a way that it is not meant to be used, he could be creating problems for his body. These problems may show up immediately or may not appear for several years. Christians need to remember that their bodies are temples of God (I Cor. 6:19) and do not belong to them. Everything that a Christian does should be to the honor and glory of God (I Cor. 10:31).

QUICK CHECK
1. Why are the hypothalamus and pituitary gland important to the interaction of the nervous and endocrine systems?
2. What is a disorder?
3. How can legal drugs sometimes be abused?

349

Direct an ACTIVITY

Provide an opportunity for students to visit patients in a nursing home or hospital (especially those patients suffering from nervous system disorders). Remind the students that even though they may not know the patients that they visit, a few minutes of cheery conversation and perhaps some gospel songs or hymns can be a good way to witness about Christ. [BAT: 5c Evangelism and missions]

Discussion:
pages 348–49

➤ Why do some athletes take steroids or other similar drugs? *{to make themselves stronger or more muscular}*

➤ Steroids and other similar drugs can temporarily improve a person's athletic performance. What are some side effects that may occur after taking drugs like these? *{seizures, heart attacks}*

💡 Why might some people decide to use drugs? *{Possible answers: peer pressure; thinking it won't hurt them; curiosity; rebelling against authority}*

📖 Is taking drugs a wise choice? *{no}* Why not? *{because God commands us to take care of our bodies, and drug abuse destroys the body}* [BAT: 3d Body as a temple]

➤ Besides the fact that a Christian's body is the temple of God, why would using drugs be dishonoring to God? *{because everything we do should glorify God}*

➤ Drug abuse is not limited to cocaine, heroin, or steroids. Non-prescription medicines and other things that are not even medicines are sometimes abused by being used in ways that they were not intended to be used.

Answers
1. Neurons in the hypothalamus regulate the pituitary gland, and the pituitary gland produces hormones that control all the other glands in the endocrine system.
2. A disorder is a condition that occurs when the body does not function properly.
3. Legal drugs, such as cold and fever medicines, can be abused any time a person uses them in excess or in ways that they are not meant to be used.

Activity Manual
Reinforcement—page 229

Study Guide—page 230
This page reviews Lessons 146–47.

Assessment
Test Packet—Quiz 14-B
The quiz may be given any time after completion of this lesson.

Lesson 148

TE pages A41–A42

Objectives

- Identify some common categories of drugs
- Explain how some types of drugs affect the nervous system
- List some biblical reasons for not taking drugs

Materials

- copy of TE pages A41–A42, *Effects of Drug Abuse* and *Keeping God's Temple Clean*, for each student

Note: This lesson is designed as an interactive discussion alerting the student to the effects and dangers of abusing drugs or other substances. The focus of the lesson should be the truth that the body of a Christian is the temple of God and that as such, the body should be kept pure.

The material discussed is not all-inclusive. You may choose to incorporate your own thoughts and additional Scripture or resources as appropriate for your student.

Introduction

Drugs can be beneficial or deadly. Most drugs are used to treat diseases or control pain. Christian young people may face temptation to experiment with drugs. Peer pressure and lack of knowledge about the seriousness of drug abuse could lead a person to submit to temptation.

Discussion:
TE pages A41–A42

💡 **What does *abuse* mean?** *{Answers will vary, but elicit that abuse can mean to use something in a way other than for its intended purpose.}*

Effects of Drug Abuse

Name _____

Category	Common drugs	Effects of drug	Possible results of abuse
Stimulants	Cocaine Crack Methamphetamine Caffeine Nicotine	Speeds up the central nervous system Interferes with the sending of nerve messages Increases alertness and energy	Shortness of breath, nausea, and seizures Decreased learning and memory Increased risk of heart attack, high blood pressure, internal bleeding, and nervous system disorders Death
Depressants	Heroin Morphine Alcohol Codeine	Slows down (depresses) the central nervous system May bring a pleasurable feeling Blocks pain message	Shortness of breath, sleepiness, and loss of appetite Increased anxiety Increased blood pressure Decreased memory and physical coordination Decreased decision making and judgment abilities Deterioration of brain, liver, heart, or other organs Death
Hallucinogens	LSD Marijuana PCP Mescaline Psilocybin	Changes the way neurons in the brain send and receive information, resulting in hallucinations	Increased blood pressure Decreased coordination and alertness Decreased memory, concentration, learning, and problem-solving abilities Increased emotions, such as fear and anger Risk of flashback reactions long after use has ended Affects the brain stem and involuntary actions such as sneezing, breathing, and coughing Death
Steroids	Prescription medications Can be found in some dietary supplements, and some herbal remedies that claim to be "all natural"; muscle builders	Affects the hypothalamus and other nerve cells Alters the function of organs and the genetic material of individual cells Disrupts hormones in the endocrine system	Headaches, nausea, seizures, and central nervous system disorders Increased anger and aggression Reduced appetite and growth Decreased learning and memory Weakened immune system Increased risk of heart attack, high blood pressure, tumors, cancer, and liver problems Death
Inhalents	Ordinary products that are inhaled contrary to their intended purpose	Prevents oxygen from reaching the brain Can cause abnormalities in the brain and nerves Can deteriorate the myelin sheaths on some nerves Inhaled drugs reach the brain faster and can cause greater damage than oral or injected drugs	Seizures Blurred vision Uncontrolled shaking Decreased memory, concentration, learning, and problem solving abilities Increased risk of blindness, deafness, or liver damage Death

Appendix — For use with Lesson 148 — A41

Use *Effects of Drug Abuse* as you discuss the types of drugs and their effects.

💡 **One group of drugs is called *stimulants*. What would you expect these drugs to do to your body?** *{Elicit that stimulant sounds like stimulate and that stimulant drugs speed up some body processes.}*

➤ **Read the first line of the chart. Which two drugs in the stimulant category are legal?** *{caffeine and nicotine}*

💡 **What are some common sources of caffeine?** *{Possible answers: soft drinks, tea, coffee, chocolate}*

💡 **How is nicotine usually taken into the body?** *{smoking or chewing tobacco}*

➤ **What are some possible results of stimulant abuse?** *{Possible answers: nausea, seizures, heart attack, high blood pressure, internal bleeding, and death}*

💡 **Another group of drugs is called *depressants*. What would you expect a depressant to do to your body?** *{Elicit that depress gives the idea of slowing down. These drugs slow down some body processes.}*

➤ **Read the line of the chart about depressants. How do depressants affect the nervous system?** *{Depressants slow down the central nervous system, sometimes block pain messages, and cause the brain to feel pleasure.}*

💡 **Cough medicine with codeine carries a warning that the user should not operate machinery when taking the medication. Why do you think it would carry this warning?** *{Possible answer: As a depressant, codeine reduces alertness, which may cause accidents.}*

➤ **What other depressant listed is a legal drug that is commonly abused?** *{alcohol}*

➤ **Read the line on the chart about *hallucinogens*. How do these drugs affect the brain?** *{They change how the brain sends and receives information.}*

➤ **What are some possible results of hallucinogen abuse?** *{Possible answers: decreased coordination, alertness, memory, and concentration; increased blood pressure; flashback reactions; increased risk of cancer}*

368   Lesson 148

Keeping God's Temple Clean   Name

A. Answer the following questions.

Which type of drug changes the way the brain receives information from the senses? _____

Which type of drug slows down the central nervous system? __depressants__

Which type of drug affects the hypothalamus and the endocrine system? _____

Which type of drug speeds up the central nervous system? __stimulants__

Which type of drug prevents oxygen from reaching the brain? _____

B. Use your Bible and the word bank to complete the puzzle.

body	destroy	obey	slavery
Christ	glory	power	sin
death	God	sacrifice	temple

Across
1. A Christian's desire should be for ___ to be magnified and exalted in his body (Phil. 1:20).
3. Everything a Christian does should be done to the ___ of God (I Cor. 10:31).
6. If someone knows the right thing to do and does not do it, it is ___ (James 4:17).
7. If someone destroys his body, God will ___ him (I Cor. 3:17).
8. The body of a Christian is the ___ of God (I Cor. 3:16).
10. We are slaves of the person or thing that we ___ (Rom. 6:16).

Down
2. A Christian should present his body as a living ___ to God (Rom. 12:1).
3. A Christian is not his own. He belongs to ___ (I Cor. 6:19).
4. Like false prophets, the temptation to use drugs may promise liberty and freedom; but actually, it results in ___ (II Pet. 2:19).
5. A Christian should honor and glorify God with his ___ (I Cor. 6:20).
7. The way that seems right to man leads to ___ (Prov. 14:12).
9. A Christian should not be under the ___ of anyone or anything other than God (I Cor. 6:12b).

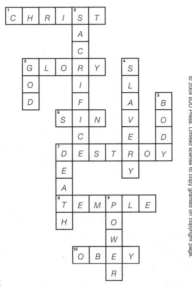

A42 For use with Lesson 148

Discussion:
TE pages A41–A42

▶ **Read the line on the chart about *steroids*. What parts of the body do these drugs affect?** *{Accept any part of the body. Steroids affect the endocrine system, which produces hormones that affect every part of the body.}*

Some steroids are used as medicine to treat conditions such as asthma, eye disease, lupus, or rheumatoid arthritis.

Emphasize to the student the importance of reading and following the directions, dosages, and warnings of medicines he takes. Using any medication in a way other than as prescribed can lead to abuse.

▶ **Have you ever been in a closed room that has just been painted?** *{Answers will vary.}* **Did the smell give you a headache or make you feel sick or dizzy?** *{Answers will vary.}*

▶ **Read the line on the chart about inhalants. Why do you think inhalants are a particularly bad problem among some young people?** *{Possible answers: Inhalants are easily accessible, and most have no medicinal purpose.}*

▶ **Why do inhalants sometimes cause greater damage than other drugs?** *{Inhaled drugs reach the brain faster.}*

▶ **How do inhalants affect the nervous system?** *{Possible answers: They prevent oxygen from reaching the brain, cause abnormalities in the brain and nerves, and deteriorate myelin sheaths on some nerves.}*

▶ **What are some common side effects of inhalant abuse?** *{Possible answers: seizures, blurred vision, shaking, decreased memory and concentration, increased risk of blindness, deafness, liver damage}*

📖 **Do you think that being a Christian prevents a person from temptations to abuse drugs? Why?** *{Not necessarily; a Christian still has a sinful nature.}*

▶ **What should a Christian do if he or she is tempted to abuse drugs?** *{Possible answers: pray; remember verses about Christians' bodies being God's temple; flee the place of temptation}*

As time allows, complete *Keeping God's Temple Clean* together and discuss the biblical principles. Use the information to reinforce that a Christian should honor and glorify God in what he does with his body.

SCIENCE BACKGROUND

Illegal drugs—Some drugs that are now considered illegal were once produced and used for medication or as additives in food and beverages. These drugs seemed helpful at first. But as their side effects were observed, doctors and scientists determined that the risks outweighed the benefits of using them, and they were classified as illegal.

Classification—Some drugs fit into more than one category. For example, nicotine is a stimulant. However, if a person is a heavy smoker, larger amounts of nicotine can act like a depressant.

Nicotine—Nicotine is now recognized as a deadly drug. Research estimates that more than 400,000 Americans die each year from nicotine-related diseases.

Caffeine—Caffeine is found in many foods, beverages, and medicines. Although most people do not normally ingest enough caffeine to be fatal to them (10 g), research has indicated that even ingesting as little as 250 mg of caffeine daily (approximately 2–3 cups of coffee or 3–4 cola drinks) can have a negative impact on one's health.

Alcohol—Many people die each year from alcohol-related traffic accidents. Even more die from alcohol-related diseases. Alcoholism is sometimes called a disease rather than drug abuse. However, an alcoholic must choose to take each drink. Alcoholism is caused by an addiction to a powerful drug.

Chapter 14   Lesson 148

Lesson 149

Student Text page 350
Activity Manual pages 231–32

Objectives

- Recall concepts and terms from Chapter 14
- Apply knowledge to everyday situations

Introduction

Material for the Chapter 14 Test will be taken from Activity Manual pages 220 and 230–32. You may review any or all of the material during the lesson.

You may choose to review Chapter 14 by playing "One, Two, Three" or a game from the *Game Bank* on TE pages A56–A57.

Diving Deep into Science

Questions similar to these may appear on the test.

Answer the Questions

Information relating to the questions can be found on the following Student Text pages:

1. page 327
2. page 336
3. page 346

Solve the Problem

In order to solve the problem, the student must apply material he has learned. The answer for this Solve the Problem is based on the material on Student Text page 349. The student should attempt the problem independently. Answers will vary and may be discussed.

Activity Manual

Study Guide—pages 231–32

These pages review some of the information from Chapter 14.

Assessment

Test Packet—Chapter 14 Test

The introductory lesson for Chapter 15 has been shortened so that it may be taught following the Chapter 14 test.

Diving Deep into Science

Answer the Questions

1. If a person has difficulty with muscle movement along his left side, which hemisphere of his brain might doctors check first? Why?
   Doctors would check the right hemisphere because the right side of the brain controls the left side of the body.

2. Why is it not completely accurate to say that ears hear?
   Ears only receive vibrations. The brain must interpret the vibrations in order for a person to hear.

3. How are the nervous system and endocrine system alike? How are they different?
   Accept any reasonable answer. They are alike because they control other parts of the body. They are different because the nervous system uses electrical impulses and the endocrine system uses chemical messengers.

Solve the Problem

Yesterday Sara found out that one of her Christian friends is taking steroids. Her friend insists that she needs to take this drug in order to build up her endurance as a softball pitcher. She insists that it is not a dangerous drug like cocaine and that she won't use the drug after softball season. What are some reasons her friend should not take this drug? What are some Scripture verses that Sara could share with her friend?

Steroids interfere with the way that the nerve cells in the brain function. The side effects may be noticeable immediately, or they may not appear until after Sara's friend is an adult. Sara can share some Bible verses, such as I Cor. 3:19–20; 10:31; and Phil. 1:20 with her friend. She needs to remind her friend that Christians should be controlled by the power of God and not by the influence of drugs.

Review Game

One, Two, Three

Divide the class into teams of three students. Give each team an object to pass. Each team represents a neuron.

Alternate questions between teams. When a review question is answered correctly by one team, the object is passed to the next person on that team. The team receives a point for each question answered correctly. When the third person answers correctly, the team receives 5 bonus points, and the object returns to the first person. If after two tries a student cannot answer his question, he may pass the object, but the team does not receive points. The student must answer a question correctly before he can pass the object to the next person and receive points. Continue playing until all review questions have been used. The team with the most points wins.

Variation: "Neurons" could be linked together to allow a message to travel from the "arm" to the "brain" and back again.

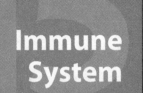

15 Immune System

GREAT & MIGHTY Things

What would you think if you were told that moonlight and swamp vapors made you get sick? Before people understood how diseases spread from person to person, they had many different ideas of how diseases spread or how they could be cured. At one time, some people believed that bathing weakened the body and caused people to get sick more often. Sometimes people practiced blood letting, allowing the "bad blood" to drain from a person's body. Unfortunately, the person sometimes bled to death. But God, the Great Physician, has always known how to heal diseases. Although much suffering and disease remain in this world, God is still in control. He is much more powerful than any disease (Matt. 9:35).

Student Text page 351
Activity Manual page 233

Lesson 150

Objectives
- Demonstrate knowledge of concepts taught in chapter 14
- Recognize that man's inferences are sometimes inaccurate

Introduction

Note: This introductory lesson has been shortened so that it may be taught following the Chapter 14 test.

➤ If you got a cold the day after having taken a bath, would you conclude that the bath caused the cold? *{Answers will vary.}*

➤ People have always tried to find the causes of diseases. Some of their conclusions were based on incomplete information or faulty reasoning. Some of their solutions were comical, but others were very unhealthy.

Chapter Outline

I. Diseases
   A. Communicable Diseases
   B. Noncommunicable Diseases
   C. Epidemiology

II. The Immune System
   A. Defensive Barriers
   B. Inflammatory Response
   C. Immune Response
   D. The Immune System at Work
   E. Immunity
   F. Antibodies and Antibiotics
   G. Malfunctions of the Immune System

Activity Manual

Preview—page 233

The purpose of this Preview page is to generate student interest and determine prior student knowledge. The concepts are not intended to be taught in this lesson.

Lesson 151

Student Text pages 352–55
Activity Manual page 234

Objectives

- Recognize that disease is a consequence of Adam's sin
- Explain how diseases are classified
- Identify four common pathogens
- List some diseases caused by each pathogen

Vocabulary

communicable pathogen
noncommunicable

Introduction

➤ What are some health and cleanliness rules that you often hear? *{Possible answers: wash your hands; brush your teeth; take a bath or shower daily; avoid eating food that has been left out too long}*

➤ Why are cleanliness rules important? *{Possible answers: to help prevent disease; to keep you healthy}*

➤ Today we will be studying about some diseases and their causes.

Purpose for reading:
Student Text pages 352–55

➤ Why is Louis Pasteur important?

➤ How is a virus different from other pathogens?

Discussion:
pages 352–53

 Why did pain, disease, and death enter the world? *{because of Adam and Eve's disobedience; as a result of sin}*

➤ Although God often uses doctors and medicines to heal people, is God limited by these methods? *{no}* Explain. *{Possible answers: God is more powerful than disease. The Bible records examples of God using miracles to heal illnesses and disease.}*

What is a *miracle*? *{Possible answer: a supernatural event that is outside the usual way in which God works}*

➤ Why did God give the Israelites some of the laws mentioned in the Old Testament? *{to promote good health habits; to protect them from dangerous diseases}*

Diseases

In the beginning, Adam and Eve enjoyed a perfect world. God talked to them directly each day. They did not experience pain, disease, or death. But when Adam and Eve chose to disobey God, everything changed. Not only did they experience spiritual death, but also they began to die physically. Disease and pain became part of their lives.

Although disease is a consequence of sin, God is more powerful than disease. The Bible records many instances where God miraculously healed people of diseases. He also controlled disease by sending and taking away plagues and diseases. In the Old Testament times, God gave to the Jewish people laws that promoted good health habits. These laws protected the Jewish people from many dangerous diseases.

Louis Pasteur

1796 Edward Jenner — Doctor Discovers Smallpox Vaccine

1847 Dr. Ignaz Semmelweisse — Dirty Hands Spread Disease

1854 Florence Nightingale — Nurse Cleans Hospitals, Saving Lives

1867 Joseph Lister — Surgical Doctor Uses Disinfectants to Kill Microorganism

1868 Louis Pasteur — Killing Microorganisms Stops Communicable Diseases

352

 Louis Pasteur (1822–1895) began his scientific career as a chemist, but his research led him from one discovery to another. His diverse accomplishments include the following: vaccinations and treatments for rabies, anthrax, and chicken cholera, silkworm diseases, fermentation and pasteurization (heating food in order to destroy harmful microbes), and showing the inaccuracy of the theory of spontaneous generation.

 The Greek word *pathos* means "suffering." The Greek suffix *-gen* means "producer."

➤ Why is the word *pathogen* an appropriate term for what a pathogen is? *{The word means producer of suffering, and that describes what a pathogen is.}*

➤ Since the suffix *-ology* means "the study of," what do you think the study of diseases is called? *{pathology}*

Communicable Diseases

For many years, people thought that evil spirits, witches, magical spells, or bad luck caused diseases. Most people did not consider cleanliness to be important or essential.

Louis Pasteur was one of the first scientists to identify the fact that diseases can be caused by organisms too small to be seen. He thought that if these microorganisms could be killed, then the disease would not spread farther. Pasteur's germ theory of disease changed the way people thought about and treated diseases. Today doctors and scientists know even more about diseases. They classify diseases as either **communicable**, contagious, or **noncommunicable**, noncontagious.

Communicable diseases spread from person to person by pathogens. Scientists define a **pathogen** (PATH uh juhn) as anything that causes a disease. Pathogens can cause diseases when they invade and attack the cells in your body. Some pathogens interfere with the normal function of your body's cells. Other pathogens produce a *toxin*, or poison, that harms the cells.

Protozoans, fungi, bacteria, and viruses are the four most common kinds of pathogens. However, there are many different types of each kind of pathogen. For example, scientists have identified more than 200 different kinds of cold viruses. Yet each pathogen can cause only one disease. Different diseases attack different cells in the body.

Discussion:
pages 352–53

▶ What is a *microorganism*? {an organism too small to be seen}

▶ Who was one of the first scientists to realize that microorganisms can cause disease? {Louis Pasteur}

▶ What did Pasteur think would prevent the spread of diseases? {killing the microorganisms that caused the diseases}

▶ How do doctors and scientists classify diseases today? {communicable or noncommunicable}

▶ What is the difference between communicable and noncommunicable diseases? {Communicable diseases are contagious, and noncommunicable diseases are not contagious.}

▶ How do communicable diseases spread from person to person? {by pathogens}

▶ What is a *pathogen*? {anything that causes a disease}

▶ What are some different ways that pathogens cause diseases? {Possible answers: invading and attacking cells in the body; interfering with the normal function of cells; producing toxins that harm the cells}

▶ What are the four most common kinds of pathogens? {protozoans, fungi, bacteria, and viruses}

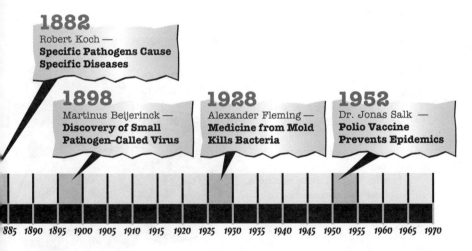

 Refer the student to the timeline.

▶ Which doctor realized that dirty hands can spread disease? {Dr. Ignez Semmelweisse}

At this time in history, many doctors went from patient to patient without washing their hands. Doctors that followed Dr. Semmelweisse's hand washing ideas experienced fewer patient deaths. Today hand washing is a standard procedure, but at that time, many doctors strongly disagreed with Dr. Semmelweisse.

▶ Which nurse saved many soldiers' lives by insisting on clean hospitals? {Florence Nightingale}

▶ Viruses were the last of the four common pathogens to be identified. In what year were viruses discovered? {1898}

▶ For what is Edward Jenner remembered? {the smallpox vaccine}

💡 A virus causes polio. How many years passed between the discovery of virus pathogens and the polio vaccine? {54}

 The timeline on Student Text pages 352–53 gives a possible news magazine headline about each person mentioned. This timeline will be used in the exploration on Student Text page 371. The student may begin researching and writing his magazine article for the exploration any time after this lesson is taught.

 In the past, washing hands was not the simple task it is today. Before indoor plumbing and running water, any water used for bathing had to be drawn and carried from its natural source. Even in the early days of running water, if warm water was desired, it had to be heated on a stove or other heat source. The ideas of Dr. Semmelweisse and Florence Nightingale were not only revolutionary, but also they were time-consuming hard work.

Discussion:
pages 354–55

➤ Which organisms are the largest known pathogens? {protozoans}

➤ Name a disease caused by a protozoan. {malaria}

➤ What are some diseases caused by fungi? {athlete's foot, ringworm}

➤ Through what part of a fungus is disease spread? {spores}

💡 What is the function of spores? {reproduction}

💡 Why do you think a fungal disease such as athlete's foot occurs in warm and moist places on the body? {Possible answer: The warmth and moisture are needed for the spores to grow and spread.}

➤ Which pathogen causes the most infections? {bacteria}

💡 Why do you think bacteria can cause so many infections? {because they reproduce quickly and can be found almost anywhere}

➤ What are some infectious diseases caused by bacteria? {Possible answers: leprosy, conjunctivitis, strep throat, tetanus}

➤ What is another name for conjunctivitis? {pink eye}

➤ Which part of the body do tetanus bacteria attack? {the nervous system}

Types of pathogens
Protozoans and fungi

Protozoans are members of the kingdom Protista. These single-celled organisms are the largest known pathogen. Many protozoans live in unpurified water, such as in streams and creeks. Drinking unclean water can cause severe illness. A tropical disease, malaria, is usually associated with mosquitoes. However, malaria is actually caused by a protozoan that infects a certain type of mosquito.

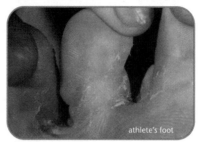
athlete's foot

Some *fungi* cause uncomfortable diseases, such as athlete's foot and ringworm. These infections spread by direct contact with spores produced by the fungi.

Bacteria

Most infections are caused by bacteria. *Bacteria* are single-celled microorganisms that can reproduce quickly and can be found almost everywhere.

Harmful bacteria cause infections, such as leprosy, conjunctivitis, strep throat, and tetanus. Leprosy bacteria often infect the skin and peripheral nerves. The bacteria that can cause conjunctivitis (kuhn JUHNG tuh VY tihs), also called pink eye, are very similar to bacteria that cause ear infections. Strep throat bacteria attack the cells of the throat. The tetanus, or lockjaw, bacteria produce toxins that attack the nervous system. For this reason, if you step on a nail or get bitten by an animal, a doctor may give you a shot to prevent tetanus.

girl with leprosy

Types of Bacteria

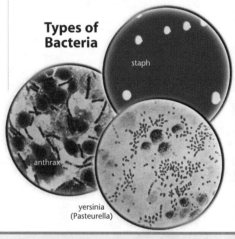

354

SCIENCE BACKGROUND

Viruses—The genetic material in viruses can be either RNA or DNA. RNA acts like a genetic messenger in the cell as it carries copies of the DNA chemical code throughout the cell. The fastest virus can replicate in less than thirty minutes. Some viruses exit the host cell gradually, but others cause the cell to burst and release many viruses at once.

Other pathogens—Other pathogens include parasitic animals, such as lice and tapeworms, as well as pathogens smaller than viruses. These smaller pathogens include *viroids, virusoids,* and *prions.*

SCIENCE MISCONCEPTIONS

Students may think that all protozoans, fungi, and bacteria are harmful. But many are harmless, and some are actually beneficial to man. Yeast, penicillin, lactobacilli found in yogurt, and other bacteria found in cheese and sauerkraut are generally not harmful to man.

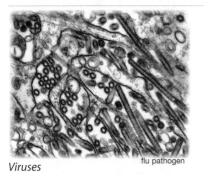

Viruses

flu pathogen

Viruses

Unlike bacteria, protozoans, and fungi, a *virus* is not a living organism. It is not made of cells. A virus usually has a protective coat and its own genetic material. It cannot move on its own and can reproduce only in cells of living organisms. A virus invades a cell by injecting its genetic material into it. Once inside the host cell, the virus tricks the cell into reproducing the virus. Sometimes the new viruses are released gradually. At other times, the increasing number of viruses causes the host cell to explode. These new viruses then attack other cells and repeat the process.

Viruses can reproduce quickly; yet they cannot attack every cell. A specific virus attacks only a specific type of cell. Viruses are some of the smallest known pathogens. They can cause colds, chickenpox, flu, rabies, hepatitis, and many other diseases.

QUICK CHECK
1. Into what two groups do scientists classify diseases?
2. What is a pathogen?
3. What are the four common types of pathogens?

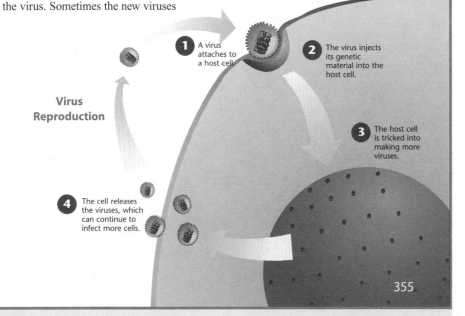

Virus Reproduction

1. A virus attaches to a host cell.
2. The virus injects its genetic material into the host cell.
3. The host cell is tricked into making more viruses.
4. The cell releases the viruses, which can continue to infect more cells.

BIBLE Most Bible scholars agree that when the Bible describes "leprosy" the term includes leprosy, other skin diseases, some molds, and other fungi. The Old Testament laws helped prevent the spread of highly contagious diseases. Leprosy can be compared to sin, since leprosy comes upon the body, spreads, and cripples, just like sin does in our spiritual lives.

Discussion
pages 354–55

➤ **How are viruses different from bacteria, protozoans, and fungi?** *{Possible answer: A virus is not a living organism. It is not made of cells, and it cannot move on its own.}*

💡 **Why do we say that a virus cannot be a living organism?** *{A virus is not made of cells. According to the cell theory, a living organism must be made up of cells.}*

➤ **What does a virus need in order to reproduce?** *{cells of a living organism}*

➤ **How do most viruses reproduce?** *{A virus invades a cell by injecting its genetic material into the cell. Then the virus tricks the cell into reproducing the virus. The viruses are then released from the cell and attack other cells.}*

➤ **What are some diseases caused by viruses?** *{Possible answers: colds, chickenpox, flu, rabies, hepatitis}*

Answers
1. communicable and noncommunicable
2. anything that causes a disease
3. protozoans, fungi, bacteria, and viruses

Activity Manual
Reinforcement—page 234

Lesson 152

Student Text pages 356–59
Activity Manual pages 235–36

Objectives

- List several ways that pathogens are spread
- Differentiate between communicable diseases and noncommunicable diseases
- Explain some of the jobs of an epidemiologist

Vocabulary

vector
epidemic
airborne pathogen
contact
food-borne pathogen
water-borne pathogen
epidemiologist

Introduction

➤ What are some diseases that you are familiar with? *{Answers will vary.}*

➤ Which of these diseases are communicable, and which are noncommunicable? *{Answers will vary.}*

➤ How do you think people get diseases? *{Answers will vary.}*

➤ Today we will study how diseases spread and about scientists that track diseases.

Purpose for reading:
Student Text pages 356–59

➤ What are some ways that pathogens are spread?

➤ What is an *epidemiologist*?

How pathogens are spread

Several hundred years ago, millions of people died from the bubonic plague. Doctors blamed swamp vapors, heat, and baths for causing the disease. Some people thought that the plague was a punishment from God. Some of them beat themselves, hoping to earn God's forgiveness. But the plague kept spreading.

Many years later, doctors learned that bacteria caused the bubonic plague. Infected rats and fleas spread these pathogens to people. A flea that bit an infected rat would become a carrier of the bacteria. Then when the flea bit a person, the bacteria would infect the person.

Insects and other animals that carry pathogens are called **vectors**. Some of the most common vectors include mosquitoes, fleas, flies, lice, and ticks.

Mosquitoes can spread many diseases, such as malaria, dengue (DENG gee) fever, the West Nile virus, and yellow fever. Even tiny sand flies can spread viruses that cause sandfly fever. The tsetse fly can carry a protozoan that causes a disease called sleeping sickness. Lice and ticks can spread many bacterial diseases, such as trench fever, Rocky Mountain spotted fever, and Lyme disease.

Epidemics (EP ih DEM ihks) happen whenever a disease spreads to a great number of people in a short time. In 1918 about half a million people in America died from a type of influenza called the "Spanish Flu." As a result of this epidemic, doctors learned how influenza and other viruses spread.

Influenza can spread through the air, so it is called an **airborne** pathogen. When a sick person coughs or sneezes,

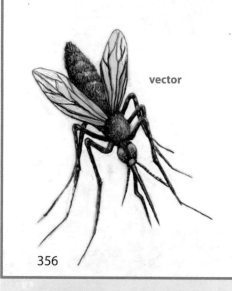

vector

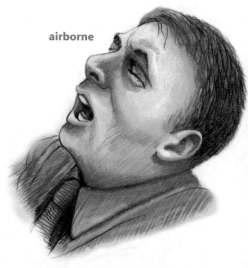

airborne

SCIENCE BACKGROUND

Vehicle—The term *vehicle* is sometimes used to refer to food and water that carry pathogens.

Poliovirus—The poliovirus usually affects the digestive system. Most cases of polio in the digestive system are mild, with symptoms such as a sore throat and intestinal upset. However, if the poliovirus enters the bloodstream, it usually attacks the nervous system, often causing paralysis and sometimes death. Another name for this type of polio is *infantile paralysis*. Some polio survivors experience muscle and joint difficulties many years after having had polio. Today both the Sabin and Salk vaccines are used to vaccinate against polio. The vaccinations have helped prevent further epidemics of this disease.

SCIENCE MISCONCEPTIONS

Vectors such as mosquitoes and ticks are not actually infected with the disease that they spread. They carry the disease without becoming ill. Vectors also include other arthropods, snails, and some clams.

 In 2002 a deadly new disease called Severe Acute Respiratory Syndrome (SARS) appeared. The epidemic started in China and spread quickly to other countries. The spread of the disease was traced to infected travelers. The World Health Organization went into action quickly to determine the cause and treatment of SARS. Epidemiologists concluded that SARS is caused by a virus. The disease appears to be spread through contact and through the air.

he expels infected droplets into the air. The pathogen is transmitted from one person to another when another person breathes in the droplets.

Viruses also spread by **contact.** Touching a sick person or something that a sick person has touched can spread the pathogens. Many common viruses, such as cold viruses, are spread both through the air and by contact.

Other illnesses, such as typhoid (TY foyd) and cholera (KOL uhr uh), are also spread by **food-borne** and **water-borne** pathogens. Contamination occurs when something infected with a pathogen touches water or food. People that drink the infected water or eat the contaminated food often become sick with the disease. Today, water treatment systems help keep drinking water clean and pathogen-free. In areas without water treatment systems, some people purify their drinking water by boiling it or by using special filtering systems to eliminate pathogens.

Epidemics help people learn more about how diseases spread. Between 1893 and 1955, many cities and towns in America had polio epidemics. These epidemics caused great anxiety and fear. People were not sure how the polio was spreading, because the poliovirus did not act like other pathogens. Scientists observed that the virus did not appear to be spread through the air or by vectors. After much observation and research, scientists learned that polio usually spreads through contaminated food and water. Sometimes contact with an infected person may also spread the disease.

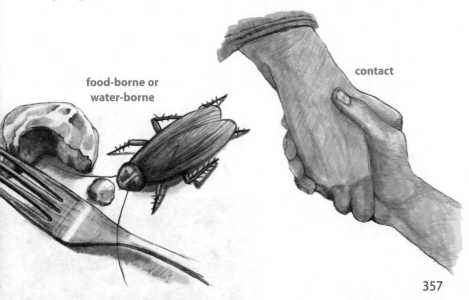

food-borne or water-borne

contact

357

Direct a DEMONSTRATION

Materials: a small plastic bag of flour or chalkboard erasers with yellow chalk dust

Lightly clap the erasers or squeeze the open bag of flour. Notice how the air carries the chalk dust or flour particles. Explain that even though we cannot see airborne pathogens, the air carries them in much the same way.

Direct an ACTIVITY

Materials: glitter or hand lotion with glitter

Allow one or two students to cover one hand in glitter. Direct the "infected" students to shake hands with other students or to touch surfaces that other students would normally touch. Notice how the glitter spreads through contact.

➤ Pathogens spread in much the same way as the glitter does.

Discussion:
pages 356–57

➤ What caused the bubonic plague? {bacteria}

➤ How did the bacteria spread to people? {through infected fleas and rats}

➤ What is a *vector?* {any insect or animal that carries pathogens}

➤ Name some of the most common vectors. {Possible answers: mosquitoes, fleas, flies, lice, and ticks}

➤ Name some diseases that are spread by vectors. {Possible answers: malaria, dengue fever, West Nile virus, yellow fever, sand fly fever, sleeping sickness, trench fever, Rocky Mountain spotted fever, Lyme disease}

➤ What term is used to describe a disease that has spread to a great number of people in a short time? {epidemic}

💡 Why would an epidemic give scientists an opportunity to learn more about how that disease is spread? {Possible answer: Since the disease spreads quickly, scientists can more easily trace the previous activities of infected people.}

💡 Why is it important for doctors and scientists to know more about the spread of disease? {The information can be used to educate people and possibly prevent epidemics.}

➤ Which epidemic helped doctors realize that some pathogens spread through the air? {the influenza epidemic}

➤ Viruses such as the influenza virus can be airborne pathogens. What is another way that viruses can spread? {by contact}

➤ What are some examples of contact? {touching a sick person or something that a sick person has touched}

💡 Why do you think you do not always become sick when you are around someone who has a communicable disease, such as a cold? {Possible answers: The sick person might have been careful not to spread the pathogen. You may not have come in contact with enough of the pathogen to make you sick. You may not be susceptible to that disease.}

➤ What did scientists learn from the polio epidemics? {The poliovirus can be spread through contact and through contaminated food or water.}

➤ What are some other diseases that can be spread by food-borne and water-borne pathogens? {Possible answers: typhoid, cholera}

Discussion:
pages 358–59

➤ How are noncommunicable diseases different from communicable diseases? {Noncommunicable diseases are not contagious and cannot be spread by contact, contamination, animals, or the air.}

➤ What are some possible causes of noncommunicable diseases? {Possible answers: health habits, heredity, environment, or a combination of these or other factors}

➤ When someone develops a lifelong disease, does that mean that God is not in control? {no} How do we know? {The Bible tells us that all things work together for good to those who love God. (Rom. 8:28)} [Bible Promise: I. God as Master]

➤ What is it called when a disease seems to disappear? {remission}

➤ What is an *epidemiologist*? {a scientist who studies the causes and spread of diseases}

➤ What is one of the first goals of an epidemiologist? {to keep diseases from spreading to more people}

➤ Why do epidemiologists try to find the source of a disease? {Possible answers: to prevent it from infecting anyone else; to inform people of how to avoid and prevent the spread of the disease}

➤ What agency was started as a training center and a response unit for epidemics? {The Epidemic Intelligence Service}

➤ What organization is this agency under? {Centers for Disease Control and Prevention}

Suppose a small town has just hosted its annual festival and barbecue. The day after the festival, many of the townspeople became sick with the same illness. What kinds of things might the epidemiologists want to find out? {Possible answers: Who is sick? Were the sick people at the festival? Did they eat at the barbecue? What exactly did they eat and drink?}

Noncommunicable Diseases

Not all diseases are contagious. A person cannot catch diabetes, arthritis, or heart disease. Noncommunicable diseases do not spread by contact, contamination, animals, or the air. Scientists do not know exactly what causes some noncommunicable diseases. A person's health habits, genes, environment, or any combination of these factors may be part of the cause. Some noncommunicable diseases, such as Huntington's disease, are not usually evident until adulthood. Other diseases, such as cancer and heart disease, may occur in both children and adults.

Once someone develops a noncommunicable disease, he usually has it for the rest of his life. Doctors can treat the symptoms of the disease, but most noncommunicable diseases cannot be cured. Some of these diseases may go into *remission*, or seem to disappear but then come back. However, scientists continue searching for cures and new treatments for diseases. Good health habits can also help to prevent many noncommunicable diseases. When a person develops a lifelong disease, it is often hard for him to understand God's purpose. However, Christians should remember that God works all things together for good (Rom. 8:28). God is sometimes glorified most in situations that seem tragic to us.

Epidemiology

Do you enjoy detective mysteries? Some scientists are actually disease detectives. **Epidemiologists** (EP ih DEE mee AHL uh jists) are scientists who study the causes and spread of diseases through communities. They look for ways to prevent and control diseases. One of their first goals is to keep a disease from spreading to more people. Epidemiologists try to track a disease's progress to find the source, or cause, of the disease. Once they find out how the disease began and spread, they can teach people how to avoid having another outbreak of that disease.

The Epidemic Intelligence Service started in 1951 as part of the Centers for Disease Control and Prevention. These scientists travel all over the world trying to solve disease mysteries. Epidemiologists study illnesses that can cause epidemics, such as meningitis (MEN ihn JY tihs), influenza, and tuberculosis (too BUHR kyuh LO sihs). At times, they also research diseases caused by contaminated food and water.

epidemiologists

SCIENCE BACKGROUND

Epidemiologists—When the SARS epidemic started in the winter of 2002, hundreds of epidemiologists began working to identify the pathogen that causes the disease and find out how it is transmitted from person to person. Other epidemiologists study diseases such as Ebola, AIDS, and the West Nile Virus. State and local governments also have epidemiologists who keep track of smaller disease outbreaks, such as food poisoning cases and influenza.

Some epidemiologists concentrate on health issues such as strokes and heart diseases. These scientists may also be involved with nutritional issues, studying how the food that people eat affects their health. Other epidemiologists specialize in a specific disease, such as cancer. Epidemiologists may also study types of diseases, like genetic diseases or infectious diseases.

Epidemiologists often work for state and local health departments or for federal government health agencies. Some teach at universities and may work for individual research programs. Others may be part of international health agencies. These medical detectives help identify diseases and trace the causes of those diseases.

QUICK CHECK
1. Name some ways that pathogens are spread.
2. How are noncommunicable diseases different from communicable diseases?
3. What is an epidemiologist?

Meet the SCIENTIST — DR. JOHN SNOW

During the 1800s, epidemics of cholera, a severe intestinal disease, were common in many countries. An English doctor, John Snow (1813–1858), was one of the first people to realize that cholera was spread through contaminated water. During Britain's second cholera epidemic, he was able to prove his theory. Dr. Snow drew maps of the London streets and marked the homes of people who had died from cholera. Through careful research he was able to trace the outbreak of cholera to a specific water pump. Dr. Snow found that more than ¾ of the people who had become ill drank water from that pump. He used a microscope to look at a water sample from the pump and noticed that the water showed some contamination. Because of Dr. Snow, England's water systems were improved, and there were fewer cases of cholera. Epidemiologists today still use some of the research methods that Dr. Snow used.

359

GEOGRAPHY Provide a blank world map that can be colored or marked. Direct the student to research to find the health and immunization requirements needed to enter ten to twelve countries around the world. Choose countries from various latitudes and continents. Tell the student to mark countries that require immunizations or preventative medications and to compare the locations of the countries with their requirements.

➤ Do the countries that require immunizations have similar latitudes? {Answers will vary.}

➤ Is there a particular continent(s) that has a higher rate of requirements? {Answers will vary.}

➤ Why do you think these similarities exist? {Possible answers: The similarities of climates and economic development may lead to similar diseases.}

WRITING Direct the student to research and write a paragraph about a disease in a third-world country. The paragraph should include information about how the disease is spread, possible vectors, and what is done to prevent the spread in that country.

TEACHER HELPS Resources to use for the Geography Link and the Writing Link can include the Centers for Disease Control and Prevention, the World Health Organization, and the *Yellow Book of Health Information for International Travel*.

Discussion:
pages 358–59

➤ Other than epidemics, what else might an epidemiologist study? {Possible answers: diseases caused by contaminated food and water; strokes; heart diseases; nutritional issues; cancer; genetic diseases; infectious diseases}

➤ What are some places that employ epidemiologists? {Possible answers: state and local health departments; federal government health agencies; universities; international health agencies}

Refer the student to *Meet the Scientist*.

➤ What is one way that cholera is spread? {by contaminated water}

➤ Who was one of the first people to realize this? {Dr. John Snow}

➤ How did he prove to others that cholera could be spread by water? {He drew maps to trace the outbreak back to a specific water pump.}

➤ How did Dr. Snow influence present-day epidemiologists? {Today, epidemiologists still use some of his research methods to track diseases.}

Answers

1. vectors, air, contact, food, and water
2. Noncommunicable diseases are not contagious and cannot be spread by animals, contact, contamination, or air. They also are usually permanent rather than temporary illnesses.
3. a scientist who identifies diseases and traces the causes of those diseases

Activity Manual
Reinforcement—page 235

Study Guide—page 236
This page reviews Lessons 151–52.

Assessment
Test Packet—Quiz 15-A
The quiz may be given any time after completion of this lesson.

Lesson 153

Student Text pages 360–61

Objectives
- Recognize how quickly pathogens can spread
- Infer the source of contamination

Materials
- sodium hydroxide solution (1.2 mL [1/4 tsp] granular lye mixed with 1 liter distilled water)
- See Student Text page
- buckets or large containers to dispose of liquids

Introduction

Infectious diseases can spread rapidly! Epidemiologists not only try to identify a disease, but also try to trace that disease back to its source. This is not always an easy job. In today's activity, you will create an epidemic and then work together to determine its source.

Purpose for reading:
Student Text pages 360–61

The student should read the pages before beginning the activity.

Before the Activity

Pour 150 mL of the sodium hydroxide solution (contaminant) into one cup. Pour 150 mL of distilled water into each of the remaining cups. Do not let students know which cup has the contaminant.

During the Activity

The following procedure may help the students exchange solutions in an orderly way:

1. Divide the students into two groups lined up and facing each other on either side of a table.
2. Distribute the safety supplies, eyedroppers, note cards, and cups of liquid.
3. Direct each student to use the eyedropper to add liquid to the cup of the student across from him.
4. Direct the line along one side of the table to remain in the same place throughout the activity. Direct those in the other line to move down two places. Tell the two people at the end to come to the front of the line.
5. Repeat Steps 3 and 4 until the students have exchanged liquids four times.

After the indicator is added, direct the students to mark the cards as directed. Tell each student to write the color of his solution on his own card before disposing of the liquid.

380 Lesson 153

ACTIVITY: Of Epidemic Proportions

Epidemics often catch people by surprise. Because pathogens are so small, a person can easily spread a disease without knowing it. In this activity, you will create an epidemic. When the epidemic is over, you will work as an epidemiologist and search for the source of the contamination.

Process Skills
- Making and using models
- Observing
- Inferring
- Recording data
- Communicating

Problem
Which cup contained the original chemical solution—the source of contamination?

Procedures
Note: This activity uses a chemical solution to create the epidemic. Use safety precautions and be careful not to spill any liquid. If any liquid splashes onto your skin, stop and wash the skin thoroughly. Do not try to pour the liquid when exchanging solutions. Do not drink any of the liquid.

Materials:
- goggles or safety glasses
- foam cups
- prepared "epidemic solution"
- eye dropper
- distilled water
- red cabbage juice indicator solution
- 3 × 5 note card
- latex gloves (optional)

1. Get a cup of liquid, an eyedropper, and a note card from your teacher. Your cup may have distilled water, or it may have the contaminated solution. Keep the cup away from your face. Write your name at the top of your note card.

2. Using the eyedropper, draw up one dropperful of liquid from your cup. When instructed, add one dropperful of your liquid to someone else's cup. On your note card record the name of the person that you gave liquid to.

3. Repeat step 2 three more times. After exchanging liquid four times, stop and wait for your teacher to add the red cabbage juice indicator solution.

4. Observe the color of the red cabbage juice indicator before and after it is added to your liquid. If your solution is not contaminated, the color of the cabbage juice will remain the same. If your solution is contaminated, the color will change.

360

TEACHER HELPS

Epidemic solution—Mix the lye and water in a bottle. Clearly label the bottle.

The solution is a weak sodium hydroxide solution. It is colorless and odorless and can be poisonous if ingested. Any liquid splashed on the skin should be washed off immediately.

Do not use liquid lye or other drain cleaners in the solution.

Prepared solutions are available from science supply companies.

Indicator—Directions for preparing cabbage juice indicator are found on TE page 188.

Eyedropper—If eyedroppers are not available, plastic spoons may be substituted.

Students—This activity works best with at least ten students. If there are less than ten students, give some students an additional cup and index card. If there are more than twenty students, you may want to add a second cup of contaminant.

Chart—Prepare a chart with a column for each student.

Under each student's name, list the names from that student's card in the same order he has them written.

You may choose to give each student a copy of the chart and allow him to work on his own to determine the source of the contamination.

Solution colors—The darkest yellow will be the source of contamination. Solutions with colors closer to yellow were most likely infected during the first two exchanges. Solutions closer to purple were most likely infected in the later exchanges.

5. If your solution is infected, circle your name at the top of your card.
6. Use the note cards to make a chart that traces the spread of the infection.

Conclusions
- What is the ratio of infected to uninfected people?
- Can you determine who had the original contaminated solution?

Follow-up
- Try the activity again with a different number of people or change the number of times you exchange liquid. Notice how this affects the ratio of infected to uninfected people.

SCIENCE PROCESS SKILLS

Inferring—*Inferring* is taking evidence or a result and drawing conclusions without having seen the origin or process leading to that result. Evidence is often gathered through observations.

➤ Why would you not consider any of the purple solutions at the end of the activity as a possible source of the contamination? *{If the solution is not contaminated at the end, it could not have been contaminated earlier.}*

➤ Look at partners for the first exchange. What inference can be made about the source of the infection if one is infected at the end and the other is not? *{Since one remained uncontaminated until the end, neither could have been the original source of contamination.}*

➤ What other solutions could you infer are not the source? *{any solution that did not contaminate other solutions in any of its four exchanges}* Why? *{If a solution remains uncontaminated until the end, none of the solutions in any of its exchanges could have been the original source of contamination.}*

➤ What can be inferred about the degree of contamination based on the colors of the solutions? *{Possible answer: The closer the color is to purple, the less contamination is in the solution.}*

➤ What can be inferred about a yellowish-green solution compared to a green or blue solution? *{The yellowish-green solution contains more contamination.}*

After the Activity
Discuss the chart. Guide the students in finding the source of the contamination. Mark the names of those who ended with an infected solution. Working from the bottom to the top of the chart, track the exchange activity of each of the marked students. Eliminate those who received solution from an uninfected source.

Conclusions

➤ How many people ended with a contaminated solution? *{Answers will vary.}*

➤ How many people did not have a contaminated solution? *{Answers will vary.}*

➤ How would you write the number of contaminated and noncontaminated results as a ratio? *{Answers will vary.}*

It may not be possible to trace the exact source. The chart should help the students eliminate all but two possible sources. Epidemiologists often have unexpected difficulties as they track diseases. The diseases they study are not as contained and controlled as this mock epidemic.

After students have discussed the chart and tried to trace the source, discuss the color variations of the solutions.

➤ What color were the solutions that did not become contaminated? *{purple}*

➤ What are some of the colors of the contaminated solutions? *{Possible answers: blue, green, yellow}*

💡 How much contamination do you think a blue solution contains—a lot or a little? *{a little}* Why? *{Blue is close to purple on the color wheel, and purple is not contaminated.}*

💡 Which color solution do you think shows the source of the infection? *{yellow}* Why? *{Yellow is the farthest color from purple on the color wheel.}*

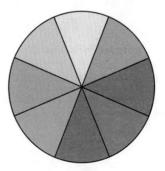

Assessment

Rubric—TE pages A50–A53

Select one of the prepared rubrics, or design a rubric to include your chosen criteria.

Lesson 154

Student Text pages 362–65

Objectives

- Identify several defensive barriers of the body
- List two of the body's nonspecific defenses
- Identify the body's specific defense against pathogens
- Explain some functions of white blood cells during the immune response

Vocabulary

defensive barrier inflammatory response
cilia immune response

Introduction

Whenever a soldier goes to battle, he wears special gear and carries equipment for protection against an enemy's attacks. Just like a soldier, the body has some special defenses against enemy pathogens. Today we will be studying how the body reacts to pathogens.

Purpose for reading:
Student Text pages 362–65

➤ How do cilia help protect air passages?
➤ What is the body's specific defense against pathogens?

Discussion:
Pages 362–63

 There are pathogens all around us. Why are we not always sick? {*God created our bodies with many defenses against pathogens.*}

➤ What are some of your body's natural barriers to pathogens? {*skin, scabs, sweat, mucus membranes, cilia, tears, earwax*}

➤ What are some ways that pathogens enter your body? {*skin openings, nose, eyes, throat, ears*}

➤ What is one purpose for scabs? {*They cover open wounds and prevent pathogens from entering your body through the wounds.*}

➤ What are some other ways that skin protects itself from pathogens? {*sweat, natural body oils*}

The Immune System

Pathogens are always around us. Many times we are exposed to pathogens without even realizing that they are there. But though we are surrounded by pathogens, we do not always become sick. God created our bodies with many defenses against disease-causing pathogens. Scientists have grouped these defenses into three main categories. Each category of defense has a job in the fight against disease.

Defensive Barriers

Ancient warriors often dug moats to surround their castles or fortresses. These barriers helped protect the castles from the enemy's attack. In a similar manner, your **defensive barriers** help to keep pathogens out of your body. Your skin keeps many pathogens from entering into your body. However, pathogens can enter through cuts and scrapes. Scabs, although not beautiful, are part of your body's defenses. Scabs cover open wounds and help prevent pathogens from entering your body through the wounds. Some pathogens, though, attack skin cells directly. Your sweat and natural body oils contain chemicals that help to kill these pathogens.

Pathogens can also enter through natural body openings, such as your nose and your eyes. The mucus membranes in your air passages produce sticky mucus that traps pathogens that enter your nose and throat. Tiny hair-like projections called **cilia** (SIL ee uh) line your air passages. Cilia help to filter out pathogens and sweep any trapped pathogens back up the air passages toward the throat and away from the lungs. Your body then expels some of these mucus-trapped pathogens by coughing or sneezing. Pathogens that are swallowed are usually killed by the *hydrochloric acid* in your stomach.

Tears help to protect your eyes from pathogens. Not only do tears wash dust and dirt away from the surface of your eyes, but they also contain a special substance that can kill some bacteria. Earwax helps to protect your ears from dust and pathogens. God has designed these barriers to work together to help prevent pathogens from gaining entrance into your body.

SCIENCE BACKGROUND

Defenses of the body—The body's defense system includes several nonspecific responses and a specific response. Nonspecific responses include defensive barriers, such as skin, the inflammatory response, and special white blood cells called macrophages. Macrophages are often found in the lymph nodes, but they are also found in other parts of the body, such as the liver. These white blood cells protect the body by surrounding pathogens and eating them. During the inflammatory response, macrophages increase in number.

The body's specific response, called the immune response, is pictured on Student Text pages 364–65. Nonspecific responses and the immune response can happen at the same time.

White blood cell carriers—Although many white blood cells are carried by the lymph fluid, the blood is the primary carrier of white blood cells.

Lymph fluid—Lymph fluid originates from blood plasma. It is a colorless liquid containing water, white blood cells, nutrients, cell debris and waste, and some pathogens. The lymph fluid flows through channels that help drain excess fluid from body tissues. Nodes and glands help filter the pathogens from lymph fluid.

Inflammatory Response

If enemy soldiers managed to cross the castle moat, they still faced direct attacks. Soldiers high on the castle wall could respond to an attack by shooting arrows, dropping rocks, or pouring boiling oil on the enemy. Your body has many ways to defend itself against any pathogens that make it past the first line of defense. These other defenses are divided into two categories: *nonspecific* and *specific*.

One nonspecific defense is often called the **inflammatory response.** Symptoms of the inflammatory response include swelling, redness, heat, and pain. If a pathogen does infect you, your body increases the supply of blood to the area of infection. This increased supply of blood often makes the infected area swollen and painful. Sometimes, too, your brain signals your body to increase its temperature. Higher temperatures can kill some pathogens. A fever can be helpful as long as the body temperature does not get too high.

Another nonspecific defense involves special white blood cells that fight pathogens by surrounding and "eating" them. These white blood cells increase in number during the inflammatory response, and they also protect your body before infection.

Immune Response

While enemy soldiers outside the castle dodge boiling oil and arrows, some enemy soldiers may actually manage to enter the castle. These soldiers engage in hand-to-hand combat with the soldiers inside the castle. Your body's specific defense, also called the **immune response,** is similar to the defending soldiers. The *lymphatic* (lim FAT ik) *system* includes special tissues and organs, such as your tonsils, appendix, and spleen. A transparent fluid called *lymph* moves throughout this system. Your blood and the lymph fluid carry many different types of white blood cells throughout the body. Other white blood cells remain in *lymph nodes,* the tiny masses of tissue found throughout the lymphatic system. All of these white blood cells act as soldiers, identifying and fighting the pathogens. Each type of white blood cell has its own special mission in the battle.

Discussion:
pages 362–63

▶ **What traps pathogens that enter the nose and throat?** {mucus from the mucus membranes}

▶ **What are *cilia*?** {hairlike projections lining the air passages}

▶ **What do cilia do?** {filter out pathogens and sweep trapped pathogens away from the lungs}

▶ **What does the body produce that can kill any pathogen that gets swallowed?** {hydrochloric acid in the stomach}

▶ **How do tears help protect the eyes?** {Possible answers: They wash dust and dirt away from the surface of the eyes. They contain a special substance that can kill some bacteria.}

▶ **What helps protect your ears from dust and pathogens?** {earwax}

▶ **How are the other defenses of the body categorized?** {nonspecific and specific}

▶ **Into which category does the inflammatory response fit?** {nonspecific}

▶ **What are some symptoms of the inflammatory response?** {Possible answers: swelling, redness, heat, pain}

▶ **Why are some fevers helpful?** {The higher temperature can kill some pathogens.}

▶ **What is a second nonspecific defense?** {special white blood cells that can surround and "eat" pathogens}

▶ **These special white blood cells are always present in your body. When does your body make more of them?** {during the inflammatory response}

▶ **What is another name for the body's specific defense?** {the immune response}

▶ **Which system includes special tissues and organs that are often involved with the immune response?** {the lymphatic system}

▶ **What are some special tissues and organs that are part of the lymphatic system?** {Possible answers: tonsils, appendix, spleen}

💡 **If a person has his spleen removed, why does he need to be more careful to avoid exposure to pathogens?** {Possible answer: The spleen is part of the body's immune response. Without it the body's defenses are weakened.}

▶ **How are white blood cells carried throughout the body?** {by the blood and the lymph fluid}

📖 Although disease is a result of Adam's sin, the immune system demonstrates God's mercy. Without the immune system, man would not be able to live to glorify God.

Discussion:
pages 364–65

 Refer the student to the diagram on Student Text pages 364–65 as the questions are discussed.

➤ How soon does the body respond to pathogens? {very quickly—often in just seconds}

➤ Which pathogen is pictured here as attacking the immune system? {bacteria}

➤ The diagram shows the response of many kinds of white blood cells to a pathogen. What type of white blood cell can surround and eat a pathogen? {macrophage}

➤ How do macrophages help other white blood cells identify the pathogen? {Little pieces of the pathogen that macrophages surrounded and "ate" are displayed on the outside of the macrophage.}

➤ Which kind of white blood cell receives the identification from the macrophage and sends chemical messages to other white blood cells? {a Helper T-cell}

➤ What do memory cells do? {store information about the pathogen}

➤ Why do memory cells store information about the pathogen? {so that the immune system can respond faster if this type of pathogen enters the body again}

💡 Do you think that the immune system responds more quickly to exposure to a new pathogen or to a pathogen that it has been exposed to before? {to a pathogen it has been exposed to before} Why? {Memory cells have been produced to help identify and produce antibodies for that pathogen.}

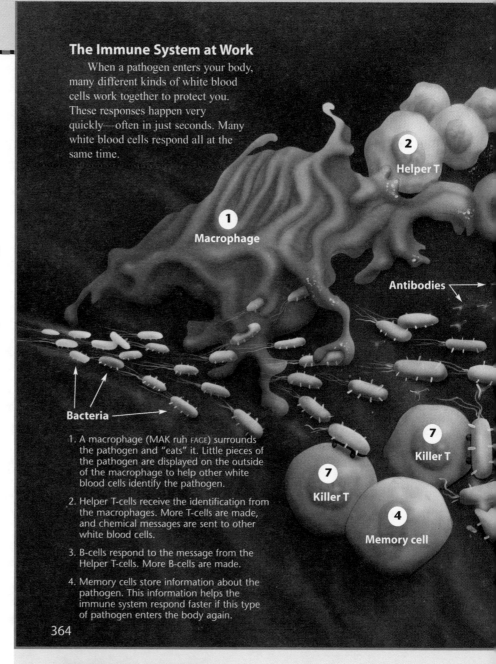

The Immune System at Work

When a pathogen enters your body, many different kinds of white blood cells work together to protect you. These responses happen very quickly—often in just seconds. Many white blood cells respond all at the same time.

1. A macrophage (MAK ruh FAGE) surrounds the pathogen and "eats" it. Little pieces of the pathogen are displayed on the outside of the macrophage to help other white blood cells identify the pathogen.
2. Helper T-cells receive the identification from the macrophages. More T-cells are made, and chemical messages are sent to other white blood cells.
3. B-cells respond to the message from the Helper T-cells. More B-cells are made.
4. Memory cells store information about the pathogen. This information helps the immune system respond faster if this type of pathogen enters the body again.

SCIENCE BACKGROUND

The immune system reaction—The immune system reaction varies slightly from the cycle illustrated when the pathogen is a virus.

B-cells and T-cells—B-cells mature in bone marrow. They produce antibodies to fight against the invading pathogen. T-cells mature in the thymus. They help produce antibodies but also are responsible to recognize and kill cells that are foreign to the body or that have been infected by a pathogen.

Macrophages—Some macrophages are free macrophages, moving throughout the body cleaning up pathogens and cellular debris. Other macrophages are fixed and remain in places such as the lymph nodes, liver, spleen, and red bone marrow.

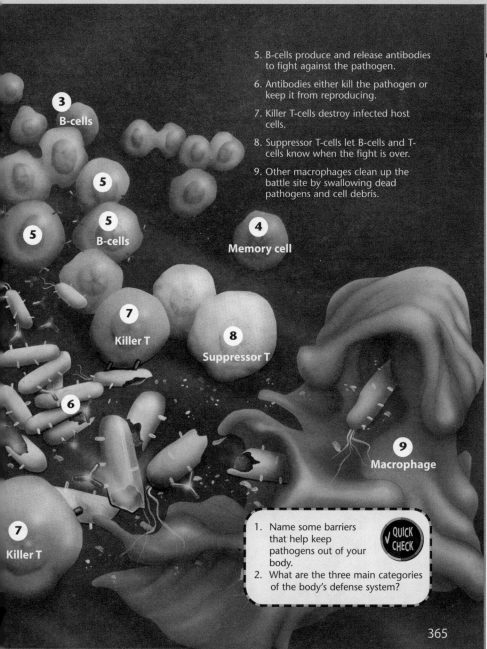

5. B-cells produce and release antibodies to fight against the pathogen.
6. Antibodies either kill the pathogen or keep it from reproducing.
7. Killer T-cells destroy infected host cells.
8. Suppressor T-cells let B-cells and T-cells know when the fight is over.
9. Other macrophages clean up the battle site by swallowing dead pathogens and cell debris.

QUICK CHECK
1. Name some barriers that help keep pathogens out of your body.
2. What are the three main categories of the body's defense system?

365

 AIDS (Acquired Immune Deficiency Syndrome) is caused by the Human Immunodeficiency Virus (HIV). This virus attacks the Helper T-cells and prevents them from functioning properly within the immune system. Scientists do not know where the virus and disease originated, but the term AIDS began to be used in 1982.

Discussion:
pages 364–65

➤ Which type of white blood cells produces antibodies? {B-cells}

➤ What are two ways that antibodies can affect the pathogen? {They either kill it or keep it from reproducing.}

📖 Read Hebrews 4:12.

➤ Just as antibodies are powerful enough to kill invading pathogens, God's Word is powerful to fight and control sin. [BAT: 8b. Faith in the power of the Word of God]

➤ What are two other types of T-cells involved in the immune response? {Killer T-cells and Suppressor T-cells}

📖 Why do you think it is necessary for the body to have Suppressor T-cells? {Answers will vary. Elicit that God designed the Suppressor T-cells to end the body's battle against infection. Without these cells, the overabundance of white blood cells might cause other problems.}

💡 Why do you think Killer T-cells must destroy any host cells infected by a pathogen? {Possible answer: Some pathogens, such as viruses, use host cells to reproduce. If the host cell is not destroyed, the virus may still be able to reproduce.}

💡 What machine or occupation can a macrophage be compared to? {vacuum cleaner, garbage collector}

💡 Why is the immune response known as a specific response? {It identifies and fights against a specific pathogen.}

➤ Which type of white blood cell helps to clean up the battle site? {macrophages}

Answers

1. skin, scabs, sweat, natural body oils, mucus membranes, cilia, tears, earwax

2. defensive barriers, nonspecific defenses, and a specific defense

Lesson 155

Student Text pages 366–69
Activity Manual pages 237–38

Objectives

- Explain three ways that the body can obtain immunity
- Compare and contrast antibiotics and antibodies
- Identify some problems that occur when the immune system malfunctions

Vocabulary

antibody
memory cell
immunity
vaccine
antibiotic
allergen
autoimmune disease

Introduction

➤ Have you ever had to get a tetanus shot because of an injury? {Answers will vary.}

➤ Why do you think tetanus shots and other vaccines are needed? {Accept reasonable answers.}

Purpose for reading:
Student Text pages 366–69

➤ What are some ways that the body can receive immunity from pathogens?

➤ How are antibiotics different from antibodies?

➤ How is the immune system involved in an organ transplant?

Discussion:
pages 366–67

➤ What are *antibodies*? {special proteins made by white blood cells that can destroy pathogens}

➤ Why does the body usually respond to a pathogen faster the second time it is exposed to that specific pathogen? {Memory cells have stored information about the pathogen and the antibody that is needed to defeat it.}

➤ What is *immunity*? {special protection against disease}

➤ What kind of immunity does the body have after a person has had a disease, such as chickenpox? {active immunity}

Immunity

Perhaps you know someone who has had chickenpox. An airborne virus causes this disease. However, if someone has had chickenpox before, he usually cannot get the disease again. Once a pathogen has entered the body, certain white blood cells make antibodies. These **antibodies** are special proteins that can destroy pathogens. The immune system can then react faster the next time it is exposed to a certain pathogen. It will usually be able to defeat that pathogen before you get sick because some white blood cells store information about that pathogen. **Memory cells** are white blood cells that remember the enemy and the specific antibody needed to defeat it.

These white blood cells provide your body with **immunity,** special protection against disease. Immunity can happen in several different ways. For example, after a person has had the chickenpox, his body remembers the chickenpox pathogen. This *active immunity* allows his body to resist the disease if it meets that pathogen again.

Active immunity can also be provided through vaccines. A doctor can give you a **vaccine,** or a shot that contains dead or weakened pathogens. Your immune system reacts to the pathogens in the vaccine and stores information about them. A person who has had the chickenpox vaccine usually does not become ill when exposed to chickenpox.

A baby receives *passive immunity,* or temporary protection, from his mother. The antibodies produced by his mother's immune system are shared with the baby's immune system. This protects him until his own immune system begins to work. The baby will develop active immunity as his immune system begins to produce its own antibodies.

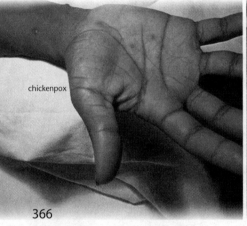
chickenpox

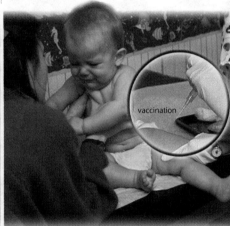
vaccination

SCIENCE BACKGROUND

Vaccines—Some vaccines, such as the measles vaccine, can provide immunity for a lifetime. Other vaccines, such as the tetanus vaccine, provide immunity for only a few years. Booster shots are then needed to keep the body protected from that disease.

Antibodies and Antibiotics

Your immune system produces antibodies that help destroy pathogens. These antibodies are able to destroy all types of pathogens, including viruses and bacteria. However, the white blood cells produce different antibodies for different pathogens. For example, the antibodies that can kill the chickenpox virus are not effective against the measles virus.

Antibiotics, though, are different. They also help your immune system destroy pathogens. However, your body does not make them. **Antibiotics** are chemicals made by microorganisms, such as fungi, that are able to destroy other microorganisms. The first antibiotic was discovered by accident. In 1929 Alexander Fleming noticed mold growing in a dish that contained some bacteria. This mold, penicillin, killed the bacteria. Later other scientists found ways to use the penicillin as a medicine. Today, many more antibiotics have been discovered. Some are synthetic, or manmade. Others are either made naturally from fungi and certain types of bacteria or are part synthetic and part natural.

Alexander Fleming

Antibiotics can work only against bacterial infections and some types of fungi. They cannot kill viruses. Some antibiotics are able to fight against many different types of bacteria. For example, the same antibiotic that fights strep throat may also be effective in helping to heal an infected cut. Other antibiotics can fight only certain types of bacteria. Each antibiotic destroys pathogens differently.

Doctors may prescribe antibiotics for certain bacterial diseases, such as bronchitis, pneumonia, and some ear infections. It is important to take all of each antibiotic that a doctor prescribes, even if you are feeling better. Failing to take all of the medicine can allow some of the bacteria to survive.

penicillin

Although at one time antibiotics were chemicals made only by microorganisms, scientists have learned to produce many antibiotics synthetically. The microorganisms are no longer needed for these antibiotics.

Discussion: pages 366–67

➤ What is another way to obtain active immunity? {vaccines}

➤ What is *passive immunity*? {The temporary protection that a baby receives from his mother's immune system until his own immune system begins to produce antibodies.}

💡 Why do you think young children tend to get sick more often than older children? {Answers will vary. Elicit that the body of a young child has not had the exposure necessary to develop antibodies to fight sickness.}

➤ How are antibodies and antibiotics similar? {They both help the immune system destroy pathogens.}

➤ What are *antibiotics*? {chemicals made by microorganisms that are able to destroy other microorganisms}

➤ How were antibiotics discovered? {Alexander Fleming discovered that a mold growing in a dish with bacteria was able to kill the bacteria. Later other scientists found ways to use this mold as a medicine.}

➤ Which are not made by the body—antibodies or antibiotics? {antibiotics}

➤ Can the body produce antibodies to fight all types of pathogens? {yes}

➤ Against which type of pathogens are antibiotics usually not effective? {viruses}

➤ Do all antibiotics work against only one pathogen? {No; most work against more than one pathogen.}

➤ Why is each type of antibody effective only against a specific pathogen? {White blood cells produce each antibody based on the body's response to a specific pathogen.}

➤ Why is it important to always take all of an antibiotic that a doctor prescribes? {Taking less may allow the bacteria to survive.}

Discussion:
pages 368–69

- Do all immune system malfunctions cause serious problems? {no} Explain. {Some are minor, but others can be life threatening.}
- What causes an allergic reaction? {Special white blood cells mistakenly identify a harmless foreign particle as a pathogen, causing the immune system to attack.}
- What is this type of attack called? {an allergic reaction}
- What is an *allergen*? {anything that causes the immune system to have an allergic reaction}
- Do all people have the same reactions to each allergen? {no}
- What are some common allergens? {Possible answers: dust, smoke, pollen, specific foods, insects, poison ivy}
- Some people have such severe reactions to allergens that they must keep an allergy kit of medication available.
- 💡 What is a *blood transfusion*? {the transfer of blood from one person to another}
- Why is it necessary for the blood types to match during a blood transfusion? {The immune system would attack unmatched donated blood.}
- What two things must match in order for an organ transplant to be successful? {the blood type and the tissue type}
- Are all organ transplants successful? Why? {No; the immune system will usually treat the new organ as an enemy and attack it.}
- What can be done to help prevent this from happening? {Medicine can be taken that suppresses, or limits, the immune system.}
- 💡 Why is it dangerous for a person with a transplant to get an illness? {The person with a transplant takes medication to suppress the immune system. A suppressed immune system may not be able to fight the illness.}
- 📖 When a person accepts Christ as Savior, God gives that person a new spiritual heart. The old nature is like an allergen or pathogen that continues to fight and attack the new nature. Only God can strengthen the new nature so that it can have victory over the old nature. [BATs: 8b Faith in the power of the Word of God; 8c Fight]
- If the body does not have very many white blood cells, how is the immune system affected? {The immune system is weakened and may lose the battle against an attacking pathogen.}

388 Lesson 155

Malfunctions of the Immune System

The immune system is a very important part of your body. It protects your body from pathogens that might attack it. Its main goal is to destroy the enemy pathogens before they can make you sick. Sometimes, though, the immune system malfunctions, or breaks down. Some malfunctions cause annoying problems, such as itching or congestion. Other malfunctions, however, can cause life-threatening situations.

Allergies

A special type of white blood cell is responsible for identifying any foreign particle that enters your body. If the particle is a pathogen, these white blood cells signal the rest of the immune system to attack. Occasionally, these white blood cells make a mistake. They might identify harmless foreign particles, such as pollen, as an enemy. The immune system attacks those particles, causing an *allergic reaction*. The pollen does not make you sick. It is the allergen that triggers your allergic reaction.

An **allergen** (AL uhr juhn) is anything that causes the immune system to have an allergic reaction. Common allergens include dust, smoke, mold, and pollen. Some people may have an allergic reaction to specific foods, such as milk or peanuts. Other people may have allergic reactions to bee stings, insect bites, or poison ivy. Allergic reactions can be mild, such as a runny nose or watery eyes. However, for some people, allergic reactions can be severe. Severe allergic reactions usually require medical treatment.

Transfusions and transplants

Sometimes the immune system needs to accept something that the body has not made naturally. For example, a person who has lost too much blood may need a blood transfusion. Healthy people can donate blood to give to people who need blood transfusions. However, the immune system accepts donated blood only if the blood types match. If the blood types do not match exactly, the immune system attacks the donated blood. These attacks usually create more problems for the person receiving the blood transfusion.

Because of illness or injury, some people need organ transplants. If the blood and tissue

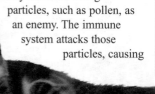
pollen

368

SCIENCE BACKGROUND

Blood transfusions—It is always better to give blood that matches. In an emergency, Type O can be given to anyone (universal donor), and someone with Type AB can receive any type (universal recipient). Certain precautions must be taken to reduce the risk of transfusion reaction, but it can be done successfully.

Cancer—With some cancers there is a minimal or no immune response. In these cases the body continues to identify the cancer cells as normal cells and does not attack them.

LANGUAGE

The prefix *trans-* means "across," "beyond," or "through."

- Which words other than *transfusion* and *transplant* begin with this prefix? {Possible answers: transcontinental, transfer, transform, transgress}

types match, organ transplants can be successful. The diseased or damaged organ is removed, and a donated healthy organ is put in its place. Usually the person's immune system treats the new organ as an enemy and attacks it. To prevent this from happening, a person who has received an organ transplant takes medicines that suppress, or limit, the reaction of the immune system. As a result, a transplant patient must be more careful to avoid being exposed to germs.

Immune deficiency

A weak immune system is unable to fight pathogens very well. It is deficient, or weak, because not enough white blood cells exist in the body. A low white blood cell count means that the immune system may lose the battle against a disease. Some diseases, such as certain cancers, kill the white blood cells that would identify the pathogen as an enemy. The remaining white blood cells do not attack the cancer or other pathogens because they have not been identified.

Autoimmune (AW toh ih MYOON) **diseases** happen when the immune system malfunctions and attacks the healthy cells that it should protect. Autoimmune diseases, such as multiple sclerosis, can affect the nervous system. Others affect certain endocrine glands. Some, like rheumatoid arthritis, affect the body's joints. These diseases are noncommunicable. Some may be inherited or may result from other major illnesses. Scientists do not know a cause for every autoimmune disease.

Many of the things that scientists know about our bodies and diseases are results of careful observation and teamwork. Great discoveries may be credited to only one or two people, but those discoveries were made possible by many observations that came before.

As we learn about God's creation, we also learn more about our Creator. His greatness should lead us to exclaim with Job, "Who knoweth not in all these that the hand of the Lord hath wrought this? In whose hand is the soul of every living thing, and the breath of all mankind" (Job 12:9–10).

1. How are antibodies and antibiotics similar? How are they different?
2. What is an allergen?
3. What is an autoimmune disease?

369

Discussion:
pages 368–69

➤ **Why does the body sometimes not recognize certain kinds of cancers as pathogens?** *{Cancer cells may have killed the white blood cells that would identify them.}*

➤ **What are *autoimmune diseases?*** *{malfunctions of the immune system that cause it to attack healthy cells}*

➤ **What are some examples of autoimmune diseases?** *{Possible answers: multiple sclerosis, arthritis}*

➤ **How are autoimmune diseases different from diseases like chickenpox or the flu?** *{They are noncommunicable and are not caused by a pathogen.}*

The word *disease* is sometimes used in place of *disorder*. A disease must be caught. A disorder is a malfunction. Autoimmune diseases are actually disorders.

Why do none of the new discoveries of scientists surprise God? *{God created all things and knows all things that exist.}*

Answers

1. Both antibodies and antibiotics help the immune system destroy pathogens. Antibodies are made by the body and can work against all types of pathogens. Different antibodies are made for each pathogen. Antibiotics are chemicals from microorganisms, and they only work against bacteria and fungi, not against viruses. Some antibiotics are effective against more than one pathogen.
2. anything that causes the immune system to have an allergic reaction
3. a malfunction of the immune system that causes it to attack healthy cells

Activity Manual
Study Guide—page 237
This page reviews Lesson 155.

Bible Integration—page 238
This page discusses bitterness as a spiritual disease.

Assessment
Test Packet—Quiz 15-B
The quiz may be given any time after completion of this lesson.

 In 1967, Dr. Christian Bernard made medical history by performing the first heart transplant. The patient lived twenty-one days before dying of pneumonia. Early organ transplant patients often died from infections because the drugs that helped suppress the body's rejection of the transplanted organ also suppressed the body's ability to deal with pathogens.

In the 1980s cyclosporin was introduced as an antirejection drug. It proved to be a beneficial drug that suppressed rejection while allowing the body to maintain a high degree of immunity. Researchers continue to develop better drugs for use by transplant patients.

In the 2000s surgeons began experimenting with transplanting an organ and bone marrow from the donor. The idea is that the donor bone marrow (which produces blood cells) would combine with the patient's bone marrow so that the patient's body would not produce antibodies against the new organ.

 Part of the last sentence on page 368 was inadvertently deleted from the first printing of the Student Text. The sentence should read "If the blood and tissue types match, organ transplants can be successful." Your student will need this information to complete Activity Manual page 237.

Lesson 156

Student Text page 370

Objectives
- Model the interaction between the immune system and pathogens

Materials
- prepared cards, TE pages A43–A48, *Blue Identity Cards* and *Red Identity Cards*
- See Student Text page

Introduction

➤ Throughout this chapter we have discussed how the immune system battles against pathogens. In today's activity, you will model some of the interaction between the immune system and pathogens.

Purpose for reading:
Student Text page 370

The student should read the page before beginning the activity.

Before the Activity
Display the blue identity cards.

➤ Which group do these cards represent—the immune system or pathogens? *{immune system}*

💡 Why do you think the spleen, bone marrow, and thymus have high point values? *{They are important in fighting pathogens. The spleen is part of the lymphatic system. White blood cells mature in the bone marrow and thymus.}*

During the Activity
Only a 60-point card can capture an ambulance.

Any 1-point card can capture a 60-point card.

The ambulance has no point-value and cannot capture another card.

The game ends when either ambulance is captured or when a set time limit has expired.

The home bases are used only for the collection of captured identities, not as a safe zone.

After the Activity
➤ Do you think this activity accurately models the warfare between pathogens and the immune system? Why? *{Accept reasonable answers. Some of the similarities may include: Pathogens can attack most parts of the body. Not all pathogens are the same strength. Some pathogens can defeat the immune system.}*

ACTIVITY: Defend and Capture

Process Skills
- Observing
- Communicating
- Defining operationally

The cells and organs of the immune system have many different functions. They work together to protect the body from pathogens. However, when any one part of the immune system is not working properly, pathogens can invade the body. In this activity, you will experience some of the "battle" between the immune system and pathogens.

Procedures

1. Divide the participants into two groups. One group is the immune system, and the other is the pathogens. Each group decides on a home base.

2. Get an identity card from your teacher. To participate in the game, you must have at least one identity card in your possession. At your teacher's signal, begin chasing your opponents. Lightly tag an opponent. Both you and your opponent must show an identity card.

3. A higher number captures a lower number. The person with the higher numbered card receives the opponent's identity card. If you both have the same number, no one is captured, and you may both continue playing the game. However, a 1-point white blood cell may capture a 60-point virus, and a 1-point bacteria may capture the 60-point spleen.

4. Do not tag the same person two times in a row. If you have lost your identity card, check your home base to see if there are any identity cards for your team that have not been used. If there are, you may continue playing. If no identity cards are left, you must sit out for the rest of the game.

5. Put captured identity cards in your team's shoebox. The game ends when one team captures the ambulance card of the other team. Add up the value of the captured identity cards. The team with the most points wins the game.

Materials:
red identity cards
blue identity cards
two shoeboxes or other containers

370

Teacher Helps

The red cards are the pathogens, and the blue cards are the immune system.

Establish the boundaries for the area that you will use to do this activity. A playground or gymnasium works best.

Choose a location away from the home bases for students to sit once they have lost their identity cards.

You may choose to play multiple rounds and keep a cumulative score.

Students may discover the advantage of protecting their 60-point card and ambulance.

SCIENCE PROCESS SKILLS

Using models—Models help us examine and experience things that are too large, small, or inconvenient to handle.

➤ What are some types of models that we have studied in science this year? *{Possible answers: earthquake, volcano, cell}*

➤ The earthquake, volcano, and cell were models that you could hold. Although this game cannot be held, it is still a type of model.

➤ What does the game model? *{the warfare between pathogens and the immune system}*

Explorations: Extra, Extra, Read All About It!

As a new reporter for a well-known magazine, you have recently been assigned to cover breaking medical news. Your editor insists on accuracy while reminding you that it is important for your magazine to be the first to publish articles about any new discoveries.

Competing reporters are racing to get the big scoop. Don't forget the basics of reporting—who, what, where, when, why, and how—and go cover that story!

What to do

1. Look back at the timeline on pages 352–53. Choose a person and headline that you would like to investigate.
2. Research the information available about your choice. Why is he or she important? What is so important about this discovery? What do other people think about this discovery?
3. You already know your headline, so go ahead and write your article. Check your dates, names, and places when you finish. Do be sure that you have not mentioned anything that would not be appropriate to the time era of this discovery!
4. If possible, try to include a picture of the person involved or some other pictures significant to your article.
5. Publish your article by presenting it to your classmates.

 Before researching, the student should prepare a sheet of paper to record his findings. He should list who, what, where, when, why, and how, leaving several lines after each to record information and notes.

You may choose for the student to write a newspaper article rather than a magazine article. If so, discuss the basic differences between the styles of writing found in a magazine and the styles found in a newspaper.

 Locate the places where the different discoveries took place.

Student Text page 371

Lesson 157

Objectives
- Research and write an article about a medical discovery

Materials
- encyclopedia

Introduction

➤ What is it like to hear about a new discovery for the first time?

➤ Does it sometimes seem too good to be true?

➤ Do you believe that the discovery is credible, or does your mind automatically start to question whether or not it is true?

➤ In today's exploration you will write an article announcing a new scientific or medical discovery as if you lived in the time era in which it happened.

Purpose for reading:
Student Text page 371
Timeline on pages 352–53

The student should read all the pages before beginning the exploration.

Discussion:
page 371

➤ If you were the first journalist to cover a great scientific discovery, what types of information would you include in your article? {Elicit that a good reporter or journalist answers the questions who, what, where, when, why, and how.}

➤ Think how you would use your article to convince the public of the importance of the discovery.

Assessment
Rubric—TE pages A50–A53

Select one of the prepared rubrics, or design a rubric to include your chosen criteria.

Lesson 158

Student Text page 372
Activity Manual pages 239–40

Objectives

- Recall concepts and terms from Chapter 15
- Apply knowledge to everyday situations

Introduction

Material for the Chapter 15 Test will be taken from Activity Manual pages 236–37 and 239–40. You may review any or all of the material during the lesson.

You may choose to review Chapter 15 by playing *Pathogen Attack* or a game from the *Game Bank* on TE pages A56–A57.

Diving Deep into Science

Questions similar to these may appear on the test.

Answer the Questions

Information relating to the questions can be found on the following Student Text pages:

1. page 356
2. pages 353, 367
3. page 368

Solve the Problem

In order to solve the problem, the student must apply material he has learned. The answer for this Solve the Problem is based on the material on Student Text pages 362–63. The student should attempt the problem independently. Answers will vary and may be discussed.

Activity Manual

Study Guide—pages 239–40

These pages review some of the information from Chapter 15.

Assessment

Test Packet—Chapter 15 Test

Lesson 159

This lesson is reserved for the last test and any activities or review that you may choose.

Diving Deep into Science

Answer the Questions

1. In tropical areas, people often sleep under netting to avoid being bitten by mosquitoes. Why is this a good practice?
   Answers will vary but should include the fact that some mosquitoes carry disease.

2. Why is it useless for a doctor to give someone an antibiotic if that person has a simple head cold?
   Viruses cause head colds, and antibiotics are ineffective against viruses.

3. Every spring the pollen causes you to get a stuffy nose and watery eyes. It also causes you to sneeze a lot. Why does your body react this way to pollen?
   For you pollen is an allergen. Your body is mistakenly trying to destroy the pollen as if it were a pathogen.

Solve the Problem

Your friend Timothy fell and skinned his knee while roller-skating. Most of the injury has now scabbed over, but some of the skin around the scab still looks a little swollen and red. Timothy usually likes to pick at his scabs and often scrapes them off. Explain to Timothy how the scabs, swelling, and redness are related to his immune system working.

Scabs provide a protective barrier for cuts. The redness and swelling are part of the inflammatory response to pathogens that gained entrance through the cuts.

Review Game
Pathogen Attack

Use the cards from Lesson 156 to play a review game. In addition to the cards, make two copies of the following chart.

1	25	5	40
20	60	10	1
40	15	45	25
55	50	25	30
25	50	1	10
35	40	5	30
1	20	40	5
55	25	15	1

Divide the students into two teams. Give one team the red pathogen cards and give the other team the blue immune system cards. Display a chart for each team.

Alternate asking review questions between the two teams. After each correct answer, the student draws the top card from his stack and marks the corresponding number on his team's chart. Return the card to the bottom of the pile. The first team to completely fill in the chart wins.

Variation: Prepare questions with varying levels of difficulty to match the card numbers. The student draws a card and then answers a corresponding question.

Appendix Contents

 Technology Lesson Plans

 Reproducibles

 Rubrics

 Games
Materials List

 Bible Action Truths
Bible Promises

Index
Photo Credits

TECHNOLOGY
lesson plans

A Technology Lesson correlates with each unit. Each lesson explores a current or future use of technology in a field covered in the unit.

Each of these lessons may be taught as part of its corresponding unit or with the other Technology Lessons at the end of the course.

The technology lessons are optional. The material covered in these lessons is not included on the quizzes or chapter tests.

Lesson 160

Activity Manual pages 241–42

Objectives

- Explain what an autonomous underwater vehicle is
- Explain the advantages of an underwater observatory
- Identify some ways that using an AUV may benefit studying the ocean

Materials

- 5 meters of rope or yarn
- a book

Introduction

Note: This technology lesson may be used in Unit 1. The Activity Manual pages for this lesson were updated after its first printing. The pages pictured in this TE contain the most recent edition of the text, so they may not match your student's copy exactly. We regret any inconvenience this may cause.

Tie the rope to the book. Place the book on the floor. Holding the end of the rope, drag the book as you walk between the desks until the rope tangles.

➤ **What problem did the book have when following my lead?** *{Possible answer: The book could not move once the rope became tangled.}*

➤ **Even if there were no obstacles, could I change the direction of the book easily?** *{no}*

➤ **Some underwater vehicles are attached, or tethered, to a survey ship similar to the way the book is attached to me. A vehicle that is tethered is limited in its ability to move. This lesson discusses underwater vehicles that are not attached to ships.**

Purpose for reading:
Activity Manual pages 241–42

➤ What is an autonomous underwater vehicle?
➤ What are some uses for autonomous underwater vehicles?

Discussion:
Activity Manual pages 241–42

➤ As scientists study the ocean, what are they using instead of dolphins in many situations? *{AUVs, or autonomous underwater vehicles}*

💡 What does *autonomous* mean? *{independent or self-contained}*

Autonomous Underwater Vehicles

Name _____

Scientists have used dolphins to map the ocean floor and locate mines in the sea. However, dolphins have some disadvantages. Even with high-tech equipment attached to them, they still become tired and hungry, and they cannot go everywhere. What if there was a dolphin that did not get tired, hungry, or uncooperative? Scientists have been working to create artificial "dolphins." These dolphins are called AUVs, or **autonomous underwater vehicles**. An AUV has no pilot and controls itself.

There are many different kinds of AUVs. Some short and thick AUVs look like blimps. Some have several propellers and look like three small submarines linked together. Some long and streamlined AUVs look like torpedoes. Most run on battery power, but scientists are developing solar-powered AUVs. AUVs are not built for speed, but they are built for endurance. They will go on missions that last for months or even years, slowly collecting information as they glide through the ocean depths. They are not tethered or attached to a survey ship, so they can go places that surface ships and manned submarines cannot go, such as in caves.

AUVs are able to operate by themselves because lists of instructions, or *mission scripts*, are programmed into them. Sometimes an AUV is not in contact with the survey ship after launch until the mission is complete. Sometimes an AUV receives instructions from a ship by acoustic signals, which travel through the water.

AUV arctic launch

Answer the questions.

1. What does the acronym (group of letters) AUV mean? _autonomous underwater vehicle_
2. What is a *mission script*? _a list of instructions that tells the AUV what to do_
3. Scientists are working on solar-powered AUVs. If solar energy were needed to recharge an AUV's batteries, how would the AUV be affected? _Answers will vary. The AUV would have to return to the surface periodically to recharge its batteries._

Lesson 160
Technology Lesson   241

➤ **What is an AUV?** *{an underwater vehicle that controls itself}*

💡 **An AUV is not tethered to the survey ship. How is this an advantage?** *{Possible answers: Since it is untethered, an AUV can go places that tethered vehicles cannot go.}*

➤ **How does an AUV know where to go?** *{A list of instructions, called a mission script, is programmed into the AUV.}*

➤ **When does an AUV contact the ship?** *{Some AUVs maintain contact throughout the mission. Others will not contact the ship until the mission is over.}*

➤ **How does an AUV stay in contact with the survey ship?** *{through acoustic signals}*

💡 **What are acoustic signals?** *{sound signals}*

➤ **How are AUVs powered?** *{battery power, although scientists are developing solar-powered AUVs}*

A4   Lesson 160

Grade 6 Science
Lesson 89

Inventors
(Use with Edison segment)

Name of Inventor _____

Country of birth _____

Where he lived _____

Family background _____
_____
_____

Most famous invention _____

How did his invention improve people's lives? _____
_____
_____
_____

How did his invention use electricity and magnetism? _____
_____
_____
_____
_____

Other Inventions _____
_____
_____

© 2005 Bob Jones University. All rights reserved.

AUVs are used to record information from the ocean. AUVs can be programmed to map the ocean floor, observe seismic activity, locate mines in a harbor, check undersea oil lines, and explore shipwrecks. AUVs can travel under the ice in the Arctic also. The only other sea vehicle that can go there is a nuclear submarine. The AUV is much less expensive and easier to work with.

AUVs have also been used by the military. AUVs can detect mines, take photographs, and operate very quietly. They can also be used for surveillance, similar to high-flying spy planes. An AUV named Remus was used by the U.S. Navy to detect mines in Operation Iraqi Freedom.

Seaglider is a small AUV owned by the University of Washington. *Seaglider* uses its wings and a buoyancy reservoir containing oil to move to higher or lower depths. All of the AUV's sensing and traveling equipment is housed in its torpedo-like body. *Seaglider* has an antenna on the back, which sends and receives information, such as instructions, findings, and GPS (Global Positioning System) location.

Scientists plan to build unmanned underwater observatories that will use AUVs to gather information. The Monterey Accelerated Research System (MARS) will be located near the Monterey Canyon in the Pacific Ocean off the coast of California. MARS will monitor the ocean in many differet ways and will be connected to land by fiber-optic and power cables that will send data directly to the Internet.

Autonomous underwater vehicles and underwater observatories are the doorways to discovering much more about what lies under the oceans. As technology continues to advance, scientists learn more and more about the mysteries of God's creation.

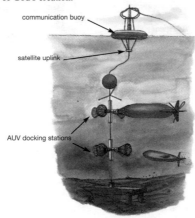

Answer the questions.

4. What are some uses of the AUV? _mapping the ocean floor, observing seismic activity, detecting mines, checking undersea oil lines, exploring shipwrecks, and taking photographs_

5. Why is an AUV a good vehicle to use to explore the Arctic? _Possible answers: It can stay underwater for long periods of time. It is less expensive and easier to work with than a nuclear submarine._

6. What other explorations might AUVs undertake? _Possible answers: biological research, chemical and mineral analysis of ocean water, measuring earthquakes, studying the migration of marine life, studying how the ocean regulates climate_

Lesson 160
242 Technology Lesson

Discussion:
Activity Manual pages 241–42

➤ How is *Seaglider* able to move up and down? *{Seaglider moves up and down by its reservoir of oil.}*

💡 Why do you think *Seaglider* must come near the surface to use the GPS satellites? *{Possible answer: GPS signals do not travel through water.}*

➤ What kinds of missions can an AUV be programmed to do? *{Possible answers: map the ocean floor; observe seismic activity; locate mines; check undersea oil lines; explore shipwrecks; travel under the ice in the arctic}*

📖 An AUV follows the commands it is given. Christians, too, should follow the leading of the Holy Spirit and the commands given to them in the Bible. [BATs: 6a Bible Study; 6c Spirit-filled]

➤ What advantages does an AUV have over a nuclear submarine? *{An AUV is less expensive and easier to work with than a nuclear submarine.}*

➤ What services will underwater observatories provide? *{Possible answer: monitor the ocean continually}*

➤ How will man receive data from an underwater observatory? *{through fiber-optic cables}*

➤ What is the name of the underwater observatory off the California coast? *{MARS, or Monterey Accelerated Research System}*

Activity Manual
Technology Lesson—pages 241–42

SCIENCE BACKGROUND

Acoustic signals—Because radio waves do not transmit well through water, sound waves are used even though they are not as fast. Data is compressed so that the AUV can transmit more data at one time. You may want to refer to Lesson 31 for further information about a GPS navigation system.

Underwater observatories—In addition to the Monterey Accelerated Research System, an even more complex underwater observatory is being developed. NEPTUNE is a joined project by the United States and Canada to observe seismic activity at the Juan de Fuca Plate, a volcanic area in the Pacific Ocean.

Lesson 161

Activity Manual pages 243–44

Objectives

- Explain how the spicules of a Rossella sponge are like optic fibers
- Identify ways that studying a Rossella sponge may improve fiber optic technology
- Recognize man's duplication of God's creation

Materials

- a world map or globe

Introduction

Note: This technology lesson may be used in Unit 2.

➤ Some of man's engineering accomplishments are based on organisms that God created. What do you think birds have shown us? *{how to fly}* hummingbirds? *{how to hover}* beavers? *{how to build a dam}*

➤ Even organisms in unlikely places give man new ideas. Today we will learn about a sponge that grows to be about one meter in length. It lives 100 to 200 meters below the ocean's surface.

Locate the Ross Sea on a world map or globe. The Ross Sea is located in the Antarctic Ocean just south of New Zealand.

Purpose for reading:
Activity Manual pages 243–44

➤ What engineering achievement can be seen in a glass sponge?

➤ What can scientists learn from the Rossella sponge?

Discussion:
Activity Manual pages 243–44

➤ What do glass sponges have in common with glass? *{Both consist of silicon dioxide, or silica.}*

➤ How does the sponge use silica in the seawater? *{It makes glassy spines.}*

➤ What are these glassy spines called? *{spicules}*

A spicule (SPIHK yool) is about as thick as an office staple and grows to be 10 to 20 centimeters long.

➤ What three functions do the sponge's spicules have? *{support the sponge, protect it, and act like optical fibers}*

💡 What are *optical fibers*? *{slender filaments of glass that transport information in the form of light pulses}*

Fiber Optic Sponges   Name_____

Hundreds of meters deep in the frigid waters of Antarctica's Ross Sea lives a sponge whose body can conduct light. The *Rossella racovitzae* belongs to a class of sponges called glass sponges. These sponges have skeletons made mostly of silicon dioxide, or silica, the major component of glass.

God has designed this sponge to take silica from the seawater that flows through its tube-shaped body. The sponge then makes glassy spines that stick out from its body. These thin glassy spines, called *spicules*, support and protect the sponge, but they also act like optical fibers. Light enters these slender glass strands and travels through them to the base of the spicules. At the base of the spicules live tiny unicellular organisms. Scientists think that these organisms use the light for photosynthesis and thus provide food for the sponge.

In the 1990s scientists began to study the ability of this sponge to gather light. Biologists beamed the light from a red laser down a sponge's spicule. They observed the smooth transmission of the light through the thin fiber. They found that the spicules could gather light from many different angles. This ability is particularly important because of the sponge's dark environment. It allows the sponge to use almost any light in its deep habitat. Direct light is not necessary.

Scientists also found a small cross-shaped cap at the top of each spicule. This four-pointed cross appears to allow the sponge to gather more light. This complex system of guiding light is similar to commercial fiber optics designed by trained engineers.

sponge spicules

Answer the questions.

1. What do a glass sponge's spicules and optical fibers have in common? <u>They both conduct light.</u>

2. How do unicellular organisms and sponges both benefit from the light-transmission ability of the sponges? <u>The unicellular organisms receive the light they need for photosynthesis and then produce food for the sponge.</u>

3. How does the ability of the light-guiding glass sponge point to an all-wise Creator? <u>God has given the glass sponge a complex system that, in the commercial world, has to be specially designed by trained engineers.</u>

4. What feature of the sponge's spicule appears to help it gather more light? <u>a four-pointed, cross-shaped cap at the end of the spicule</u>

SCIENCE BACKGROUND

Venus Flower Basket—Another glass sponge being researched for its fiber optic similarities is the *Euplectella*, or the Venus Flower Basket. Found in the ocean in the tropics, this sponge grows to about 15 cm (6 in.) in length.

Fiber optic cables—Optical fibers are long filaments of glass that are about the same thickness as a human hair. The fibers are arranged in bundles called *optic cables*, which pass light signals over long distances. Compared to conventional copper wire, optical fibers are lower-priced, thinner, better suited for carrying digital information, non-flammable, and better able to carry a signal. Because optical fibers are thinner than wire, more fibers can be bundled together.

You may choose to have a guest speaker address the use of fiber optics in his profession. Some of the fields that use fiber optics are medicine, telecommunications, computer technology, plumbing, and aircraft inspections.

Until recently scientists were not sure what they could learn from the Rossella sponge. Its abilities were interesting, but they were not very useful. Laboratory tests showed that man-made optical fibers worked better than the sponge did. However, some scientists now think that the Rossella sponge may help improve the design of optical fibers. They point out that the sponge probably performs better in its natural surroundings. Now scientists are very interested in the sponge's ability to conduct light in the extreme cold of the Ross Sea. Man-made fibers need special chemical treatments to work there.

The spicules of the Rossella sponge have a different structure and chemical makeup than man-made fibers have. The spicules are tough but flexible. Bending optical fibers or spicules may cause small cracks. A crack in a man-made fiber often causes the whole fiber to break. However, the Rossella sponge is layered. Its spicules stop cracks after only a few layers are penetrated. The spicules can bend and still gather light. Scientists think that layers might improve the strength and flexibility of optical fibers.

Sponge spicules can collect light from many angles. If engineers could copy this ability, they might be able to lower the cost of fiber optics. They might also make fiber optics more useful in places with limited light, such as in outer space. Scientists can once again go back to God's creation to make man's inventions better.

Rosella racovitzae (sponge)

Answer the questions.

5. Why are scientists interested in how the glass sponge works in its natural habitat? _Possible answers: It may conduct light better there than it does in a laboratory; the sponge performs in cold water, but man-made optical fibers need chemical treatments to work there._

6. What are some ways a sponge has a better structure than man-made fibers? _Its layered structure can stop cracks better; its chemical makeup allows it to conduct light in cold conditions._

7. What ability of the sponge could lower the cost of fiber optics if successfully copied in man-made fibers? _the ability to collect and conduct light from many different angles_

8. What are some situations in which optical fibers could be used if engineers could produce them to be like the sponge spicules? _Answers will vary but may include space exploration and deep-sea research._

Direct a DEMONSTRATION

Demonstrate how an optical cable transmits light

Materials: 3 cardboard tubes, tape, black paper, flashlight, 2 small mirrors

  Tape the tubes together as illustrated.

Hold the sheet of black paper 10 cm from one end. Shine the flashlight into the other end.

▶ **Does the light shine onto the black paper?** *{yes}*

Bend the tube as illustrated. Hold the black paper and shine the flashlight as before.

▶ **Does the light shine onto the black paper? Why?** *{No; because light travels in a straight line.}*

Tape a mirror at each bend in the tube as illustrated. Hold the black paper and shine the flashlight as before.

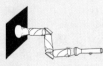

▶ **Does the light shine onto the black paper? Why?** *{Yes; because the mirrors reflect the light.}*

An optical fiber or cable works in much the same way. The light travels through the core of the fiber (the paper tube) by constantly bouncing from the material surrounding the core (mirrors). In this way the light wave can travel great distances, because it is not absorbed.

Discussion:
Activity Manual pages 243–44

Fiber optics are used in telecommunications, computer networks, and medical technology. They are also used to inspect pipes and engines.

▶ **What can be found at the base of the spicules?** *{unicellular organisms that use the light to produce food for the sponge}*

▶ **The spicules of a glass sponge are able to collect light from many different angles. Why is this ability so important?** *{This ability enables the sponge to use almost any light available in its dark environment.}*

▶ **What part of the spicule seems to help the sponge gather more light?** *{the cross-shaped cap at the end of the spicule}*

📖 **The Rossella sponge glows with light in a dark environment. Matthew 5:14–16 tells Christians to shine as lights for Christ in a dark, sinful world.**

💡 **Why do you think the laboratory tests showed that the man-made optical fibers worked better than the sponge?** *{Possible answer: The sponge probably performs better in its natural habitat.}*

▶ **What do man-made optical fibers need in order to work in the sponge's cold environment?** *{special chemical treatments}*

▶ **How do the spicules of the Rossella sponge differ in structure from man-made fibers?** *{The spicules are layered, causing them to be tough and flexible. The spicules are able to stop cracks from penetrating more than a few layers. Cracks in man-made fibers, however, often cause the whole fiber to break.}*

▶ **How do scientists think adding layers might benefit man-made fibers?** *{The layers might improve the strength and flexibility of the fibers.}*

▶ **How would man-made fibers become more beneficial if engineers could imitate the sponge's ability to collect light from many angles?** *{Fiber optics might become less expensive. They might also be able to be used in places with limited light.}*

📖 **Man has benefited from this amazing sponge. Who designed and created the sponge?** *{God}*

Activity Manual
Technology Lesson—pages 243–44

Lesson 162

Activity Manual pages 245–46

Objectives

- Explain how electromagnets are used in maglev trains
- Identify some ways a maglev train may benefit the environment and transportation

Introduction

Note: This technology lesson may be used in Unit 3.

In the 1890s the fastest way to travel was by the train called *Empire State Express*. The engineer, Charles Hogan, took the steam engine along a 36-mile straightaway, reaching a speed of 112.5 mi/h, the fastest a human had gone up to that time.

➤ Do you think the engine could go that fast while pulling a load? *{probably not}*

➤ What happened in Kitty Hawk, NC, in the early 1900s? *{Orville and Wilbur Wright flew the airplane.}*

➤ How has airplane travel progressed? *{Possible answer: from the first plane in North Carolina to passenger planes to the stealth fighter}*

As highways and airports become busier, people look for more efficient ways of traveling that will have little negative impact on the environment.

Purpose for reading:
Activity Manual pages 245–46

➤ What kind of train can be suspended above its track?

➤ How does this type of train work?

Discussion:
Activity Manual pages 245–46

➤ What kind of train is suspended above a track? *{a maglev train}*

➤ What does *maglev* stand for? *{magnetic levitation}*

➤ What kind of magnets does a maglev train use? *{electromagnets}*

➤ How does an electromagnet work? *{A coil of wire with a core produces magnetism when a current of electricity passes through it.}*

➤ What does *levitate* mean? *{to lift}*

➤ How does a maglev train levitate? *{Electromagnets in the guideway repel the magnets on the bottom of the train.}*

➤ 💡 What principle of magnetic attraction does the train use? *{Like poles of magnets repel each other.}*

Lesson 162

Magnetic Levitation   Name_____

When you think of trains, what comes to your mind? Perhaps you see smoke billowing from a smokestack or hear the loud clacking of wheels turning on a track. Maybe you think of a caboose or of an engineer up front blowing the whistle. Perhaps you can feel the vibrations of a subway rushing into an underground station. Or maybe you picture a train quietly whooshing by, suspended above a specially designed track.

Suspended above a track! What kind of a train would do that? Magnetic levitation trains, or maglev trains, do exactly that. The maglev train does not use a conventional track. Instead, electromagnets propel it along a guideway. Because the train does not touch the track, it runs very quietly.

The maglev train works by using guide rails lined with metal coils that are connected to a power supply. An electric current flows through the coils, thus creating an electromagnet. There are also magnets located on various parts of the train cars. These magnets have many uses. Some of the magnets stabilize the train so that it will not tip over. Other magnets propel the train forward. The magnetized coil in the guideway repels the magnets on the bottom of the train, causing the train to lift, or levitate, about ten millimeters off the track. The train floats above the track on this cushion of air.

Opposite magnetic principles can be used to propel the train. One train uses a traveling magnetic field's repulsion to push the train forward. Other trains use the magnetic attraction to pull the train along. Some trains even use a combination of the two principles.

Maglev Guideway

Answer the questions.

1. How high does the train levitate off the track? _about 10 mm_
2. What does levitate mean? _to lift_
3. What principle of magnets causes the train to "float" above the track? _Answer should include that since like poles repel each other, the large wire coils repel the magnets on the underside of the train._
4. What benefits to the environment might this train provide? _Possible answers: The train uses magnets to power it instead of burning fossil fuels. Electromagnets for the train can be obtained through renewable resources. The relatively quiet trains do not disturb the surrounding animals and people._

Lesson 162 Technology Lesson 245

SCIENCE BACKGROUND

United States—The "Transportation Equity Act for the 21st Century," passed by Congress in 1998, created a National Magnetic Levitation Transportation Technology Deployment Program. The program was designed to show high-speed maglev technology in commercial service.

Old Dominion University—On its campus located in Norfolk, VA, Old Dominion University has built a maglev train guideway that is 1036 m (3,400 ft) long. The 14 m (45 ft) car is slightly bigger than a bus and can carry 100 students at about 64 km/h (40 mi/h).

Types of systems—Different systems of electromagnets exist. The Japanese use supercooled, superconducting electromagnets for maglev trains. This type of electromagnet allows trains to run without electric power. The German version's electromagnets must have power to operate.

Japan—Japan has built two maglev lines, the first in the 1960s and a second in 1996. One train north of Mt. Fuji runs on U-shaped guideways. It reached a speed of 550 km/h (334 mi/h) in January 1998.

Germany—The German system uses the Transrapid Levitation system in Emsland, Germany. Thousands of passengers have used the system. The Transrapid system can also be used commercially.

China—The first commercially operated maglev train in the world travels between the Shanghai and Pudong airports at a speed of 430 km/h (267 mi/h) along a 30 km route. The Transrapid Shanghai uses German technology.

A8

The maglev train can move at very high speeds. It floats on a cushion of air, so there is no friction with the track to slow it down. Also, its aerodynamic design reduces wind resistance. These factors and others allow maglev trains to go much faster than regular trains. They have been recorded going more than 500 km/h (310 mi/h)!

Maglev train cars are designed to carry large amounts of freight as well as passengers. To help reduce the amount of traffic on roads in the future, tractor-trailers will be able to be parked, fully loaded, in one of the cars and taken to another city. Carrying trucks on trains would clear up traffic on roads and save on gasoline. If a maglev train left New York City and had a straight track to Dallas, Texas, it would arrive in just over 5 hours. That is a 26-hour car ride! With these trains, you would be able to drive your car onto the train, enjoy the dining car or the sleeper cars, and get off at Grandma's house with your car.

Because the trains have no moving parts, there is less maintenance to do. There are no wheels to break down or axles to fix. This makes the cost of maintaining the trains relatively low. However, because of the complicated guiderails, the trains are very expensive to build. Also, the trains can run only on the guiderails built specifically for them. They cannot use railroads previously built for other trains.

Although maglev trains are not in common use yet, China, Japan, the United States, and Germany are developing them. However, the high cost has slowed the development of these trains severely. But someday maglev trains may be more popular than airplanes for long-distance, high-speed travel.

TRANSRAPID INTERNATIONAL

Answer the questions.

5. What two things that hinder the speed of other trains are eliminated in the design of the maglev train? _friction with the track and wind resistance_

6. How fast has the train been recorded going? _more than 500 km/h (310 mi/h)_

7. How many kilometers is it from Dallas, Texas, to New York City? (Hint: Multiply the kilometers per hour by how long it takes to get there by maglev train.) _about 2,500 kilometers_

8. What are two disadvantages of the maglev train? _Maglev trains require expensive guideways to travel on. Also, they cannot use existing railroad tracks._

Lesson 162
246 Technology Lesson

Science 6 Activity Manual

Discussion:
Activity Manual pages 245–46

📖 In the same way as magnets attract and repel, a Christian should be attracted to Christ and godly living and should repel sin. [Bible Promise: D. Identified in Christ; BAT: 1c Separation from the world]

▸ How high does a maglev train levitate? {about 10 mm (0.39 in.)}

Different trains levitate different amounts. Some trains are not raised when they are in the stop position. Others maintain their levitation at all times.

▸ How does the train move forward? {Magnets attract and repel to move the train.}

〰️ Look at the diagram *Maglev Guideway*. After the train is raised, electrical power flows through the coils in the guideway to create magnetic fields. The current alternates constantly to change the polarity in the magnetized coils. This change causes the magnetic field in the front of the train to pull forward. At the same time the magnetic field behind the train adds more thrust.

💡 What three things does a maglev train need to operate? {a source of electricity, metal coils in the guideway, and magnets underneath the train}

💡 What do you think happens if the power, or electricity, goes out? {Answers will vary.}

Some maglev trains have an emergency battery power supply. The maglev train in Japan also has rubber wheels as a backup. It uses its wheels until liftoff speed (100 km/h or 62 mi/h) is attained.

▸ What kind of track does a maglev use? {a guiderail}

▸ Can a maglev train run on traditional railroad tracks? {no}

▸ What factors help the maglev train go faster than regular trains? {no friction; its aerodynamic design}

💡 How would a maglev train affect the freight industry? {Possible answers: reduce the time for a truck to drive to its destination; not as much wear and tear on the vehicles; not as much fuel used}

💡 How do you think maglev trains might benefit the environment? {Possible answers: No fossil fuels are used. They are not as noisy as conventional trains.}

Activity Manual
Technology Lesson—pages 245–46

Direct a DEMONSTRATION
Demonstrate magnetic levitation

Materials: shoebox, masking tape, twelve disc-shaped magnets

Prepare the shoebox before the lesson. Cut one short end from the shoebox. Trim all the sides from the lid so that the remaining piece of cardboard will lay inside the box without touching the sides. Tape 3 magnets along both long edges of the bottom of the box. Tape 6 magnets in matching positions on the underside of the lid. Place the lid, magnet side down, gently on the bottom of the box. Adjust the magnets as needed until each pair repels.

▸ Why does the lid not touch the bottom of the box? {The magnets are repelling each other, causing the lid to float.}

▸ What is it called when something floats? {levitation}

▸ How is this model like a maglev train? {Possible answer: Both use magnets to levitate. The sides of the box act as guiderails.}

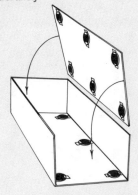

Appendix

Lesson 162

A9

Lesson 163
Activity Manual pages 247–48

Objectives
- Describe some types of inflatable spacecraft
- Understand the basics of inflatable technology
- Explain the advantages of inflatable spacecraft

Materials
- package of balloons

Introduction

Note: This technology lesson may be used in Unit 4. The Activity Manual pages for this lesson were updated after its first printing. The pages pictured in this TE contain the most recent edition of the text, so they may not match your student's copy exactly. We regret any inconvenience this may cause.

Display the package of balloons.

➤ **If you were asked to bring 25 balloons to a party, how would you carry them—blow them up at home or take the balloons to the party and then blow them up?** *{take the balloons to the party and then blow them up}*

Inflate one balloon. Compare the amount of space that the entire package takes up compared to just one inflated balloon.

Purpose for reading:
Activity Manual pages 247–48

➤ How are scientists applying what we know about balloons to travel in outer space?
➤ What types of spacecraft use this technology?

Discussion:
Activity Manual pages 247–48

➤ **What is one of the main reasons for space research today?** *{to help cut down high costs for space equiment}*
➤ **What are scientists beginning to use to help with the expense of space travel?** *{inflatable materials}*
➤ 📖 **We are to be good stewards of all God has provided for us. How are these scientists being good stewards?** *{They are trying to reduce the cost of space exploration.}*
➤ 💡 **What is an *acronym*?** *{a word formed from the initial letters or parts of a series of words}*
➤ **What does the acronym NASA represent?** *{National Aeronautics and Space Administration}*
➤ **What words does the acronym ARISE represent?** *{Advanced Radio Interferometry between Space and Earth}*

Lesson 163

Inflatable Spacecraft   Name_____

How would you like to live in space, or maybe even on Mars, inside something that looks like a giant balloon? Though it sounds strange, this may be a reality in the future because of research at NASA.

Researchers spend billions and billions of dollars on space equipment. To help cut down these high costs, scientists research ways to make space equipment cheaper. One way they have discovered that will help make spacecraft less expensive is to make some of the equipment inflatable. The equipment will be lifted into orbit uninflated, but once in orbit it will fill with gas and can become huge.

Inflatable spacecraft has many advantages. It is smaller in size and lighter than other types of spacecraft, which makes it cheaper to launch into orbit. Also, inflatable equipment takes up less space. One of the first inflatable objects in space was an inflatable telescope. In 1996 the Space Shuttle *Endeavor* deployed a telescope that inflated and could be seen from earth. The telescope cost much less than a metal one would have cost. NASA plans to use inflatable telescopes for the Advanced Radio Interferometry between Space and Earth (ARISE) project. Interferometry is the use of antennae at various locations to produce a three-dimensional image. The ARISE project will use several telescopes on Earth and in space to view distant objects in space, such as black holes, with a very high resolution. This new technology will allow man to see even more of God's majestic creation.

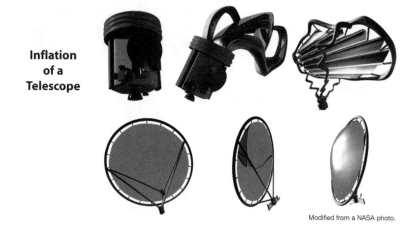

Inflation of a Telescope

Modified from a NASA photo.

Answer the questions.

1. Why have scientists turned to inflatable spacecraft? _Inflatable spacecraft are easier to use and less expensive to assemble than metal spacecraft._
2. What does ARISE stand for? _Advanced Radio Interferometry between Space and Earth_

Lesson 163
Technology Lesson   247

Inflatable technology will be used for more than just telescopes. NASA plans to explore other planets using inflatable materials. For example, NASA is developing an all inflatable planetary rover. Also, they want to use inflatable balloons powered by the Sun's heat to fly around the atmosphere of other planets.

NASA even plans to build inflatable habitats for astronauts to live in as they do their research. They will work the same way an astronaut's space suit works, only on a larger scale. Scientists are developing inflatable space habitats for Mars, the Moon, and space stations. One of the ideas for an inflatable habitat is called TransHab. TransHab would have three stories, including a kitchen, computers, and exercise equipment. The inflatable outside walls would be made of thick, bulletproof material which could withstand small meteors.

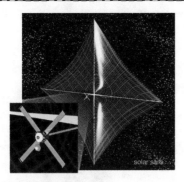
solar sails

inflatable space habitat

Another example of inflatable space equipment is inflatable solar sails. Solar sails are powered by the Sun, requiring no engines. The inflatable solar sail would be sent into orbit in a small container that could fit in the trunk of a car but then would inflate to a huge size—maybe as big as a football field! Then the solar sail would travel through the galaxy, collecting information and pictures and sending them to Earth.

As more spacecraft and space equipment become inflatable, the possibilities for studying God's universe increase greatly. Inflatable spacecraft give men the opportunity to travel farther and for longer periods than ever before.

Answer the questions.

3. How large might some solar sails be? _as big as a football field_
4. Why do the walls of inflatable habitats need to be thick and strong? _to withstand small meteors_

5. Which kind of inflatable do you think would be the hardest to design and build? Why?
Answers will vary, but the inflatable living quarters would probably be the hardest to design because of safety concerns and because many items needed in the living quarters would be difficult to make inflatable.

Discussion:
Activity Manual pages 247–48

➤ **What is *interferometry?*** {the use of antennae at various locations to produce a three-dimensional image}

➤ **What is a *black hole?*** {an area of extreme gravitational force that may result from the collapse of a massive supergiant star}

➤ **How large can a solar sail be?** {the size of a football field; 4,900 m^2, or 52,900 ft^2}

➤ **How will solar sails work?** {They are powered by the Sun.}

➤ **What is the name of the planned space habitat?** {TransHab}

➤ **Of what kind of material will TransHab be made?** {a bulletproof substance}

➤ **How many stories would TransHab have?** {three}

➤ **Would you like to go to space and live in an inflatable space habitat such as TransHab?** {Answers will vary.}

Activity Manual
Technology Lesson—pages 247–48

MATH

According to a 2002 study by a company called Futron, the cost of transporting 454 g (1 lb) of cargo into orbit is about $10,000. Direct the student to calculate the estimated cost of a student going to space as cargo.

Since people are not really cargo, encourage the student to research to find the weight of a space suit, food, water, and other essential supplies to include in the cost of transporting the student into space.

Discuss ways the load could be lightened.

Direct the student to calculate the cost of carrying other items into space, such as the total weight of the materials used in one of the experiments in SCIENCE 6.

Note: Mass is usually used when calculating payload. For convenience, weight is used for this activity.

Lesson 164

Activity Manual pages 249–50

Objectives

- Identify characteristics of the thale cress
- Explain how genetic engineering produces glowing plants
- Recognize that scientists use the same basic methods that Mendel used

Introduction

Note: This technology lesson may be used in Unit 5.

➤ **Plants cannot communicate in the same ways that humans do. How might a gardener be able to determine the health of his plants?** *{Possible answers: by checking the condition of the leaves and stems; checking whether or not the plant bears fruit or produces seeds; checking the roots; checking the condition of the soil}*

➤ **The plants discussed in today's lesson will be located far away from people. Yet scientists will be able to monitor the health of these plants as well as their environmental conditions. These special plants will be able to communicate information to the scientists.**

Purpose for reading:
Activity Manual pages 249–50

➤ Why is thale cress useful for genetic research?

➤ Genes from what two animals have been added to thale cress to cause the plants to glow?

Discussion:
Activity Manual pages 249–50

➤ **What weed have some scientists been using for their genetic research?** *{Arabidopsis thaliana, or thale cress}*

➤ **Why has this plant been useful for genetic research?** *{Possible answers: Thale cress is similar to many other plants, has a simple DNA structure, is small, matures quickly, and produces many seeds.}*

➤ **How many genes does a thale cress plant have?** *{25,000}*

💡 **These genes are made up of millions of DNA base pairs. Why do you think scientists consider this plant to have a small amount of DNA?** *{Other organisms have many more genes and base pairs.}*

➤ **Scientists have added different genes to thale cress. What did these genes cause the plant to do?** *{glow}*

➤ **What insect's gene caused the thale cress to glow when the weather was too cold for the plant to grow?** *{firefly}*

A12 Lesson 164

Glowing Space Plants

Name _____

Weeds are not usually considered useful to humans, but scientists have found one weed from the mustard plant family that works well for genetic research. Genetics researchers call *Arabidopsis thaliana*, or thale cress, the "model plant" because it is very similar to many other plants. It has a simple DNA structure, matures quickly, and produces many seeds. In the year 2000, scientists were able to map its genome (JEE nohm), or genetic information. Thale cress has 25,000 genes and about 140 million base pairs of DNA. Because scientists know exactly what DNA the weed has, they can insert special marker genes from other organisms into its genetic structure. Scientists have discovered that adding these genes can make the plants glow under certain conditions.

Arabidopsis thaliana

Scientists inserted a gene from a firefly into the thale cress to make the plant glow when it was cold. The harder it was for the plant to grow because of the cold, the more it glowed. Scientists took the plants that glowed the least and cross-pollinated them to produce a plant that resists the cold. Because scientists have mapped the whole genome, they can now isolate the cold-resistant gene and put it into other plants.

Scientists have also altered some thale cress plants so that they glow when they are sick. The scientists inserted the firefly's gene where the plant's defensive response genes were located. Then they infected a group of genetically altered plants with a virus and watched. The plants that glowed the least were pollinated with each other to make a more disease-resistant plant.

Answer the questions.

1. What weed have scientists been using for genetic research? _Arabidopsis thaliana, or thale cress_
2. Why is thale cress good for genetic experimentation? _It is similar to other plants, matures quickly, produces many seeds, and has a simple DNA structure. Its genetic structure has been mapped._
3. List two conditions that caused the thale cress to glow. _coldness, sickness_
4. How is what scientists do to produce cold- or disease-resistant plants similar to Mendel's experiments with pea plants? _Mendel cross-pollinated plants that had certain traits to produce plants with the traits that he wanted. Today's scientists also cross-pollinate the plants to produce plants with traits that they want. (Note: Related information about Mendel's experiments is found in Chapter 13 of the Student Text.)_

Lesson 164
Technology Lesson 249

SCIENCE BACKGROUND

Arabidopsis thaliana plants have been in space before. Thirty-six plants were sent to space aboard the space shuttle *Columbia* in 1999. These plants were genetically engineered with a gene that would cause a stain on any plant that experienced stress. The scientists, a team of researchers from the University of Florida, used the plants to test stresses such as carbon dioxide levels, temperature, and extreme light.

The study of *Arabidopsis thaliana* will not be limited to Earth. Scientists hope to put several varieties of these plants on Mars in 2007. Scientists plan to insert more marker genes into the plants to see if the plants have difficulty growing in the soil on Mars. One of the marker genes comes from a Pacific coast jellyfish that glows with a soft green color. With these marker genes, any plant that is stressed or experiencing difficulty will begin to glow.

When the plants land on Mars, they will not be exposed to the actual surface of Mars. The spacecraft's robot will collect a sample of soil. It will then plant the seeds in soil in a small greenhouse attached to the spacecraft. A camera will be mounted near them so scientists on Earth can watch for signs of trouble and adaptation. Groups of plants will be engineered to glow under different conditions. Some plants will glow if they do not receive enough oxygen. Others will glow if they detect metals in the soil or if there are certain chemical compounds in the environment. Scientists will use the information gained from the plants to adjust the soil to the plant's needs.

an artist's concept of a futuristic greenhouse on Mars

These tests are the first steps toward building a livable environment in space, and especially on Mars. These little plants can tell scientists all about an environment without saying a word. All they have to do is glow.

Answer the questions.

5. What ocean animal provides one of the marker genes for the Mars plants? _a jellyfish_
6. How will the plants be set up on Mars? _A robot will collect soil from Mars. The seeds will be planted in the soil in a small greenhouse attached to the spacecraft._
7. What are some reasons that *Arabidopsis thaliana* would glow on Mars? _not enough oxygen, metals in the soil, certain chemical compounds in the environment_
8. How will these plants make progress toward establishing a livable environment on Mars? _These plants will help scientists determine what type of environment is currently on Mars and see how the plants will adapt to the environment._

Lesson 164
250 Technology Lesson

Direct an Activity

Materials: 6 similar plants (such as petunias or marigolds), 6 pots, soil, sand, gravel

Plant four plants in pots with soil. Label the pots as dry, moist, cold, and dark. Plant another plant in sand and the last plant in gravel. Label these pots as sand or gravel.

Place the plants labeled sand, gravel, dry, and moist in a warm, bright location. Place the plant labeled cold in a refrigerator and the plant labeled dark in a cupboard.

Water all the plants except the one labeled dry. Keep these plants moist throughout the activity. Observe the plants for 2–4 weeks. Record observations. Compare results.

➤ Which plant or plants thrived?
➤ Which conditions seem the most ideal for that type of plant?

Discussion:
Activity Manual pages 249–50

➤ What was another reason that the plants with the firefly's gene glowed? *{The plants were sick with a disease.}*

📖 The health of a plant is reflected by its appearance and response to the environment. The strength of a Christian's walk with the Lord is reflected in his attitudes and responses to the circumstances of life (1 Cor. 15:58).

💡 How is what scientists do to produce cold- or disease-resistant plants similar to Mendel's experiments with pea plants? *{Possible answer: Mendel cross-pollinated plants that had certain traits in order to produce plants with the traits that he wanted. Today's scientists also cross-pollinate plants to produce plants with the traits that they want.}*

💡 How is their experimentation different from Mendel's experiment? *{Possible answer: The marker genes cause the plants to glow, so today's scientists can determine quickly whether or not the plant possesses the desired traits. Mendel had to continue cross-pollinating throughout several generations before he could be certain that the plants had the traits he wanted.}*

➤ Other than on Earth, where else do scientists hope to be able to grow these plants? *{on Mars}*

➤ What ocean animal provides one of the marker genes for the plants going to Mars? *{a jellyfish}*

💡 In what conditions on Earth might plants have difficulty surviving? *{Possible answers: extreme heat or drought in deserts; high altitude in mountains; in extremely cold weather}*

💡 Are the conditions on Mars more or less extreme than the most severe conditions on Earth? *{more extreme}*

➤ What are some reasons that the plants on Mars might glow? *{Possible answers: not enough oxygen; metals in the soil; certain chemical compounds in the environment}*

💡 How will these plants make progress toward man's establishing a livable environment on Mars? *{Possible answers: These plants will help scientists determine what type of environment is currently on Mars. They can also see how the plants will adapt to that environment. If plants can survive, then they will produce oxygen for future human exploration.}*

Activity Manual
Technology Lesson—pages 249–50

Appendix

Lesson 164 A13

Lesson 165

Activity Manual pages 251–52

Objectives

- Compare robotic surgery with traditional surgery
- Describe some advantages and disadvantages of long-distance robotic surgery

Introduction

Note: This technology lesson may be used in Unit 6.

➤ **How do you paint or draw a picture on the computer?** *{Answers will vary but should include the idea that you move the mouse and the computer shows your movements as lines and colors on the computer screen.}*

➤ **Some computer programs and games use controls other than the keys and mouse. What are other types of computer controls you have seen or used?** *{Possible answers: joystick, foot pedal, stylus}*

Purpose for reading:
Activity Manual pages 251–52

➤ How does a surgeon control the movements of the robot during surgery?

➤ What are some advantages of long-distance surgery using a robot?

Discussion:
Activity Manual pages 251–52

➤ **When you think of a robot performing surgery, what picture comes into your mind?** *{Answers will vary.}*

➤ **How are the pictures on these pages different from what you thought of?** *{Answers will vary.}*

➤ **What kinds of medical instruments might a robot use during surgery?** *{a camera, a scalpel, a laser}*

💡 **Doctors use similar instruments during traditional surgeries. What is one way that these robotic medical instruments are different than those used in traditional surgeries?** *{Possible answer: They are smaller.}*

➤ **What does the surgeon look at to direct the robotic arms?** *{a 3-D image from the camera inside the patient}*

A14 Lesson 165

Robotic Surgery   Name_____

Imagine that your doctor says you are in urgent need of surgery. He tells you that he will perform open-heart surgery on you without opening your chest. Then he tells you that his new surgical assistant stands six feet tall and has four arms. No, your doctor has not lost his mind. Robotic surgeries like these are happening today in hospitals around the world.

Robots can help doctors perform delicate surgeries. The arms of the robot insert small surgical instruments through tiny incisions. These instruments might include a camera, a scalpel, and a laser. Up to four different robotic arms might be used. The surgeon looks into a viewfinder that shows him a 3-D image from the camera inside the patient. The surgeon directs the robotic arms with joystick-like controls. Each time he moves the joystick, the robotic arms move the instruments inside the patient. For every few centimeters that the surgeon moves, the surgical robot can be programmed to move only a few millimeters. Thus, the robot can make smaller movements than the surgeon can.

Several different robotic operating systems have been developed. Currently these robotic surgeons are being used for gallbladder removal, heart bypass surgery, cancer surgery, and other similar procedures.

Answer the questions.

1. What are some of the instruments available on a robot? _camera, scalpel, and laser_

2. Why can the robot make smaller movements than a human surgeon can? _Answers will vary, but elicit that a human's physical makeup limits how small he can make movements. A robot does not have these limitations, and it can be programmed to make very precise, tiny movements._

3. What types of surgery do robots perform? _gallbladder removal, heart bypass surgery, cancer surgery, and other similar procedures_

Lesson 165
Technology Lesson   251

SCIENCE BACKGROUND

Types of robotic surgery—Several different robotic operating systems have been developed. In addition to those described on Activity Manual page 251, some use foot pedals, voice activated software, or hand signals.

Robotic surgery has many benefits. After robotic surgery, patients usually heal more quickly and spend less time in the hospital. In traditional heart surgery, the surgeon cuts open the skin, muscles, and bone with a long incision. Robotic heart surgery, though, requires only three or four small incisions. Each incision is less than a centimeter wide.

Robots also make long surgeries less tiring for the surgeon. The doctor sits at a control console instead of standing over a patient for hours at a time. The robot can be programmed to ignore small shakes of the surgeon's hands and keep the instruments steady.

Robotic surgery also allows the surgeon to be in a different location than the patient. The surgeon could even be across the ocean from the patient. Nurses and assistant surgeons are with the patient while the master surgeon controls the robot from somewhere else.

Robotic surgery has some drawbacks. Robots are very expensive—over one million dollars—so only a few hospitals can afford them. Another problem is that most robotic surgeries take twice as long as traditional surgeries. However, the surgery time should decrease as surgeons gain more experience using robots. Finally, the robot is unable to feel the patient's tissues. Doctors often make decisions based on their sense of touch. A surgeon using robotic technology must operate without feeling the tissue himself.

What does the future hold for robotic surgery? Someday millions of surgeries may be done this way. As robots become smaller, we may see a day when a surgeon inserts a robot into the patient and lets the robot do all of the work.

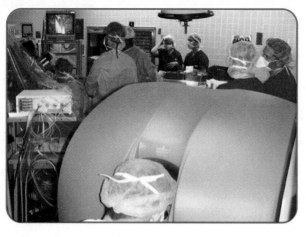

Answer the questions.

4. What are some benefits of robotic surgery? _Possible answers: Patients heal more quickly and spend less time in the hospital; doctors do not get as tired; steadier hands; long-distance surgery_

5. What are some drawbacks of robotic surgery? _Possible answers: The robots are very expensive to buy; surgery takes longer; and the surgeon does not use his sense of touch._

6. What would be the advantage of a surgeon performing a robotic operation from a different location? _Accept any reasonable answer, but elicit that expert surgeons could perform surgery on people in different locations without having to travel to the patients._

© 2004 BJU Press. Reproduction prohibited.

252 Lesson 165
Technology Lesson

HISTORY

Long-distance surgery—Today's telecommunication systems are continually updated to transmit data faster. These developments have allowed long-distance surgery, also called *telesurgery*, to become possible.

First transatlantic surgery—On September 7, 2001, a surgeon in New York successfully removed a gallbladder. The patient, however, was in France! The New York surgeon sent commands via computer to the robot in the operating room in France. The surgery took 54 minutes and was successful.

First robotic brain telesurgery—On September 18, 2002, neurosurgeons in Halifax, Nova Scotia, Canada, removed a cancerous brain tumor from a patient 400 km away in New Brunswick.

Appendix

Discussion:
Activity Manual pages 251–52

💡 Why do you think the robot can make smaller movements than a human surgeon can? *{Possible answers: A human's physical makeup limits how small and precise his movements can be. A robot does not have those limitations and can be programmed to make very precise and tiny movements.}*

➤ Why would the robot only require three or four small incisions for heart surgery instead of a traditional long incision? *{Possible answers: The tiny camera allows the surgeon to see under the skin without having to make an extensive incision. More precision results in smaller cuts.}*

➤ What types of surgeries can the robots perform? *{Possible answers: gallbladder removal, heart bypass surgery, cancer surgery, and other similar procedures}*

➤ What are some benefits of robotic surgery? *{Possible answers: Patients heal more quickly and spend less time in the hospital. Doctors do not get as tired. Incisions can be smaller. Long-distance surgery is possible.}*

➤ What are some disadvantages of robotic surgery? *{Possible answers: expense; surgery sometimes takes longer; surgeon cannot use his sense of touch}*

💡 What are some advantages for long-distance robotic surgeries? *{Possible answers: The surgeon can perform surgery without having to travel to the patient. The patient might be in an emergency situation and not able to reach the doctor in time. The patient might be in an area that is remote or inaccessible to a skilled surgeon.}*

📖 Because God tells us to love one another, man should continue to research and develop technology that helps people.

➤ Do you think there will ever be a time when a tiny robot can be inserted into a person and programmed to find the problem and perform the surgery? Why? *{Answers will vary.}*

Activity Manual
Technology Lesson—pages 251–52

Lesson 165 A15

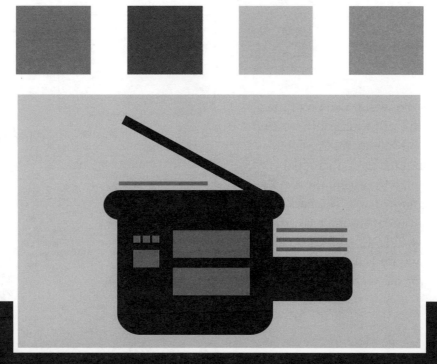

REPRODUCIBLES

Devils Postpile National Monument

Name_____

Let's take a hike along the western slope of the Sierra Nevada Mountains in California. Here we find the Devils Postpile National Monument. President Theodore Roosevelt declared the park a national monument in 1911. He wanted to protect the Devils Postpile formation and Rainbow Falls. The park is small but rich in natural resources. In the park, the San Joaquin (wah KEEN) River winds through forests and small canyons. The meadows are full of flowers and grasses that provide food for the wildlife.

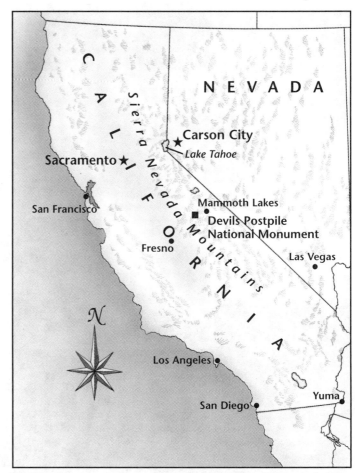

The 800-acre monument is near the resort community of Mammoth Lakes.

Volcanic Formation

Years ago Devils Postpile formed from basalt lava. **Basalt** is a dark-gray, fine-grained rock containing feldspar crystals. This lava flowed from a vent and filled the valley. The molten lava cooled slowly and evenly. Shrinkage of the cooling lava caused stress on the formation. Surface cracks formed when the shrinkage was greater than the lava's strength. Each crack then branched. A hexagonal pattern of cracks formed on the surface of the hardened lava. Gradually the cracks deepened, creating the six-sided columns you see today. This type of geologic formation is called **columnar basalt** and forms only under ideal conditions. It requires a slow cooling time and an even mineral composition of the lava. Devils Postpile met these conditions. It is one of the world's finest examples of columnar basalt.

Columns of Devils Postpile National Monument

Glacier Erosion

Glaciers eroded most of the lava flows at Devils Postpile. The slow-moving ice cut away one side of the formation. The cutting away exposed a sheer wall of columns 18.3 m (60 ft) high. **Glacier polish** is evident on the tops of the columns. The polishing happened as glacial ice carrying silt scraped and smoothed off the upper part of the lava flow. The glaciers also left behind deep grooves called **striations** (stry AY shunz).

Glacier polish looks like floor tiles on top of the columns.

Rainbow Falls

Just downstream from Devils Postpile is Rainbow Falls. The San Joaquin River drops 30.8 m (101 ft) over a cliff of volcanic rock. When the sun is overhead, rainbows highlight the falls. Rainbow Falls was once called "a gem unique and worthy of its name." A stairway and short trail lead to the bottom of the falls.

When the lava in the upper layer cooled, it fractured horizontally. Geologists call these rocks **rhyodacite.** Rhyodacite is visible in the cliffs that surround Rainbow Falls. As water rushes over the cliff, it erodes the rhyodacite more easily. A small cavern begins to form at the base of the falls. The overlying rock loses its support and collapses. This process is known as **undercutting.** It causes the waterfall to recede upstream slowly. Rainbow Falls has receded about 152.4 m (500 ft) due to undercutting.

Rhyodacite is evident in the cliffs of Rainbow Falls.

How to Read a Flow Chart

Begin at the top of the chart and follow the boxes as each decision is made.

Here is a possible thought process for when your Mom tells you to make sure the dog is fed.

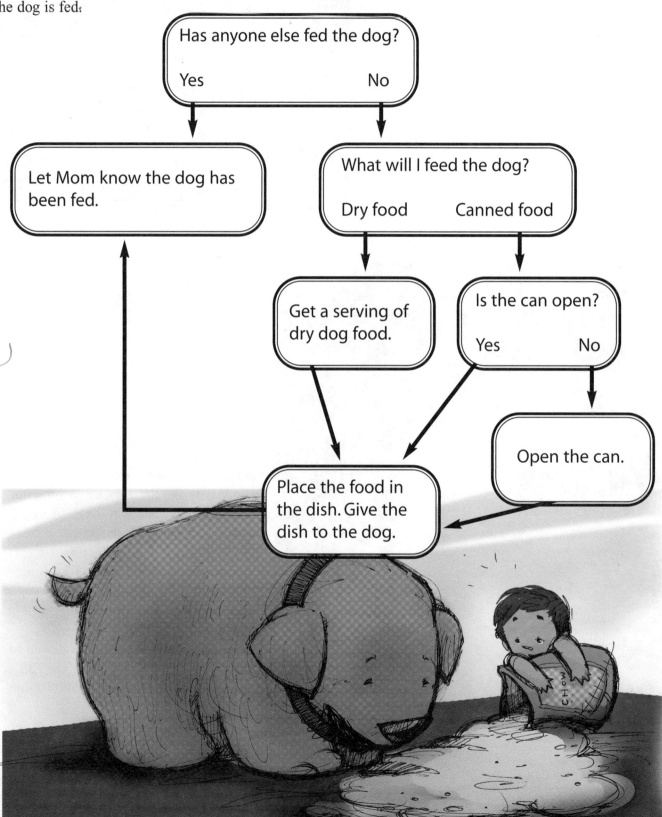

A20 For use with Lesson 20

Water in Israel

Name_____

Because some of the country of Israel is hot and dry, water is an especially important resource. The country's usable water, from both rivers and aquifers, is already being used to its limit. As the country's population grows, the stress on its water system grows as well. Consequently, Israel has a number of desalination plants to help supplement its fresh water supply.

Conflicts Over Water

Israel shares many water sources with its neighboring countries—Syria, Lebanon, and Jordan—as well as with its own Palestinian occupied areas. Conflicts sometimes develop over how much water each area is allotted. But disputes over water in the Middle East are not new. As far back as Abraham's time, wells were often a source of conflict. Digging wells in Bible times was an important but difficult task. Wells were considered the property of the person who dug them, and usually they were passed on as an inheritance just as land was. To take over a well that someone else had dug was considered stealing. In Genesis 21 Abraham reproves the leader of the Philistines, Abimelech, because Abimelech's servants had violently taken over a well that Abraham's servants had dug. Abraham and Abimelech made a covenant to establish Abraham's ownership of the well. Abraham called the place of the well Beersheba.

After Abraham died, the Philistines filled with dirt the wells that he had dug. When Abraham's son Isaac opened each of the wells and found water, the herdsmen of the area fought to gain control of the wells. Isaac moved on to another well that was fought over also. Finally he came to Beersheba. At Beersheba, Isaac's servants again dug a well and found water. Abimelech recognized that Isaac was blessed by God and reestablished a covenant with Isaac as he had with Abraham.

Modern Middle East

Sources of Water

Wells were the cleanest sources of water in ancient Israel. A short wall of limestone or stone often surrounded wells. The wall helped protect people and animals from falling into the well. Often the wells would have a stone cover that would have to be removed to draw water from the well. Because the wells were often used for watering flocks, they usually had a trough of wood or stone into which water could be poured to allow the flocks to drink.

Usually the women in a household were responsible for drawing water for the family. Typically, women would draw water early in the day and toward evening. Water was drawn by dropping a vase or waterskin attached to a rope into the well. Some wells were dug into the limestone and had steps descending into them.

Another source of water in ancient Israel was cisterns. A cistern is a large basin made

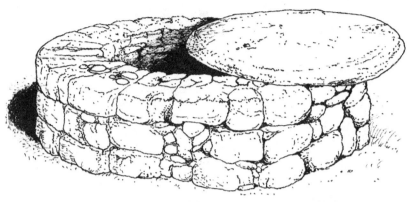

A large stone protected the well from dirt and smaller stones.

of stone. The cistern collected rainwater that ran off rooftops during the rainy months. Because of the dirt that rainwater collects, water from cisterns was not suitable for drinking or cooking. However, in ancient Israel, cisterns provided water for washing and bathing.

Living Water

Probably one of the most well-known Bible stories concerning a well is found in John 4. Jesus met a Samaritan woman at the well. He asked her to draw water for him to drink. Surprised that a Jew would ask a Samaritan for water, she questioned Jesus. He described for her living water that she could obtain. She would never thirst again. What was this water? The gift of salvation that God offers to all men. No conservation is needed for the living water. It is freely available in abundance for all those who seek for Christ.

The Microscope

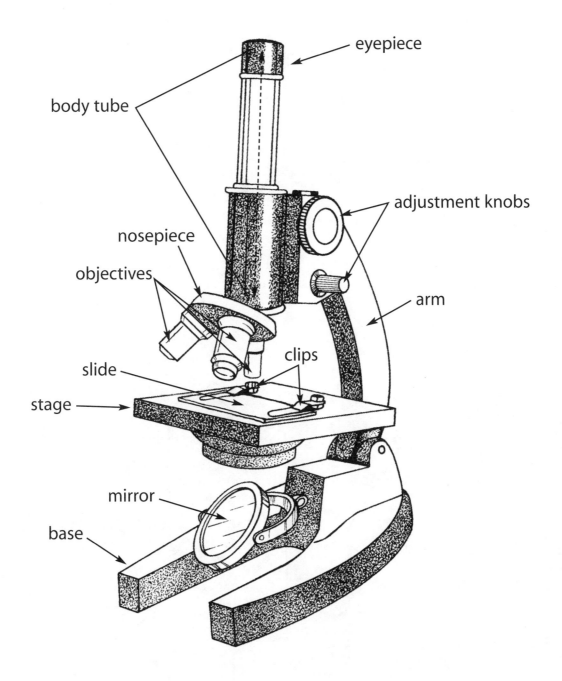

APPENDIX • For use with Lesson 39 • A23

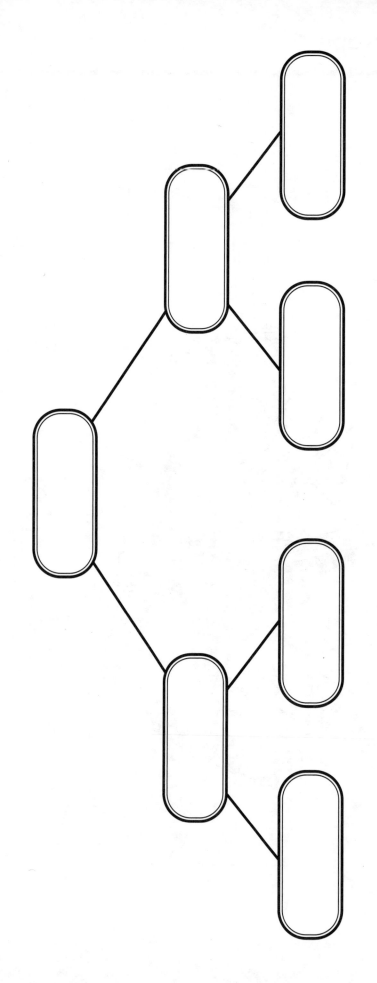

Flow Chart

Observation Record

Name_____

Date	Observation

Magnetic Personalities and Shocking Discoveries

Years	Person	Key discoveries
1745–1827	Alessandro Volta	battery; volt
1775–1836	Andre-Marie Ampere	flow of a current
1790–1845	John Frederic Daniell	battery for telegraph systems
1791–1867	Michael Faraday	electric motor, generator, and transformer
1791–1872	Samuel F. B. Morse	telegraph
1797–1878	Joseph Henry	electromagnet
1818–1889	James Joule	conductors; mechanical advantage of heat; Joule's law
1847–1922	Alexander Graham Bell	telephone
1847–1931	Thomas Edison	light bulb
1856–1943	Nikola Tesla	alternating current
1864–1930	Sebastian Ferranti	high voltage generation stations; alternating current
1873–1961	Lee de Forest	vacuum tube; sound amplifier
1876–1958	Charles Franklin Kettering	electric ignition for autos
1905–1995	S. Joseph Begun	magnetic recording
1906–1968	Chester F. Carlson	electrophotography (photocopying)
1914–	Bessie Blount	device to help disabled veterans
1916–1995	Marvin Camras	magnetic recording
1919–	Dr. Wilson Greatbatch	implantable pacemaker
1920–1982	Otis Boykin	resistor; device for guided missiles and computers
1936–	Raymond Damadian	magnetic resonance imaging (MRI) machine

Pop Can Constellations

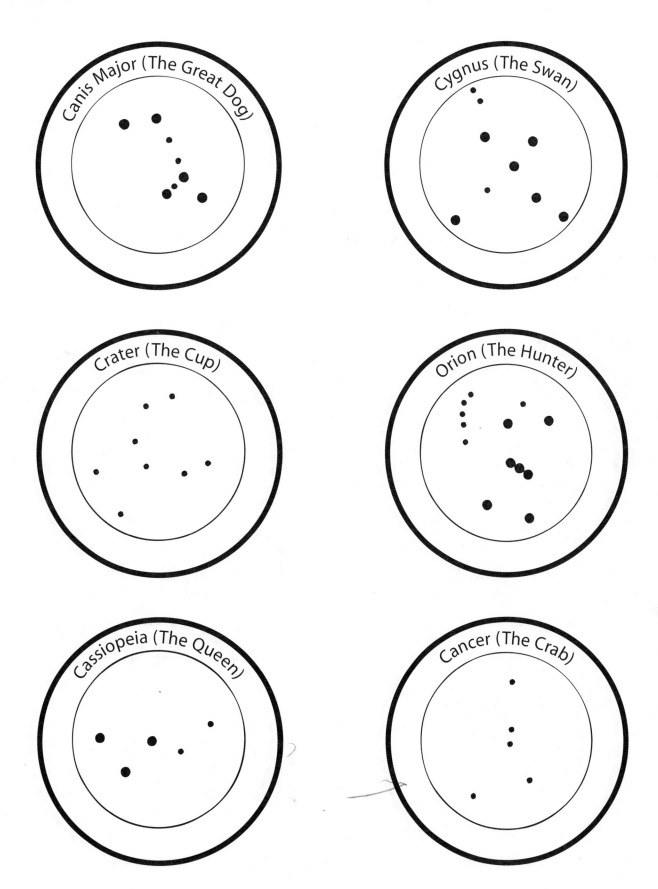

Appendix   For use with Lesson 106   A27

Pop Can Constellations

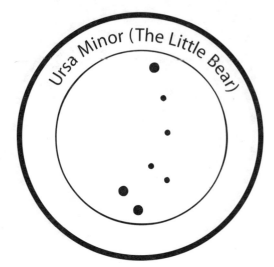

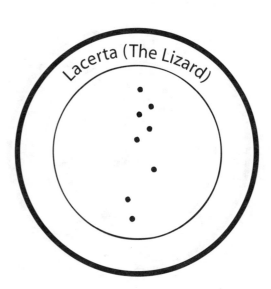

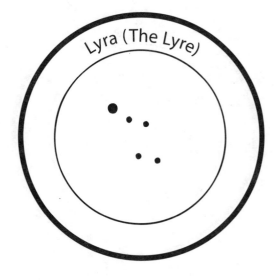

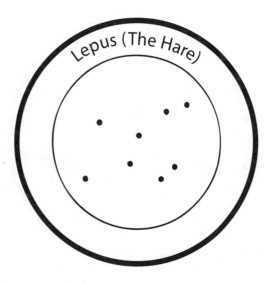

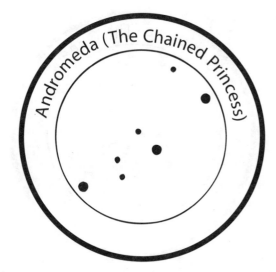

© 2004 BJU Press. Limited license to copy granted on copyright page.

A28   For use with Lesson 106

Star Coordinates

Andromeda

(x, y) coordinates   **Length**

A (2, 12)   8 cm
B (7, 6)   5 cm
C (9, 6.5)   14 cm
D (11, 4.5)   11 cm
E (13, 3.5)   18 cm
F (13, 22)   15 cm
G (7, 3)   19 cm

y coordinates   1 square = 30 light years

Cancer

(x, y) coordinates   **Length**

A (4, 5)   19 cm
B (8, 7)   13 cm
C (9, 6)   11 cm
D (8, 17)   5 cm
E (15, 7.5)   21 cm

y coordinates   1 square = 25 light years

Star Coordinates

Cassiopeia

(x, y) coordinates	Length
A (2, 15)	9 cm
B (7, 2)	12 cm
C (11, 21)	12 cm
D (12, 3.5)	15 cm
E (16, 1)	13 cm

y coordinates   1 square = 40 light years

Crater

(x, y) coordinates	Length
A (1, 18)	12 cm
B (4, 7.5)	13 cm
C (6, 17)	5 cm
D (10, 23.5)	6 cm
E (10, 5.5)	12 cm
F (11, 5)	10 cm
G (13, 4)	16 cm
H (16, 7.5)	13 cm

y coordinates   1 square = 15 light years

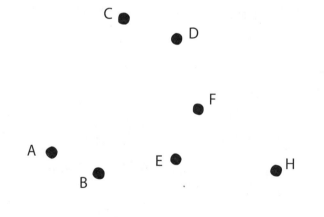

A30   For use with Lesson 107

Star Coordinates

Cygnus

(x, y) coordinates **Length**

- A (1, 1)   24 cm
- B (4, 0.5)   21 cm
- C (5, 21)   12 cm
- D (7, 10)   16 cm
- E (10, 1)   20 cm
- F (12, 1)   12 cm
- G (14, 0.75)   7 cm
- H (14, 2.5)   26 cm
- I (15, 0.75)   5 cm

y coordinates   1 square = 150 light years

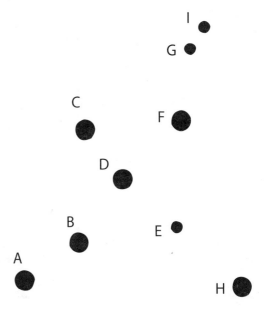

Lyra

(x, y) coordinates **Length**

- A (5, 4)   16 cm
- B (8, 16)   8 cm
- C (10, 4)   14 cm
- D (12, 2)   7 cm
- E (16, 0.5)   6 cm

y coordinates   1 square = 50 light years

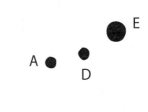

Appendix   For use with Lesson 107   A31

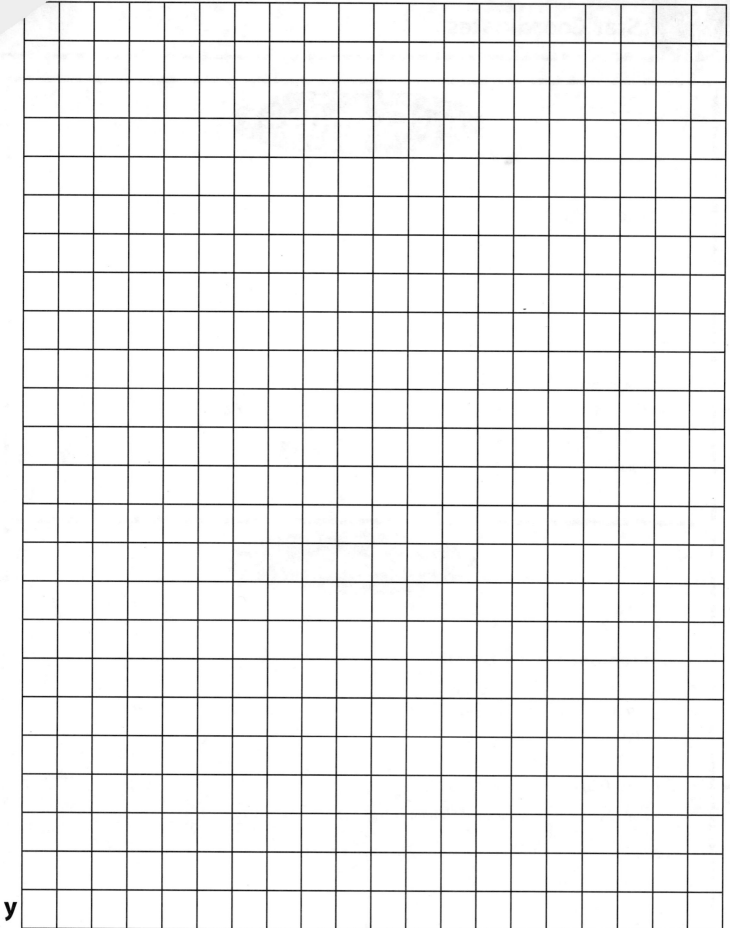

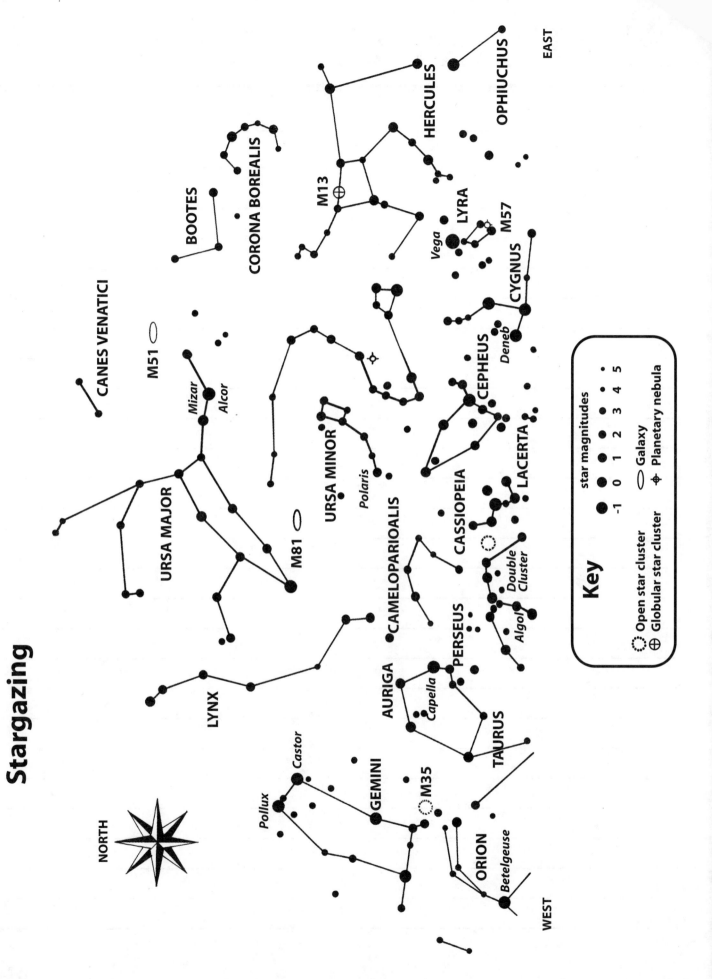

PLANT CITY TIMES

_____

by _____

Coded Instructions

MORSE CODE

A ·—	B —···	C —·—·	D —··	E ·	F ··—·	G ——·
H ····	I ··	J ·———	K —·—	L ·—··	M ——	N —·
O ———	P ·——·	Q ——·—	R ·—·	S ···	T —	U ··—
V ···—	W ·——	X —··—	Y —·——	Z ——··		

Decode the message written in Morse code.

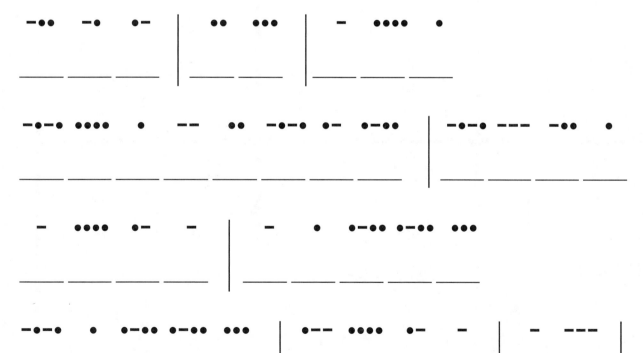

Parental Genotype Cards

Card	Genotype
1 (round face, pointed ears)	Face Color: Bb / Face Shape: ff / Eye Shape: RR / Ear Shape: Ee
2 (square face, pointed ears)	Face Color: Bb / Face Shape: FF / Eye Shape: Rr / Ear Shape: Ee
3 (square face, round ears)	Face Color: BB / Face Shape: Ff / Eye Shape: Rr / Ear Shape: ee
4 (round face, round ears)	Face Color: Bb / Face Shape: ff / Eye Shape: Rr / Ear Shape: ee
5 (square face, pointed ears)	Face Color: BB / Face Shape: Ff / Eye Shape: rr / Ear Shape: EE
6 (square face, round ears)	Face Color: BB / Face Shape: FF / Eye Shape: rr / Ear Shape: ee
7 (round face, pointed ears)	Face Color: Bb / Face Shape: ff / Eye Shape: rr / Ear Shape: Ee
8 (round face, round ears)	Face Color: Bb / Face Shape: ff / Eye Shape: rr / Ear Shape: ee
9 (square face, pointed ears)	Face Color: BB / Face Shape: FF / Eye Shape: RR / Ear Shape: Ee

A36 For use with Lesson 137

© 2004 BJU Press. Limited license to copy granted on copyright page.

Science 6 Teacher's Edition

Parental Genotype Cards

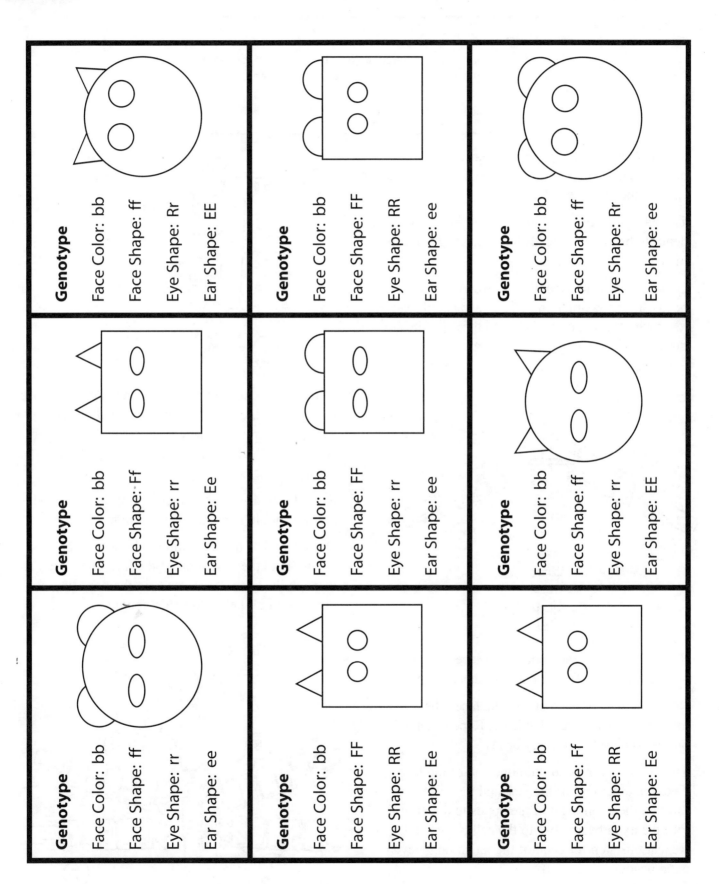

Appendix

For use with Lesson 137   A37

Bead Neuron

Name_____

Materials:
65 beads: Make each part a different color—1 bead for synaptic terminal (color #1); 12 beads for axon (color #2); 10 beads for cell body (color #3); 42 beads for dendrites (color #4)

120 cm plastic lace or flexible wire

1. Thread one dendrite bead on the lace until the bead is exactly in the middle. Make sure that the ends of the lace are even.

2. Thread both ends of the lace through five more dendrite beads. Pull beads together snugly. Keep the ends of the lace even.

3. On one end of the lace, thread two cell body beads. Take the other end of the lace and thread it through the same two beads in the opposite direction. Pull beads together snugly.

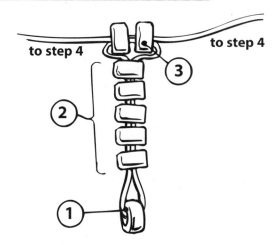

4. Thread six dendrite beads onto one end of the lace. Using the same end, skip the last bead placed on the string and thread the lace back through the other five beads. Push the dendrite beads up close to the cell body beads.

5. Repeat step 4 with the other end of the lace. Your neuron should now have three dendrites.

6. On one end of the lace, thread three cell body beads. Take the other end of the lace and thread it through the same three beads in the opposite direction. Pull beads together snugly.

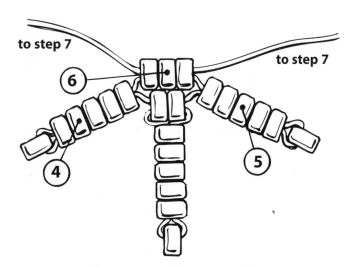

7. Repeat step 4 for each end of the lace. You should now have five dendrites. Repeat step 6.

8. Repeat step 4 for each end of the lace. The neuron now has seven dendrites. Repeat step 6, but use only two cell body beads.

9. Thread one axon bead onto one end of the lace. Take the other end of the lace and thread it through the axon bead in the opposite direction. Pull the axon bead close to the cell body beads. Repeat this step for each of the remaining eleven axon beads.

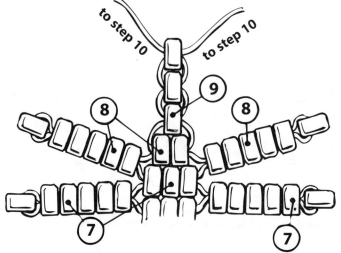

A38   For use with Lesson 142

Bead Neuron

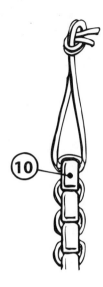

10. Repeat step 9 for the synaptic terminal bead. This is where neurotransmitters are stored and released. Thread both laces in opposite directions through the same bead again. This will help fasten the end of the neuron. Tie the two ends of the lace together to form a small loop. Cut off any excess.

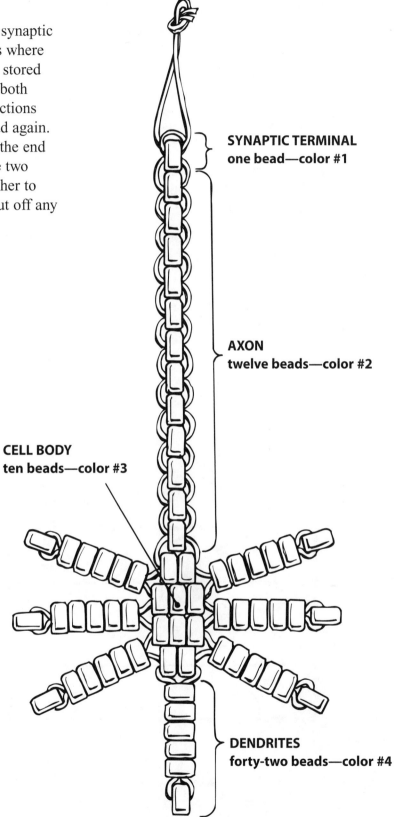

SYNAPTIC TERMINAL
one bead—color #1

AXON
twelve beads—color #2

CELL BODY
ten beads—color #3

DENDRITES
forty-two beads—color #4

Appendix   For use with Lesson 142   A39

Touch Tester

1. Cut out parts A and B.
2. Fold down tabs on part A. Tape both ends of each tab within the area marked with lines.
3. Center a toothpick along the edge of part A. Tape in place.
4. Align a toothpick with the 0 mark on part B. Tape in place.
5. Insert part B through both tabs on part A so both toothpicks are next to each other. Adjust the tabs so part B slides easily.
6. Adjust the position of the toothpicks as needed so they both extend the same distance below the Touch Tester.

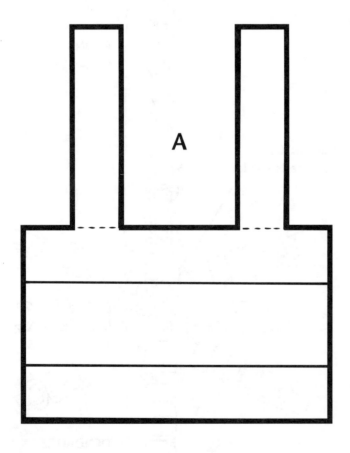

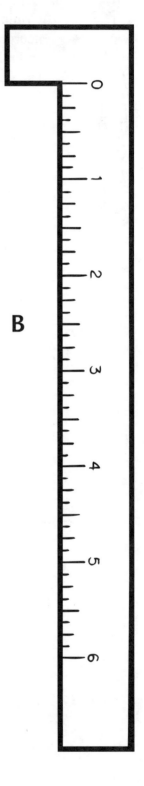

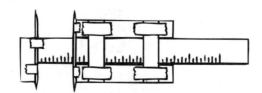

Effects of Drug Abuse

Name_____

Category	Common drugs	Effects of drug	Possible results of abuse
Stimulants	Cocaine Crack Methamphetamine Caffeine Nicotine	Speeds up the central nervous system Interferes with the sending of nerve messages Increases alertness and energy	Shortness of breath, nausea, and seizures Decreased learning and memory Increased risk of heart attack, high blood pressure, internal bleeding, and nervous system disorders Death
Depressants	Heroin Morphine Alcohol Codeine	Slows down (depresses) the central nervous system May bring a pleasurable feeling Blocks pain message	Shortness of breath, sleepiness, and loss of appetite Increased anxiety Increased blood pressure Decreased memory and physical coordination Decreased decision making and judgment abilities Deterioration of brain, liver, heart, or other organs Death
Hallucinogens	LSD Marijuana PCP Mescaline Psilocybin	Changes the way neurons in the brain send and receive information, resulting in hallucinations	Increased blood pressure Decreased coordination and alertness Decreased memory, concentration, learning, and problem-solving abilities Increased emotions, such as fear and anger Risk of flashback reactions long after use has ended Affects the brain stem and involuntary actions such as sneezing, breathing, and coughing Death
Steroids	Prescription medications Can be found in some dietary supplements, and some herbal remedies that claim to be "all natural"; muscle builders	Affects the hypothalamus and other nerve cells Alters the function of organs and the genetic material of individual cells Disrupts hormones in the endocrine system	Headaches, nausea, seizures, and central nervous system disorders Increased anger and aggression Reduced appetite and growth Decreased learning and memory Weakened immune system Increased risk of heart attack, high blood pressure, tumors, cancer, and liver problems Death
Inhalents	Ordinary products that are inhaled contrary to their intended purpose	Prevents oxygen from reaching the brain Can cause abnormalities in the brain and nerves Can deteriorate the myelin sheaths on some nerves Inhaled drugs reach the brain faster and can cause greater damage than oral or injected drugs	Seizures Blurred vision Uncontrolled shaking Decreased memory, concentration, learning, and problem solving abilities Increased risk of blindness, deafness, or liver damage Death

Keeping God's Temple Clean

Name_____

A. Answer the following questions.

Which type of drug changes the way the brain receives information from the senses? _____

Which type of drug slows down the central nervous system? _____

Which type of drug affects the hypothalamus and the endocrine system? _____

Which type of drug speeds up the central nervous system? _____

Which type of drug prevents oxygen from reaching the brain? _____

B. Use your Bible and the word bank to complete the puzzle.

body	destroy	obey	slavery
Christ	glory	power	sin
death	God	sacrifice	temple

Across

1. A Christian's desire should be for ___ to be magnified and exalted in his body (Phil. 1:20).

3. Everything a Christian does should be done to the ___ of God (I Cor. 10:31).

6. If someone knows the right thing to do and does not do it, it is ___ (James 4:17).

7. If someone destroys his body, God will ___ him (I Cor. 3:17).

8. The body of a Christian is the ___ of God (I Cor. 3:16).

10. We are slaves of the person or thing that we ___ (Rom. 6:16).

Down

2. A Christian should present his body as a living ___ to God (Rom. 12:1).

3. A Christian is not his own. He belongs to ___ (I Cor. 6:19).

4. Like false prophets, the temptation to use drugs may promise liberty and freedom; but actually, it results in ___ (II Pet. 2:19).

5. A Christian should honor and glorify God with his ___ (I Cor. 6:20).

7. The way that seems right to man leads to ___ (Prov. 14:12).

9. A Christian should not be under the ___ of anyone or anything other than God (I Cor. 6:12b).

A42   For use with Lesson 148

Blue Identity Cards

Make 2 copies of this page on blue paper.

WHITE BLOOD CELL 1	WHITE BLOOD CELL 15	LYMPH NODE 40
WHITE BLOOD CELL 5	WHITE BLOOD CELL 20	TONSIL 45
WHITE BLOOD CELL 1	WHITE BLOOD CELL 10	LYMPH NODE 40

Appendix

For use with Lesson 156   A43

Blue Identity Cards

Make 1 copy of this page on blue paper.

BRAIN 25	LEFT LUNG 25	STOMACH 35
WHITE BLOOD CELL 5	LIVER 25	KIDNEY 30
WHITE BLOOD CELL 1	HEART 25	RIGHT LUNG 25

A44   For use with Lesson 156

Blue Identity Cards

Make 1 copy of this page on blue paper.

BONE MARROW 55	LIVER 50	SPLEEN 60
AMBULANCE	APPENDIX 50	THYMUS 55

Appendix · For use with Lesson 156 · A45

Red Identity Cards

Make 2 copies of this page on red paper.

BACTERIA 1	PROTOZOAN 15	BACTERIA 40
VIRUS 5	BACTERIA 20	BACTERIA 45
BACTERIA 1	FUNGI 10	BACTERIA 40

Red Identity Cards

Make 1 copy of this page on red paper.

BACTERIA 25	BACTERIA 25	BACTERIA 35
VIRUS 5	BACTERIA 25	VIRUS 30
BACTERIA 1	BACTERIA 25	BACTERIA 25

Red Identity Cards

Make 1 copy of this page on red paper.

VIRUS 50	BACTERIA 55	VIRUS 60
AMBULANCE	VIRUS 50	BACTERIA 55

© 2004 BJU Press. Limited license to copy granted on copyright page.

A48 For use with Lesson 156

Science 6 Teacher's Edition

RUBRICS

A rubric is a scoring device that helps the teacher assess activities. It allows the teacher to establish criteria and evaluate performance based on those criteria. The rubric can also be used by the student to self-assess his performance.

Three prepared rubrics are provided. They are designed for general use with the experiments, activities, presentations, and projects included in SCIENCE 6 *for Christian Schools*. A blank rubric grid is included for use as needed.

The style of the rubrics provided includes a column for establishing the level of importance of each category listed. For example, if an experiment emphasizes measuring, the number written as Level of Importance for measurement should be higher than the numbers written for the other categories. On the other hand, although spelling and grammar are important, their importance on a research activity may be rated lower than the actual research.

To use a rubric, write the level of importance for each category. Then assign points to each category using the Scoring Key. Multiply the points by the level of importance to get the total. If the Level of Importance column is not used, record and total the points in the Total Points column.

Science Rubric

Name _____

_____
Title

Category	Points	X	Level of Importance	=	Total Points
Instructions Student followed instructions					
Materials Student handled materials properly					
Measurements Student took accurate and appropriate measurements					
Recorded information Student recorded necessary information					
Group participation (optional) Student participated in the group and made a worthwhile contribution					
				Total	

Comments:

Scoring Key
- **4 points** correct, complete, detailed
- **3 points** partially correct, complete, detailed
- **2 points** partially correct, partially complete, lacks some detail
- **1 point** incorrect or incomplete, needs assistance

A50   Rubrics

Science Rubric— Oral Presentation

Name_____

_____
Title

Category	Points X	Level of Importance =	Total Points
Instructions Student followed instructions			
Research Student researched thoroughly and accurately			
Organization Student presented information in a logical sequence			
Presentation Student demonstrated good presentation techniques • eye contact • posture • pronunciation • volume			
Group presentation (optional) Student participated in the group and made a worthwhile contribution			
Visual aid (optional) Student used a visual aid that reinforced the presentation			
		Total	

Comments:

Scoring Key
- **4 points** correct, complete, detailed
- **3 points** partially correct, complete, detailed
- **2 points** partially correct, partially complete, lacks some detail
- **1 point** incorrect or incomplete, needs assistance

Appendix   Rubrics   A51

Science Rubric— Written Presentation

Name _____

_____
Title

Category	Points	X	Level of Importance	=	Total Points
Instructions Student followed instructions					
Research Student researched thoroughly and accurately					
Organization Student displayed organization of thought in the project					
Grammar & spelling Student's completed presentation reflected proper use of grammar and spelling					
Project design Student presented the material in a neat and visually appealing manner					
				Total	

Comments:

Scoring Key
- **4 points** correct, complete, detailed
- **3 points** partially correct, complete, detailed
- **2 points** partially correct, partially complete, lacks some detail
- **1 point** incorrect or incomplete, needs assistance

Name _____

_____
Title

Category	Points X	Level of Importance =	Total Points
		Total	

Comments:

Scoring Key
- **4 points** correct, complete, detailed
- **3 points** partially correct, complete, detailed
- **2 points** partially correct, partially complete, lacks some detail
- **1 point** incorrect or incomplete, needs assistance

Science Fair Judging Form

Name _____

_____
Title

	Outstanding	Commendable	Acceptable	Unsatisfactory	Incomplete	Comments
Display						
Content **Scientific merit** Creative experimental design that includes a definition of the problem, a hypothesis, procedures, results, and conclusion						
Records and research Logbook record thorough, including a step-by-step description of the plans and procedures used to accomplish the project. Written report included with the project						
Bible integration Biblical references, truths, and applications related to the scientific study						
Layout Display well constructed, neat, well laid out, and attractive with an appropriate amount of work done by student						
Presentation of information sequential and complete						
Good spelling and grammar (records, logs, display) evident						
Interview with Judge						
Knowledge and understanding of concepts evident						
Conclusions and applications from findings logical						
Discussion of the topic communicated with confidence and clarity						
Overall Rating						

A54   Science Fair Judging Form

RESOURCES
game bank • materials list

Game Bank

The following games can be used to review any lesson or chapter.

Baseball Challenge

Identify four areas in the classroom to use as bases. Divide the class into two teams, and flip a token to determine which team bats first. Ask a question to the first student, or "batter." If the batter gets the question right, he may proceed to first base. Players advance to the next base each time a batter answers a question correctly. Correct answers to difficult questions could allow the batter to advance more bases.

If the batter answers incorrectly, someone from the opposing team (the outfield) may answer the question. If the outfield answers correctly, the batter is out. If the outfield's answer is incorrect, the batter receives a second chance. Three incorrect answers from the batter equal an out. When a team has received three outs, switch sides.

Basketball

Divide the class into two teams. Each team should choose a spokesperson. Give the teams time to make up several questions about the lesson or chapter. They must know the correct answers to the questions.

The teams will take turns asking each other their questions. The team may discuss the answer to a question, but the final answer should come from the team spokesperson. If the team answers correctly, they receive two points. If desired, the team may also get a chance to make a "basket" by shooting a foam ball into a trash can or other container.

Ducks and Decoys

This game is best used with a multiple choice review. Designate the four corners of the classroom as A, B, C, and D. Count out enough blank index cards so that you have one for each student. Label three-fourths of the index cards "Duck" and the remaining cards "Decoy." Give one card to each student. The student should not tell others whether he is a duck or a decoy. As the review questions are read, ducks should go to the corner that corresponds with the answer they believe is correct. Decoys should intentionally pick a corner that does not have the correct answer. This will encourage students to think of the answer for themselves and not to "follow the flock."

Football

Display a picture of a football field. Divide the students into two teams. Use a token or marker to represent each team. Decide which team goes first. This team receives the first question. If they answer correctly, their team marker should advance ten yards. The opposing team's marker also moves to the same position. If the answer is incorrect, the markers stay in the same place. When two consecutive questions are answered incorrectly, the ball switches to the other team. The second team now receives the questions.

When a team has advanced its marker to the end zone, a touchdown is scored (6 points). The team also has the option of receiving an additional question for an extra point while in the end zone. After a touchdown has been scored, place the team markers back at the 50-yard line. The team who did not score the touchdown should now receive the questions.

Jump Start

Divide the class into teams. Provide two "jump" chairs in the front of the room—one for each team. One person from each team should sit in his team's chair, keeping both feet flat on the floor and his back against the back of the chair. After the question is read, the seated students who know the answer should jump to their feet and remain standing. The first student to stand and give the correct answer receives points for his team.

Mix and Match

Write several questions on strips of paper. Write the answers to those questions on separate strips of paper. Place the questions in one container and the answers in another container. Mix up the papers. Draw a question from the container and read it aloud. Choose a student to draw an answer. The student should determine whether the answer matches the question he heard. If it does match, the student receives a point for his team. If it does not match, he has the option of stating the correct answer. If he can give the correct answer, he receives a point for his team. Questions and answers should be placed back into their respective containers before the next question is drawn.

Puzzling Questions

Divide the class into teams. Write several questions on strips of paper. Write the answers on separate strips of paper. Mix up the papers and give each team a set of questions and corresponding answers. At a given signal, each team should start organizing its papers and matching up the questions with the correct answers. The first team to display all the questions and answers correctly is the winner.

Science Four in a Row

Display a grid of two horizontal lines and three vertical lines. Divide the class into two teams. As team members answer the review questions correctly, they may place their team symbol in a section of the grid. The first team to get four symbols in a row wins.

Materials List

Standard science supplies

- balance scale
- calculator
- goggles
- magnifying glass
- meter stick
- metric beakers
- metric measuring cups
- metric measuring spoons
- protractor
- spring scale
- stopwatch
- thermometer
- timer

Basic supplies

- bucket
- colored pencils
- construction paper
- containers
- crayons
- dishpan
- glue
- newspaper
- paper towels
- scissors
- tape

Materials by chapter

Materials marked with a pink X are used only for enrichment demonstrations and activities.
The "T" column indicates materials needed for the optional Technology Lessons found on TE pages A4–A15.

Materials	1	2	3	4	5	6	7	8	9	10	11	12	13	14	15	T
aerosol can	X															
aluminum foil		X						X		X	X					
ammonia					X		X									
antacids, commercial							X									
apple or potato slices					X											
appliances, several small								X								
bag of candy									X							
bag, black plastic													X			
bag, plastic sandwich		X							X	X						
bags, resealable		X			X		X									
baking soda	X						X									
baking soda solution							X									
ball, golf										X						
ball, ping pong										X						
ball, rubber										X						

Materials	1	2	3	4	5	6	7	8	9	10	11	12	13	14	15	T
balloons				X			X			X	X					X
basketball											X					
batteries, C- or D-cell								X								
batteries, various sizes								X								
battery, 6-volt								X								
BBs									X		X					
bead, small											X					
bead, tiny											X					
blocks, toy					X					X						
board, 30 cm x 80 cm										X						
Borax							X									
bowl					X											
buckets		X	X		X											
Bunsen burner										X						
butter		X														
calculator						X										
can, soft drink										X						
candle										X				X		
candle holder										X						
cardboard box											X					
cardboard or card stock											X					
cardboard or meat tray, 20 cm x 30 cm										X						
cardboard or wooden base	X															
carnation, white						X										
carpenter's level	X															
carrot top														X		
checkers			X				X									
cheesecloth		X	X		X											
chocolate milk mix										X						
clay, dry		X														

Appendix   Materials List   A59

Materials	1	2	3	4	5	6	7	8	9	10	11	12	13	14	15	T
clay, modeling	X	X														
clothespin, clip style												X				
coins								X	X							
construction paper, black					X			X			X	X				
construction paper, blue													X			
construction paper, green													X			
construction paper, orange													X			
construction paper, white				X	X					X	X	X				
construction paper, yellow													X			
containers		X	X			X						X		X		
containers, glass						X										
cooking oil			X													
cooking spray		X														
copper, small piece								X								
cotton balls			X													
cotton swabs						X										
craft sticks												X				
crayons													X			
crispy rice cereal		X														
cups or containers, clear plastic		X	X	X		X	X	X	X			X	X			
cups, foam	X	X									X				X	
dead plant or flower			X													
dehydrated food											X					
dirt		X			X											
dishpan					X											
distilled water							X							X		
drawing compass													X			
droppers, medicine or eye	X		X				X							X		
earthworm					X											
eggs, raw														X		

A60 Materials List

Materials	1	2	3	4	5	6	7	8	9	10	11	12	13	14	15	T
electric fan or hair dryer			X													
electric pencil sharpener				X												
electronic devices, small											X					
encyclopedia					X	X					X				X	
family tree													X			
fern						X										
fettuccine noodles	X															
film containers with lids		X														
fishbowl, glass														X		
fishbowl or glass jar					X											
fishing line											X					
flashlight										X	X					X
flour										X	X				X	
flower pots																X
flower, large													X			
flowering plants																X
foam ball										X						
foam block, 20 cm x 25 cm	X															
foil baking pans		X	X													
foil loaf pans		X														
foil pan, deep, or dishpan										X						
foil pie plate											X					
food coloring			X			X										
freezer		X														
fruits, assorted dry and fleshy													X			
funnel	X															
glitter															X	
globe												X				
glue, white							X						X	X		
goggles or eye protection	X						X								X	

Appendix   Materials List   A61

Materials	1	2	3	4	5	6	7	8	9	10	11	12	13	14	15	T
grass tufts			X													
grasshopper or picture of grasshopper					X											
gravel																X
hammer			X							X						
helmet, bicycle (or other protective helmet)											X			X		
hot plate	X	X	X													
hydrogen peroxide							X									
ice		X			X											
information on roller coasters									X							
iron filings								X								
jar, large glass		X			X											
jars, plastic with screwable lids														X		
keys, assorted														X		
knife, small												X				
labels, blank										X						
lamp					X					X	X					
lamp, 3-way										X						
leaves			X													
leaves, dried				X												
lemon juice							X									
lenses, concave and convex										X						
levers, assorted									X							
light bulb, 3-way										X						
light bulb, 40-watt										X						
light bulbs with sockets								X								
lights, Christmas tree								X								
liquid detergent or dish soap				X									X			
list of numbers														X		
litmus paper, red and blue							X									
magnet, horseshoe									X							

A62 Materials List

Materials	1	2	3	4	5	6	7	8	9	10	11	12	13	14	15	T
magnets, bar			X		X			X								
magnets, disc-shaped																X
magnifying glass				X	X					X		X				
map, road													X			
map, world	X	X	X		X					X						X
marble, large											X					
marble, small											X					
marker, permanent										X						
marshmallows, large	X	X		X							X					
mass scale		X	X							X						
matches										X						
mealworms					X											
measuring wheel											X					
meat tenderizer													X			
metals, samples of			X													
meter stick		X			X				X	X	X					
metric beakers or measuring cups		X			X	X										
metric measuring spoons							X						X			
microscope				X												
microscope slides, prepared				X												
milk							X									
milk of magnesia solution							X									
mirror					X											X
model 3-stage rocket (optional)												X				
mushroom cap													X			
nail, common										X						
nail, finishing		X								X						
needle										X						
newspaper										X	X					
note card									X	X					X	

Appendix   Materials List   A63

Materials	1	2	3	4	5	6	7	8	9	10	11	12	13	14	15	T
notepad, plastic cover					X											
oatmeal or wheat bran					X											
overhead projector										X						
overhead transparency								X								
padlock with key														X		
paint											X					
paintbrushes											X					
pan, large	X	X														
paper							X									
paper clip, large metal								X			X					
paper fasteners											X					
paper hole punches		X														
paper towels			X			X		X			X					
paper, stiff											X					
paper, unlined					X											
pasta, 8 varieties of uncooked				X												
peas, dry											X					
pebbles			X	X												
pen										X						
pencils, colored					X		X	X			X					
pennies	X								X		X					
pH indicator paper							X									
picture—bison				X												
picture—car				X												
picture—flying gecko or flying squirrel					X											
picture—hang glider					X											
picture—jellyfish					X											
picture—mole					X											
picture—mother cat and kittens													X			
picture—mountain lion				X												

Materials	1	2	3	4	5	6	7	8	9	10	11	12	13	14	15	T
picture—mountain range										X						
picture—octopus					X											
picture—porcupine					X											
picture—rhinoceros					X											
picture—sea star					X											
picture—snail					X											
picture—solar system											X					
picture—Statue of Liberty						X										
picture—swan or duck					X											
picture—the night sky											X					
picture—whale					X											
picture—woodpecker					X											
pictures—common plants						X										
pictures—flowers						X										
pictures—Grand Canyon		X														
pictures—Yellowstone National Park	X															
pie plate, glass	X															
piece of clear plastic or glass					X											
pine cone, male												X				
pine cone, mature female												X				
pine cone, young female												X				
pinwheel			X													
plastic lacing															X	
plastic wrap											X					
pony beads, assorted colors															X	
popcorn, 2 different brands														X		
posterboard, black														X		
posterboard, white							X							X		
posterboard, assorted colors														X		
potato chips, 2 different brands														X		

Appendix   Materials List   A65

Materials	1	2	3	4	5	6	7	8	9	10	11	12	13	14	15	T
potato with sprouts												X				
potting soil		X	X									X				X
prism										X						
protractor		X	X									X				
pulley									X							
quilt batting					X											
raw wheat germ													X			
red cabbage juice							X								X	
resource books—animals					X											
resource books—planets											X					
resource books—plants						X										
resource books—tree field guide						X										
ringstand									X							
rocks, assorted small		X														
rocks, igneous	X															
rope																X
rubber bands		X			X					X	X	X				
rubber spatula					X											
rubbing alcohol													X			
ruler, centimeters and inches		X		X	X	X				X	X	X	X	X		
salt								X	X							
sand		X										X				X
scissors, heavy duty		X	X									X				
scissors, school														X	X	
seed catalogs or seed packets		X				X										
seed, mustard											X					
seeds, carrot													X			
seeds, lima bean													X			
seeds, poppy											X					
self-stick white dots										X						

A66 Materials List

Materials	1	2	3	4	5	6	7	8	9	10	11	12	13	14	15	T
shoebox																X
shortening					X											
simple machines									X							
small bottle or container	X															
snail					X											
soccer ball											X					
soda bottle		X					X									
sodium hydroxide or lye mixture solution															X	
soft drink, colorless, carbonated							X									
soft drinks, 2 different brands														X		
softball											X					
sponge, man-made	X				X											
sponge, natural					X											
spoon, large		X														
spoons			X				X									
spray bottle		X														
spring scale									X							
spring toy	X															
staples, bar of								X								
star charts										X						
sticks		X		X	X	X										
stopwatch or clock with second hand			X		X		X				X			X		
straw, drinking											X					
string	X					X			X		X					
suction cups						X										
sugar		X		X												
sunglasses											X					
sunscreen											X					
tape measure						X				X						
tape, clear														X		

Appendix   Materials List   A67

Materials	1	2	3	4	5	6	7	8	9	10	11	12	13	14	15	T
tape, electrical								X								
tape, masking										X	X	X				X
tea containing ginkgo						X										
tea kettle			X													
thermometers					X		X									
thread									X	X						
thread spool, empty									X							
tissue tube				X												X
toothpick					X							X	X	X		
toy animal or doll				X												
toy car				X					X							
trail mix							X									
travel brochures												X				
tubing, plastic									X							
umbrella, dark												X				
various objects that are conductors								X								
various objects that are insulators								X								
vinegar	X				X		X	X								
voltmeter									X							
watering can or large salt shaker		X	X													
wire screen					X											
wire, insulated								X								
yeast				X			X									

Bible Action Truths
Bible Promises

Bible Action Truths

The quality and consistency of a man's decisions reflect his character. Christian character begins with justification, but it grows throughout the lifelong process of sanctification. God's grace is sufficient for the task, and a major part of God's gracious provision is His Word. The Bible provides the very "words of life" that instruct us in salvation and Christian living. By obeying God's commands and making godly decisions based on His Word, Christians can strengthen their character.

Too often Christians live by only vague guidance—for instance, that we should "do good" to all men. While doing good is desirable, more specific guidance will lead to more consistent decisions.

Consistent decisions are made when man acts on Bible principles—or Bible Action Truths. The thirty-seven Bible Action Truths (listed under eight general principles) provide Christians with specific goals for their actions and attitudes. Study the Scriptures indicated for a fuller understanding of the principles in Bible Action Truths.

Thousands have found this format helpful in identifying and applying principles of behavior. Yet, there is no "magic" in this formula. As you study the Word, you likely will find other truths that speak to you. The key is for you to study the Scriptures, look for Bible Action Truths, and be sensitive to the leading of the Holy Spirit.

1. **Salvation-Separation Principle**
Salvation results from God's direct action. Although man is unable to work for this "gift of God," the Christian's reaction to salvation should be to separate himself from the world unto God.
   a. **Understanding Jesus Christ** (Matthew 3:17; 16:16; I Corinthians 15:3–4; Philippians 2:9–11) Jesus is the Son of God. He was sent to earth to die on the cross for our sins. He was buried but rose from the dead after three days.
   b. **Repentance and faith** (Luke 13:3; Isaiah 55:7; Acts 5:30–31; Hebrews 11:6; Acts 16:31) If we believe that Jesus died for our sins, we can accept Him as our Savior. We must be sorry for our sins, turn from them, confess them to God, and believe that He will forgive us.
   c. **Separation from the world** (John 17:6, 11, 14, 18; II Corinthians 6:14–18; I John 2:15–16; James 4:4; Romans 16:17–18; II John 10–11) After we are saved, we should live a different life. We should try to be like Christ and not live like those who are unsaved.

2. **Sonship-Servant Principle**
Only by an act of God the Father could sinful man become a son of God. As a son of God, however, the Christian must realize that he has been "bought with a price"; he is now Christ's servant.
   a. **Authority** (Romans 13:1–7; I Peter 2:13–19; I Timothy 6:1–5; Hebrews 13:17; Matthew 22:21; I Thessalonians 5:12–13) We should respect, honor, and obey those in authority over us. (attentiveness, obedience)
   b. **Servanthood** (Philippians 2:7–8; Ephesians 6:5–8) Just as Christ was a humble servant while He was on earth, we should also be humble and obedient. (attentiveness, helpfulness, promptness, teamwork)
   c. **Faithfulness** (I Corinthians 4:2; Matthew 25:23; Luke 9:62) We should do our work so that God and others can depend on us. (endurance, responsibility)
   d. **Goal setting** (Proverbs 13:12, 19; Philippians 3:13; Colossians 3:2; I Corinthians 9:24) To be faithful servants, we must set goals for our work. We should look forward to finishing a job and going on to something more. (dedication, determination, perseverance)
   e. **Work** (Ephesians 4:28; II Thessalonians 3:10–12) God never honors a lazy servant. He wants us to be busy and dependable workers. (cooperativeness, diligence, initiative, industriousness, thoroughness)
   f. **Enthusiasm** (Colossians 3:23; Romans 12:11) We should do *all* tasks with energy and with a happy, willing spirit. (cheerfulness)

3. **Uniqueness-Unity Principle**
No one is a mere person; God has created each individual a unique being. But because God has an overall plan for His creation, each unique member must contribute to the unity of the entire body.
   a. **Self-concept** (Psalm 8:3–8; 139; II Corinthians 5:17; Ephesians 2:10; 4:1–3, 11–13; II Peter 1:10) We are special creatures in God's plan. He has given each of us special abilities to use in our lives for Him.
   b. **Mind** (Philippians 2:5; 4:8; II Corinthians 10:5; Proverbs 23:7; Luke 6:45; Proverbs 4:23; Romans 7:23, 25; Daniel 1:8; James 1:8) We should give our hearts and minds to God. What we do and say really begins in our minds. We should try to think of ourselves humbly as Christ did when He lived on earth. (orderliness)
   c. **Emotional control** (Galatians 5:24; Proverbs 16:32; 25:28; II Timothy 1:7; Acts 20:24) With the help of God and the power of the Holy Spirit, we should have control over our feelings. We must be careful not to act out of anger. (flexibility, self-control)
   d. **Body as a temple** (I Corinthians 3:16–17; 6:19–20) We should remember that our bodies are the dwelling place of God's Holy Spirit. We should keep ourselves pure, honest, and dedicated to God's will.
   e. **Unity of Christ and the church** (John 17:21; Ephesians 2:19–22; 5:23–32; II Thessalonians 3:6, 14–15) Since we are saved, we are now part of God's family and should unite ourselves with others to worship and grow as Christians. Christ is the head of His church, which includes all believers. He wants us to work together as His church in carrying out His plans, but He forbids us to work in fellowship with disobedient brethren.

4. **Holiness-Habit Principle**
   Believers are declared holy as a result of Christ's finished action on the cross. Daily holiness of life, however, comes from forming godly habits. A Christian must consciously establish godly patterns of action; he must develop habits of holiness.

   a. **Sowing and reaping** (Galatians 6:7–8; Hosea 8:7; Matthew 6:1–8) We must remember that we will be rewarded according to the kind of work we have done. If we are faithful, we will be rewarded. If we are unfaithful, we will not be rewarded. We cannot fool God. (thriftiness)

   b. **Purity** (I Thessalonians 4:1–7; I Peter 1:22) We should try to live lives that are free from sin. We should keep our minds, words, and deeds clean and pure.

   c. **Honesty** (II Corinthians 8:21; Romans 12:17; Proverbs 16:8; Ephesians 4:25) We should not lie. We should be honest in every way. Even if we could gain more by being dishonest, we should still be honest. God sees all things. (fairness)

   d. **Victory** (I Corinthians 10:13; Romans 8:37; I John 5:4; John 16:33; I Corinthians 15:57–58) If we constantly try to be pure, honest, and Christlike, with God's help we will be able to overcome temptations.

5. **Love-Life Principle**
   We love God because He first loved us. God's action of manifesting His love to us through His Son demonstrates the truth that love must be exercised. Since God acted in love toward us, believers must act likewise by showing godly love to others.

   a. **Love** (I John 3:11, 16–18; 4:7–21; Ephesians 5:2; I Corinthians 13; John 15:17) God's love to us was the greatest love possible. We should, in turn, show our love for others by our words and actions. (courtesy, compassion, hospitality, kindness, thankfulness to men, thoughtfulness)

   b. **Giving** (II Corinthians 9:6–8; Proverbs 3:9–10; Luke 6:38) We should give cheerfully to God the first part of all we earn. We should also give to others unselfishly. (hospitality, generosity, sharing, unselfishness)

   c. **Evangelism and missions** (Psalm 126:5–6; Matthew 28:18–20; Romans 1:16–17; II Corinthians 5:11–21) We should be busy telling others about the love of God and His plan of salvation. We should share in the work of foreign missionaries by our giving and prayers.

   d. **Communication** (Ephesians 4:22–29; Colossians 4:6; James 3:2–13; Isaiah 50:4) We should have control of our tongues so that we will not say things displeasing to God. We should encourage others and be kind and helpful in what we say.

   e. **Friendliness** (Proverbs 18:24; 17:17; Psalm 119:63) We should be friendly to others, and we should be loyal to those who love and serve God. (loyalty)

6. **Communion-Consecration Principle**
   Because sin separates man from God, any communion between man and God must be achieved by God's direct action of removing sin. Once communion is established, the believer's reaction should be to maintain a consciousness of this fellowship by living a consecrated life.

   a. **Bible study** (I Peter 2:2–3; II Timothy 2:15; Psalm 119) To grow as Christians, we must spend time with God daily by reading His Word. (reverence for the Bible)

   b. **Prayer** (I Chronicles 16:11; I Thessalonians 5:17; John 15:7, 16; 16:24; Psalm 145:18; Romans 8:26–27) We should bring all our requests to God, trusting Him to answer them in His own way.

   c. **Spirit-filled** (Ephesians 5:18–19; Galatians 5:16, 22–23; Romans 8:13–14; I John 1:7–9) We should let the Holy Spirit rule in our hearts and show us what to say and do. We should not say and do just what we want to, for those things are often wrong and harmful to others. (gentleness, joyfulness, patience)

   d. **Clear conscience** (I Timothy 1:19; Acts 24:16) To be good Christians, we cannot have wrong acts or thoughts or words bothering our consciences. We must confess them to God and to those people against whom we have sinned. We cannot live lives close to God if we have guilty consciences.

   e. **Forgiveness** (Ephesians 4:30–32; Luke 17:3-4; Colossians 3:13; Matthew 18:15–17; Mark 11:25–26) We must ask forgiveness of God when we have done wrong. Just as God forgives our sins freely, we should forgive others when they do wrong things to us.

7. **Grace-Gratitude Principle**
   Grace is unmerited favor. Man does not deserve God's grace. However, after God bestows His grace, believers should react with an overflow of gratitude.

   a. **Grace** (I Corinthians 15:10; Ephesians 2:8–9) Without God's grace we would be sinners on our way to hell. He loved us when we did not deserve His love and provided for us a way to escape sin's punishment by the death of His Son on the cross.

   b. **Exaltation of Christ** (Colossians 1:12–21; Ephesians 1:17–23; Philippians 2:9–11; Galatians 6:14; Hebrews 1:2–3; John 1:1–4, 14; 5:23) We should realize and remember at all times the power, holiness, majesty, and perfection of Christ, and we should give Him the praise and glory for everything that is accomplished through us.

   c. **Praise** (Psalm 107:8; Hebrews 13:15; I Peter 2:9; Ephesians 1:6; I Chronicles 16:23–36; 29:11–13) Remembering God's great love and goodness toward us, we should continually praise His name. (thankfulness to God)

   d. **Contentment** (Philippians 4:11; I Timothy 6:6–8; Psalm 77:3; Proverbs 15:16; Hebrews 13:5) Money, houses, cars, and all things on earth will last only for a little while. God has given us just what He meant for us to have. We should be happy and content with what we have, knowing that God will provide for us all that we need. We should also be happy wherever God places us.

   e. **Humility** (I Peter 5:5–6; Philippians 2:3–4) We should not be proud and boastful but should be willing to be quiet and in the background. Our reward will come from God on Judgment Day, and men's praise to us here on earth will not matter at all. Christ was humble when He lived on earth, and we should be like Him.

8. **Power-Prevailing Principle**
   Believers can prevail only as God gives the power. "I can do all things through Christ." God is the source of our power used in fighting the good fight of faith.

   a. **Faith in God's promises** (II Peter 1:4; Philippians 4:6; Romans 4:16–21; I Thessalonians 5:18; Romans 8:28; I Peter 5:7; Hebrews 3:18; 4:11) God always remains true to His promises. Believing that He will keep all the promises in His Word, we should be determined fighters for Him.

   b. **Faith in the power of the Word of God** (Hebrews 4:12; Jeremiah 23:29; Psalm 119; I Peter 1:23–25) God's Word is powerful and endures forever. All other things will pass away, but God's Word shall never pass away because it is written to us from God, and God is eternal.

   c. **Fight** (Ephesians 6:11–17; II Timothy 4:7–8; I Timothy 6:12; I Peter 5:8–9) God does not have any use for lazy or cowardly fighters. We must work and fight against sin, using the Word of God as our weapon against the Devil. What we do for God now will determine how much He will reward us in heaven.

   d. **Courage** (I Chronicles 28:20; Joshua 1:9; Hebrews 13:6; Ephesians 3:11–12; Acts 4:13, 31) God has promised us that He will not forsake us; therefore, we should not be afraid to speak out against sin. We should remember that we are armed with God's strength.

Bible Promises

A. **Liberty from Sin**—Born into God's spiritual kingdom, a Christian is enabled to live right and gain victory over sin through faith in Christ. (Romans 8:3-4—"For what the law could not do, in that it was weak through the flesh, God sending his own Son in the likeness of sinful flesh, and for sin, condemned sin in the flesh: that the righteousness of the law might be fulfilled in us, who walk not after the flesh, but after the Spirit.")

B. **Guiltless by the Blood**—Cleansed by the blood of Christ, the Christian is pardoned from the guilt of his sins. He does not have to brood or fret over his past because the Lord has declared him righteous. (Romans 8:33—"Who shall lay any thing to the charge of God's elect? It is God that justifieth." Isaiah 45:24—"Surely, shall one say, in the Lord have I righteousness and strength: even to him shall men come; and all that are incensed against him shall be ashamed.")

C. **Basis for Prayer**—Knowing that his righteousness comes entirely from Christ and not from himself, the Christian is free to plead the blood of Christ and to come before God in prayer at any time. (Romans 5:1-2—"Therefore being justified by faith, we have peace with God through our Lord Jesus Christ: by whom also we have access by faith into this grace wherein we stand, and rejoice in hope of the glory of God.")

D. **Identified in Christ**—The Christian has the assurance that God sees him as a son of God, perfectly united with Christ. He also knows that he has access to the strength and the grace of Christ in his daily living. (Galatians 2:20—"I am crucified with Christ: nevertheless I live; yet not I, but Christ liveth in me: and the life which I now live in the flesh I live by the faith of the Son of God, who loved me, and gave himself for me." Ephesians 1:3—"Blessed be the God and Father of our Lord Jesus Christ, who hath blessed us with all spiritual blessings in heavenly places in Christ.")

E. **Christ as Sacrifice**—Christ was a willing sacrifice for the sins of the world. His blood covers every sin of the believer and pardons the Christian for eternity. The purpose of His death and resurrection was to redeem a people to Himself. (Isaiah 53:4-5—"Surely he hath borne our griefs, and carried our sorrows: yet we did esteem him stricken, smitten of God, and afflicted. But he was wounded for our transgressions, he was bruised for our iniquities: the chastisement of our peace was upon him; and with his stripes we are healed." John 10:27-28—"My sheep hear my voice, and I know them, and they follow me: and I give unto them eternal life; and they shall never perish, neither shall any man pluck them out of my hand.")

F. **Christ as Intercessor**—Having pardoned them through His blood, Christ performs the office of High Priest in praying for His people. (Hebrews 7:25—"Wherefore he is able also to save them to the uttermost that come unto God by him, seeing he ever liveth to make intercession for them." John 17:20—"Neither pray I for these alone, but for them also which shall believe on me through their word.")

G. **Christ as Friend**—In giving salvation to the believer, Christ enters a personal, loving relationship with the Christian that cannot be ended. This relationship is understood and enjoyed on the believer's part through fellowship with the Lord through Bible reading and prayer. (Isaiah 54:5—"For thy Maker is thine husband; the Lord of hosts is his name; and thy Redeemer the Holy One of Israel; The God of the whole earth shall he be called." Romans 8:38-39—"For I am persuaded, that neither death, nor life, nor angels, nor principalities, nor powers, nor things present, nor things to come, nor height, nor depth, nor any other creature, shall be able to separate us from the love of God, which is in Christ Jesus our Lord.")

H. **God as Father**—God has appointed Himself to be responsible for the well-being of the Christian. He both protects and nourishes the believer, and it was from Him that salvation originated. (Isaiah 54:17—"No weapon that is formed against thee shall prosper; and every tongue that shall rise against thee in judgment thou shalt condemn. This is the heritage of the servants of the Lord, and their righteousness is of me, saith the Lord." Psalm 103:13—"Like as a father pitieth his children, so the Lord pitieth them that fear him.")

I. **God as Master**—God is sovereign over all creation. He orders the lives of His people for His glory and their good. (Romans 8:28—"And we know that all things work together for good to them that love God, to them who are the called according to his purpose.")

Index

Bold indicates definition. Italic indicates illustration.

A

abrasion, 32–33, *32*
abortion, 312
acceleration, **218,** 222
accelerometer, 218–19
acid, **184**–91
acid rain, 22, 34–35
active volcano, 18
adrenal gland, *364,* 365
aerial roots, 161
agents of erosion, **46**
AIDS, 385
alchemy, 168
Aldebaran, *242, 249, 249*
Aldrin, Edwin "Buzz," 276, *276*
algae, 102–**3,** *102,* 143
alkalis, **184**
allergen, **388**
allergic reaction, 388
Alzheimer's disease, **366**
amnesia, 361
amoeba, 102, *102,* 314
ampere, **201**
amphibian, 126–27, 310–11
angiosperm, **148,** 152–56, 298
animal, 105–7, 109–42, 308–14
annelid, 116–17
annual, **153**
antacid, 186–87, 190–91
antibiotic, **387**
antibody, *384–85,* 385, **386**–**87**
aquifer, **76**–77, *76*
arachnid, **119**
Aristarchus, 265
Armstrong, Neil, 276, *276*
arthropod, **118**–21
asexual reproduction, **314**–15, *314–15*
ash, 16, **17**–19, 60
asterism, 252
asteroid, **258,** 270
asteroid belt, 270–71, *270–71*
astrology, 248–**49**
atomic bonding, **180**–81
atomic mass, **169,** 172, *173*
atomic number, **170**
atomic theory, **171**
atomic weight, 172–**73**
atoms, **168**–81, *169*
aurora, **267,** *267*
astronomy, **249**
Aswan High Dam, 48
autoimmune disease, **389**
autonomic nervous system, 350

autonomous underwater vehicle, **A4**–A5, *A4–A5*
avalanche, 19, 47
axon, **348**–49, *348–49,* 366

B

Babbage, Charles, 212
bacteria, **101**–2, *101,* 312, 374, *374,* 384–85, 387
Barringer Crater, 258–59, *258–59*
Barringer, Daniel, 259
base, **184**–89
bats, 134, 299
battery, **202**–3, *202–3*
bedrock, 42, *42*
Bernard, Dr. Christian, 389
Betelgeuse, *242,* 248, *249*
biennial, **153**
bilateral symmetry, 116
binary fission, 314, *314*
binary number system, **210**
binary star, **254**
bird, 130–31, 310–11
birdfoot delta, 49, *49*
bivalve, 112, *112*
black hole, 246–47, *247*
block and tackle, **230**
blubber, **136,** 138–39
Bohr, Niels, 170, *170*
brain, **344**–51, *345–47,* 354–57, 360–70
brain stem, 345–**46,** *346*
Braun, Wernher von, 286
budding, **314,** *314*

C

cambium, **158**–59, *158*
Cape Hatteras, 52, *52*
carbonic acid, **34**
Carlsbad Caverns, 36, *36*
carnivore, 129, 135–36
cave, 38–39
cell membrane, **94,** *94–95*
cell reproduction, 98–99
cells, **89**–99, *94–95,* 104–5
cell theory, **89**
cell wall, *94–95,* **95**
centipede, 119
central nervous system, **344**–47
cephalopod, 113
cerebellum, **346,** *346*
cerebrospinal fluid, 344–45, 347
cerebrum, 344, **345**–46, *345–46*
chemical families, 174–75

chemical formula, **178**
chemical reaction, **177,** 179, 182–83, 202
chemical symbol, **172**
chemistry, **168**
Chernobyl, 61
chlorophyll, **95,** 104
chloroplast, **95,** *95,* 104
chromosome, **94,** *94–95,* 98–99, 308, 320–23, *320,* 328, *328,* 337–38
chromosphere, **266,** *266*
cilia, 102, 382
cinder, **18**
cinder cone volcano, 18, *18*
circuits
    closed circuit, **196,** *196, 198,* 210–11
    integrated circuit, 211, *211*
    open circuit, **196,** *196,* 210–11
    parallel circuit, 198–99, **200,** *200*
    series circuit, 198–99, **200,** *200*
    short circuit, **196**
circumpolar constellations, **248**
classification
    animals, 110–37, *96*
    cells and organisms, 100–107
    elements, 173–75
    plants, 144–61
clay, 38–42, **40,** *40,* 53
cloning, 338
club mosses, *146–47,* 147, 156
coal, 58, **60**–51, *60,* 64
cochlear implants, 354
codominance, **329**
cold-blooded, 124–29
Collins, Michael, 276, *276*
colony, **101,** 103
comet, **259,** *259*
communicable disease, **373**
complete metamorphosis, **121,** *121*
composite cone volcano, 18, *18*
compound machine, **233**
compounds, **177**–81
computer, 212–13
conductor, **196**
cone (of a volcano), **17**
conifer, 148–51, *148,* 156, 304–5, *305*
conservation, 70–71
constellation, **248**–49, *248–49,* 252–53, *253,* 255, 260
contour plowing, **71,** *71*
corona, **266,** *266*
cotyledon, **154**–55, 301, *301*
covalent bond, **180**
CPU, **212**–13, *212*

Appendix   Index   A73

crater, meteorite, 270–73
crater, volcano, **16**, *16*
Crick, Francis, 323, *323*
crocodilian, 129
cross-pollination, **299**
crude oil, 58–59
crustacean, **118**
current electricity, **196**–97
cystic fibrosis, **337**
cytologist, **96**
cytoplasm, **94**, *94–95*

Dalton, John, 170
da Vinci, Leonardo, 215
debris flow, **22**
deflation, **53**
delta, **49**, *49*
dendrite, 348–49, *348–49*
deposition, **46**
depressants, 368–69
dicotyledon, **154**, *154*
dissolved load, 48
distance, **217**, 234
distillation tower, **59**
DNA, 94, **320**, *320, 322,* 323–25, *325*
dominant gene, 328–32
dominant trait, **327**–29, 332
dormant volcano, 19
double helix, *32,* 323, *323*
Douglas fir, 150–51
Down syndrome, 337
drawdown, **76**
drip curtain, 37, *37*
drug abuse, 367–69
dust storm, 53
dwarf star, 242–43

ear, *354*
Earth, 4, 30–31, 266–77, *269–71,* 274, *277*
earth flow, 47
earthquake, **6**–15
echinoderm, 115
echolocation, 134, 136
eclipse, 277
eclipsing variable stars, 244, *244*
Edison, Thomas, 194, *194,* 197
effort force, **228**–30
eggs, 310–11, *310–11*
electrical signal, **210**
electric cell, 202–3
electricity, 183–203
electroencephalograph (EEG), 363
electrolyte, **202**–3
electromagnet, **205**–6, *205,* 208–9
electron, **169**, 174, 180, 194–97, 202–3

electronic device, **216**–18
electron shell, 169, 180–81
element, **168**, 172–76
embryo, **300**–301, 308–10
endocrine glands, **364**
endocrine system, 364–65, *364*
endoplasmic reticulum, **94**, *94–95*
endothermic reaction, 182
energy, 58–61, 64–67, **88,** 216
environment, 87
epicenter, **8**
epidemic, **376**–81
epidemiologist, **378**–80, *378*
epidemiology, 378–79
epilepsy, **366**
equinox, **268**, *269*
erosion, **46**–55, *46–47,* 70–73
eruptions, 19, 24
euthanasia, 312
Ewing, Ella, 365
exfoliation, 32, *32*
exoskeleton, **118**
exothermic reaction, 182
extinct volcano, 19
Exxon *Valdez,* 63
eye, *355*

facula, **266**, *266*
fallow, **70**
Faraday, Michael, 204
faults, **6**–7
  normal, 6–7, *6*
  reverse, 6, *6*
  strike-slip, 7, *7*
fern, *146–47,* 147, 306, *306*
fertilization, **299**, 308
fiber optic sponges, A6–A7, *A7*
fibrous roots, **160**–61, *160*
fiddlehead, **147**
filter feeder, **115**, 124–25
fish, 124–25
flatworm, 116, *116*
Fleming, Alexander, 387, *387*
floodplain, 48
flower, 152–55, 298–300, 302–3
  parts of a, 298, *298*
focus, **8**
force, 219–35
fossil fuel, **58**–60
fragmentation, **314**, *314*
Franklin, Rosalind, 323, *323*
free-living, **116**
friction, 197, **219**, 221, 232, 233
Fritz, Charles, 67
frond, **147**, *147,* 306
frost heaving, 32, *32*

frost wedging, 32
fruit, 299–**300**
fruiting body, **307**
fulcrum, **228**–29
fungi, 103, 307, 374, 387

galaxy, **256**–57, *256–57*
Galilei, Galileo, 220, 250, 280
Galvani, Luigi, 193
gastropod, 112–13
gene, **320**–39, *320, 328*
generator, **206**
genetic engineering, **338**–39
genetics, 319–40
genotype, **328**–30, 332, 334
geothermal energy, **65**
germinate, **301**
gestation, **308**–9
geyser, **24,** *24*
giant star, 242–43
glacier, **54**–55
globular star cluster, **255**, *255*
Goddard, Robert, 286
GPS, 71
gravity, 46–47, 216, **220**–21, 274, 289
Great Barrier Reef, 111
ground cover, **71**
ground water, **76**
gymnosperm, **148**–51, 156, 304–5

hallucinogens, 368–69
heliculture, 112
hemophilia, 333
Henry, Joseph, 204
herbaceous, **159**
herbivore, **128**
heredity, 319–40, **320**
Hipparchus, 241
hippocampus, 361
hoofed mammal, 134
Hooke, Robert, 89
horizons, soil, **42**, *42*
hormones, **364**–65
hornworts, 144
horsetails, 146, *146*
hosts, 116
hot spots, **17**
hot spring, **24,** *24, 102*
Hubble Space Telescope, 250, 270, 288
humidity, 77
humus, **40**–42
Huntington's disease, 337, 378
hybrid, 326–30, **327**
hydrochloric acid, 184, 382
hydroelectric energy, **64**

A74 Index

hydrosphere, **74**
hypothalamus, **364**–65, *364*

iceberg, **78**
ice sheet, **78**–79
ice shelf, **78**
igneous rock, **23**, 30, *31*
inhalents, 368–69
immune response, **383**–85
immune system, 371–90, *384–85*
immunity, 386
impulse, **348**–51
inclined plane, **232**, *232*, 234–35
incomplete dominance, **329**
incomplete metamorphosis, **120**, *120*
indicator, **186**, 188–91
inertia, 220–21, 226
inflammatory response, 383
inflatable spacecraft, A10–A11, *A10–A11*
insect, 120–23, *120*, 299, 376
insectivore, **133**
instantaneous speed, **217**
insulator, **197**, 210–11
integrated circuit, **211**, *211*
invertebrate, **110**–23
ion, **181**
ionic bond, **181**

Jansen, Zacharias, 90
joule, **227**
Jupiter, 270–71, *270–71*, 280, *280*

Kilby, Jack, 210
kinetic energy, **216**
kingdoms, 101–7
    kingdom Animalia, 105
    kingdom Archaebacteria, 102
    kingdom Eubacteria, 101
    kingdom Fungi, 103
    kingdom Plantae, 104
    kingdom Protista, 102–3

larva, 121–23
Latimer, Lewis, 197, *197*
lava, **16**, 22–23
laws of motion, 220–25
Leeuwenhoek, Anton van, 90
lever, **228**–29, *229*, *232*
life cycle, 86
    amphibian, *126*
    conifer, *305*
    fern, *306*
    insect, 120–21

life span, **86**
light-year, 243, 256
Linnaeus, Carolus, 106–7
lithosphere, **4**, *4*, 16
liverworts, 145, *145*
lizard, 128
load, **48**
loam, **40**
lobe (of the brain), 344–**45**, *345*
Local Group (stars), **257**
Lowell, Percival, 273
lunar eclipse, 277
Lyme disease, 118–19
lymphatic system, 383
lymph fluid, 382

macrophage, 382, 384–85, *384–85*
maglev train, A8–A9, *A9*
magma, 7, **16**, 17, 23, 30–31
magma chamber, **16**
magnetic field, **204**–6, *204*, 267
magnets, **204**–6
magnitude
    absolute (stars), 240–**41**
    apparent (stars), **241**, *241*
    earthquake, **10**
mammal, 132–37
marine mammal, 136
Mars, 270–71, *270–71*, 273, *273*, 288, A13
marsupial, 132–33, **309**
mass movement, **47**, *47*
Mauna Loa, 18
mealworm, 122–23
mechanical advantage, **230**, 232
mechanical energy, **216**
meiosis, 98–**99**, *99*, 308, 320
memory, **360**–61
memory cells, 384–86, *384*
Mendel, Gregor Johann, 326–30, *326*
Mendeleev, Dmitri, 173
meninges, 347
Mercury, 270, 272, *270–72*
metal, 68–69
metallic bonds, 180
metamorphic rock, **30**, *31*
metamorphosis, **120**–21, *120–21*, 126
meteor, **258**
meteorite, **258**
meteoroid, **258**
microscope, **89**–91, *90–91*
    parts of, 91
Milky Way, 256–57
millipede, 119
mineral, **68**
mitochondria, **94**, *94–95*, 96
mitosis, 98–**99**, *98*, 308, 320

mole, 133
molecules, **177**
mollusk, **112**–13
molt, **118**
momentum, **219**, 226
monocotyledon, *154*, **155**
monotreme, 132
Moon, 274–77, *274–77*, 290
    phases, 275, *275*
moraine, 54–55, *55*
moss, 144–45, *144*, 307, *307*
motion, **216**, 220–23
motor neurons, **348**
Mt. St. Helens, 11, *11*, 18
mudflow, 47
mud pot, **24**, *24*
multicellular organism, **93**
multiple sclerosis, 366, 389
multiple star group, **254**, *254*
myelin sheath, **348**–49, *348*

natural gas, **60**
natural resources, **57**–79
nebula, 245–246, *246*
nematocysts, **110**–11
Neptune, 270, *271*, 282, *282–83*
nervous system, 343–70
neuron, **348**–57, *348–49*
neutral, **185**, 194–95
neutralize, **187**, 190–91
neutron, **169**–70
neutron star, **246**–47
Newton, Sir Isaac, 220, 222–25, 250, 265, 268
Nightingale, Florence, 372
noble gases, 180
nocturnal, 134
noncommunicable disease, **373**, 378
nonrenewable resources, **58**–61
nonvascular plant, **144**–45
nova, 244–45, *245*
Noyce, Robert, 210
nuclear energy, 61
nucleus (atom), **169**–71
nucleus (cell), **94**, *94–95*
nudibranch, 113
nymph, 120

ocean, 75
Oersted, Hans Christian, 204
Old Faithful, 24
omnivore, **128**
open star cluster, **255**, *255*
ore, **68**
organelles, **94**–95

organism, **86**
organs, **93**, 388–89
oxidation, 34

P

pancreas, 364–65, *364*
Pangaea, 5
parallax, 243, *243*
paramecium, 102, *102*
parasite, **116**
Parkinson's disease, **366**
particle accelerator, **171**, *171*
passive immunity, 386
Pasteur, Louis, 372–73, *372*
pathogen, **373**–90
pedigree, **332**
pedologist, 57
perennial, **153**
periodic table of the elements, **173**–175, *174–75*
peripheral nervous system, 348–50, 354–57
petrochemicals, **59**
petroleum, **58**–59
phenotype, **328**–30
phloem, **158**, *158*
photosphere, **266**, *266*
photosynthesis, 74, 95, **104**
phrenology, 343
pH scale, 184–**85**, *184–85*
phytoplankton, **75**, *75*
pinniped, 135–36
pituitary gland, *364,* 365
placental mammal, **309**
planets, 270–83
plants, 104, 143–64, 298–307, 315, 326–330, 338–39, A12–A13
plate boundaries, **5**–7
Pleiades, 249, *249*
plucking, **54**
Pluto, 270–71, *271,* 282–83, *282*
Polaris (North Star), 248, *248*
poliovirus, 376
pollination, 299, 304–5
pollution, 77
Pompeii, 17, 22
potential energy, **216**
precious metals, 68
pressure release, 32
primary root, **160**, *301*
primate, 136–37
probe (space), 289
proton, **169**–70
protozoan, **102**–3, 374
pulley, **230**–31, *230*
pulsar, **247**, *247*
pulsating variable star, 244, *244*

Punnett, Reginald, 330
Punnett square, 330–31, *330–31*
pupa, 121
purebred, **326**–30
pyroclastic flow, **19**

Q

Queen Victoria, 333, *333*
quillworts, 146

R

rabbit, 133
radial symmetry, 115
radio telescope, 251, *251*
RAM, **213**, *212*
receding glacier, 55
recessive gene, 328–32
recessive trait, **327**–29, 332
recycling, **79**
Redi, Francesco, 297
redshift, **251**
reference point, **216**
refinery, **58**–59, *59*
reflecting telescope, **250**, *250*
reflex, **351**
refracting telescope, **250**, *250*
regeneration, **314**, *314*
regolith, 42
REM sleep, 362–63
renewable energy resources, 64–67
renewable resources, **58**, 64–67, 70–78
reproduction, 298–315
reptile, 128–29
reservoir, 64
resistance force, **228**
resistor, **197**, 210
revolution, **268**
rhizoid, **145**, *145*
rhizome, **146**, *147*
ribosome, **94**, *94–95*
Richter scale, 10, *10*
Rigel, 238, *242,* 249
Ring of Fire, *5,* 16–**17**
robotic surgery, A14–A15, *A14–A15*
rock cycle, 30–31, *30–31*
rocket, 286–87, 292
rock flour, 54
rockslide, 47
Rocky Mountain spotted fever, 119, 376
rodent, 133
ROM, 212–**13**
roots, 154–55, *154,* 160–61, *160–61*
Ross Ice Shelf, 78, *78,* A6
rotation, **268**
roundworm, 116, *116*
Rutherford, Ernest, 170

S

saline, 75
salt, **187**
sand, **40**–41, *40*
sandstorm, 53
SARS, 376, 378
satellite, **274**, 288, *288*
Saturn, 270–71, *271,* 280–81, *280–81*
science process skills
   classifying, 100, 156
   communicating, 96
   defining operationally, 235
   experimenting, 199, 225
   hypothesizing, 191, 292
   identifying and controlling variables, 15, 353
   inferring, 73, 279, 381
   measuring/using numbers, 51, 163, 303
   observing, 123, 189, 317
   predicting, 45, 209, 359
   recording data, 139, 183, 263, 322
   using models, 21, 63, 252, 335, 390
scientific name, 107
screw, **233**, *233*
sea-floor spreading, 7
sea ice, 78
seasons, 268–69, *269*
sediment, **46**–49, 52–55
sedimentary rock, 30, *31*
seed, 148, 154–55, *154,* 300–301, *301,* 304–5
seed coat, 39
seedless vascular plant, 146–47
segmented worm, 116–17, *117*
seismic waves, **8**–11
seismograph, **9**, *9*
seismologist, **8**–9
seismoscope, 8
self-pollination, 299
semiconductor, 210–**11**
Semmelweisse, Dr. Ignez, 373
senses, 354–57, *354–56*
sensory neurons, **348**
setae, 117
sex-linked trait, **333**
sexual reproduction, **99**
shield volcano, 18, *18*
sickle cell anemia, **336**, *336*
silicon, 210
silt, **40**, *40*
simple machine, 227–33
sleep, 362–63
smell, **356**
smelting, 68, *68*
snake, 129
Snow, John, 379, *379*

soil, **40**–45, 70–71
soil creep, 47
soil horizons, **42**
solar eclipse, 277
solar energy, **66**, 67
solar field, 67
solar flare, 266–67, *266*
solar oven, 278–79
solar prominence, *266*, 267
solar storm, 266–67
solar wind, 267
solstice, 268, *269*
somatic nervous system, 350
space exploration
    Apollo 13, 278–79, 287
    Challenger, 287
    Columbia, 287
    International Space Station, 287, 289
    Magellan, 273
    Mariner 2, 273
    Mariner 10, 272
    Mars Climate Orbiter, 288
    Mars Exploration Rovers
        Opportunity, 289
        Spirit, 289
    Mars Global Surveyor, 289
    Mars Pathfinder, 273
    Project Apollo, 276
    Saturn series rockets, *286*, 287
    Sojourner, 273, *273*
    Sputnik, 288
    Venera, 273
    Voyagers 1 and *2*, 283, *283*, 289
space shuttle, 287, *287*
spectroscope, **251**, *251*
speed, **217**
speleothems, 34
spelunkers, 37
spike mosses, 146
spinal column, 347, *347*
spinal cord, 344, **347**, *347*, 351
sponge, 110
spores, **306**–7
stalactite, **36**, *37*
stalagmite, **36**, *37*
stars, 240–61, *242*
static electricity, 194–**95**

stem, 158–59, *301*
stimulants, 368–69
Stratton, Charles (Tom Thumb), 365
Sturgeon, William, 205
subsoil, 42, *42*
Sun, 240, 266–70, *266–70*, 277, *277*, 280–82
sunspots, 266–**67**
supergiant (star), *242*, 243
supernova, **246**
suspended load, 48
switch, **196**
synapse, **349**, *349*
systems (of the body), 93

taproot, **160**–61, *160*
target cells, **364**
taste, **356**
taxonomy, 156
telescope, 250–51, *250–51*, A10, *A10*
tephra, **18**
terrestrial planet, **270**
texture, **40**–41, 43
theory of plate tectonics, **4**
Thomson, J. J., 170
thyroid gland, *364*, 365
tissue, **93**
topsoil, 42, *42*
totality, **277**
touch, **356**, 358–59
traits, **320**–22, 326–33
transfusion, 388
transplant (organ), 388–89
triangulation, 11
tsunami, **11**
tuber, **159**
turtle, 128

unicellular, **93**
Uniformitarianism, 30
univalve, 112
uranium, 60
Uranus, 270–71, *271*, *281*

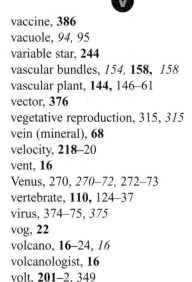

vaccine, **386**
vacuole, *94*, 95
variable star, **244**
vascular bundles, *154*, **158**, *158*
vascular plant, **144**, 146–61
vector, **376**
vegetative reproduction, 315, *315*
vein (mineral), **68**
velocity, **218**–20
vent, **16**
Venus, 270, *270–72*, 272–73
vertebrate, **110**, 124–37
virus, 374–75, *375*
vog, **22**
volcano, **16**–24, *16*
volcanologist, **16**
volt, **201**–2, 349
Volta, Alessandro, 193

Wadlow, Robert, 365
warm-blooded, **131**–32
water cycle, **74**, *74*
Watson, James, 323, *323*
watt, **201**
weathering, 30–37
    chemical, **30**, 34–37
    mechanical, **30**–33
wedge, **232**, *232*
Welles, Orson, 273
wheel and axle, **231**, *231*
White, Jim, 36
white blood cells, 383–89
Wilkins, Maurice, 323
wind energy, **66**
wind farm, 66, *66*
work, **227**–28, 234–35
worms, 116–17

xylem, **158**, *158*

zygote, **299**–301, 308

Photo Credits

These page numbers refer to the Student Text.

The following agencies and individuals have furnished materials to meet the photographic needs of this textbook. We wish to express our gratitude to them for their important contribution.

Suzanne Altizer
American Leprosy Missions
American Iron and Steel Institute
J. Ramon Arrowsmith
Auburn University
Philip Baird
Tom Barnes
Bat Conservation International
BJU Press Files
Bureau of Reclamation
3dCafe.Com
Carolina Biological
David A. Caron
Cedar Fair L.P./Cedar Point
Centers for Disease Control
John Charlton
Comstock
Bill Cook
Corbis
Dr. Thomas Coss
Dr. Stewart Custer
Terry Davenport
Carl Dennis
Department of Energy
Department of Natural Resources
Egyptian Tourist Authority
Fred Espenak
Evergreen Photo Alliance
Donna Fare
Kim Fennema
Forestry Images
Joyce Garland
Gary Gauler
Getty Images
Philip Gladstone
Ann Glenn
Phillip Greenspun
Greenville Police Department
Hemera Technologies, Inc.
Dr. Joseph Henson
Holiday Film
Robert Hooke
imagequest3D.com
Brian Johnson
Breck Kent
Joyce Landis
Library of Congress
Kerrie Ann Lloyd
George Loun
Luray Caverns
Ron Magill
David Malin
Fred Mang, Jr.
Meteor Crater, Northern Arizona, USA

Michigan State University Extension
Mark S. Mohlman
Greg Moss
George Musil
NASA
National Library of Medicine
National Mining Association
The National Museum of Health and Medicine, Armed Forces Institute of Pathology
National Oceanic and Atmospheric Administration (NOAA)
National Park Service
Christine Nichols
Boyd Norton
Ohaus
Dr. A. A. Padhye
David Parker
Susan Perry
J. S. Peterson
Dave Powell
Queens Borough Public Library
Dr. Margene Ranieri
Sue Renault
Richmond–HRR
Bruce Roberts
Paul Roberts
Wendy Searles
Science Photo Library
Roger Steene
Stem Labs, Inc.
The Telegraph
Merlin D. Tuttle
United States Geological Services (USGS)
University of Pennsylvania Library
University of Southern California
Unusual Films
USDA
USDA Forest Service
Eric Vallery
Visuals Unlimited
Ward's Natural Science Establishment
Paul Wray
Wrigley Institute for Environmental Studies
Jim Zimmerlin

Front Matter
PhotoDisc/Getty Images iv (background scene, fish); David A. Caron, Wrigley Institute for Environmental Studies, University of Southern California iv (microbes)

Unit One
PhotoDisc/Getty Images 1

Chapter 1
Unusual Films 3 (both), 12–13, 18, 19; © Philip Baird/www.anthroarcheart.org 6 (top); Boyd Norton/Evergreen Photo Alliance 6 (bottom); J. Ramon Arrowsmith 7; Paul Roberts 9; PhotoDisc/Getty Images 11, 16 (bottom), 20, 22 (top); Dr. Stewart Custer 15, 21 (bottom); USGS 16 (top, center); Susan Perry 21 (top); Dr. Joseph Henson 21 (center); Joyce Garland 22 (center); Joyce Landis 22 (bottom); NASA 23

Chapter 2
Unusual Films 25 (both), 39, 44, 45, 48 (top); PhotoDisc/Getty Images 27 (center), 28 (bottom), 30, 31 (bottom), 40 (background, bottom), 42 (top), 46 (bottom), 47 (both); Dr. Stewart Custer 27 (top, bottom); Kerrie Ann Lloyd 28 (top); John Charlton, Kansas Geological Survey, University of Kansas 29, 40 (top), 48 (bottom); Brian Johnson 31 (top); Greg Moss 31 (center); Holiday Films 32; Luray Caverns 33; Ward's Natural Science Establishment 40 (center); Egyptian Tourist Authority 43 (top); NASA 43 (center, bottom); Bruce Roberts 46 (top); USDA Forest Service 49

Chapter 3
Unusual Films 51 (both), 56, 63 (top left, center left), 66–67; PhotoDisc/Getty Images 52, 53, 55, 60–61 (both), 62 (both), 63 (top right), 64, 73; National Mining Association 54 (all); Bureau of Reclamation 58; Department of Energy 59; Wendy Searles 63 (center right, bottom right); USDA 65 (center); Department of Natural Resources 65 (bottom); David A. Caron, Wrigley Institute for Environmental Studies, University of Southern California 69 (both); NOAA 72

Unit Two
PhotoDisc/Getty Images 75

Chapter 4
Unusual Films 77 (both), 90 (all), 92 (center right, bottom right); Joyce Landis 78 (top left); PhotoDisc/Getty Images 78 (top right), 79 (both), 92 (bottom left), 94 (inserts), 95 (all); Robert Hooke 81; The National Museum of Health and Medicine,

A78 Photo Credits

Armed Forces Institute of Pathology 82; Carolina Biological/Visuals Unlimited 84; Stem Labs, Inc. 91; Dr. Thomas Coss 92 (top right); Susan Perry 93 (top); Suzanne Altizer 93 (bottom); Dr. Margene Ranieri 94 (plant cell); Corbis 96

Chapter 5
Unusual Films 99 (both), 106 (top), 112 (all), 113, 128, 129; PhotoDisc/Getty Images 102, 114–15 (all), 116, 118 (center), 120, 121 (left, center), 123 (top), 124 (all), 125 (top, middle), 126 (bottom), 127 (bottom); Roger Steene/imagequest3D.com 103; Breck Kent 105, 107, 116–17 (bottom), 117 (top), 119 (both), 122 (bottom); Ward's Natural Science Establishment 106 (bottom); Corbis 109; Ron Magill 118 (bottom); © 2003 Hemera Technologies, Inc., All Rights Reserved 121 (right); Visuals Unlimited 122 (top); Terry Davenport 123 (bottom); Guillaume Dargaud 125 (bottom); NOAA 126 (top); Phillip Greenspun 127 (top)

Chapter 6
Unusual Films 133 (both), 143 (insets), 150 (center, right); Suzanne Altizer 134 (all), 135; © J. S. Peterson@PlantsDatabase 136 (bottom left); Tom Barnes 136 (inset); USDA Forest Service 137 (top left); Ann Glenn 137 (top right); Dave Powell/Forestry Images 137 (bottom); George Loun/Visuals Unlimited 138 (top); Donna Fare 138 (bottom, inset); Fred Mang, Jr., National Park Service 139; Bill Cook, Michigan State University Extension 140 (top left, top center, top right), 140–41 (bottom), 141 (top right, inset); Paul Wray 140 (inset); Dr. Stewart Custer 141 (top left); PhotoDisc/Getty Images 142–43 (top), 142 (far left, right, far right), 143 (top, bottom), 144, 145 (all), 147 (bottom left), 148–49 (bottom), 149 (both), 150 (left); © 2003 Hemera Technologies, Inc. All Rights Reserved 142 (center left), 147 (bottom right, bottom center, bottom left); Suzanne Altizer 142 (center right); Susan Perry 147 (top); Kim Fennema/Visuals Unlimited 151

Unit 3
PhotoDisc/Getty Images 155

Chapter 7
Unusual Films 157 (both), 167 (bottom left), 172–73, 177 (both), 178–79, 181; Edgar Fahs Smith Collection, University of Pennsylvania Library 160; Department of Energy 161; PhotoDisc/Getty Images 162; Susan Perry 167 (bottom right, bottom center), 176 (both); PhotoDisc/Getty Images 170

Chapter 8
Unusual Films 183 (both), 187 (top left, top right), 188, 189, 191, 194 (all), 199 (both); Edgar Fahs Smith Collection, University of Pennsylvania Library 184; Queens Borough Public Library 187 (bottom); Susan Perry 200, 203; PhotoDisc/Getty Images 201

Chapter 9
Unusual Films 205 (both), 223 (top, center), 225; PhotoDisc/Getty Images 206, 211 (both), 213 (top), 214; Susan Perry 207, 218, 221 (top, bicycle gear), 223 (bottom); Brian Johnson 213 (bottom); Greenville Police Department 215; Cedar Fair L.P./Cedar Point 216; Ohaus 217; © 2003 Hemera Technologies, Inc. All Rights Reserved 221 (screwdriver)

Unit 4
PhotoDisc/Getty Images 227

Chapter 10
Unusual Films 229 (both), 251 (both); NASA 230 (top), 231 (Sun, Moon, Venus), 235, 236, 237, 240, 244, 245 (both), 246 (center), 248 (top), 249 (top); PhotoDisc/Getty Images 230–31 (background); David Parker, Science Photo Library 241; Wendy Searles 242; David Malin 246–47 (top), 246 (bottom), 247 (bottom); Meteor Crater, Northern Arizona, USA 248–49 (bottom)

Chapter 11
Unusual Films 253 (both), 266, 268, 277; Jan Curtis 255; NASA 258–59 (planets), 260–61 (all), 262 (bottom), 263 (all), 264 (bottom), 265 (top inset), 268–69 (all), 270–71 (all), 272–75; PhotoDisc/Getty Images 262 (top), 264 (top); © 2003 by Fred Espenak, www.mreclipse.com 265 (bottom); Wendy Searles 266–67 (both)

Unit 5
PhotoDisc/Getty Images 279

Chapter 12
Unusual Films 281 (both), 286–87 (bottom), 287 (top), 291, 296 (bottom left), 297 (top right, bottom left, center left, top left), 298, 299; PhotoDisc/Getty Images 283 (top), 292 (both), 294 (center); Carl Dennis, Auburn University. www.insect-images.com 283 (center); Merlin D. Tuttle, Bat Conservation International 283 (bottom); © 2003 Hemera Technologies, Inc. All Rights Reserved 288 (top, top inset); Eric Vallery/USDA Forest Service 288 (bottom right, bottom center, bottom left); Gary Gauler/Visuals Unlimited 291 (inset); Jim Zimmerlin 293 (left); Breck Kent 293 (right), 294–95 (top, bottom); Philip Gladstone 296 (top); Ward's Natural Science Establishment, Inc. 296 (center right); Mark S. Mohlman 296 (bottom); George Musil/Visuals Unlimited 296 (center left); BJU Press Files 297 (center right)

Chapter 13
Unusual Films 301 (both), 302, 303 (all); Susan Perry 307, 316 (both), 317 (all); National Library of Medicine 308; Richmond–HRR Pseudoisochromatic Plates 4th edition 315 (left); Library of Congress 315 (right); Stem Labs, Inc. 318 (insets); Suzanne Altizer 319; Centers for Disease Control 320; American Iron and Steel Institute 321

Unit 6
PhotoDisc/Getty Images 323

Chapter 14
Unusual Films 325 (both), 333 (right), 335, 340–41, 343; PhotoDisc/Getty Images 326–27, 333 (left); Model provided by 3dCafe.com 327; www.comstock.com 344; Library of Congress 347 (left); The Telegraph 347 (right); Susan Perry 349

Chapter 15
Unusual Films 351 (both), 361, 370; Edgar Fahs Smith Collection, University of Pennsylvania Library 352; Sue Renault/Christine Nichols/American Leprosy Missions 354 (right); Dr. A.A. Padhye/Centers for Disease Control 354 (left); Centers for Disease Control 354 (bottom right), 355, 358, 366 (both), 368 (center), 369; Visuals Unlimited 367 (both); PhotoDisc/Getty Images 368 (top right, top left, bottom)

These page numbers refer to the Teacher's Edition.

Front Matter
PhotoDisc/Getty Images iii

Appendix
Cartesia Software A18 (top), A21; National Park Service; photographer, Wymond Eckhardt A18 (bottom); photograph by C. D. Miller in 1980/US Geological Survey A19 (top); Mammoth Lakes Visitors Bureau/Jim Stroup A19 (bottom)

Working Together

Whether you have been teaching for many years or are just getting started, your comments are vital in helping us maintain our standard of excellence. In fact, most of the improvements in our materials started with good advice from consumers. So after you have put our products to the test, please give us your thoughtful comments and honest assessment.

And thanks for your valuable help!

Book Title _____ Grade level _____

Material was ☐ used in classroom.   ☐ used in home school.   ☐ examined only.

How did you hear about us?
_____
_____
_____

I liked
_____
_____
_____

I'd like it better if
_____
_____
_____

How did our material compare with other publishers' materials?
_____
_____
_____

Other comments?
_____
_____
_____

(OPTIONAL)
☐ Dr. ☐ Miss ☐ Mrs. ☐ Mr. _____
School _____
Street _____
City _____ State ____ ZIP _____
Phone (____) _____
E-mail _____

Fold and tape. DO NOT STAPLE
Mailing address on the other side.

BJU PRESS
Greenville, SC 29614

Return address:

TAPE SHUT–DO NOT STAPLE

NO POSTAGE
NECESSARY IF
MAILED IN THE
UNITED STATES

BUSINESS REPLY MAIL
FIRST-CLASS MAIL   PERMIT NO. 344   GREENVILLE, SC

POSTAGE WILL BE PAID BY ADDRESSEE

BJU PRESS
TEXTBOOK DIVISION
1700 WADE HAMPTON BLVD.
GREENVILLE, SC 29609-9971

please fold

please fold